George Anson, Richard Walter

A voyage round the world in the years 1740-44

George Anson, Richard Walter

A voyage round the world in the years 1740-44

ISBN/EAN: 9783744722285

Printed in Europe, USA, Canada, Australia, Japan

Cover: Foto ©Andreas Hilbeck / pixelio.de

More available books at **www.hansebooks.com**

A VOYAGE ROUND THE WORLD

BY

SIR FRANCIS DRAKE

AND

WILLIAM DAMPIER

According to the Text of the Original Narratives

EDITED, WITH NOTES, ETC., BY

D. LAING PURVES

Special Edition.

LONDON:
PUBLISHED BY THE LI-QUOR TEA COMPANY,
5 GEORGE STREET, TOWER HILL.
1879.

CONTENTS.

DRAKE'S VOYAGE ROUND THE WORLD.

DRAKE'S LAST VOYAGE.

DAMPIER'S VOYAGE ROUND THE WORLD.

BIOGRAPHICAL NOTICES

OF

SIR FRANCIS DRAKE & WILLIAM DAMPIER.

AMONG the voyagers and naval heroes flourishing in the times of Queen Elizabeth, a high place must be given to Sir Francis Drake for his courage and persevering bravery, displayed in almost every enterprise, successful or unsuccessful, with which he was identified. His father is said to have been a poor yeoman, inhabiting a humble cottage at Tavistock, Devonshire, where his son, Francis, was born in or about the year 1539 or 1541; but having embraced the Protestant religion, he was compelled to fly into Kent, where, for some time, the family are said to have inhabited the hull of a ship on the sea coast. His family being in poor circumstances, and Francis being the eldest of twelve sons, he was early inured to hardihood, and was trained as a sailor from earliest youth. He was apprenticed to the master of a bark trading on the coast, and making occasional voyages between Zealand and France, in which employment he is said to have proved himself so painstaking and diligent, that his old master, having no children of his own, at his death bequeathed to him his bark and all its belongings. He continued this coasting trade for some time, but these narrow seas proving insufficient for his adventurous spirit, and fired with the idea of the possible wealth to be gained from an expedition to the New World, he gladly took advantage of an opportunity to join Captain John Hawkins in a voyage to the Spanish Main. Selling his vessel, he embarked his fortune and his person in this expedition at Plymouth in 1567. Drake's ship was called the Judith (50 tons), and which, notwithstanding the perils of the voyage, by skilful seamanship, he brought safely home. This expedition proving unfortunate, and losing all he had, he returned with an enlarged experience, and an increased and growing hatred towards the Spaniards. On the 24th of May 1572, he sailed from Plymouth in command of the Pasha, of 70 tons, and the Swan, of 25 tons, the latter commanded by his brother. In the end of July he came in sight of Santa Martha, and a few days afterwards was unexpectedly reinforced by another English bark, the Lion, commanded by Captain Rouse, and with thirty men on board. The Lion willingly joined the Pasha and the Swan, when they sailed together for Nombre de Dios. Leaving his ships in charge of Rouse, he selected three-and-fifty men, and with these he landed under cover of the night of July 22d, and made his attack upon the town, which proved successful.

After this voyage his thoughts were directed to the best means of realising his dream of ploughing the Pacific Ocean with English keels. While gathering help and enlisting supporters, he served with the Earl of Essex in an Irish campaign; and his tactics and brilliant valour secured him the patronage of Queen Elizabeth. He was thus enabled, towards the close of 1577, to sail from Plymouth, with five vessels, the largest of which was 100 and the smallest 15 tons. This was destined to prove his great voyage of circumnavigation, which occupied about two years and ten months. The execution of Thomas Doughty in the course of this voyage has been regarded as one of the most doubtful acts in Drake's life, although he is represented as being perfectly honest and straightforward in the act, regretting Doughty's death, but looking upon it as necessary for the safety of the expedition. On his return he was knighted by Queen Elizabeth.

Drake was next employed as commander-in-chief of the great fleet despatched in September 1585 against the Spanish West Indies. They made a successful attack on San Domingo, and, after a desperate struggle, carried Carthagena. Then, after doing infinite damage, and securing immense booty, Drake brought back his fleet to England in perfect safety. At this time he is said to have visited Virginia, and it is stated by Camden with regard to this voyage, that he was the first to bring tobacco to England, though Raleigh was the first to make its use popular. On the same authority it is stated that from the books, papers, and charts which were taken from an East India ship which he captured off the coast of Spain in 1587, originated the first suggestion for undertaking our East Indian trade, and suggested an application to the Queen for liberty to establish an East India Company. Drake played a high and honourable part in the defeat of the "Invincible Armada" of Spain.

In April 1589 he took the command of the naval portion of a joint expedition against Spain. Corunna was captured, but owing to disease appearing among the land forces, little else was done or attempted, and Drake returned to England. For some time he addressed himself to civil pursuits, and in 1592-3 sat in Parliament as the representative of Plymouth. In 1594 he was again called to active service, Queen Elizabeth's government having determined on a new expedition against the Spanish colonies. It consisted of six royal and twenty private ships, and Drake and Hawkins were associated in the command.

The narrative of this ill-fated expedition, written by Thomas Maynarde, is given in the present work. When they had taken and plundered in succession Rio de la Hacha, Santa Martha, and Nombre de Dios, it became evident that Drake's career was nearly ended. He was seized with a severe illness, which, acting fatally on an already weakened frame, terminated in his death on the 20th December 1596. His body was placed in a leaden coffin, the solemn service of the Church was read over it, and then it was lowered into the deep.

In Stowe's Annals Drake is described as "low of stature, of strong limbs, broad breasted, rounde headed, broune hayre, full bearded; his eyes round, large, and clear, well favoured, fayre, and of a cheerfull countenance. His name was a terror to the French, Spaniard, Portugal, and Indians."

The incursions of the buccaneers on the Spanish settlements in the South Seas, though undertaken in the first place for gain and plunder, helped to familiarise our English seamen with the geography of the South American coast, and the other islands in the South Seas. The derivation of the word "buccaneer" is ascribed to the method which prevailed in Cuba at that time of killing, and curing the flesh of the cattle, according to the Carib method, on hurdles raised a few feet above the fire. This apparatus, the meat, and also the method of preparing it, the Indians called *boocan*, and hence those sailors who were engaged in supplying it to the cruisers and others were called *buccaneers*. Many of these adventurers were Englishmen, carrying on a smuggling trade both by sea and land. They all, without exception, plundered the Spaniards, and under this bond and unity of aim, they were sometimes called the *brethren of the coast*. Those who did their plundering on shore were called freebooters, and those who mainly cruised against the Spaniard were called buccaneers. If, in the case of a war with Spain, a commission could be obtained, these buccaneers became privateers. The ordinary buccaneer set propriety at defiance by dirtiness and negligence in dress. Every buccaneer leader had a mate, who was heir to all his money, and in some cases they held a community of property.

Among the more notable of the buccaneers who have left a record of their doings in the South Seas, we must place the name of Captain William Dampier, whose Voyage Round the World is given in the present volume. While reading this narrative, we feel that he was one of the most acute of observers, readily remarking anything which at that time would be counted new or extraordinary, although these details at the present time, with our amazingly increased facilities of travel, and increased familiarity with the places and people described, may be ranked as more commonplace. In the "author's account of himself" we have a concise narrative of his early training and way of life, with a graphic sketch of logwood cutting in Campeachy Bay, till the date of his joining with the buccaneers in 1679. His Voyage Round the World may be considered as a natural continuation of the story of his life, as it deals with all the public and personal affairs in which he was concerned up till the date of his return to England on September 16th, 1691.

Dampier having recommended himself very favourably to public attention by the publication of his "Voyage Round the World," at the instance of the Earl of Pembroke he was given the command of an expedition, ordered by King William in 1699 for the discovery of new countries, and the examination of New Holland and New Guinea. A vessel called the Roebuck was equipped for this purpose. After visiting New Holland, he sailed for New Guinea, which he descried on January 1st, 1700. He had explored the west and north-west coasts of Australia, and gave his name to a small archipelago, east of North-west Cape. After exploring the coasts of New Guinea, New Britain, and New Ireland, he returned. In the homeward voyage the Roebuck sprang a leak off the Island of Ascension. Dampier and his men were forced to stay ten weeks on the island, but they were eventually picked up by three English ships of war, and conveyed to England.

Although his last voyage had been partially unfortunate as far as the loss of the vessel was concerned, we find that he was next given command of the St George, a vessel of 26 guns, which, with the Cinque Ports of 16 guns, had been fitted out by English merchants on a privateering expedition to the South Seas. He did not shine as a commander, being, it is said, at times too familiar with his men, at other times using injudicious severity with frequent bursts of ill temper. The story of the crew of this somewhat mutinous expedition contains the incidents in the life of Alexander Selkirk, which form the groundwork of De Foe's world-famous "Robinson Crusoe."

Little is known of Dampier's personal history after this voyage, although he remained at sea up till 1711. After forty years' wandering over the world, he seems to have sunk into obscurity, as no record remains of how or when he died.

THE VOYAGE ABOUT THE WORLD

BY SIR FRANCIS DRAKE.

TO

THE TRULY NOBLE

ROBERT EARL OF WARWICK.

RIGHT HONOURABLE,

FAME and envy are both needless to the dead because unknown; sometimes dangerous to the living when too well known; reason enough that I rather choose to say nothing, than too little, in praise of the deceased author, or of your Lordship my desired fautor.[1] COLUMBUS did neatly check his emulators, by rearing an egg without assistance. Let the slighter of this voyage apply. If your Lordship vouchsafe the acceptance, 'tis yours; if the reader can pick out either use or content, 'tis his; and I am pleased. Example being the public, and your Lordship's favour the private, aim of

Your humbly devoted,

FRANCIS DRAKE.[2]

EVER since Almighty God commanded Adam to subdue the earth, there have not wanted in all ages some heroical spirits which, in obedience to that high mandate, either from manifest reason alluring them, or by secret instinct enforcing them, thereunto, have expended their wealth, employed their time, and adventured their persons, to find out the true circuit thereof.

Of these, some have endeavoured to effect this their purpose by conclusion and consequence, drawn from the proportion of the higher circles to this nethermost globe, being the centre of the rest. Others, not contented with school points and such demonstrations (for that a small error in the beginning groweth in the progress to a great inconvenience) have added thereunto their own history and experience. All of them in reason have deserved great commendation of their own ages, and purchased a just renown with all posterity. For if a surveyor of some few lordships, whereof the bounds and limits were before known, worthily deserve his reward, not only for his travel, but for his skill also in measuring the whole and every part thereof, how much more, above comparison, are their famous travels by all means possible to be eternized, who have bestowed their studies and en-

[1] Favourer, patron.

[2] Nephew of "the General," as Drake is called throughout Mr Fletcher's narrative.

deavour to survey and measure this globe, almost unmeasurable? Neither is here that difference to be objected which in private possessions is of value: "Whose land survey you?" forasmuch as the main ocean is by right the Lord's alone, and by nature left free for all men to deal withal, as very sufficient for all men's use, and large enough for all men's industry.

And therefore that valiant enterprise, accompanied with happy success, which that right rare and thrice worthy captain, Francis Drake, achieved, in first turning up a furrow about the whole world, doth not only overmatch the famous Argonauts, but also outreacheth in many respects that noble mariner, Magellan, and by far surpasseth his crowned victory. But hereof let posterity judge. It shall for the present be deemed a sufficient discharge of duty to register the true and whole history of that his voyage, with as great indifference of affection as a history doth require, and with the plain evidence of truth, as it was left recorded by some of the chief and divers other actors in that action.

The said Captain Francis Drake, having in a former voyage, in the years 1572 and 1573 (the description whereof is already imparted to the view of the world[1]), had a sight, and only a sight, of the South Atlantic; and thereupon, either conceiving a new, or renewing a former, desire of sailing on the same in an English bottom, he so cherished, thenceforward, this his noble desire and resolution in himself, that notwithstanding he was hindered for some years, partly by secret envy at home, and partly by public service for his Prince and country abroad (whereof Ireland, under Walter Earl of Essex, gives honourable testimony), yet, against the year 1577, by gracious commission from his sovereign, and with the help of divers friends adventurers, he had fitted himself with five ships:

1. The Pelican, Admiral, burthen one hundred tons, Captain-General Francis Drake.

2. The Elizabeth, Vice-Admiral, burthen eighty tons, Captain John Winter.

3. The Marigold, a bark of thirty tons, Captain John Thomas.

4. The Swan, a fly-boat of fifty tons, Captain John Chester.

5. The Christopher, a pinnace of fifteen tons, Captain Thomas Moon.

These ships he manned with one hundred and sixty-four able and sufficient men, and furnished them also with such plentiful provision of all things necessary, as so long and dangerous a voyage did seem to require; and, amongst the rest, with certain pinnaces ready framed, but carried aboard in pieces, to be new set up in smoother water when occasion served. Neither had he omitted to make provision also for ornament and delight, carrying to this purpose with him expert musicians, rich furniture (all the vessels for his table, yea, many belonging even to the cook-room, being of pure silver), and divers shows of all sorts of curious workmanship, whereby the civility and magnificence

---

[1] It was written by Philip Nichols, preacher, and subsequently published by the navigator's nephew, heir, and godson, Sir Francis Drake. In the course of an expedition to intercept a convoy of treasure from Panama to Nombre de Dios, Drake was conducted by a friendly native chief to a "great and goodly tree" upon the ridge of the hills, from a bower or look-out in the top of which both the Atlantic and the Pacific could be seen. When Drake had beheld that sea, "of which he had heard such golden reports, he besought Almighty God of His goodness to give him life and leave to sail once in an English ship in that sea." Calling up his men, he acquainted them, John Oxenham especially, with his resolve, which all approved. Oxenham, indeed, more than kept his promise to follow his chief, for two years later, crossing the Isthmus with a devoted band, he built a pinnace, launched it on the South Sea, and took two Spanish ships; but being made prisoner on his return, he was executed at Lima.

of his native country might, amongst all nations whithersoever he should come, be the more admired.

Being thus appointed, we set sail out of the Sound of Plymouth [1] about five o'clock in the afternoon, November 15, of the same year 1577, and running all that night SW., by the morning were come as far as the Lizard, where meeting the wind at SW. (quite contrary to our intended course), we were forced, with our whole fleet, to put into Falmouth. The next day, towards evening, there arose a storm, continuing all that night and the day following (especially between ten of the clock in the forenoon and five in the afternoon) with such violence, that though it was in a very good harbour, yet two of our ships—the Admiral, wherein our General himself went, and the Marigold —were fain to cut their mainmasts by board; and for the repairing of them, and many other damages in the tempest sustained (as soon as the weather would give leave), to bear back to Plymouth again, where we all arrived the thirteenth day after our first departure thence [November 28]. Whence, having in a few days supplied all defects, with happier sails we once more put to sea, December 13, 1577. As soon as we were out of sight of land, our General gave us occasion to conjecture in part whither he intended, both by the directing of his course, and appointing the rendezvous, if any should be severed from the fleet, to be the Island Mogador. And so sailing with favourable winds, the first land we had sight of was Cape Caulin [2] in Barbary, December 25, Christmas Day, in the morning. The shore is fair white sand, and the inland country very high and mountainous; it lies in 32° 30′ N. latitude: and so coasting from hence southward about eighteen leagues, we arrived the same day at Mogador, the island before named.

This Mogador lies under the dominion of the King of Fesse,[3] in 31° 40′, about a mile off from the shore, by this means making a good harbour between the land and it. It is uninhabited, of about a league in circuit, not very high land, all overgrown with a kind of shrub breast high, not much unlike our privet, very full of doves, and therefore much frequented of goshawks and such-like birds of prey, besides divers sorts of sea-fowl very plenty. At the south side of this island are three hollow rocks, under which are great store of very wholesome but very ugly fish to look to. Lying here about a mile from the main, a boat was sent to sound the harbour, and finding it safe, and in the very entrance on the north side about five or six fathoms' water (but at the south side it is very dangerous), we brought in our whole fleet, December 27, and continued there till the last day of the month, employing our leisure the meanwhile in setting up a pinnace, one of the four brought from home in pieces with us. Our abode here was soon perceived by the inhabitants of the country, who coming to the shore, by signs and cries made show that they desired to be fetched aboard, to whom our General sent a boat, into which two of the chief of the Moors were presently received, and one man of ours, in exchange, left aland, as a pledge for their return. They that came aboard were right courteously entertained with a dainty banquet, and such gifts as they seemed to be most glad of, that they might thereby understand that this fleet came in peace and friendship, offering to traffic with them for such commodities as their country yielded, to their own content. This offer they seemed most gladly to accept, and promised the next day to resort again, with such

[1] To throw the Spaniards off their guard, the destination of the fleet was given out as Alexandria; and to give countenance to the report the course first steered was towards the Straits of Gibraltar.

[2] In lat. 32° N., long. 10° W.

[3] Fez, the northern portion of the Empire of Morocco.

things as they had, to exchange for ours. It is a law amongst them to drink no wine, notwithstanding by stealth it pleaseth them well to have it abundantly, as here was experience. At their return ashore, they quietly restored the pledge which they had stayed; and the next day at the hour appointed returning again, brought with them camels, in show laden with wares to be exchanged for our commodities, and calling for a boat in haste, had one sent them, according to order which our General (being at this present absent) had given before his departure to the island. Our boat coming to the place of landing, which was among the rocks, one of our men, called John Fry, mistrusting no danger nor fearing any harm pretended by them, and therefore intending to become a pledge, according to the order used the day before, readily stepped out of the boat and ran aland; which opportunity (being that which the Moors did look for) they took the advantage of, and not only they which were in sight laid hands on him to carry him away with them, but a number more, who lay secretly hidden, did forthwith break forth from behind the rocks, whither they had conveyed themselves, as it seems, the night before, forcing our men to leave the rescuing of him that was taken as captive, and with speed to shift for themselves.

The cause of this violence was a desire which the King of Fesse had to understand what this fleet was, whether any forerunner of the King of Portugal's[1] or no, and what news of certainty the fleet might give him. And therefore, after that he was brought to the King's presence, and had reported that they were Englishmen, bound to the Straits under the conduct of General Drake, he was sent back again with a present to his Captain, and offer of great courtesy and friendship, if he would use his country. But in this meantime the General, being aggrieved with this show of injury, and intending, if he might, to recover or redeem his man, his pinnace being ready, landed his company, and marched somewhat into the country, without any resistance made against him, neither would the Moors come nigh our men to deal with them any way; wherefore having made provision of wood, as also visited an old fort built sometime by the King of Portugal but now ruined by the King of Fesse, we departed, December 31, towards Cape Blanco, in such sort that when Fry returned he found to his great grief that the fleet was gone; but yet, by the King's favour, he was sent home into England not long after, in an English merchant ship.

Shortly after our putting forth of this harbour, we were met with contrary winds and foul weather, which continued till the 4th of January; yet we still held on to our course, and the third day after fell with Cape de Guerre,[2] in 30°, where we lighted on three Spanish fishermen called caunters, whom we took with our new pinnace, and carried along with us till we came to Rio del Oro,[3] just under the Tropic of Cancer, where with our pinnace also we took a carvel. From hence till the fifteenth day we sailed on towards Cape Barbas, where the Marigold took a carvel more, and so onward to Cape Blanco till the next day at night. This cape lies in 20° 30′, showing itself upright like the corner of a wall, to them that come towards it from the north, having between it and Cape Barbas, low, sandy, and very white land all the way. Here we observed the South Guards, called the Croziers,[4] 9° 30′ above the horizon. Within the Cape we took one Spanish ship more riding at anchor (all her men being fled ashore in the boat save

[1] King Sebastian was then preparing that expedition into Mauritania, the calamitous result of which, on the fatal day of Alcazar-Seguer, will afterwards appear.

[2] Cape Ghir, in about latitude 31°; it marks the end of the Atlas mountain-chain towards the Atlantic.

[3] Rio do Ouro.

[4] The constellation of the Southern Cross.

two), which, with all the rest we had formerly taken, we carried into the harbour, three leagues within the Cape.[1]

Here our General determined for certain days to make his abode, both for that the place afforded plenty of fresh victuals for the present refreshing of our men, and for their future supply at sea (by reason of the infinite store of divers sorts of good fish which are there easy to be taken, even within the harbour, the like whereof is hardly to be found again in any part of the world), as also because it served very fitly for the despatching of some other businesses that we had. During the time of our abode at this place, our General, being ashore, was visited by certain of the people of the country, who brought down with them a woman, a Moor (with her babe hanging upon her dry dug, having scarce life in herself, much less milk to nourish her child), to be sold as a horse, or a cow and calf by her side; in which sort of merchandise our General would not deal. But they had also ambergris, with certain gums of some estimation, which they brought to exchange with our men for water, whereof they have great want; so that coming with their *alforges*[2] (they are leathern bags holding liquor) to buy water, they cared not at what price they bought it, so they might have it to quench their thirst. A very heavy judgment of God upon the coast! The circumstances whereof considered, our General would receive nothing of them for water, but freely gave it to them that came to him, yea, and fed them also ordinarily with our victuals, in eating whereof their manner was not only uncivil[3] and unsightly to us, but even inhuman and loathsome in itself.[4]

And having washed and trimmed our ships, and discharged all our Spanish prizes except one caunter (for which we gave to the owner one of our own ships, the Christopher) and one carvel, formerly bound to St Iago, which we caused to accompany us hither, where she also was discharged; after six days' abode here, we departed, directing our course for the Islands of Cape Verd, where (if anywhere) we were of necessity to store our fleet with fresh water, for a long time, for that our General from thence intended to run a long course, even to the coast of Brazil, without touch of land. And now having the wind constant at NE. and ENE., which is usual about those parts, because it blows almost continually from the shore, January the 27th we coasted Buenavista, and the next day after we came to anchor under the western part, towards St Iago, of the island Mayo; it lies in 15° high land, saving that the northwest part stretches out into the sea the space of a league, very low; and is inhabited by subjects to the King of Portugal. Here landing, in hope of traffic with the inhabitants for water, we found a town, not far from the water-side, of a great number of desolate and ruinous houses, with a poor naked chapel or oratory, such as small cost and charge might serve and suffice, being to small purpose, and as it seems only to make a show, and that a false show, contrary to the nature of a scarecrow, which feareth birds from coming nigh; this enticeth such as pass by to haul in and look for commodity,[5] which is not at all to be found there; though in the inner parts of the island it is in great abundance.

For when we found the springs and wells which had been there (as appeared) stopped up again, and no other water to purpose to be had to serve our need, we marched up to

[1] Probably the Bay du Levrier, which runs up into the land northwards behind the peninsula-promontory that Cape Blanco forms.

[2] Spanish, "Alforja," a saddle-bag.

[3] Barbarous, uncivilised.

[4] These people were worshippers of the sun; they never quitted their abodes until he had mounted above the horizon, and then paid their adoration prostrate, or kneeling upon a hillock or stone.

[5] Accommodation, convenience of supply, etc.

seek some more convenient place to supply our want, or at least to see whether the people would be dealt withal to keep us therein. In this travelling, we found the soil to be very fruitful, having everywhere plenty of fig trees, with fruit upon most of them. But in the valleys and low ground, where little low cottages were built, were pleasant vineyards planted, bearing then ripe and most pleasant grapes. There were also tall trees, without any branch till the top, which bare the cocoa nuts. There were also great store of certain lower trees, with long and broad leaves, bearing the fruit which they call plantains in clusters together like puddings, a most dainty and wholesome fruit. All of these trees were even laden with fruit,—some ready to be eaten, others coming forward, others overripe. Neither can this seem strange, though about the midst of Winter with us, for that the Sun doth never withdraw himself farther off from them, but that with his lively heat he quickeneth and strengtheneth the power of the soil and plant; neither ever have they any such frost and cold as thereby to lose their green hue and appearance.

We found very good water in divers places, but so far off from the road,[1] that we could not with any reasonable pains enjoy it. The people would by no means be induced to have any conference with us, but keeping in the most sweet and fruitful valleys among the hills, where their towns and places of dwelling were, gave us leave without interruption to take our pleasure in surveying the island, as they had some reason not to endanger themselves, where they saw they could reap nothing sooner than damage and shame, if they should have offered violence to them which came in peace to do them no wrong at all. This island yieldeth other great commodities, as wonderful herds of goats, infinite store of wild hens, and salt without labour (only the gathering it together excepted), which continually in a marvellous quantity is increased upon the sands by the flowing of the sea, and the heat of the sun kerning[2] the same. So that of the increase thereof they keep a continual traffic with their neighbours in the other adjacent islands. We set sail thence the 30th day [of January].

Being departed from Mayo, the next day we passed by the island of St Iago, ten leagues west of Mayo, in the same latitude, inhabited by the Portugals and Moors together. The cause whereof is said to have been in the Portugals themselves, who, continuing long time lords within themselves in the said island, used that extreme and unreasonable cruelty over their slaves, that (their bondage being intolerable) they were forced to seek some means to keep themselves and to lighten that so heavy a burthen; and thereupon chose to fly into the most mountainous parts of the island; and at last, by continual escapes, increasing to a great number, and growing to a set strength, do now live with that terror to their oppressors, that they now endure no less bondage in mind than the *forcatoz* did before in body; besides the damage that they daily suffer at their hands in their goods and cattle, together with the abridging of their liberties in the use of divers parts of the fruitful soil of the said island, which is very large, marvellous fruitful (a refuge for all such ships as are bound towards Brazil, Guinea, the East Indies, Binny,[3] Calicut, etc.), and a place of rare force, if it were not for the cause afore recited, which hath much abated the pride and cooled the courage of that people, who under pretence of traffic and friendship at first making an entrance, ceased not practising upon the poor islanders (the ancient re-

[1] The roadstead.

[2] Granulating, forming into corns or kernels.

[3] Apparently Benin, on the west coast of Africa, is meant, though in the list of places it is geographically out of order.

mainder of the first planters thereof, as it may seem from the coast of Guinea), until they had excluded them from all government and liberty, yea almost life. On the south-west of this island we took a Portugal, laden the best part with wine, and much good cloth, both linen and woollen, besides other necessaries, bound for Brazil, with many gentlemen and merchants in her.[1] As we passed by with our fleet, in sight of three of their towns, they seemed very joyful that we touched not with their coast; and seeing us depart peaceably, in honour of our fleet and General, or rather to signify that they were provided for an assault, shot off two great pieces into the sea, which were answered by one given them again from us.

South-west from St Iago, in 14° 30′, about twelve leagues distant, yet by reason of the height seeming not above three leagues, lies another island, called of the Portugals "Fogo,"—the burning island, or fiery furnace—in which rises a steep upright hill, by conjecture at least six leagues, or eighteen English miles, from the upper part of the water; within the bowels whereof is a consuming fire, maintained by sulphury matters, seeming to be of a marvellous depth, and also very wide. The fire showeth itself but four times in an hour, at which times it breaketh out with such violence and force, and in such main abundance, that besides that it giveth light like the moon a great way off, it seemeth that it would not stay till it touch the heavens themselves. Herein are engendered great store of pumice-stones, which being in the vehement heat of the fire carried up without the mouth of that fiery body, fall down, with other gross and slimy matter, upon the hill, to the continual increasing of the same; and many times these stones falling down into the sea are taken up and used, as we ourselves had experience by sight of them swimming on the water. The rest of the island is fruitful, notwithstanding, and is inhabited by Portugals, who live very commodiously therein, as in the other islands thereabout.

Upon the south side, about two leagues off this Island of Burning, lieth a most sweet and pleasant island; the trees thereof are always green and fair to look on, the soil almost set full of trees, in respect whereof it is named the Brave Island,[2] being a storehouse of many fruits and commodities, as figs always ripe, cocoas, plantains, oranges, lemons, cotton, etc. From the banks into the sea do run in many places the silver streams of sweet and wholesome water, which with boats or pinnaces may easily be taken in. But there is no convenient place or road for ships, neither any anchoring at all. For after long trial, and often casting of leads, there could no ground be had at any hand, neither was it ever known, as is reported, that any line would fetch ground in any place about that island. So that the top of Fogo burneth not so high in the air, but the root of Brava (so is the island called) is buried and quenched as low in the seas. The only inhabitant of this island is a hermit, as we suppose, for we found no other houses but one, built as it seemed for such a purpose; and he was so delighted in his solitary living, that he would by no means abide our coming, but fled, leaving behind him the relics of his false worship; to wit, a cross with a crucifix, an altar with his[3] superaltar, and certain other idols of wood of rude workmanship.

---

[1] Command of this prize was given to Thomas Doughty, who afterwards figures so prominently in the narrative; but being found appropriating to his own use the propitiatory presents made by the prisoners, he was superseded by Thomas Drake, brother of the Admiral. The pilot of the "Portugal" ship, Nuno da Silva—an expert mariner and well acquainted with the coast of Brazil—was detained by Drake, and afterwards liberated at Guatulco.

[2] Ilha Brava.

[3] Its.

Here we dismissed the Portugals taken near St Iago,[1] and gave them in exchange of their old ship our new pinnace built at Mogador, with wine, bread, and fish for their provision, and so sent them away, February 1.

Having thus visited, as is declared, the Islands of Cape Verd, and provided fresh water as we could, the 2d of February we departed thence, directing our course towards the Straits,[2] so to pass into the South Sea; in which course we sailed sixty-three days without sight of land (passing the Line Equinoctial the 17th day of the same month) till we fell with the coast of Brazil, the 5th of April following. During which long passage on the vast gulf, where nothing but sea beneath us and air above us was to be seen, as our eyes did behold the wonderful works of God in His creatures, which He hath made innumerable both small and great beasts, in the great and wide seas: so did our mouths taste, and our natures feed on, the goodness thereof in such fulness at all times, and in every place, as if He commanded and enjoined the most profitable and glorious works of His hands to wait upon us, not only for the relief of our necessities, but also to give us delight in the contemplation of His excellence, in beholding the variety and order of His providence, with a particular taste of His fatherly care over us all the while. The truth is, we often met with adverse winds, unwelcome storms, and, to us at that time, less welcome calms; and being as it were in the bosom of the burning zone, we felt the effects of sweltering heat, not without the affrights of flashing lightnings, and terrifyings of often claps of thunder; yet still with the admixture of many comforts. For this we could not but take notice of, that whereas we were but badly furnished (our case considered) of fresh water, having never at all watered, to any purpose, or that we could say we were much the better for it, from our first setting forth out of England till this time, nor meeting with any place where we might conveniently water, till our coming to the River of Plate, long after—continually, after once we were come within four degrees of the Line on this side, viz., after February 10, and till we were past the Line as many degrees towards the south, viz., till February 27, there was no one day went over us but we received some rain, whereby our want of water was much supplied. This also was observable, that of our whole fleet, being now six in number, notwithstanding the uncouthness[3] of the way, and whatever other difficulties, by weather or otherwise, we met withal, not any one, in all this space, lost company of the rest; except only our Portugal prize for one day, who, March 28, was severed from us, but the day following, March 29, she found us again, to both her own and our no little comfort. She had in her twenty-eight of our men, and the best part of all our provision for drink; her short absence caused much doubting and sorrow in the whole company, neither could she then have been finally lost without the overthrow of the whole voyage.

Among the many strange creatures which we saw, we took heedful notice of one, as strange as any, to wit, the flying fish, a fish of the bigness and proportion of a reasonable or middle sort of pilchards; he hath fins, of the length of his whole body, from the bulk to the top of the tail, bearing the form and supplying the like use to him that wings do to other creatures. By the help of these fins, when he is chased of the *Bonits*, or great mackerel (whom the *Aurata* or dolphin likewise pursueth), and hath not strength to escape by swimming any longer, he lifteth up himself

[1] Except the pilot Nuna da Silva, who willingly stayed with Drake when he learned that the voyage was to be prosecuted into Mare del Zur, or the South Sea.

[2] Of Magellan.

[3] Strangeness, unknown character,

above the water, and flieth a pretty height, sometimes lighting into boats or barks as they sail along. The quills of their wings are so proportionable, and finely set together, with a most thin and dainty film, that they might seem to serve for a much longer and higher flight; but the dryness of them is such, after some ten or twelve strokes, that he must needs into the water again to moisten them, which else would grow stiff and unfit for motion. The increase of this little and wonderful creature is in a manner infinite, the fry whereof lies upon the upper part of the waters, in the heat of the sun, as dust upon the face of the earth; which being in bigness of a wheat straw, and in length an inch more or less, do continually exercise themselves in both their faculties of nature; wherein, if the Lord had not made them expert indeed, their generation could not have continued, being so desired a prey to many which greedily hunt after them, forcing them to escape in the air by flight when they cannot in the waters live in safety. Neither are they always free, or without danger, in their flying; but as they escape one evil by refusing the waters, so they sometimes fall into as great a mischief by mounting up into the air, and that by means of a great and ravening fowl, named of some a Don or Spurkite, who feeding chiefly on such fish as he can come at by advantage, in their swimming in the brim of the waters, or leaping above the same, presently seizes upon them with great violence, making great havoc, especially among these flying fishes, though with small profit to himself. There is another sort of fish which likewise flies in the air, named a Cuttill; it is the same whose bones the goldsmiths commonly use, or at least not unlike that sort, a multitude of which have at one time in their flight fallen into our ships among our men.

Passing thus, in beholding the most excellent works of the Eternal God upon the seas, as if we had been in a garden of pleasure, April 5 we fell in with the coast of Brazil, in 30° 30′ towards the Pole Antarctic,[1] where the land is low near the sea, but much higher within the country having in depth not above twelve fathoms three leagues off from the shore; and being descried by the inhabitants we saw great and huge fires made by them in sundry places. Which order of making fires, though it be universal as well among Christians as heathens, yet it is not likely that many use it to that end which the Brazilians do: to wit, for a sacrifice to devils, whereas they intermix many and divers ceremonies of conjugations, casting up great heaps of sand, to this end, that if any ships shall go about to stay upon their coasts, their ministering spirits may make wreck of them, whereof the Portugals by the loss of divers of their ships have had often experience.

In the reports of Magellan's voyage, it is said that this people pray to no manner of thing, but live only according to the instinct of nature; which if it were true, there should seem to be a wonderful alteration in them since that time, being fallen from a simple and natural creature to make gods of devils. But I am of the mind that it was with them then as now it is, only they lacked then the like occasion to put it in practice which now they have; for then they lived as a free people among themselves, but now are in most miserable bondage and slavery, both in body, goods, wife, and children, and life itself, to the Portugals, whose hard and most cruel dealings against them force them to fly into the more unfruitful parts of their own land, rather there to starve, or at least live miserably, with liberty, than to abide such intolerable bondage as they lay upon them; using the aforesaid practices with devils both for a revenge against their oppressors, and also for a defence, that they have no further entrance into the country. And supposing indeed that no others

[1] That is, in latitude South of the Line.

had used travel by sea in ships, but their enemies only, they therefore used the same at our coming; notwithstanding, our God made their devilish intent of none effect; for albeit there lacked not, within the space of our falling with this coast, forcible storms and tempests, yet did we sustain no damage, but only the separating of our ships out of sight for a few days. Here our General would have gone ashore, but we could find no harbour in many leagues. And therefore coasting along the land towards the south, April 7, we had a violent storm for the space of three hours, with thunder, lightning, and rain in great abundance, accompanied with a vehement south wind directly against us, which caused a separation of the Christopher (the caunter which we took at Cape Blanco in exchange for the Christopher, whose name she thenceforward bore) from the rest of the fleet. After this we kept on our course, sometimes to the seaward, sometimes toward the shore, but always southward, as near as we could, till April 14, in the morning, at which time we passed by Cape St Mary, which lies in 35°, near the mouth of the River of Plate; and running within it about six or seven leagues, along by the main, we came to anchor in a bay under another cape, which our General afterwards called Cape Joy, by reason that the second day after our anchoring here the Christopher, whom we had lost in the former storm, came to us again.

Among other cares which our General took in this action,[1] next the main care of effecting the voyage itself, these were the principal and chiefly subordinate: to keep our whole fleet, as near as possibly we could, together; to get fresh water, which is of continual use; and to refresh our men, wearied with long toils at sea, as oft as we should find any opportunity of effecting the same. And for these causes it was determined, and public notice thereof given at our departure from the Islands of Cape Verd, that the next rendezvous, both for the re-collecting of our navy if it should be dispersed, as also for watering and the like, should be the River of Plate; whether we were all to repair with all the convenient speed that could be made, and to stay one for another if it should happen that we could not arrive there all together; and the effect we found answerable to our expectations, for here our severed ship (as hath been declared) found us again, and here we found those other helps also so much desired. The country hereabout is of a temperate and most sweet air, very fair and pleasant to behold, and, besides the exceeding fruitfulness of the soil, it is stored with plenty of large and mighty deer. Notwithstanding that in this first bay we found sweet and wholesome water, even at pleasure, yet the same day, after the arrival of the caunter, we removed some twelve leagues farther up into another, where we found a long rock, or rather island of rocks, not far from the main, making a commodious harbour, especially against a southerly wind; under them we anchored and rode till the 20th day at night, in which mean space we killed divers seals, or sea-wolves as the Spaniard calls them, which resorted to these rocks in great abundance. They are good meat, and were an acceptable food to us for the present and a good supply of our provision for the future. Hence, April 20, we weighed again and sailed yet farther up into the river, even till we found but three fathoms' depth, and that we rode with our ships in fresh water; but we stayed not there, nor in any other place of the river, because that the winds being strong, the shoals many, and no safe harbour found, we could not without our great danger so have done. Hauling therefore to seaward again, the 27th of the same month, after that we had spent a just fortnight in the river to the great comfort of the whole fleet, we passed by the south side thereof into the main. The land here lies SW., and NNE., with shoal water

[1] Enterprise, expedition.

some three or four leagues off into the sea; it is about 36° 20′ and somewhat better S. latitude.

At our very first coming forth to sea again, to wit, the same night, our fly-boat, the Swan, lost company of us: whereupon, though our General doubted nothing of her happy coming forward again to the rest of the fleet, yet because it was grievous to have such often losses, and that it was his duty as much as in him lay to prevent all inconveniences besides that might grow, he determined to reduce the number of his ships, thereby to draw his men into less room, that both the fewer ships might the better keep company, and that they might also be the better appointed with new and fresh supplies of provisions and men, one to ease the burthen of another: especially for that he saw the coast (it drawing now towards winter here) to be subject to many and grievous storms. And therefore he continued on his course to find out a convenient harbour for that use; searching all the coast from 36° to 47°, as diligently as contrary winds and sundry storms would permit, and yet found none for the purpose. And in the meantime—viz., May 8, by another storm the caunter also was once more severed from us. May 12 we had sight of land in 47°, where we were forced to come to anchor in such road as we could find for the time. Nevertheless our General named the place Cape Hope; by reason of a bay discovered within the headland, which seemed to promise a good and commodious harbour. But by reason of many rocks lying off from the place, we durst not adventure with our ships into it without good and perfect discovery beforehand made. Our General, especially in matters of moment, was never wont to rely on other men's care, how trusty or skilful soever they might seem to be; but always contemning danger, and refusing no toil, he was wont himself to be one, whosoever was a second, at every turn where courage, skill, or industry, was to be employed; neither would he at this time entrust the discovery of these dangers to another's pains, but rather to his own experience, in searching out and sounding of them. A boat being therefore hoisted forth, himself with some others the next morning, May 13, rowed into the bay; and being now very nigh the shore, one of the men of the country showed himself unto him, seeming very pleasant, singing and dancing, after the noise of a rattle which he shook in his hand, expecting earnestly his landing.

But there was suddenly so great an alteration in the weather, into a thick and misty fog, together with an extreme storm and tempest, that our General, being now three leagues from his ship, thought it better to return than either to land or make any other stay; and yet the fog thickened so mightily, that the sight of the ships was bereft them; and if Captain Thomas, upon the abundance of his love and service to his General, had not adventured with his ship to enter the bay in this perplexity, where good advice would not suffer our ships to bear in while the winds were more tolerable and the air clearer, we had sustained some great loss, or our General had been further endangered. Who was now quickly received aboard his ship;[1] out of which, being within the bay, they let fall an anchor, and rode there (God be praised) in safety; but our other ships, riding without, were so oppressed with the extremity of the storm, that they were forced to run off to sea for their own safeguard, being in good hope only of the success of the ship which was gone in to relieve our General. Before this storm arose, our caunter, formerly lost, was come in the same day unto us in the road, but was put to sea again, the same evening, with the rest of the fleet.

The next day, May 14, the weather being fair and the winds moderate, but the fleet out of sight, our General determined to go ashore, to this end, that he might, by making of fires,

[1] Captain Thomas's ship, the Marigold.

give signs to the dispersed ships to come together again into that road; whereby at last they were all assembled, excepting the Swan, lost long time before, and excepting our Portugal prize, called the Mary, which, weighing in this last storm the night before, had now parted company, and was not found again in a long time after. In this place (the people being removed up into the country, belike for fear of our coming) we found near unto the rocks, in houses made for that purpose, as also in divers other places, great store of ostriches, at least to the number of fifty, with much other fowl, some dried and some in drying, for their provision, as it seemed, to carry with them to the place of their dwellings. The ostriches' thighs were in bigness equal to reasonable legs of mutton. They cannot fly at all; but they run so swiftly, and take so long strides, that it is not possible for a man in running by any means to take them, neither yet to come so nigh them as to have a shot at them either with bow or piece; whereof our men had often proof on other parts of the coast, for all the country is full of them. We found there the tools or instruments which the people use in taking them. Among other means they use in betraying these ostriches, they have a great and large plume of feathers, orderly compact together upon the end of a staff, in the forepart bearing the likeness of the head, neck, and bulk of an ostrich, and in the hinder part spreading out very large, sufficient (being held before him) to screen the most part of the body of a man. With this, it seemeth, they stalk, driving them into some strait or neck of land close to the seaside, where spreading long and strong nets, with their dogs which they have in readiness at all times, they overthrow them, and make a common quarry. The country is very pleasant, and seemeth to be a fruitful soil. Being afterwards driven to fall with this place again, we had great acquaintance and familiarity with the people, who rejoiced greatly in our coming, and in our friendship, in that we had done them no harm. But because this place was no fit or convenient harbour for us to do our necessary business, neither yet to make much provision of such things as we wanted, as water, wood, and the like, we departed thence the 15th of May.

At our departure hence, we held our course South and by West, and made about nine leagues in twenty-four hours, bearing very little sail, that our fleet might the easier get up with us, which by reason of contrary winds were cast astern of us. In 47° 30′ we found a bay which was fair, safe, and beneficial to us, very necessary for our use, into which we hauled, and anchored May 17; and the next day we came further into the same bay, where we cast anchor, and made our abode full fifteen days. The very first day of our arrival here, our General having set things in some order, for the despatch of our necessary business, being most careful for his two ships which were wanting, sent forth to the southward Captain Winter in the Elizabeth, Vice-admiral, himself in the Admiral going forth northward into the sea, to see if happily they might meet with either of them; at which time, by the good providence of God, he himself met with the Swan, formerly lost at our departure, from the River of Plate, and brought her into the same harbour the same day; where being after wards unladen and discharged of he freight, she was cast off, and, her iron-work and other necessaries being saved for the better provision of the rest, of the remainder was made firewood and other implements which we wanted. But all this while of the other ship, which we lost so lately in our extremity, we could have no news.

While we were thus employed, after certain days of our stay in this place, being on shore in an island nigh unto the main, where at low-water was free passage on foot from the one to the other, the people of the country did show themselves unto us with leaping, dancing, and holding up their hands, and making outcries after their manner; but, being then high water, we could not g over to them on foot. Wherefore the General caused immediately a boat to be in readiness, and sent unto them such things as he thought would delight them, as knives, bells, bugles, etc. Whereupon they, being assembled together upon a hill, half an English mile from the water-side, sent down two of their company, running one after the other with a great pace, traversing their ground, as it seemed after the manner of their wars, by degrees descending towards the water's side very swiftly. Notwithstanding, drawing nigh unto it, they made a stay, refusing to come near our men: which our men perceiving, sent such things as they had, tied with a string upon a rod, and stuck the same up a reasonable distance from them, where they might see it. And as soon as our men were departed from the place, they came and took those things, leaving instead of them, as in recompense, such feathers as they use to wear about their heads, with a bone made in manner of a toothpick, carved round about the top, and in length about six inches, being very smoothly burnished. Whereupon our General, with divers of his gentlemen and company, at low water, went over to them to the main. Against his coming they remained still upon the hill, and set themselves in a rank, one by one, appointing one of their company to run before them from the one end of the rank to the other, and so back again, continually East and West, with holding up his hands over his head, and yielding forward his body in his running towards the rising and setting of the Sun, and, at every second or third turn at the most, erected his body against the midst of the rank of the people, lifting himself vaulting-wise from the ground towards the Moon, being then over our heads: signifying thereby, as we conceived, that they called the Sun and Moon (which they serve for gods) to witness that they meant nothing towards us but peace. But when they perceived that we ascended the hill apace, and drew nigh unto them, they seemed very fearful of our coming. Wherefore our General, not willing to give them any way any occasion to mislike or be discomfited, retired his company; whereby they were so allured, and did so therein confirm themselves of us that we were no enemies, neither meant them harm, that without all fear divers came down with all speed after us, presently entering into traffic with our men. Notwithstanding, they would receive nothing at our hands, but the same must first be cast upon the ground, using this word, "Zussus," for exchange, "Tóytt," to cast upon the ground. And if they misliked anything, they cried "Coróh! Coróh!" speaking the same with rattling in the throat. The wares we received from them were arrows of reeds, feathers, and such bones as are afore described.

This people go naked, except a skin of fur, which they cast about their shoulders when they sit or lie in the cold; but having anything to do, as going or any other labour, they use it as a girdle about their loins. They wear their hair very long; but lest it might trouble them in their travel, they knit it up with a roll of ostrich feathers, using the same rolls and hair together for a quiver for their arrows, and for a store-house, in which they carry the most things which they carry about them. Some of them, within these rolls, stick on either side of their heads (for a sign of honour in their persons) a large and and plain feather, showing like horns afar off; so that such a head upon a naked body—if devils do appear with horns—might very nigh resemble devils. Their whole bravery and setting out themselves standeth[1] in painting their bodies with divers colours, and such works as they can devise. Some wash[2] their faces with sulphur, or some such like substance;

[1] Consists.

[2] Dye their faces, "or give them a wash," to use a modern phrase.

some paint their whole bodies black, leaving only their necks before and behind white, much like our damsels that wear their squares,[1] their necks and breasts naked. Some paint one shoulder black, another white; and their sides and legs interchangeably, with the same colours, one still contrary to the other. The black part hath set upon it white Moons, and the white part black Suns, being the marks and characters of their gods, as is before noted. They have some commodity[2] by painting of their bodies, for the which cause they use it so generally; and that I gather to be, the defence it yieldeth against the piercing and nipping cold. For the colours being close laid on upon their skin, or rather in their flesh, as by continual renewing of these juices which are laid on, soaked into the inner part thereof, doth fill up the pores so close that no air or cold can enter, or make them once to shrink.

They have clean, comely, and strong bodies; they are swift of foot, and seem very active. Neither is anything more lamentable, in my judgment, than that so goodly a people, and so lively creatures of God, should be ignorant of the true and living God. And so much the more is this to be lamented, by how much they are more tractable, and easy to be brought to the sheepfold of Christ; having, in truth, a land sufficient to recompense any Christian Prince in the world for the whole travail and labour, cost and charges, bestowed in their behalf: with a wonderful enlarging of a kingdom, besides the glory of God by increasing the Church of Christ. It is wonderful to hear, being never known to Christians before this time, how familiar they became in short space with us; thinking themselves to be joined with such a people as they ought rather to serve than offer any wrong or injury unto; presuming that they might be bold with our General as with a father, and with us as brethren and their nearest friends; neither seemed their love less towards us. One of the chiefest among them having on a time received a cap off the General's head, which he did daily wear, removing himself but a little from us, with an arrow pierced his leg deeply, causing the blood to stream out upon the ground: signifying thereby how unfeignedly he loved him, and giving therein a covenant of peace. The number of men which did here frequent our company was about fifty persons. Within, in the southernmost part of this bay, there is a river of fresh water, with a great many profitable islands; of which some have always such store of seals, or sea-wolves, as were able to maintain a huge army of men. Other islands, being many and great, are so replenished with birds and fowl, as if there were no other victuals: a wonderful multitude of people might be nourished by the increase of them for many posterities. Of these we killed some with shot, and some with staves,[3] and took some with our hands, from men's heads and shoulders, upon which they lighted. We could not perceive that the people of the country had any sort of boat or canoe to come to these islands. Their own provisions which they ate, for ought we could perceive, was commonly raw: for we should sometimes find the remnants of seals, all bloody, which they had gnawn with their teeth like dogs. They go all of them armed with a short bow, of about an ell in length, in their hands, with arrows of reeds, and headed with a flint stone, very cunningly cut and fastened.

This bay, by reason of the plenty of seals therein found, insomuch that we killed two hundred in the space of one hour, we called Seal Bay. And having now made sufficient provision

[1] Square-bodied dresses; the reproduction of which is only one of the signs of the fatigue of fashionable inventions which have fallen to the present period.

[2] They gain some convenience.

[3] Or, possibly, by misreading of the text, "stones."

of victuals and other necessaries, as also happily finished all our businesses, on June 3 we set sail from thence; and coasting along towards the Pole Antarctic, on June 12 we fell in with a little bay, in which we anchored for the space of two days, spent in the discharging of our caunter, the Christopher, which we here laid up. The 14th day we weighed again, and kept on our course southward till the 17th, and then cast anchor in another bay, in 50° 20′, lacking but little more than one degree of the mouth of the Straits through which lay our so much desired passage into the South Sea. Here our General, on good advice, determined to alter his course, and turn his stern to the northward again, if haply God would grant that we might find our ship [1] and friends whom we lost in the great storm, as is before said. Forasmuch as, if we should enter the Straits without them in our company, it must needs go hard with them; and we also in the mean time, as well by their absence as by the uncertainty of their state, must needs receive no small discomfort. And therefore, on June 18 in the morning, putting to sea again, with hearty and often prayers we joined watchful industry to serve God's good providence, and held on our purpose to run back towards the Line into the same height [2] in which they were first dissevered from us. The 19th day of June, towards night, having sailed within a few leagues of Port St Julian, we had our ship in sight, for which we gave God thanks with most joyful minds. And forasmuch as the ship was far out of order, and very leaky, by reason of extremity of weather which she had endured, as well before her losing company as in her absence, our General thought good to bear into Port St Julian with his fleet, because it was so nigh at hand, and so convenient a place; intending there to refresh his wearied men, and cherish them who had in their absence tasted such bitterness of discomfort, besides the want of many things which they sustained.

Thus the next day, the 20th of June, we entered Port St Julian, which stands in 49° 30′, and has on the south side of the harbour peaked rocks like towers, and within the harbour many islands, which you may ride hard aboard of, but in going in you must borrow of the north shore. Being now come to anchor, and all things fitted and made safe aboard, our General with certain of his company—viz., Thomas Drake his brother, John Thomas, Robert Winter, Oliver the master-gunner, John Brewer, and Thomas Hood—on June 22 rowed farther in with a boat to find out some convenient place which might yield us fresh water, during the time of our abode there, and furnish us with supply for provision to take to sea with us at our departure; which work, as it was of great necessity, and therefore carefully to be performed, so did not he think himself discharged of his duty if he himself bestowed not the first travail therein, as his use was at all times in all other things belonging to the relieving of our wants and the maintenance of our good estate, by the supply of what was needful. Presently upon his landing he was visited by two of the inhabitants of the place, whom Magellan named "Patagous," or rather "Pentagours," from their huge stature and strength proportionable. These, as they seemed greatly to rejoice at his arrival, so did they show themselves very familiar, receiving at our General's hands whatsoever he gave them, and taking great pleasure in seeing Mr Oliver, the master-gunner of the Admiral, shoot an English arrow—trying with him to shoot at length, but came nothing near him.

Not long after came one more of

---

[1] The "Portugal prize," the Mary; which had on board most or all of their provision of liquor for the voyage.

[2] Latitude; the word is frequently used in this and in other old voyagers' narrations, to signify the amount of ascendant, on one side or the other, towards the plane of the Equator.

the same cast, but of a sourer sort; for he, misliking of the familiarity which his fellows had used, seemed very angry with them, and strove earnestly to withdraw them, and to turn them to become our enemies. Which our General, with his men, not suspecting in them, used them as before, and one Mr Robert Winter, thinking of pleasure to shoot an arrow at length, as Mr Oliver had done before, that he who came last also might have a sight thereof, the string of his bow broke; which, as before it was a terror unto them, so now, broken, it gave them great encouragement and boldness, and, as they thought, great advantage in their treacherous intent and purpose, not imagining that our calivers,[1] swords, and targets, were any munition or weapon of war. In which persuasion—as the General and his company were quietly, without any suspicion of evil, going down towards the boat—they suddenly, being prepared and gotten by stealth behind them, shot their arrows, and chiefly at him which had the bow, not suffering him to string the same again, which he was about to have done, as well as he could; but being wounded in the shoulder at the first shot, and turning about, was sped by an arrow, which pierced his lungs, yet he fell not. But the Master Gunner, being ready to shoot off his caliver, which took not fire in levelling thereof,[2] was presently slain outright. In this extremity, if our General had not been both expert in such affairs, able to judge and to give present direction in the danger thereof, and had not valiantly thrust himself into the dance against these monsters, there had no one of our men, that there were landed, escaped with life. He therefore, giving order that no man should keep any certain ground, but shift from place to place, encroaching still upon the enemy, using their targets and other weapons for the defence of their bodies, and that they should break so many arrows as by any means they could come by, being shot at them, wherein he himself was very diligent, and careful also in calling upon them, knowing that their arrows being once spent, they should have these enemies at their devotion and pleasure, to kill or save; and this order being accordingly taken, himself, I say, with a good courage and trust in the true and living God, taking and shooting off the same piece which the Gunner could not make to take fire, despatched the first beginner of the quarrel, the same man who slew our Master Gunner. For the pieces being charged with a bullet and hail-shot, and well aimed, tore out his belly and guts, with great torment, as it seemed by his cry, which was so hideous and horrible a roar, as if ten bulls had joined together in roaring; wherewith the courage of his partners was so abated, and their hearts appalled, that notwithstanding divers of their fellows and countrymen appeared out of the woods on each side, yet they were glad, by flying away, to save themselves, quietly suffering our men either to depart or stay. Our General chose rather to depart, than to take farther revenge of them, which now he might, by reason of his wounded man, whom for many good parts he loved dearly, and therefore would rather have saved him than slain a hundred enemies; but being past recovery, he died the second day after his being brought on board again. That night, our Master Gunner's body being left ashore, for the speedier bringing of the other aboard, our General himself the next day, with his boat well-appointed, returned to the shore to find it likewise; which they found lying where it was left, but stripped of his uppermost garment, and having an English arrow

[1] The same word as "the modern calibre;" both, by old philologists, derived from "equilibrium." Caliver, strictly, not merely means a gun, but the shot, of whatever weight, which the gun propels.

[2] That is, though he aimed his piece, it missed fire, or flashed in the pan.

stuck in his right eye. Both of these dead bodies were laid together in one grave, with such reverence as was fit for the earthen tabernacles of immortal souls, and with such commendable ceremonies as belong unto soldiers of worth in time of war, which they most truly and rightfully deserved.

Magellan was not altogether deceived in naming them Giants, for they generally differ from the common sort of men, both in stature, bigness, and strength of body, as also in the hideousness of their voice; but yet they are nothing so monstrous or giantlike as they were reported, there being some Englishmen as tall as the highest of any that we did see: but peradventure the Spaniards did not think that ever any Englishman would come thither to reprove them, and thereupon might presume the more boldly to lie; the name "*Pentagones*," "Five Cubits," viz., seven feet and a half, describing the full height, if not somewhat more, of the highest of them. But this is certain, that the Spanish cruelties there used have made them more monstrous in mind and manners than they are in body, and more inhospitable to deal with any strangers that shall come hereafter. For the loss of their friends (the remembrance whereof is assigned and conveyed over from one generation to another among their posterity) breedeth an old grudge, which will not easily be forgotten with so quarrelsome and revengeful a people. Notwithstanding, the terror which they had conceived of us did henceforward so quench their heat, and take down their edge, that they both forgot revenge, and, seeming by their countenance to repent them of the wrong they had offered us that meant them no harm, suffered us to do what we would the whole space of two months after this, without any interruption or molestation by them; and it may be perhaps a means to breed a peace in that people towards all that may, hereafter this, come that way.

To this evil, thus received at the hands of infidels, there was adjoined and grew another mischief, wrought and contrived closely amongst ourselves; as great, yea, far greater, and of far more grievous consequence, than the former, but that it was by God's providence detected and prevented in time; which else had extended itself not only to the violent shedding of innocent blood by murdering our General, and such others as were most firm and faithful to him, but also to the final overthrow of the whole action intended, and to divers other most dangerous effects.[1] These plots had been laid before the voyage began, in England: the very model of them was showed and declared to our General in his garden at Plymouth before his setting sail: which yet he either would not credit as true or likely of a person whom he loved so dearly, and was persuaded of to love him,[2] likewise unfeignedly; or thought by love and benefits to remedy it, if there were any evil purposes conceived against him. And therefore he did not only continue to this suspected and accused person, all countenance, credit, and courtesies which he was wont to show and give him; but increased them, using him in a manner as another himself; giving him the second place in all companies, in his presence; leaving in his hand the state, as it were, of his own person in his absence; imparting unto him all his counsels; allowing him free liberty in all things that were reasonable; and bearing often, at his hands great infirmities: yea, despising that any private inquiry should break so firm a friendship as he meant towards him. And therefore was he oftentimes not

[1] Without entering here on the much-debated question as to Drake's conduct in the trial and execution of Doughty—which has been well called the most dubious act in the life of the great navigator—it may be briefly said, that the balance both of testimony and of character is decidedly in Drake's favour. The matter has been more fully handled in the Introduction.

[2] That is, "and who he was persuaded loved him."

a little offended even with those who, upon conscience of their duty, and knowledge that otherwise they should indeed offend, disclosed from time to time unto him how the fire increased that threatened his own together with the destruction of the whole action.[1]

But at length, perceiving that his lenity of favours did little good, in that the heat of ambition was not yet allayed, nor could be quenched, as it seemed, but by blood; and that the manifold practices[2] grew daily more and more, even to extremities; he thought it high time to call these practices into question before it were too late to call any question of them into hearing. And therefore setting good watch over him, and assembling all his captains and gentlemen of his company together, he propounded to them the good parts which were in the gentleman, the great good will and inward affection, more than brotherly which he had ever since his first acquaintance borne him, not omitting the respect which was had of him among no mean personages in England; and afterwards delivered the letters which were written to him, with the particulars from time to time which had been observed, not so much by himself as by his good friends; not only at sea, but even in Plymouth; not bare words, but writings; not writings alone, but actions, tending to the overthrow of the service in hand, and making away of his[3] person. Proofs were required and alleged, so many and so evident, that the gentleman himself, stricken with remorse of his inconsiderate and unkind dealing, acknowledged himself to have deserved death, yea many deaths; for that he conspired, not only the overthrow of the action, but of the principal actor also, who was not a stranger or ill-willer, but a dear and true friend unto him; and therefore in a great assembly openly besought them, in whose hands justice rested, to take some order for him, that he might not be compelled to enforce his own hands against his own bowels,[4] or otherwise to become his own executioner.

The admiration[5] and astonishment hereat in all the hearers, even those who were his nearest friends and most affected him, was great, yea, in those who for many benefits received from him had good cause to love him; but yet the General was most of all distracted, and therefore withdrew himself, as not able to conceal his tender affection, requiring them that had heard the whole matter to give their judgments as they would another day answer it unto their Prince and unto Almighty God, judge of all the earth. Therefore they all, above forty in number, the chiefest of place and judgment in the whole fleet, after they had discussed diversly of the case, and had alleged whatsoever came in their minds, or could be there produced by any of his other friends, with their own hands, under seal, adjudged that "He had deserved death: and that it stood by no means with their safety to let him live: and therefore they remitted the matter thereof, with the rest of the circumstances, to the General." This judgment, and as it were assize, was held aloud, in one of the islands of the port, which afterwards, in memory hereof, was called the Island of "True Justice and Judgment."

Now after this verdict was thus returned unto our General (unto whom, for his company, Her Majesty before his departure had committed her sword, to use for his safety, with this word: "We do account that he which striketh at thee, Drake, striketh at us"), he called for the guilty party,

---

[1] That threatened his own destruction, along with the ruin of the whole enterprise.

[2] "Practice," in the time of Drake, was used generally in an ill sense—and is thus a rare specimen of a word which has improved, or at least grown less tart, by keeping.

[3] The Admiral's.

[4] A curious and literal description of the Japanese "hari-kari," or "happy despatch."

[5] Wonder.

and caused to be read unto him the several verdicts which were written and pronounced of him. Which being acknowledged for the most part (for none had given heavier sentence against him than he had given against himself), our General proposed unto him this choice: "Whether he would take, to be executed in this island? or to be set a-land on the main? or return into England, there to answer his deed before the Lords of Her Majesty's Council? He most humbly thanked the General for his clemency, extended towards him in such ample sort; and craving some respite to consult thereon, and so make his choice advisedly, the next day he returned this answer: "That albeit he had yielded in his heart to entertain so great a sin, whereof now he was justly condemned; yet he had a care, and that excelling all other cares, to die a Christian man, that whatsoever did become of his clay body, he might yet remain assured of an eternal inheritance in a far better life. This he feared, if he should be set a-land among Infidels, how he should be able to maintain this assurance; feeling, in his own frailty, how mighty the contagion is of lewd custom." And therefore he besought the General most earnestly, "That he would yet have a care and regard of his soul, and never jeopard it amongst heathen and savage Infidels. If he should return into England, he must first have a ship, and men to conduct it, besides sufficient victuals; two of which, although they were had, yet for the third, he thought that no man would accompany him, in so bad a message, to so vile an issue, from so honourable a service. But if that there were who could induce their minds to return with him, yet the very shame of the return would be as death, or grievouser, were that possible: because he should be so long a-dying, and die so often. Therefore he professed, that with all his heart he did embrace the first branch of the General's proffer, desiring only this favour, that they might receive the Holy Communion once again together before his death, and that he might not die other than a gentleman's death."

Though sundry reasons were used by many to persuade him to take either of the other ways, yet when he remained resolute in his former determination, both parts of his last request were granted; and the next convenient day a communion was celebrated by Mr Francis Fletcher, preacher and pastor of the fleet at that time. The General himself communicated at this Sacred Ordinance, with this condemned penitent gentleman, who showed great tokens of a contrite and repentant heart, as who was more deeply displeased with his own act than any man else. And after this holy repast they dined, also at the same table together, as cheerfully in sobriety, as ever in their lives they had done aforetime: each cheering up the other, and taking their leave, by drinking each to other, as if some journey only had been in hand. After dinner, all things being brought in readiness by him that supplied the room of the Provost Marshall, without any dallying, or delaying the time, he came forth and kneeled down, preparing at once his neck for the axe, and his spirit for Heaven; which having done without long ceremony, as who had before digested this whole tragedy, he desired all the rest to pray for him, and willed the executioner to do his office, not to fear nor spare.

Thus having by the worthy manner of his death being much more honourable by it than blamable for any other of his actions) fully blotted out whatever stain his fault might seem to bring upon him, he left unto our fleet a lamentable example of a goodly gentleman who, in seeking advancement unfit for him, cast away himself; and unto posterity a monument of I know not what fatal calamity,[1]

[1] The context shows that these words would have been better reversed; the Narrator plainly referring to the "calamitous fatality" of the place, where both Drake and

as incident to that port, and such like actions, which might haply afford a new pair of Parallels to be added to Plutarch's: In that the same place, near about the same time of the year, witnessed the execution of two gentlemen, suffering both for the like cause, employed both in like service, entertained both in great place, endued both with excellent qualities, the one fifty-eight years after the other. For on the main our men found a gibbet, fallen down, made of a spruce mast, with men's bones underneath it, which they conjectured to be the same gibbet which Magellan commanded to be erected, in the year 1520, for the execution of John Carthagena,[1] the Bishop of Burgor's cousin, who by the King's order was joined with Magellan in commission, and made his Vice-admiral. In the island as we digged to bury this gentleman, we found a great grinding-stone, broken in two parts, which we took and set fast in the ground, the one part at the head, the other at the feet, building up the middle space with other stones and tufts of earth, and engraved in the stones the names of the parties buried there, with the time of their departure, and a memorial of our General's name, in Latin, that it might be the better understood of all that should come after us.

These things thus ended and set in order, our General discharged the Mary—our Portugal prize—because she was leaky and troublesome, defaced her,[2] and then left her ribs and keel upon the island where for two months together we had pitched our tents. And so having wooded, watered, trimmed our ships, despatched all our other business, and brought our fleet into the smallest number—even three only, besides our pinnaces—that we might the easier keep ourselves together, be the better furnished with necessaries, and be the stronger manned, against whatsoever need should be—August 17, we departed out of this port; and being now in great hope of a happy issue to our enterprise, which Almighty God hitherto had so blessed and prospered, we set our coast for the Straits, south-west.

August 20, we fell with the Cape near which lies the entrance into the Straits, called by the Spaniards Capo Virgin Maria,[3] appearing four leagues before you come to it, with high and steep grey cliffs, full of black stars, against which the sea beating showeth as it were the spouting of whales, having the highest of the cape like Cape Vincent in Portugal. At this cape our General caused his fleet, in homage to our Sovereign Lady the Queen's Majesty, to strike their top-sails upon the bunt,[4] as a token of his willing and glad mind to shew his dutiful obedience to her Highness, whom he acknowledged to have full interest and right in that new discovery; and withal, in remembrance of his honourable friend and favourer, Sir Christopher Hatton, he changed the name of the ship which himself went in from the Pelican to be called the Golden Hind.[5] Which ceremonies being ended, together with a sermon, teaching true obedience, with prayers and giving of thanks for Her Majesty and her most honourable Council, with the whole body of the Commonweal and Church of God, we continued our course on into the said frete,[6] where passing with land in sight on both sides, we shortly fell with

Magellan had to exercise the extremity of justice.

[1] Not Don Juan de Carthagena, but Don Luis de Mendoza—upon whom Magellan placed great reliance—suffered, with some other ringleaders in the mutiny, the fate indicated in the text; Don Juan, with several of the less guilty accomplices, being left among the Patagonians.

[2] Stripped her of her planking.

[3] Cabo de las Virgenes, or Cape Virgins, in modern maps.

[4] To lower the topsails half-way, upon the bunt or bend of the sail.

[5] Conjectured to have formed part of the Chancellor's armorial bearings.

[6] A sound or narrow sea; Latin, "fretum."

so narrow a strait, as, carrying with it much wind, often turnings, and many dangers, requireth an expert judgment in him that shall pass the same: it lies WNW. and ESE. But having left this strait astern, we seemed to be come out of a river of two leagues broad, into a large and main sea; having, the night following, an island in sight, which—being in height nothing inferior to the island Fogo, before spoken of—burneth, like it also, aloft in the air, in a wonderful sort, without intermission.

It has formerly been received as an undoubted truth, that the seas, following the course of the first mover, from East to West, have a continual current through the Strait, but our experience found the contrary; the ebbings and flowings here being as orderly—in which the water rises and falls more than five fathoms upright—as on other coasts.

The 24th of August, being Bartholomew's Day, we fell with three islands, bearing triangle-wise one from another: one of them was very fair and large and of a fruitful soil, upon which, being next unto us and the weather very calm, our General with his gentlemen and certain of his mariners then landed, taking possession thereof in Her Majesty's name, and to her use, and calling the same Elizabeth Island. The other two, though they were not so large nor so fair to the eye, yet were they to us exceeding useful, for in them we found great store of strange things,[1] which could not fly at all, nor yet run so fast as that they could escape us with their lives; in body they are less than a goose, and bigger than a mallard, short and thick set together, having no feathers, but instead thereof a certain hard and matted down; their beaks are not much unlike the bills of crows; they lodge and breed upon the land, where, making earths, as the conies do, in the ground, they lay their eggs and bring up their young; their feeding and provision to live on is in the sea, where they swim in such sort, as Nature may seem to have granted them no small prerogative in swiftness, both to prey upon others, and themselves to escape from any others that seek to seize upon them. And such was the infinite resort of these birds to these islands, that in the space of one day we killed no less than 3000, and if the increase be according to the number, it is not to be thought that the world hath brought forth a greater blessing, in one kind of creature in so small a circuit, so necessarily and plentifully serving the use of man. They are a very good and wholesome victual. Our General named these islands, the one Bartholomew, according to the day, the other Saint George's, in honour of England, according to the ancient custom there observed. In the Island of Saint George we found the body of a man, so long dead before, that his bones would not hold together, being moved out of the place whereon they lay.

From these islands to the entrance into the South Sea, the frete is very crooked, having many turnings, and as it were shuttings-up, as if there were no passage at all; by means whereof we were often troubled with contrary winds, so that some of our ships recovering a cape of land, entering another reach, the rest were forced to alter their course and come to anchor where they might. It is true which Magellan reports of this passage: namely, that there be many fair harbours and store of fresh water; but some ships had need to be freighted with nothing else besides anchors and cables, to find ground in most of them to come to anchor; which when any extreme gusts or contrary winds do come, whereunto the place is altogether subject, is a great hindrance to the passage, and carries with it no small danger. The land on both sides is very high and mountainous, having on the North and West side the continent of America, and on the South and East part nothing but islands, among which lie innumerable fretes or passages into the South Sea. The mountains arise with such tops and

[1] Penguins.

spires into the air, and of so rare a height, as they may well be accounted amongst the wonders of the world; environed, as it were, with many regions of congealed clouds and frozen meteors, whereby they are continually fed and increased, both in height and bigness, from time to time, retaining that which they have once received, being little again diminished by the heat of the sun, as being so far from reflection and so nigh the cold and frozen region. But notwithstanding all this, yet are the low and plain grounds very fruitful, the grass green and natural, the herbs, that are of very strange sorts, good and many; the trees, for the most part of them, always green; the air of the temperature of our country; the water most pleasant; and the soil agreeing to any grain which we have growing in our country: a place, no doubt, that lacketh nothing but a people to use the same to the Creator's glory and the increasing of the Church. The people inhabiting these parts made fires as we passed by in divers places.

Drawing nigh the entrance of the South Sea, we had such a shutting-up to the northwards, and such large and open fretes towards the south, that it was doubtful which way we should pass, without further discovery;[1] for which cause, our General having brought his fleet to anchor under an island, himself, with certain of his gentlemen, rowed in a boat to descry the passage: who having discovered a sufficient way toward the North, in their return to their ships met a canoe, under the same island where we rode then at anchor, having in her divers persons. This canoe, or boat, was made of the bark of divers trees, having a prow and a stern standing up, and semicirclewise yielding inward, of one form and fashion, the body whereof was a most dainty mould, bearing in it most comely proportion and excellent workmanship, in so much as to our General and us it seemed never to have been done without the cunning and expert judgment of art; and that not for the use of so rude and barbarous a people, but for the pleasure of some great and noble personage, yea, of some Prince. It had no other closing-up or caulking in the seams, but the stitching with thongs made of seal skins, or other such beast, and yet so close that it received very little or no water at all.

The people are of a mean[2] stature, but well set and compact in all their parts and limbs; they have great pleasure in painting their faces, as the others have, of whom we have spoken before. Within the said Island they had a house of mean building, of certain poles, and covered with skins of beasts, having therein fire, water, and such meat as commonly they can come by, as seals, mussels, and such like. The vessels wherein they keep their water, and their cups in which they drink, are made of barks of trees, as was their canoe, and that with no less skill (for the bigness of the thing), being of a very formal shape and good fashion. Their working tools, which they use in cutting these things and such other, are knives made of most huge and monstrous mussel shells (the like whereof have not been seen or heard of lightly by any travellers, the meat thereof being very savoury and good in eating); which after they have broken off the thin and brittle substance of the edge, they rub and grind them upon stones had for the purpose, till they have tempered and set such an edge upon them, that no wood is so hard but they will cut it at pleasure with the same; whereof we ourselves had experience. Yea, they cut therewith bones of a marvellous hardness, making of them fisgics[3] to kill fish, wherein they have a most pleasant exercise with great dexterity.

The 6th of September we had left astern of us all these troublesome islands, and were entered into the South Sea, or *Mare del Zur*,[4] at the

---

[1] Exploration.

[2] Middling, ordinary.

[3] Or fizgigs; see Note 4, page 128.

[4] Drake was the fourth person who

Cape whereof our General had determined with his whole company to have gone on shore, and there, after a sermon, to have left a monument of Her Majesty, engraven in metal, for a perpetual remembrance, which he had in a readiness for that end prepared: but neither was there any anchoring, neither did the wind suffer us by any means to make a stay. Only this by all our men's observations was concluded: that the entrance, by which we came into this Strait, was in 52°, the middle, in 53° 15′, and the going out in 52° 30′, being 150 leagues in length: at the very entry, supposed also to be about ten leagues in breadth. After we were entered ten leagues within it, it was found not past a league in breadth: farther within, in some places very large, in some very narrow; and in the end found to be no Strait at all, but all islands. Now when our General perceived that the nipping cold, under so cruel and frowning a winter, had impaired the health of some of his men, he meant to have made the more haste again towards the Line, and not to sail any further towards the Pole Antarctic, lest being further from the sun, and nearer the cold, we might haply be overtaken with some greater danger of Sickness. But God, giving men leave to purpose, reserved to himself the disposition of all things; making their intents of none effect, or changing their meaning oft-times clean into the contrary, as may best serve for his own glory and their profit.

For September 7th, the second day after our entrance into the South Sea—called by some *Mare Pacificum*, but proving to us rather to be Mare Furiosum—God by a contrary wind and intolerable tempest seemed to set himself against us, forcing us not only to alter our course and determination, but with great trouble, long time, many dangers, hard escapes, and final separating of our fleet, to yield ourselves unto his will. Yea, such was the extremity of the tempest, that it appeared to us as if he had pronounced a sentence not to stay his hand, nor to withdraw his judgment, till he had buried our bodies, and ships also, in the bottomless depths of the raging sea. In the time of this incredible storm, the 15th of September, the moon was eclipsed in Aries, and darkened about three points, for the space of two glasses; which being ended might seem to give us some hope of alteration and change of weather to the better. Notwithstanding, as the ecliptical conflict could add nothing to our miserable estate, no more did the ending thereof ease us anything at all, nor take away any of our troubles from us: but our eclipse continued still in its full force, so prevailing against us, that, for the space of full fifty-two days together, we were darkened more than the moon by twenty parts, or more than we by any means could ever have preserved or recovered light of ourselves again, if the Son of God, which laid this burthen upon our backs, had not mercifully borne it up with his own shoulders, and upheld us in it by his own power, beyond any possible strength or skill of man. Neither indeed did we at all escape, but, with the feeling of great discomforts through the same. For these violent and extraordinary flaws, such as seldom have been seen, still continuing or rather increasing, September 30th, in the night, caused the sorrowful separation of the Marigold from us; in which was Captain John Thomas, with many others of our dear friends, who by no means that we could conceive could help themselves, but by spooming along before the sea.[1] With

achieved the passage of the Straits, having been preceded by Magellan in 1520, by Loyasa in 1526, and by Juan de Ladrilleros, from the Pacific side, in 1558. The English commander had better fortune than his predecessors, in respect to weather and temperature; accomplishing in about a fortnight what had occupied months.

[1] Running straight before the wind, and with the sea; usually done in the

whom albeit we could never meet again, yet (our General having beforehand given order, that if any of our fleet did lose company the place of resort to meet again should be in 30° or thereabouts upon the coast of Peru towards the Equinoctial) we long time hoped, till experience shewed our hope was vain, that there we should joyfully meet with them: especially for that they were well provided of victuals, and lacked no skilful and sufficient men (besides their Captain) to bring forward the ship to the place appointed.

From the 7th of September, in which the storm began, till the 7th of October, we could not by any means recover any land; having in the meantime been driven so far south as to the 57° and somewhat better. On this day, towards night, somewhat to the northward of that Cape of America whereof mention is made before in the description of our departure from the Strait into this Sea, with a sorry sail we entered a harbour where hoping to enjoy some freedom and ease till the storm was ended, we received within few hours after our coming to anchor so deadly a stroke and hard entertainment that our Admiral left not only an anchor behind her, through the violence and fury of the flaw, but in departing thence also lost the company and sight of our Vice-Admiral, the Elizabeth, partly through the negligence of those who had the charge of her, partly through a kind of desire that some in her had to be out of these troubles, and to be at home again; which (as since is known) they thenceforward by all means assayed and performed. For the very next day, October 8th, recovering the mouth of the Straits again, which we were now so near unto, they returned back the same way by which they came forward, and so coasting Brazil they arrived in England June 2d the year following. So that now our Admiral, if she had retained her old name of Pelican, which she bare at our departure from our Country, she might have been now indeed said to be as a pelican in the wilderness. For albeit our General sought the rest of his fleet with great care, yet could we not have any sight or certain news of them by any means.[1]

From this Bay of Parting of Friends, we were forcibly driven back again into 55° towards the Pole Antarctic. In which height we ran in among the islands before mentioned, lying to the southward of America, through which we passed from one sea to the other, as hath been declared. Where coming to anchor, we found the waters there to have their indraught and free passage, and that through no small guts or narrow channels, but indeed through as large fretes or straits as it hath at the supposed Straits of Magellan, through which we came. Among these islands making our abode with some quietness for a very little while (viz., two days)

case of weak ships, which by lying to the sea might have their masts carried by the board. The Marigold justified the worst apprehensions of her friends, for nothing more was ever heard of her or of her company.

[1] Edward Cliffe, who narrates the voyage of the Elizabeth back to England, denies that Winter intended to desert his Admiral, and declares that some attempts were made to rejoin him. As these attempts, however, seem to have been limited to the lighting of fires on the shore *within* the narrows, just the direction in which Drake did *not* design to prosecute his voyage, they do not seem to have been either very energetic or very sincere. The Elizabeth's company, after resting and recruiting themselves in Port Health for several weeks, desired to resume the enterprise; but Captain Winter compelled them to abandon the voyage, "full sore against the mariners' minds," affirming that he now despaired of the Admiral's safety, or of being able to gain the golden shores of Peru. Winter was the first Englishman to navigate the Straits of Magellan eastward.

and finding divers good and wholesome herbs, together with fresh water; our men, who before were weak, and much impaired in their health, began to receive good comfort, especially by the drinking of one herb (not much unlike that herb which we commonly call Pennyleaf) which, purging with great facility, afforded great help and refreshing to our wearied and sickly bodies. But the winds returning to their old wont, and the seas raging after their former manner, yea everything as it were setting itself against our peace and desired rest, here was no stay permitted us, neither any safety to be looked for. For such was the present danger by forcing and continual flaws, that we were rather to look for present death than hope for any delivery, if God Almighty should not make the way for us. The winds were such as if the bowels of the Earth had set all at liberty, or as if all the clouds under heaven had been called together to lay their force upon that one place. The seas, which by nature and of themselves are heavy, and of a weighty substance, were rolled up from the depths, even from the roots of the rocks, as if it had been a scroll of parchment which by the extremity of heat runneth together; and being aloft were carried in most strange manner and abundance, as feathers or drifts of snow, by the violence of the winds, to water the exceeding tops of high and lofty mountains. Our anchors, as false friends in such a danger, gave over their holdfast, and as if it had been with horror of the thing, did shrink down to hide themselves in this miserable storm, committing the distressed ship and helpless men to the uncertain and rolling seas, which tossed them like a ball in a racket. In this case, to let fall more anchors would avail us nothing; for being driven from our first place of anchoring, so unmeasurable was the depth, that 500 fathoms would fetch no ground. So that the violent storm without intermission; the impossibility to come to anchor; the want of opportunity to spread any sail; the most mad seas; the lee shores; the dangerous rocks; the contrary and most intolerable winds; the impossible passage out; the desperate tarrying there, and inevitable perils on every side, did lay before us so small likelihood to escape present destruction, that if the special providence of God himself had not supported us, we could never have endured that woeful state, as being environed with most terrible and most fearful judgments round about. For, truly, it was more likely that the mountains should have been rent in sunder from the top to the bottom, and cast headlong into the sea, by these unnatural winds, than that we by any help or cunning of man should free the life of any amongst us.[1]

Notwithstanding, the same God of mercy which delivered Jonah out of the whale's belly, and heareth all those that call upon him faithfully in their distress, looked down from heaven, beheld our tears, and heard our humble petitions, joined with holy vows. Even God—whom not the winds and seas alone, but even the devils themselves and powers of hell obey—did so wonderfully free us, and make our way open before us, as it were by his holy angels still guiding and conducting us, that, more than the affright and amaze of this Estate, we received no part of damage in all the things that belonged to us. But escaping from these Straits and miseries, as it were through the needle's eye (that God might have the greater glory in our delivery), by the great and effectual care and travail of our General, the Lord's instrument therein; we could now no longer forbear, but must needs find some place of refuge, as well to provide water, wood, and other necessaries, as to comfort our men, thus worn and tired out by so many and so long intolerable toils; the like whereof, it is to be supposed, no traveller hath felt,

[1] Compare with this account of Drake's difficulties, that of Anson's in the same navigation.

neither hath there ever been such a tempest, that any records make mention of, so violent and of such continuance, since Noah's flood; for, as hath been said, it lasted from September 7th to October 28th, full fifty-two days.

Not many leagues, therefore, to the southward of our former anchoring, we ran in again among these islands, where we had once more better likelihood to rest in peace; and so much the rather, for that we found the people of the country travelling for their living from one island to another in their Canoes, both men, women, and young infants wrapt in skins and hanging at their mothers' backs; with whom we had traffic for such things as they had, as chains of certain shells, and such other trifles. Here the Lord gave us three days to breathe ourselves and to provide such things as we wanted, albeit the same was with continual care and troubles to avoid imminent dangers, which the troubled seas and blustering winds did every hour threaten unto us. But when we seemed to have staid there too long, we were more rigorously assaulted by the not formerly ended but now more violently renewed storm, and driven thence also with no small danger, leaving behind us the greater part of our cable with the anchor; being chased along by the winds and buffeted incessantly in each quarter by the seas, (which our General interpreted as though God had sent them of purpose to the end which ensued), till at length we fell with the uttermost part of land towards the South Pole, and had certainly discovered how far the same doth reach southward from the coast of America aforenamed.[1] The uttermost cape or headland of all these islands stands near in 56°, without[2] which there is no main nor island to be seen to the southwards, but that the Atlantic Ocean and the South Sea meet in a most large and free scope.

It hath been a dream through many ages, that these islands have been a main,[3] and that it hath been *Terra Incognita*, wherein many strange monsters lived. Indeed, it might truly before this time be called *Incognita*,[4] for howsoever the maps and general descriptions of cosmographers, either upon the deceivable reports of other men, or the deceitful imaginations of themselves (supposing never herein to be corrected), have set it down, yet it is true, that before this time it was never discovered or certainly known by any traveller that we have heard of. And here, as in a fit place, it shall not be amiss to remove that error in opinion, which hath been held by many, of the impossible return out of Mare del Zur into the West Ocean by reason of the supposed Eastern current and levant winds, which (say they) speedily carry any thither, but suffer no return. They are herein likewise altogether deceived, for neither did we meet with any such current, nor had we any such certain winds with any such speed to carry us through; but at all times, in our passage there, we found more opportunity to return back again into the West Ocean, than to go forward into Mare del Zur, by means either of current or winds to hinder us, whereof we had experience more than we wished: being glad oftentimes to alter our course, and to fall astern again with frank wind, without any impediment of any such surmised current, farther in one afternoon, than we could fetch up or recover again in a whole day, with a reasonable gale. And in that they allege the narrowness of the frete, and the want of sea-room, to be the cause of this violent current, they are herein no less deceived, than they

[1] Thus Drake accidentally discovered Cape Horn, which received its name from Schouten and Le Maire, who sailed round it for the first time in 1616.

[2] Beyond, outside.

[3] A continent or mainland of themselves.

[4] Elsewhere we read that Drake held himself warranted in changing the title of *Terra Incognita* into *Terra nunc bene Cognita*.

were in the other without reason: for besides that it cannot be said, that there is one only passage, but rather innumerable, it is most certain that, a-seaboard[1] all these islands, there is one large and main sea; wherein if any will not be satisfied, nor believe the report of our experience and eye-sight, he should be advised to suspend his judgment till he have either tried it himself by his own travel, or shall understand, by other travellers, more particulars to confirm his mind herein.

Now as we were fallen to the uttermost part of these islands, October 28th, our troubles did make an end, the storms ceased, and all our calamities (only the absence of our friends excepted) were removed; as if God, all this while, by his secret providence, had led us to make this discovery, which being made, according to his will, he stayed his hand, as pleased his majesty therein, and refreshed us as his servants. At these southerly parts we found the night in the latter end of October to be but two hours long: the sun being yet above seven degrees distant from the Tropic; so that it seems, being in the Tropic, to leave very little or no night at all in that place. There be few of all these islands but have some inhabitants, whose manners, apparel, houses, canoes, and means of living, are like unto those formerly spoken of, a little before our departure out of the Strait. To all these islands did our General give one name, to wit, Elizabethides. After two days' stay which we made in and about these islands, the 30th of October we set sail, shaping our course right North-west, to coast alongst the parts of Peru (for so the general maps set out the land to lie) both for that we might, with convenient speed, fall with the height of 30°, being the place appointed for the rest of our fleet to re-assemble; as also that no opportunity might be lost in the meantime to find them out, if it seemed good to God to direct them to us.

In this course we chanced, the next day, with two islands, being, as it were, storehouses of the most liberal provision of victuals for us, of birds; yielding not only sufficient and plentiful store for us who were present, but enough to have served all the rest also who were absent. Thence, having furnished ourselves to our content, we continued our course, November 1st, still North-west, as we had formerly done; but in going on we soon espied that we might easily have been deceived; and therefore casting about and steering upon another point we found that the general maps did err from the truth in setting down the coast of Peru for twelve degrees at least to the Northward of the supposed Strait, no less than is the NW. point of the compass different from the NE.; perceiving thereby that no man had ever by travel discovered any part of these twelve degrees; and therefore the setters forth of such descriptions are not to be trusted, much less honoured, in their false and fraudulent conjectures which they use, not in this alone, but in divers other points of no small importance.

We found this part of Peru, all alongst to the height of Lima, which is 12° south of the Line, to be mountainous and very barren, without water or wood, for the most part, except in certain places inhabited by the Spaniards, and few others, which are very fruitful and commodious. After we were once again thus fallen with the land, we continually coasted along, till we came to the height of 37° or thereabout; and finding no convenient place of abode, nor likelihood to hear any news of our ships, we ran off again with an island which lay in sight, named of the Spaniards Mucho, by reason of the greatness and large circuit thereof.[2] At this island coming to anchor November 25th, we found it to be a fruitful place, and well stored with sundry sorts of

[1] On the seaward side, to the South.

[2] It is, despite this derivation, marked in the maps as "Mocha," lying off the Chilian coast midway between Valdivia and Concepcion.

good things: as sheep and other cattle, maize (which is a kind of grain whereof they make bread), potatoes, with such other roots; besides that, it is thought to be wonderfully rich in gold, and to want no good thing for the use of man's life. The inhabitants are such Indians as by the cruel and most extreme dealing of the Spaniards have been driven to fly from the main here, to relieve and fortify themselves. With this people our General thought it meet to have traffic for fresh victuals and water; and for that cause, the very same night of our arrival there, himself with divers of his company went ashore, to whom the people with great courtesy came down, bringing with them such fruits and other victuals as they had, and two very fat sheep, which they gave our General for a present. In recompense whereof he bestowed upon them again many good and necessary things; signifying unto them that the end of his coming was for no other cause but by way of exchange, to traffic with them for such things as we needed and they could spare; and, in particular, for such as they had already brought down unto us, besides fresh water, which we desired of them. Herein they held themselves well contented, and seemed to be not a little joyful of our coming, appointing where we should have the next morning fresh water at pleasure, and withal signifying that then also they would bring us down such other things as we desired to serve our turns.

The next day therefore, very early in the morning (all things being made ready for traffic, as also vessels prepared to bring the water), our General, taking great care for so necessary provision, repaired to the shore again; and setting a-land two of his men, sent them with their barricoes[1] to the watering-place assigned the night before. Who having peaceably passed on one-half of the way, were then with no small violence set upon by those traitorous people, and suddenly slain: and to the end that our General with the rest of his company should not only be stayed from rescuing them, but also might fall, if it were possible, into their hands in like manner, they had laid closely behind the rocks an ambushment (as we guessed) of about 500 men, armed and well appointed for such a mischief. Who suddenly attempting their purpose (the rocks being very dangerous for the boat, and the sea-gate[2] exceeding great) by shooting their arrows hurt and wounded every one of our men, before they could free themselves, or come to the use of their weapons to do any good. The General himself was shot in the face, under his right eye, and close by his nose, the arrow piercing a marvellous way in under *basis cerebri*, with no small danger of his life; besides that he was grievously wounded in the head. The rest, being nine persons, in the boat, were deadly wounded in divers part of their bodies, if God almost miraculously had not given cure to the same. For our chief surgeon being dead, and the other absent by the loss of our Vice-admiral, and having none left us but a boy whose goodwill was more than any skill he had, we were little better than altogether destitute of such cunning and helps as so grievous a state of so many wounded bodies did require. Notwithstanding God, by the good advice of our General, and the diligent putting-to of every man's help, did give such speedy and wonderful cure, that we had all great comfort thereby, and yielded God the glory thereof.

The cause of this force and injury by these islanders was no other but the deadly hatred which they bear against their cruel enemies the Spaniards, for the bloody and most tyrannous oppression which they had used towards them. And therefore with purpose against them (suspecting us to be Spaniards indeed, and that the rather by occasion that, though com-

[1] Casks; Spanish, "Barrica," a keg.

[2] The force of the waves lifting the boat towards the rocks.

mand was given to the contrary, some of our men, in demanding water, used the Spanish word "Aqua") sought some part of revenge against us. Our General, notwithstanding he might have revenged this wrong with little hazard or danger, yet being more desirous to preserve one of his own men alive, than to destroy an hundred of his enemies, committed the same to God; wishing this only punishment to them, that they did but know whom they had wronged; and that they had done this injury not to an enemy, but to a friend; not to a Spaniard, but to an Englishman; who would rather have been a patron to defend them, than any way an instrument of the least wrong that should have been done unto them. The weapons which this people use in their wars, are arrows of reeds, with heads of stone very brittle and indented, but darts of a great length, headed with iron or bone.

The same day that we received this dangerous affront, in the afternoon, we set sail from thence; and because we were now nigh the appointed height wherein our ships were to be looked for, as also the extremity and crazy[1] state of our hurt men advising us to use expedition to find some convenient place of repose which might afford them some rest, and yield us necessary supply of fresh victuals for their diet; we bent our course, as the wind would suffer us, directly to run in with the main. Where falling with a bay called Philip's Bay,[2] in 32° or thereabout, November 30, we came to anchor and forthwith manned and sent our boat to discover what likelihood the place would offer to afford us such things as we stood in need of. Our boat doing her uttermost endeavour in a diligent search, yet after long travel could find no appearance of hope for relief, either of fresh victuals or of fresh water; huge herds of wild buffs[3] they might discern, but not so much as any sign of any inhabitant thereabout. Yet in their return to us they descried within the bay an Indian with his canoe, as he was a-fishing; him they brought aboard our General, canoe and all, as he was in it. A comely personage, and of a goodly stature; his apparel was a white garment, reaching scarcely to his knees; his arms and legs were naked; his hair upon his head very long; without a beard, as all the Indians for the most part are. He seemed very gentle, of mild and humble nature, being very tractable to learn the use of everything, and most grateful for such things as our General bestowed upon him. In him we might see a most lively pattern of the harmless disposition of that people, and how grievous a thing it is that they should by any means be so abused as all those are whom the Spaniards have any command or power over.

This man being courteously entertained, and his pains of coming doubly requited, after we had shewed him, partly by signs, and partly by such things as we had, what things we needed, and would gladly receive by his means, upon exchange of such things as he would desire, we sent him away with our boat and his own canoe (which was made of reed straw) to land him where he would. Who being landed, and willing our men to stay his return, was immediately met with by two or three of his friends; to whom imparting his news, and shewing what gifts he had received, he gave so great content, that they willingly furthered his purpose: so that, after certain hours of our men's abode there, he with divers others (among whom was their head or captain) made their return, bringing with them their loadings of such things as they thought would do

[1] Used in the simply physical sense of sickly or weakly.

[2] The name, conferred in honour of the natives afterwards mentioned, who guided them to Valparaiso, has not been maintained in the modern maps; probably the place was Pichidanqui Cove, rather more than a degree to the north of Valparaiso.

[3] Buffaloes, wild oxen.

us good, as some hens, eggs, a fat hog, and such like. All which, that our men might be without all suspicion of all evil to be meant or intended by them, they sent in one of their canoes, a reasonable distance from off the shore, to our boat, the sea-gate being at present very great; and their captain, having sent back his horse, would needs commend himself to the credit of our men, though strangers, and come with them to the General, without any of his own acquaintance or countrymen with him.

By his coming, as we understood that there was no means or way to have our necessities relieved in this place; so he offered himself to be our pilot to a place, and that a good harbour, not far back to the southward again, where, by way of traffic, we might have at pleasure both water and those other things which we stood in need of. This offer our General very gladly received,[1] and so much the rather, for that the place intended was near about the place appointed for the rendezvous of our fleet. Omitting therefore our purpose of pursuing the buffs formerly spoken of, of which we had otherwise determined, if possible, to have killed some, this good news of better provision, and more easy to come by, drew us away; and so the fifth day after our arrival, December 4, we departed hence, and the next day, by the willing conduct of our new Indian pilot, we came to anchor in the desired harbour. This harbour the Spaniards call Valparaiso, and the town adjoining Saint James[2] of Chili: it stands in 35° 40′[3]; where, albeit we neither met with our ships nor heard of them, yet there was no good thing which the place afforded, or which our necessities indeed for the present required, but we had the same in great abundance. Amongst other things, we found in the town divers storehouses of the wines of Chili; and in the harbour a ship called the Captain of Moriall, or the Grand Captain of the South, Admiral to the Islands of Salomon, laden for the most part with the same kind of liquors; only there was besides a certain quantity of fine gold of Baldivia, and a great cross of gold beset with emeralds, on which was nailed a god of the same metal.[4] We spent some time in refreshing ourselves, and easing this ship of so heavy a burthen; and on the 8th day of the same month—having in the meantime sufficiently stored ourselves with necessaries, as wine, bread, bacon, &c., for a long season—we set sail, returning back towards the Line, carrying again our Indian pilot with us, whom our General bountifully rewarded, and enriched with many good things, which pleased him exceedingly, and caused him by the way to be landed in the place where he desired.[5]

---

[1] By other accounts, it had been Drake's purpose to go for Valparaiso, but he oversailed that port, and Felipe—the name of the "head or captain"—undertook to pilot them back, believing them Spaniards.

[2] Santiago, the present capital of Chili.

[3] An obvious misprint for 33° 40′.

[4] Drake's men were welcomed with beat of drum by the few Spaniards on board, and asked to partake of Chili wine, under the belief that they were compatriots. The Spaniards were soon rudely undeceived; but one escaped to shore and alarmed the town, the inhabitants of which speedily took refuge inland. About 1800 jars of wine, and a quantity of gold variously stated at from 25,000 to 60,000 *pesos*, were found in the Grand Captain, when she was subsequently overhauled at sea; she was destined for Peru. Mr Fletcher touches mildly on this act of open piracy; he does not mention at all the sacrilege of which the explorers were guilty, in plundering the church of its ornaments and relics—among the former two cruets, a silver chalice, and an altar-cloth, which became by gift the property of the chaplain himself.

[5] Felipe, who had unwittingly be-

Our necessities being thus to our content relieved, our next care was the regaining—if possible—of the company of our ships so long severed from us: neither would anything have satisfied our General or us so well, as the happy meeting or good news of them. This way therefore, all other thoughts for the present set apart, were all our studies and endeavours bent, how to fit it so as that no opportunity of meeting them might be passed over. To this end, considering that we could not conveniently run in with our ship in search of them to every place where there was likelihood of being a harbour, and that our boat was too little, and unable to carry men enough to encounter the malice or treachery of the Spaniards (if we should by any chance meet with any of them) who are used to show no mercy where they may overmaster; and therefore, meaning not to hazard ourselves to their cruel courtesy, we determined, as we coasted now towards the Line, to search diligently for some convenient place where we might, in peace and safety, stay[1] the trimming of our ship, and the erecting of a pinnace, in which we might with better security than in our boat, and without endangering of our ship by running into each creek, leave no place untried, if happily we might so find again our friends and countrymen.

For this cause, December 19th, we entered a bay not far to the southward of the town of Cyppo,[2] now inhabited by the Spaniards, in 29° 30′; where, having landed certain of our men, to the number of fourteen, to search what conveniency the place was likely to afford for our abiding there, we were immediately descried by the Spaniards of the town of Cyppo aforesaid, who speedily made out 300 men at least, whereof 100 were Spaniards, every one well-mounted upon his horse: the rest were Indians, running as dogs at their heels, all naked, and in most miserable bondage.[3] They could not come anyway so closely, but God did open our eyes to see them, before there was any extremity of danger; whereby our men, being warned, had reasonable time to shift themselves as they could: first from the main to a rock within the sea, and from thence into their boat, which, being ready to receive them, conveyed them with expedition out of the reach of the Spaniards' fury, without the hurt of any man. Only one Richard Minioy, being over bold and careless of his own safety, would not be entreated by his friends, nor feared[4] by the multitude of his enemies, to take the present benefit of his own delivery; but chose either to make 300 men, by outbraving of them, to become afraid, or else himself to die in the place; the latter of which he did. Whose dead body being drawn by the Indians from the rock to the shore, was there manfully by the Spaniards beheaded, the right hand cut off, the heart plucked out; all which they carried away in our sight, and for the rest of his carcass they caused the Indians to shoot it full of arrows, made but the same day, of green wood, and so left it to be devoured by the beasts and fowls, but that we went ashore again and buried it; wherein as there appeareth a most extreme and barbarous cruelty, so doth it declare to the world in what miserable fear the Spaniard holdeth the Government of those parts; living in continual dread of foreign invasion by strangers, or secret cutting of their throats by those whom they kept under them in so shameful slavery, I mean the innocent and harmless Indians. And therefore they make sure to murder what strangers soever

trayed the Spaniards at Valparaiso, was replaced in the post of temporary pilot by a Greek, named Juan Griego, captured on board the Grand Captain, who took Drake as far as Lima.

[1] Remain so long as would suffice for.

[2] Coquimbo.

[3] Other accounts make the numbers 300 horse and 200 foot.

[4] Alarmed.

they can come by, and suffer the Indians by no means to have any weapon longer than they be in present service: as appeared by their arrows cut from the tree the same day, as also by the credible report of others who knew the matter to be true. Yea, they suppose they show the wretches great favour when they do not for their pleasures whip them with cords, and day by day drop their naked bodies with burning bacon, which is one of the least cruelties among many which they universally use against that nation and people.

This not being the place we looked for, nor the entertainment such as we desired, we speedily got hence again, and December 20th, the next day, fell with a more convenient harbour, in a bay somewhat to the northward of the forenamed Cyppo, lying in 27° 25′ South the Line. In this place we spent some time in trimming of our ship, and building of our pinnace, as we desired; but still the grief for the absence of our friends remained with us, for the finding of whom our General, having now fitted all things to his mind, intended—leaving his ship the meanwhile at anchor in the bay — with his pinnace and some chosen men, himself to return back to the southwards again, to see if happily he might either himself meet with them, or find them in some harbour or creek, or hear of them by any others whom he might meet with. With this resolution he set on, but after one day's sailing, the wind being contrary to his purpose, he was forced, whether he would or no, to return again. Within this bay, during our abode there, we had such abundance of fish, not much unlike our gurnard in England, as no place had ever afforded us the like—Cape Blanco only upon the coast of Barbary excepted—since our first setting forth of Plymouth until this time; the plenty whereof in this place was such, that our gentlemen sporting themselves day by day, with four or five hooks or lines, in two or three hours would take sometimes 400, sometimes more, at one time.

All our businesses being thus despatched, January 19th we set sail from hence; and the next place that we fell withal, January 22d, was an island standing in the same height with the north cape of the province of Mormorena. At this island we found four Indians with their canoes, who took upon them to bring our men to a place of fresh water on the foresaid cape; in hope whereof, our General made them great cheer, as his manner was towards all strangers, and set his course by their direction; but when we came unto the place, and had travelled up a long way into the land, we found fresh water indeed, but scarce so much as they had drunk wine in their passage thither. As we sailed along, continually searching for fresh water, we came a place called Tarapaca,[1] and landing there we lighted on a Spaniard who lay asleep, and had lying by him thirteen bars of silver, weighing in all about 4000 Spanish ducats: we would not, could we have chosen, have awaked him of his nap: but seeing we, against our wills, did him that injury, we freed him of his charge, which otherwise perhaps would have kept him waking, and so left him to take out, if it pleased him, the other part of his sleep in more security. Our search for water still continuing, as we landed again not far from thence we met a Spaniard with an Indian boy, driving eight lambs or Peruvian sheep: each sheep bare two leathern bags, and in each bag was 50 pounds weight of refined silver, in the whole 800 pounds weight: we could not endure to see a gentleman Spaniard turned carrier so, and therefore without entreaty we offered our service and became drovers; only his directions were not so perfect that we could keep the way which he intended, for almost as soon as he was parted from us, we with our new kind of carriages were come unto our boats.[2]

---

[1] Better known now by its port of Iquique, a few miles distant.

[2] It is somewhat amusing to notice the grim humour with which the re-

Farther beyond this cape forementioned lie certain Indian towns, from whence, as we passed by, came many of the people in certain bawses[1] made of sealskins; of which two being joined together, of a just length, and side by side, resemble in fashion or form a boat: they have in either of them a small gut, or some such thing, blown full of wind, by reason whereof it floateth, and is rowed very swiftly, carrying in it no small burthen.[2] In these, upon sight of our ships, they brought store of fish of divers sorts, to traffic with us for any trifles we would give them, as knives, margarites,[3] glasses, and such like, whereof men of sixty and seventy years old were as glad as if they had received some exceeding rich commodity, being a most simple and plain-dealing people. Their resort unto us was such as, considering the shortness of the time, was wonderful to us to behold.

Not far from this, viz., in 22° 30′, lay Mormorena,[4] another great town of the same people, over whom two Spaniards held the government; with these our General thought meet to deal, or at least to try their courtesy, whether they would, in way of traffic, give us such things as we needed or no; and therefore, January the 26th, we cast anchor here. We found them more from fear than from love, somewhat tractable, and received from them by exchange many good things, very necessary for our uses. Amongst other things which we had of them, the sheep of the country (such as we mentioned before, bearing the leathern bags) were most memorable. Their height and length was equal to a pretty[5] cow, and their strength fully answerable, if not by much exceeding their size or stature. Upon one of their backs did sit at one time three well-grown and tall men, and one boy, no man's foot touching the ground by a large foot in length, the beast nothing at all complaining of his burthen in the mean time. These sheep have necks like camels, their heads bearing a reasonable resemblance of another sheep. The Spaniards use them to great profit. Their wool is exceeding fine, their flesh good meat, their increase ordinary, and besides they supply the room of horses for burthen or travel; yea, they serve to carry over the mountains marvellous loads, for 300 leagues together, where no other carriage can be made but by them only,[6] thereabout, as also all along, and up into the country throughout the province of Cuzco, the common ground, wheresoever it be taken up, in every hun-

---

verend chaplain carried off acts that in their nature fell very little short of sheer highway robbery.

[1] Boats, "bottoms"; "bawse" may be either connected with "base," or with "buss," a box-shaped small decked vessel employed in fishery.

[2] Answering very much to the description of the Greenland boats, as given by Dr Rae, in his latest book, "The Land of Desolation," where the "women's canoes" or "Omyacks" are made of sealskins extended on a wicker frame.

[3] Beads: the original word, "marguerite" or "margarette," is used to signify a pearl by Wycliffe, and a daisy by Chaucer.

[4] Neither the town nor the province of this name survives in maps of the present day. They seem, however, generally to correspond with the districts of Atacam and Cobija, at the extreme north of Chili, and the contiguous region of Moquegua, at the extreme south of Peru. Cobija town —El Puerto de la Mar—would nearly answer to the latitude ascribed to Mormorena in the text.

[5] A somewhat small or undersized cow, like the Alderney.

[6] All later and more scientific accounts of the llama, or Peruvian sheep, only serve to corroborate Drake's description. They stand to the south American populations of the Cordillera coast, even in these days of partial railroad invasion, much in the same relation as the "ship of the desert" to the Bedaween of Sahara or the Arabian wilderness.

dred pounds weight of earth yieldeth 25s. of pure silver, after the rate of a crown an ounce. The next place likely to afford us any news of our ships (for in all this way from the height where we builded our pinnace, there was no bay or harbour at all for shipping) was the port of the town of Arica, standing in 20°,[1] whither we arrived the 7th of February. This town seemed to us to stand in the most fruitful soil that we saw all alongst these coasts, both for that it is situate in the mouth of a most pleasant and fertile valley, abounding with all good things, as also in that it hath continual trade of shipping, as well from Lima as from all other parts of Peru. It is inhabited by the Spaniards. In two barks here we found some forty and odd bars of silver, of the bigness and fashion of a brickbat, and in weight each of them about twenty pounds; of which we took the burthen on ourselves to ease them, and so departed towards Chowley,[2] with which we fell the second day following, February 9th; and in our way to Lima we met with another bark at Arequipa, which had begun to load some silver and gold, but having had (as it seemed, from Arica by land) some notice of our coming, had unladen the same again before our arrival.[3] Yet in this passage we met another bark laden with linen, some of which we thought might stand us in some stead, and therefore took it with us.

At Lima we arrived February 15th, and notwithstanding the Spaniards' forces, though they had thirty ships at that present in harbour there, whereof seventeen (most of them the especial ships in all the South Seas) were fully ready, we entered and anchored all night in the midst of them, in the Callao, and might have made more spoil amongst them in a few hours, if we had been affected to revenge, than the Spaniards could have recovered again in many years.[4] But we had more care to get up that company which we had so long missed, than to recompense their cruel and hard dealing by an even requital, which now we might have taken. This Lima stands in 12° 30′ South Latitude.[5] Here, albeit no good news of our ships could be had, yet got we the news of some things that seemed to comfort if not to countervail our travels thither; as, namely, that in the ship of one Miguel Angels there, there were 1500 bars of plate; besides some other things (as silks, linen, and in one a chest full of royals of plate), which might stand us in some stead, in the other ships, aboard whom we made somewhat bold to bid ourselves welcome. Here also we heard the report of something that had befallen in and near Europe since our departure thence; in particular of the death of some great personages, as the King of Portugal, and both the Kings of Morocco and Fesse, dead all three in one day at one battle;[6] the

[1] More nearly in 18° 30′ or 18° 40′.

[2] Either Ylo, or Yslay, both lying on the coast between Arica and Quilca, the port of Arequipa; probably Ylo is intended, that town lying within the northern sweep of the Point of Colas, which the very un-Spanish word in the text may have been meant to represent.

[3] The plate amounted to 800 bars of silver, belonging to the King of Spain.

[4] According to another narrative—that of Nuno da Silva, the Portuguese pilot taken at the Cape Verd Islands—the English, being among the ships, enquired for that which had the silver, on board; but learning that all the silver had been carried on shore, they cut the cables of all the ships and the masts of the two largest, and so left them. A ship which came in from Panama nearly fell into the hands of the English in the harbour; alarmed in time, she made her escape to sea, but was afterwards captured and plundered.

[5] Callao is in 12°.

[6] The battle of Alcazar-Seguer, fought August 4th 1578, when Sebastian of Portugal, and his ally Muley Hamet of Fez, fell in the decisive overthrow inflicted on their

death of the King of France, and the Pope of Rome,[1]—whose abominations, as they are in part cut off from some Christian kingdoms, where his shame is manifest, so do his vassals and accursed instruments labour by all means possible to repair that loss, by spreading the same the farther in these parts, where his devilish illusions and damnable deceivings are not known. And as his doctrine takes place anywhere, so do the manners that necessarily accompany the same insinuate themselves together with the doctrine. For as it is true, that in all the parts of America where the Spaniards have any government, the poisonous infection of Popery hath spread itself; so on the other side it is as true, that there is no city, as Lima, Panama, Mexico, etc., no town or village, yea, no house almost in all these provinces, wherein (amongst the other like Spanish virtues) not only whoredom, but the filthiness of Sodom, not to be named among Christians, is not common without reproof: the Pope's pardons[2] being more rife in these parts than they be in any part of Europe for these filthinesses, whereout he sucketh no small advantage. Notwithstanding, the Indians, who are nothing nearer the true knowledge of God than they were before, abhor this most filthy and loathsome manner of living; showing themselves, in respect of the Spaniards, as the Scythians did in respect of the Grecians: who, in their barbarous ignorance, yet in life and behaviour did so far excel the wise and learned Greeks, as they were short of them in the gifts of learning and knowledge. But as the Pope and anti-Christian Bishops labour by their wicked factors[3] with tooth and nail to deface the glory of God, and to shut up in darkness the light of the Gospel; so God doth not suffer His name and religion to be altogether without witness, to the reproving both of his[4] false and damnable doctrine, as also crying out against his unmeasurable and abominable licentiousness of the flesh, even in these parts. For in this city of Lima, not two months before our coming thither, there were certain persons, to the number of twelve, apprehended, examined, and condemned for the profession of the Gospel, and reproving the doctrines of men, with the filthy manners used in the city: of which twelve, six were bound to one stake and burnt, the rest remained yet in prison, to drink of the same cup within few days.

Lastly, here we had intelligence of a certain rich ship which was laden with gold and silver for Panama, that had set forth of this haven the 2d of February. The very next day, therefore, in the morning, the 16th of the same month, we set sail, as long as the wind would serve our turn, and towed our ship as soon as the wind failed; continuing our course towards Panama, making stay nowhere, but hastening all we might, to get sight if it were possible of that gallant ship the Cacafuego, the great glory of the South Sea, which was gone from Lima fourteen days before us. We fell with the port of Paita in 4° 20′, February 20th; with the port Saint Helena and the river and port of Guayaquil, February 24th. We passed the Line on the 28th, and on the 1st of March we fell with Cape Francisco, where, about mid-day, we descried a sail ahead of us, with whom, after once we had spoken with her, we lay still in the same place about six days to recover our breath again, which we had almost spent with hasty following, and to recall to mind what adventures

---

combined invading forces by Muley Moluc, the Emperor of Morocco—himself dying of a lingering malady before the fight began, and dead before it ended.

[1] This reference is somewhat bewildering. Henry III. of France reigned from 1574 to 1589. Gregory XIII. was Pope from 1572 to 1585; and it is difficult to imagine that on leaving England in 1577, Drake and his followers had not learned the death of the predecessors of these potentates.

[2] Indulgences.

[3] Agents.

[4] The Pope's.

had passed us since our late coming from Lima; but especially to do Juan de Anton a kindness, in freeing him of the care of those things with which his ship was laden. This ship we found to be the same of which we had heard, not only in the Callao of Lima, but also by divers occasions afterwards, which now we are at leisure to relate, viz., by a ship which we took between Lima and Paita; by another, which we took laden with wine in the port of Paita; by a third, laden with tackling and implements for ships, besides eighty pounds weight in gold[1] from Guayaquil; and lastly, by Gabriel Alvarez, with whom we talked somewhat nearer the Line. We found her to be indeed the Cacafuego, though before we left her she were now named by a boy of her own the Cacaplata.[2] We found in her some fruit, conserves, sugars, meal, and other victuals, and (that which was the especialest cause of her heavy and slow sailing), a certain quantity of jewels and precious stones, thirteen chests of royals of plate, eighty pounds weight in gold, twenty-six tons of uncoined silver, two very fair gilt silver drinking-bowls, and the like trifles, valued in all at about 360,000 pesos.[3] We gave the master a little linen and the like for these commodities, and at the end of six days we bade farewell and parted. He hastened, somewhat lighter than before, to Panama; we plying off to sea, that we might with more leisure consider what course henceforward were fittest to be taken.[4]

---

[1] Besides a golden crucifix, with "goodly great emeralds" set in it, of which the Reverend Mr Fletcher carefully eschews notice. Between Lima and Panama, the Viceroy, Don Francisco de Toledo, although surprised by this unexpected inroad of the English, had fairly defended his coasts from any descent, and had even put such a force to sea, that Drake judged it prudent—having richer game to stalk—to show a clean pair of heels.

[2] Or, as the jest is narrated in Hakluyt: "The pilot's boy said to our General, 'Captain, our ship shall be called no more the Cacafuego, but the Cacaplata, and your ship shall be the Cacafuego,' which pretty speech of the pilot's boy ministered matter of laughter to us both then and long after."

[3] The total value of the silver and gold alone has been estimated at £750,000 or £1,000,000 of our present money, leaving the precious stones and other booty out of account. It is narrated elsewhere that the commander of the Cacafuego so little suspected the presence of enemies in those seas, as to let the Golden Hind approach him in full security, without taking any defensive measures till the last moment, believing that she was a Spanish ship sent after him with despatches from the Viceroy; yet he did not strike his flag until one of his masts had fallen by the board and he himself was wounded. The silver bowls belonged to the pilot, to whom Drake said, "that these were fine bowls, and he must needs have one of them; to which the pilot yielded, not knowing how to help himself; but, to make this appear less like compulsion, he gave the other to the Admiral's Steward."

[4] Drake at parting gave the captain of the Cacafuego the following letter, addressed to Captain Winter, on the chance of her falling in with the Elizabeth: "Master Winter, if it pleaseth God that you should chance to meet with this ship of Sant John de Anton, I pray you use him well, according to my word and promise given unto them; and if you want anything that is in this ship of Sant John de Anton, I pray you pay them double the value for it, which I will satisfy again, and command your men not to do her any hurt; and what composition or agreement we have made, at my return into England I will by God's help perform, although I am in doubt that this letter will never come to your hands: notwithstanding I am the man I have promised to be: beseech-

And considering that now we were come to the Northward of the Line (Cape Francisco standing in the entrance of the Bay of Panama, in 1° North latitude), and that there was no likelihood or hope that our ships should be before us that way by any means: seeing that, in running so many degrees from the Southernmost Islands hitherto, we could not have any sign or notice of their passage that way, notwithstanding that we had made so diligent search and careful enquiry after them, in every harbour and creek almost, as we had done; and considering also that the time of the year now drew on wherein we must attempt, or of necessity wholly give over, that action which chiefly our General had determined, namely, the discovery of what passage there was to be found about the Northern parts of America from the South Sea into our own Ocean (which being once discovered and made known to be navigable, we should not only do our country a good and notable service, but we also ourselves should have a nearer cut and passage home; where otherwise we were to make a very long and tedious voyage of it, which would hardly agree with our good liking, we having been so long from home already, and so much of our strength separated from us), which could not at all be done if the opportunity were now neglected: we therefore all of us willingly harkened and consented to our General's advice, which was, first to seek out some convenient place wherein to trim our ship, and store ourselves with wood and water and other provisions as we could get, and thenceforward to hasten on our intended journey for the discovery of the said passage, through which we might with joy return to our longed homes.[1]

From this Cape therefore we set onward, March the 7th, shaping our course towards the Island of Cano,[2] with which we fell March 16th, setting ourselves for certain days in a fresh river, between the main and it, for the finishing of our needful businesses, as it is aforesaid. While we abode in this place, we felt a very terrible earthquake, the force whereof was such that our ship and pinnace, riding very near an English mile from the shore, were shaken and did quiver as if it had been laid on dry land. We found here many good commodities which we wanted, as fish, fresh water, wood, &c., besides alargartoes, monkeys, and the like; and in our journey hither we met with one ship more (the last we met with in all those coasts), laden with linen, China silk, and China dishes, amongst which we found also a falcon of gold, handsomely wrought, with a great emerald set in the breast

ing God, the Saviour of all the world, to have us in his keeping, to whom only I give all honour, praise, and glory. What I have written is not only to you Mr Winter, but also to Mr Thomas, Mr Charles, Mr Caube, and Mr Anthony, with all our other good friends, whom I commit to the tuition of him that with his blood redeemed us, and am in good hope that we shall be in no more trouble, but that he will help us in adversity; desiring you, for the passion of Christ, if you fall into any danger, that you will not despair of God's mercy, for he will defend you and preserve you from all danger, and bring us to our desired haven; to whom be all honour, glory, and praise, for ever and ever, Amen.—Your sorrowful Captain, whose heart is heavy for you,

"FRANCIS DRAKE."

[1] "It is not unworthy of notice," says one modern chronicler of this voyage, "that the scheme for exploring a North-eastern channel from the Pacific, thus adopted by Drake, is the same with that recommended about a century later by the celebrated Dampier." See *post*, Dampier's Voyage, Chapter IX.

[2] Off the coast of Nicaragua; it is mentioned by Dampier, who (Chap. VIII., page 172) "coasted along shore, passing by the Gulf of Nicoya, the Gulf of Dulce, and the Island Cano."

of it. From thence we parted the 24th day of the month forenamed, with full purpose to run the nearest course, as the wind would suffer us, without touch of land a long time; and therefore passed by port Papagaya: the port of the Vale, of the most rich and excellent balms of Jericho; Quantapico,[1] and divers others; as also certain gulfs hereabout, which without intermission send forth such continual and violent winds, that the Spaniards, though their ships be good, dare not venture themselves too near the danger of them. Notwithstanding having notice that we should be troubled with often calms and contrary winds if we continued near the coast, and did not run off to sea to fetch the wind, and that if we did so we could not then fall with land again when we would; our General thought it needful that we should run in with some place or other before our departure from the coast, to see if happily we could, by traffic, augment our provision of victuals and other necessaries,—that being at sea we might not be driven to any great want or necessity; albeit we had reasonable store of good things aboard us already.

The next harbour therefore which we chanced with, on April 15th, in 15° 40′, was Guatulco, so named of the Spaniards who inhabited it, with whom we had some intercourse, to the supply of many things which we desired, and chiefly bread, &c. And now having reasonably, as we thought, provided ourselves, we departed from the coast of America for the present; but not forgetting, before we got a-shipboard, to take with us also a certain pot, of about a bushel in bigness, full of royals of plate, which we found in the town, together with a chain of gold and some other jewels which we entreated a gentleman Spaniard to leave behind him as he was flying out of town.[2] From Guatulco we departed the day following, viz., April 16th, setting our course directly into the sea, whereon we sailed 500 leagues in longitude, to get a wind: and between that and June 3d, 1400 leagues in all, till we came into 42° of North latitude, wherein the night following we found such alteration of heat into extreme and nipping cold, that our men in general did grievously complain thereof, some of them feeling their healths much impaired thereby; neither was it that this chanced in the night alone, but the day following carried with it not only the marks, but the stings and force of the night going before, to the great admiration[3] of us all. For besides that the pinching and biting air was nothing altered, the very ropes of our ship were stiff, and the rain which fell was an unnatural congealed and frozen substance, so that we seemed rather to be in the Frozen Zone than any way so near unto the Sun, or these hotter climates. Neither did this happen for the time only, or by some sudden accident, but rather seems indeed to proceed from some ordinary cause, against which the heat of the Sun prevails not; for it came to that extremity in sailing but two degrees farther to the Northward in our course, that though seamen lack not good stomachs, yet it seemed a question to many amongst us whether their hands should feed their mouths, or rather keep themselves within their coverts from the pinching cold that did benumb them. Neither could we impute it to the tenderness of our

[1] Probably Tehuantepec is meant.

[2] Here the voyagers surprised a council engaged in the trial of some Indians accused of trying to burn the town, and carried culprits and judges on board together as temporary prisoners. The name of the man who pursued and plundered the wearer of the golden chain was Thomas Moon. At Guatulco the Portuguese pilot, Nuno da Silva, and all the other prisoners, were liberated; the pilot wrote a narrative of the voyage up to this point, which was sent to the Portuguese Viceroy in India, and afterwards fell into English hands.

[3] Wonder, astonishment.

bodies, though we came lately from the extremity of heat, by reason whereof we might be more sensible of the present cold; insomuch as the dead and senseless creatures were as well affected with it as ourselves: our meat, as soon as it was removed from the fire, would presently in a manner be frozen up, and our ropes and tackling in few days were grown to that stiffness, that what three men afore were able with them to perform, now six men, with their best strength and uttermost endeavour, were hardly able to accomplish: whereby a sudden and great discouragement seized upon the minds of our men, and they were possessed with a great mislike and doubting of any good to be done that way. Yet would not our General be discouraged, but as well by comfortable speeches, of the Divine Providence and of God's loving care over his children, out of the Scriptures, as also by other good and profitable persuasions, adding thereto his own cheerful example, he so stirred them up to put on a good courage, and to quit themselves like men, to endure some short extremity to have the speedier comfort, and a little trouble to obtain the greater glory, that every man was throughly armed with willingness, and resolved to see the uttermost, if it were possible, of what good was to be done that way.

The land in that part of America bearing farther out into the West than we before imagined, we were nearer on it than we were aware; and yet the nearer still we came unto it, the more extremity of cold did seize upon us. The 5th day of June, we were forced by contrary winds to run in with the shore, which we then first descried, and to cast anchor in a bad bay, the best road we could for the present meet with, where we were not without some danger by reason of the many extreme gusts and flaws that beat upon us, which if they ceased and were still at any time, immediately upon their intermission there followed most vile, thick, and stinking fogs, against which the sea prevailed nothing, till the gusts of wind again removed them, which brought with them such extremity and violence when they came, that there was no dealing or resisting against them. In this place was no abiding for us; and to go farther North, the extremity of the cold (which had now utterly discouraged our men) would not permit us; and the winds, directly bent against us, having once got us under sail again, commanded us to the Southward whether we would or no. From the height of 48° in which now we were, to 38°, we found the land, by coasting along it, to be but low and reasonably plain; every hill (whereof we saw many, but none very high), though it were in June, and the sun in his nearest approach unto them, being covered with snow. In 38° 30′ we fell with a convenient and fit harbour, and, June 17th, came to anchor therein, where we continued till the 23d day of July following. During all which time, notwithstanding it was in the height of summer, and so near the sun, yet were we continually visited with like nipping colds as we had felt before; insomuch that if violent exercises of our bodies, and busy employment about our necessary labours, had not sometimes compelled us to the contrary, we could very well have been contented to have kept about us still our winter clothes, yea (had our necessities suffered us), to have kept our beds; neither could we at any time, in whole fourteen days together, find the air so clear as to be able to take the height of sun or star.

And here having so fit occasion (notwithstanding it may seem to be beside the purpose of writing the history of this our voyage), we will a little more diligently enquire into the causes of the continuance of the extreme cold in these parts, as also into the probabilities or unlikelihoods of a passage to be found that way. Neither was it (as hath formerly been touched) tenderness of our bodies, coming so lately out of the heat, whereby the pores were opened, that made us so sensible of the colds we here felt: in this respect, as in many others, we found our God a provident Father and careful Physician for us. We lacked no outward helps nor inward comforts to restore and fortify nature, had it been decayed or weakened in us; neither was there wanting to us the great experience of our General, who had often himself proved the force of the Burning Zone, whose advice always prevailed much to the preserving of a moderate temper in our constitutions; so that even after our departure from the heat we always found our bodies, not as sponges, but strong and hardened, more able to bear out cold, though we came out of excess of heat, than a number of chamber champions could have been, who lie on their feather-beds till they go to sea, or rather, whose teeth in a temperate air do beat in their heads at a cup of cold sack and sugar by the fire. And that it was not our tenderness, but the very extremity of the cold itself, that caused this sensibleness in us, may the rather appear, in that the natural inhabitants of the place (with whom we had for a long season familiar intercourse, as is to be related), who had never been acquainted with such heat, to whom the country, air, and climate was proper, and in whom custom of cold was as it were a second nature, yet used to come shivering to us in their warm furs, crowding close together, body to body, to receive heat one of another, and sheltering themselves under a lee bank, if it were possible, and as often as they could labouring to shroud themselves under our garments also to keep them warm. Besides, how unhandsome and deformed appeared the face of the earth itself! showing trees without leaves, and the ground without greenness, in those months of June and July. The poor birds and fowls not daring (as we had great experience to deserve it) so much as once to arise from their nests after the first egg laid, till it, with all the rest, be hatched and brought to some strength of nature, able to help itself. Only this recompense hath Nature afforded them, that the heat of their own bodies being exceeding great, it perfecteth the creature with greater expedition, and in shorter time than is to be found in many places.

As for the causes of this extremity, they seem not to be so deeply hidden but that they may, at least in part, be guessed at. The chief of which we conceive to be the large spreading of the Asian and American continent, which (somewhat Northward of these parts), if they be not fully joined, yet seem they to come very near one to the other. From whose high and snow-covered mountains the North and North-west winds (the constant visitants of these coasts) send abroad their frozen Nymphs, to the infecting the whole air with this insufferable sharpness: not permitting the Sun, no, not in the pride of his heat, to dissolve the congealed water and snow which they have breathed out so nigh the Sun, and so many degrees distant from themselves. And that the North and North-west winds are here constant in June and July, as the North wind alone is in August and September, we not only found by our own experience, but were fully confirmed in the opinion thereof by the continued observations of the Spaniards. Hence comes the general squalidness and barrenness of the country; hence comes it that in the midst of their Summer the snow hardly departeth even from their very doors, but is never taken away from their hills at all; hence come those thick mists and most stinking fogs, which increase so much the more by how much higher the Pole is raised:[1] wherein a blind pilot is as good as the best director of a course. For the Sun striving to perform his natural office, in elevating the vapours out of these inferior bodies, draweth necessarily abundance of moisture out of the sea; but the nipping cold, from the former causes, meeting and opposing the Sun's endeavour, forces him to give over his work imperfect, and, instead of higher elevation, to leave in the lowest

[1] The nearer one approaches to the Pole, causing the North Star to rise apparently higher in the heavens.

region, wandering upon the face of the earth and waters, as it were a second sea, through which its own beams cannot possibly pierce, unless sometimes when the sudden violence of the winds doth help to scatter and break through it; which thing happeneth very seldom, and when it happeneth is of no continuance. Some of our mariners in this voyage had formerly been at Wardhouse,[1] in 72° of North latitude, who yet affirmed that they felt no such nipping cold there in the end of the Summer, when they departed thence, as they did here in those hottest months of June and July. And also from these reasons we conjecture, that either there is no passage at all through these Northern coasts (which is most likely); or if there be, that yet it is unnavigable. Add hereunto, that though we searched the coast diligently, even unto the forty-eighth degree, yet found we not the land to trend so much as one point in any place towards the East, but rather running on continually North-west, as if it went directly to meet with Asia; and even in that height, when we had a frank wind to have carried us through, had there been a passage, yet we had a smooth and calm sea, with ordinary flowing and reflowing, which could not have been had there been a frete;[2] of which we rather infallibly concluded, than conjectured, that there was none. But to return.

The next day after our coming to anchor in the aforesaid harbour,[3] the people of the country shewed themselves, sending off a man with great expedition to us in a canoe. Who being yet but a little from the shore, and a great way from our ship, spake to us continually as he came rowing on. And at last at a reasonable distance staying himself, he began more solemnly a long and tedious oration, after his manner: using in the delivery thereof many gestures and signs, moving his hands, turning his head and body many ways; and after his oration ended, with great show of reverence and submission returned back to shore again. He shortly came again the second time in like manner, and so the third time, when he brought with him, as a present from the rest, a bunch of feathers, much like the feathers of a black crow, very neatly and artificially[4] gathered upon a string, and drawn together into a round bundle; being very clean and finely cut, and bearing in length an equal proportion one with another: a special cognisance (as we afterwards observed) which they that guard their King's person wear upon their heads. With this also he brought a little basket made of rushes, and filled with an herb which they called "Tabáh."[5] Both which, being tied to a short rod, he cast into our boat. Our General intended to have recompensed him immediately with many good things he would have bestowed on him; but entering into the boat to deliver the same, he could not be drawn[6] to receive them by any means, save one hat, which being cast into the water out of the ship, he took up (refusing utterly to meddle with any other thing, though it were upon a board put off to him), and so presently made his return. After which time our boat could row no way, but, wondering at us as at gods, they would follow the same with admiration.

The third day following, viz., the 21st, our ship, having received a leak at sea, was brought to anchor nearer the shore, that, her goods being landed, she might be repaired; but for that we were[7] to prevent any danger that might chance against our safety, our General first of all landed his men, with all necessary provision

---

1 Wardhys, at the extreme north-east point of Norway.

2 A narrow passage or contracted channel.

3 The Bay of San Francisco, the present prosperous capital of California.

4 Cleverly, skilfully.

5 Tobacco—"tabac" in French.

6 Induced, tempted.

7 Were obliged or bound

to build tents and make a fort for the defence of ourselves and goods, and that we might under the shelter of it with more safety (whatever should befall) end our business. Which when the people of the country perceived us doing, as men set on fire to war in defence of their country, in great haste and companies, with such weapons as they had, they came down unto us, and yet with no hostile meaning or intent to hurt us; standing, when they drew near, as men ravished in their minds with the sight of such things as they never had seen or heard of before that time: their errand being rather with submission and fear to worship us as gods, than to have any war with us as with mortal men. Which thing, as it did partly show itself at that instant, so did it more and more manifest itself afterwards, during the whole time of our abode amongst them. At this time, being willed by signs to lay from them their bows and arrows, they did as they were directed; and so did all the rest, as they came more and more by companies unto them, growing in a little while to a great number, both of men and women. To the intent, therefore, that this peace which they themselves so willingly sought might, without any cause of the breach thereof on our part given, be continued, and that we might with more safety and expedition end our businesses in quiet, our General, with all his company, used all means possible gently to entreat them, bestowing upon each of them liberally good and necessary things to cover their nakedness; withal signifying unto them that we were no gods, but men, and had need of such things to cover our own shame; teaching them to use them to the same ends. For which cause also we did eat and drink in their presence, giving them to understand that without that we could not live, and therefore were but men as well as they. Notwithstanding, nothing could persuade them, nor remove that opinion which they had conceived of us, that we should be gods.

In recompense of those things which they had received of us, as shirts, linen cloth, &c., they bestowed upon our General, and divers of our company, divers things; as feathers, cauls of network, the quivers of their arrows, made of fawn-skins, and the very skins of beasts that their women wore upon their bodies. Having thus had their fill of this time's visiting and beholding of us, they departed with joy to their houses; which houses are digged round within the earth, and have from the uppermost brims of the circle clefts of wood set up, and joined close together at the top, like our spires on the steeple of a church; which being covered with earth, suffer no water to enter, and are very warm. The door in the most part of them performs the office also of a chimney to let out the smoke: it is made in bigness and fashion like to an ordinary scuttle in a ship, and standing slopewise. Their beds are the hard ground, only with rushes strewed upon it, and, lying round about the house, have their fire in the midst, which, by reason that the house is but low vaulted, round, and close, giveth a marvellous reflection to their bodies to heat the same. Their men for the most part go naked; the women take a kind of bulrushes, and kembing it[1] after the manner of hemp, make themselves thereof a loose garment, which being knit about their middles, hangs down about their hips, and so affords to them a covering of that which Nature teaches should be hidden; about their shoulders they wear also the skin of a deer, with the hair upon it. They are very obedient to their husbands, and exceeding ready in all services; yet of themselves offering to do nothing, without the consent or being called of the men. As soon as they were returned to their houses, they began amongst themselves a kind of most lamentable weeping and crying out, which they continued also a great while together, in such sort that in the place where they left us (being

[1] Combing it out, or "heckling" it.

near about three-quarters of an English mile distant from them) we very plainly, with wonder and admiration, did hear the same, the women especially extending their voices in a most miserable and doleful manner of shrieking. Notwithstanding this humble manner of presenting themselves, and awful demeanour used towards us, we thought it no wisdom too far to trust them (our experience of former Infidels dealing with us before, made us careful to provide against an alteration of their affections or breach of peace if it should happen); and therefore with all expedition we set up our tents, and entrenched ourselves with walls of stone; that so being fortified within ourselves, we might be able to keep off the enemy (if they should so prove) from coming amongst us without our good wills. This being quickly finished, we went the more cheerfully and securely afterwards about our other business.

Against the end of two days, during which time they had not been with us again, there was gathered together a great assembly of men, women, and children (invited by the report of them which first saw us, who, as it seems, had in that time of purpose dispersed themselves into the country, to make known the news), who came now the second time unto us, bringing with them, as before had been done, feathers and bags of "Tabáh" for presents, or rather indeed for sacrifices, upon this persuasion that we were gods. When they came to the top of the hill, at the bottom whereof we had built our fort, they made a stand; where one, appointed as their chief speaker, wearied both us his hearers, and himself too, with a long and tedious oration, delivered with strange and violent gestures, his voice being extended to the uttermost strength of nature, and his words falling so thick one in the neck of another, that he could hardly fetch his breath again. As soon as he had concluded, all the rest, with a reverent bowing of their bodies (in a dreaming manner, and long producing of the same) cried "Oh:" thereby giving their consents that all was very true which he had spoken, and that they had uttered their mind by his mouth unto us. Which done, the men laying down their bows upon the hill, and leaving their women and children behind them, came down with their presents; in such sort as if they had appeared before a God indeed, thinking themselves happy that they might have access unto our General, but much more happy when they saw that he would receive at their hands those things which they so willingly had presented: and no doubt they thought themselves nearest unto God when they sat or stood next to him. In the meantime the women, as if they had been desperate, used unnatural violence against themselves, crying and shrieking piteously, tearing their flesh with their nails from their cheeks in a monstrous manner, the blood streaming down along their breasts; besides, despoiling the upper parts of their bodies of those single coverings they formerly had, and holding their hands above their heads that they might not rescue their breasts from harm, they would with fury cast themselves upon the ground, never respecting whether it were clean or soft, but dashed themselves in this manner on hard stones, knobby hillocks, stocks of wood, and pricking bushes, or whatever else lay in their way, iterating the same course again and again; yea, women great with child, some nine or ten times each, and others holding out till fifteen or sixteen times, till their strength failed them, exercised this cruelty against themselves: a thing more grievous for us to see or suffer, could we have help[1] it, than trouble to them, as it seemed, to do it. This bloody sacrifice, against our wills, being thus performed, our General, with his company, in the presence of those strangers, fell to prayers; and by signs in lifting up our eyes and hands to heaven, signified unto them that that God whom we did serve, and whom they ought to worship,

[1] Helped, prevented.

was above: beseeching God, if it were His good pleasure, to open by some means their blinded eyes, that they might in due time be called to the knowledge of Him, the true and everlasting God, and of Jesus Christ whom He hath sent, the salvation of the Gentiles. In the time of which prayers, singing of Psalms, and reading of certain chapters in the Bible, they sat very attentively: and observing the end at every pause, with one voice still cried "Oh," greatly rejoicing in our exercises. Yea, they took such pleasure in our singing of Psalms, that whensoever they resorted to us, their first request was commonly this, "Gnaáh," by which they entreated that we would sing. Our General having now bestowed upon them divers things, at their departure they restored them all again, none carrying with him anything of whatsoever he had received, thinking themselves sufficiently enriched and happy that they had found so free access to see us.

Against the end of three days more (the news having the while spread itself farther, and as it seemed a great way up into the country), were assembled the greatest number of people which we could reasonably imagine to dwell within any convenient distance round about. Amongst the rest the King himself, a man of a goodly stature and a comely personage, attended with his guard of about 100 tall and warlike men, this day, June 26th, came down to see us. Before his coming, were sent two ambassadors or messengers to our General, to signify that their "Hióh," that is, their King, was coming and at hand. They in the delivery of their message, the one spake with a soft and low voice, prompting his fellow; the other pronounced the same, word by word, after him with a voice more audible, continuing their proclamation, for such it was, about half an hour. Which being ended, they by signs made request to our General to send something by their hands to their "Hióh" or King, as a token that his coming might be in peace. Our General willingly satisfied their desire; and they, glad men, made speedy return to their "Hióh." Neither was it long before their King (making as princely a show as possibly he could) with all his train came forward. In their coming forwards they cried continually after a singing manner, with a lusty courage. And as they drew nearer and nearer towards us, so did they more and more strive to behave themselves with a certain comeliness and gravity in all their actions. In the forefront came a man of a large body and goodly aspect, bearing the sceptre or royal mace (made of a certain kind of black wood, and in length about a yard and a half), before the King. Whereupon hung two crowns, a bigger and a less, with three chains of a marvellous length, and often doubled, besides a bag of the herb "Tabáh." The crowns were made of network, wrought upon most curiously with feathers of divers colours, very artificially placed, and of a formal fashion. The chains seemed of a bony substance, every link or part thereof being very little, thin, most finely burnished, with a hole pierced through the midst. The number of links going to make one chain is in a manner infinite; but of such estimation is it amongst them, that few be the persons that are admitted to wear the same; and even they to whom it is lawful to use them, yet are stinted what number they shall use, as some ten, some twelve, some twenty, and as they exceed in number of chains, so thereby are they known to be the more honourable personages.

Next unto him that bare this sceptre was the King himself, with his guard about him; his attire upon his head was a caul[1] of network, wrought upon somewhat like the crowns, but differing much both in fashion and perfectness of work; upon his shoulders he had on a coat of the skins of conies, reaching to his waist; his guard also had each coats of the same shape, but of other skins; some having cauls likewise stuck with

[1] Cowl, cap.

feathers, or covered over with a certain down, which groweth up in the country upon an herb much like our lettuce, which exceeds any other down in the world for fineness, and being laid upon their cauls, by no winds can be removed. Of such estimation is this herb amongst them, that the down thereof is not lawful to be worn, but of such persons as are about the King (to whom also it is permitted to wear a plume of feathers on their heads in sign of honour), and the seeds are not used but only in sacrifice to their gods. After these in their order did follow the naked sort of common people, whose hair, being long, was gathered into a bunch behind, in which stuck plumes of feathers; but in the forepart only single feathers like horns, every one pleasing himself in his own device. This one thing was observed to be general amongst them all, that every one had his face painted, some with white, some black, and some with other colours, every man also bringing in his hand one thing or other for a gift or present. Their train or last part of their company consisted of women and children, each woman bearing against her breast a round basket or two, having within them divers things, as bags of "Tabáh," a root which they call "Petah,"[1] whereof they make a kind of meal and either bake it into bread or eat it raw; broiled fishes, like a pilchard; the seed and down aforenamed; with such like. Their baskets were made in fashion like a deep bowl, and though the matter were rushes, or such other kind of stuff, yet was it so cunningly handled, that the most part of them would hold water: about the brims they were hung with pieces of the shells of pearls, and in some places with two or three links at a place of the chains forenamed: thereby signifying that they were vessels wholly dedicated to the only[2] use of the gods they worshipped, and besides this they were wrought upon with the matted down of red feathers, distinguished into divers works and forms.

In the mean time, our General having assembled his men together (as forecasting the danger and worst that might fall out) prepared himself to stand upon sure ground, that we might at all times be ready in our own defence, if any thing should chance otherwise than was looked for or expected. Wherefore every man being in a warlike readiness, he marched within his fenced place, making against their approach a most warlike show (as he did also at all other times of their resort), whereby if they had been desperate enemies they could not have chosen but have conceived terror and fear, with discouragement to attempt anything against us, in beholding of the same.

When they were come somewhat near unto us, trooping together they gave us a common or general salutation, observing in the mean time a general silence. Whereupon, he who bare the sceptre before the King, being prompted by another whom the King assigned to that office, pronounced with an audible and manly voice what the other spake to him in secret, continuing, whether it were his oration or proclamation, at the least half an hour. At the close whereof there was a common "Amen" in sign of approbation given by every person: and the King himself, with the whole number of men and women (the little children only remaining behind), came further down the hill, and as they came set themselves again in their former order. And being now come to the foot of the hill and near our fort, the sceptre-bearer, with a composed countenance and stately carriage, began a song, and answerable thereunto observed a kind of measures in a dance: whom the King with his guard and every other sort of person following, did in like manner sing and dance, saving only the women, who danced but kept silence. As they danced they still came on: and our General, perceiving their plain and simple meaning, gave

[1] Probably the potato.

[2] Sole.

order that they might freely enter without interruption within our bulwark. Where after they had entered, they yet continued their song and dance a reasonable time, their women also following them with their wassail bowls in their hands, their bodies bruised, their faces torn, their dugs, breasts, and other parts bespotted with blood, trickling down from the wounds which with their nails they had made before their coming After that they had satisfied or rather tired themselves in this manner, they made signs to our General to have him sit down; unto whom both the King and divers others made several orations, or rather, indeed, if we had understood them, supplications, that he would take the province and kingdom into his hand, and become their king and patron: making signs that they would resign unto him their right and title in the whole land, and become his vassals in themselves and their posterities: which that they might make us indeed believe that it was their true meaning and intent, the King himself, with all the rest, with one consent and with great reverence, joyfully singing a song, set the crown upon his head, enriched his neck with all their chains, and offering him many things, honoured him by the name of "Hióh." Adding thereunto, as it might seem, a song and dance of triumph, because they were not only visited of the gods (for so they still judged us to be) but the great and chief god was now become their god, their king and patron, and themselves were become the only happy and blessed people in the world.

These things being so freely offered, our General thought not meet to resign or refuse the same, both for that he would not give them any cause of mistrust or disliking of him (that being the only place wherein at this present we were of necessity enforced to seek relief of many things), and chiefly for that he knew not to what good end God had brought this to pass, or what honour and profit it might bring to our country in time to come. Wherefore, in the name and to the use of Her most excellent Majesty, he took the sceptre, crown, and dignity of the said country into his hand; wishing nothing more than that it had lain so fitly for Her Majesty to enjoy, as it was now her proper own, and that the riches and treasures thereof (wherewith in the upland countries it abounds) might with as great conveniency be transported, to the enriching of her kingdom here at home, as it is in plenty to be attained there; and especially that so tractable and loving a people as they shewed themselves to be might have means to have manifested their most willing obedience the more unto her, and by her means, as a Mother and Nurse of the Church of Christ, might by the preaching of the Gospel be brought to the right knowledge and obedience of the true and ever-living God. The ceremonies of this resigning and receiving of the kingdom being thus performed, the common sort, both of men and women, leaving the the King and his guard about him, with our General, dispersed themselves among our people, taking a diligent view or survey of every man; and finding such as pleased their fancies (which commonly were the youngest of us), they presently enclosing them about offered their sacrifices unto them, crying out with lamentable shricks and moans, weeping and scratching and tearing their very flesh of their faces with their nails; neither were it the women alone which did this, but even old men, roaring and crying out, were as violent as the women were. We groaned in spirit to see the power of Satan so far prevail in seducing these so harmless souls, and laboured by all means, both by showing our great dislike, and, when that served not, by violent withholding of their hands from that madness; directing them, by our eyes and hands lift up towards heaven, to the living God whom they ought to serve. But so mad were they upon their idolatry, that forcibly withholding them would not prevail; for as soon as they could get liberty to their hands again they would be as violent as they were before, till such time as they whom they worshipped were conveyed from them into the tents; whom yet, as men beside themselves, they would with fury and outrage seek to have again.

After that time had a little qualified their madness, they then began to show and make known unto us their griefs[1] and diseases which they carried about them; some of them having old aches, some shrunk sinews, some old sores and cankered ulcers, some wounds more lately received, and the like: in most lamentable manner craving help and cure thereof from us, making signs, that if we did but blow upon their griefs, or but touched the diseased places, they would be whole. Their griefs we could not but take pity on them, and to our power desire to help them; but that (if it pleased God to open their eyes) they might understand we were but men and no gods, we used ordinary means, as lotions, plasters, and unguents, most fitly, as far as our skills could guess, agreeing to the natures of their griefs; beseeching God, if it made for his glory, to give cure to their diseases by these means. The like we did from time to time as they resorted unto us. Few were the days wherein they were absent from us, during the whole time of our abode in that place; and ordinarily every third day they brought their sacrifices, till such time as they certainly understood our meaning, that we took no pleasure but were displeased with them; whereupon their zeal abated, and their sacrificing, for a season, to our good liking ceased. Notwithstanding they continued still to make their resort unto us in great abundance, and in such sort, that they oftentimes forgot to provide meat for their own sustenance, so that our General (of whom they made account as of a father) was fain to perform the office of a father towards them, relieving them with such victuals as we had provided for ourselves, as mussels, seals, and such like, wherein they took exceeding much content; and seeing that their sacrifices were displeasing unto us, yet (having gratitude) they sought to recompense us with such things as they had, which they willingly forced upon us, though it were never so necessary or needful for themselves to keep. They are people of a tractable, free, and loving nature, without guile or treachery; their bows and arrows (their only weapons, and almost all their wealth) they use very skilfully, but yet not to do any great harm with them, being by reason of their weakness more fit for children than for men, sending the arrows neither far off nor with any great force: and yet are the men commonly so strong of body, that that which two or three of our men could hardly bear, one of them would take upon his back, and without grudging carry it easily away, up hill and down hill, an English mile together. They are also exceeding swift in running, and of long continuance, the use whereof is so familiar with them, that they seldom go, but for the most part run. One thing we observed in them with admiration, that if at any time they chanced to see a fish so near the shore that they might reach the place without swimming, they would never, or very seldom, miss to take it.

After that our necessary businesses were well despatched, our General, with his gentlemen and many of his company, made a journey up into the land, to see the manner of their dwelling, and to be the better acquainted with the nature and commodities of the country. Their houses were all such as we have formerly described, and being many of them in one place, made several villages here and there. The inland we found to be far different from the shore, a goodly country and fruitful soil, stored with many blessings fit for the use of man. Infinite was the company of very large and fat deer which there we saw by

[1] Used here, of course, in the merely physical sense of pain, or wound, or sore.

thousands, as we supposed, in a herd; besides a multitude of a strange kind of conies, by far exceeding them in number. The head and body, in which they resemble other conies, are but small; his tail, like the tail of a rat, exceeding long, and his feet like the paws of a want or mole; under his chin, on either side, he hath a bag, into which he gathereth his meat, when he hath filled his belly abroad, that he may with it either feed his young, or feed himself when he lists not to travel from his burrow. The people eat their bodies, and make great account of their skins, for their King's holiday coat was made of them.[1]

This country our General named Albion,[2] and that for two causes: the one in respect of the white banks and cliffs which lie towards the sea; the other, that it might have some affinity, even in name also, with our own country, which was sometimes so called. Before we went from thence, our General caused to be set up a monument of our being there, as also of Her Majesty's and successor's right and title to that kingdom: namely, a plate of brass, fast nailed to a great and firm post, whereon is engraven her Grace's name, and the day and year of our arrival there, and of the free giving up of the province and kingdom, both by the King and people, into Her Majesty's hands; together with Her Highness' picture and arms, in a piece of sixpence current English money, showing itself by a hole made of purpose through the plate; underneath was likewise engraven the name of our General, &c. The Spaniards never had any dealing, or so much as set a foot in this country, the utmost of their discoveries reaching only to many degrees southward of this place.[3]

And now as the time of our departure was perceived by them to draw nigh, so did the sorrows and miseries of this people seem to themselves to increase upon them, and the more certain they were of our going away, the more doubtful they showed themselves what they might do: so that we might easily judge that that joy (being exceeding great) wherewith they received us at our first arrival, was clean drowned in their excessive sorrow for our departing. For they did not only lose on a sudden all mirth, joy, glad countenance, pleasant speeches, agility of body, familiar rejoicing one with another, and all pleasure whatever flesh and blood might be delighted in, but with sighs and sorrowings, with heavy hearts and grieved minds, they poured out woeful complaints and moans, with bitter tears and wringing of their hands, tormenting themselves. And as men refusing all comfort they only accounted themselves as castaways, and those whom the gods were about to forsake: so that nothing we could say or do was able to ease them of their so heavy a burthen, or to deliver them from so desperate a strait, as our leaving of them did seem to them that it would cast them into.

---

[1] Captain Beechey, in his "Voyage to the Pacific," says that the fields about San Francisco are burrowed by a small rat resembling the field-mouse, by a larger mountain rat, and by another little animal resembling a squirrel, called the "ardillo," which is excellent eating. The coney described by Drake is thought to answer most closely to the Canada pouched rat, or *Mus barsarius*.

[2] More correctly, New Albion; the whiteness of the cliffs, which suggested the name to Drake, has been noted by subsequent voyagers.

[3] This is a mistake, for Juan Rodriguez Cabrillo, a Portuguese by birth, had by command of the Viceroy of New Spain explored the same coast thirty-seven years before. Indeed, some English editors have shown a decided inclination to take a very liberal discount from Mr Fletcher's eloquent and elaborate account of the doings in California—so much in contrast with the meagre details he gives of such really important incidents as the combat with and capture of the Cacafuego.

Howbeit, seeing they could not still enjoy our presence, they (supposing us to be gods indeed) thought it their duty to entreat us that, being absent, we would yet be mindful of them; and making signs of their desires that in time to come we would see them again, they stole upon us a sacrifice, and set it on fire ere we were aware, burning therein a chain and a bunch of feathers. We laboured by all means possible to withhold or withdraw them, but could not prevail, till at last we fell to prayers and singing of Psalms, whereby they were allured immediately to forget their folly, and leave their sacrifice unconsumed, suffering the fire to go out; and imitating us in all our actions, they fell a-lifting of their eyes and hands to heaven, as they saw us do.

The 23d of July they took a sorrowful farewell of us; but, being loth to leave us, they presently ran to the top of the hills to keep us in their sight so long as they could, making fires before and behind and on each side of them, burning therein (as is to be supposed) sacrifices at our departure.

Not far without this harbour did lie certain islands (we called them the Islands of Saint James),[1] having on them plentiful and great store of seals and birds, with one of which we fell July 24th, whereon we found such provision as might competently serve our turn for a while. We departed again the day next following, July 25th. And our General now considering that the extremity of the cold not only continued, but increased, the Sun being gone farther from us, and that the wind blowing still, as it did at first, from the North-west, cut off all hope of finding a passage through these Northern parts, thought it necessary to lose no time; and therefore, with general consent of all, bent his course directly to run with the Islands of the Moluccas. And so having nothing in our view but air and sea, without sight of any land for the space of full sixty-eight days together, we continued our course through the main Ocean, till September 30th[2] following, on which day we fell in ken of certain islands lying about eight degrees to the Northward of the Line. From these islands, presently upon the discovery of us, came a great number of canoes, having in each of them in some four, in some six, in some fourteen or fifteen men, bringing with them cocoas, fish, potatoes, and certain fruits to small purpose.[3] Their canoes were made after the fashion that the canoes of all the rest of the Islands of Moluccas for the most are, that is, of one tree, hollowed within with great art and cunning, being made so smooth, both within and without, that they bore a gloss as if it were a harness most finely burnished. A prow and stern they had of one fashion, yielding inward in manner of a semicircle, of a great height, and hung full of certain white and glistening shells for bravery:[4] on each side of their canoes lay out two pieces of timber, about a yard and a half long, more or less according to the capacity of their boat. At the end whereof was fastened crosswise a great cane, the use whereof was to keep their canoes from overthrowing, and that they might be equally borne up on each side.[5]

The people themselves have the nether parts of their ears cut round or circle-wise, hanging down very low upon their cheeks, wherein they hang things of a reasonable weight. The

[1] The three Farallons, North, Middle, and South, which lie about a day's sail to the Westward of the Golden Gate.

[2] By another account, the 13th of October. The islands were doubtless some of the Caroline group, which lay in the direct track from Drake's Californian harbour—whether San Francisco or Port Sir Francis Drake under Punta de los Reyes to the North—to the Moluccas.

[3] Of little value or consequence.

[4] Adornment.

[5] Compare Dampier's minute description of similar craft at Guam; Chapter X.

nails on the fingers of some of them were at least an inch long, and their teeth as black as pitch, the colour whereof they use to renew by often eating of an herb, with a kind of powder, which in a cane they carry about them to the same purpose. The first sort and company of those canoes being come to our ship (which then, by reason of a scant wind, made little way) very subtilely and against their natures began in peace to traffic with us, giving us one thing for another very orderly, intending (as we perceived) hereby to work a greater mischief to us; entreating us by signs most earnestly to draw nearer towards the shore, that they might, if possible, make the easier prey both of the ship and us. But these passing away, and others continually resorting, we were quickly able to guess at them what they were; for if they received anything once into their hands, they would neither give recompense nor restitution of it, but thought whatever they could finger to be their own, expecting always with brows of brass to receive more, but would part with nothing. Yea, being rejected for their bad dealing, as those with whom we would have no more to do, using us so evilly, they could not be satisfied till they had given the attempt to revenge themselves because we would not give them whatsoever they would have for nothing: and having stones good store in their canoes, let fly amain of them against us. It was far from our General's meaning to requite their malice by like injury. Yet that they might know that he had power to do them harm if he had listed, he caused a great piece to be shot off, not to hurt them, but to affright them. Which wrought the desired effect amongst them; for at the noise thereof they every one leaped out of his canoe into the water, and, diving under the keel of their boats, stayed them from going any way till our ship was gone a good way from them. Then they all lightly recovered into their canoes, and got them with speed toward the shore. Notwithstanding, other new companies (but all of the same mind) continually made resort unto us. And seeing that there was no good to be got by violence, they put on a show of seeming honesty; and offering in show to deal with us by way of exchange, under the pretence they cunningly fell a-filching of what they could, and one of them pulled a dagger and knives from one of our men's girdles, and being required to restore it again, he rather used what means he could to catch at more. Neither could we at all be rid of this ungracious company, till we made some of them feel some smart as well as terror; and so we left that place, by all passengers to be known hereafter by the name of the Island of Thieves.

Till the 3d of October we could not get clear of these consorts, but from thence we continued our course within sight of land till the 16th of the same month, when we fell with four Islands standing in 7° 5′ to the Northward of the Line. We coasted them till the 21st day, and then anchored and watered upon the biggest of them, called Mindanao. The 22d of October, as we passed between two islands, about six or eight leagues south of Mindanao,[1] there came from them two canoes to be talked with us, and we would willingly be talked with them, but there arose so much wind that put us from them to the Southwards. October the 25th we passed by the island named Talao,[2] in 3° 40′. We saw to the northward of it three or four other islands, Teda, Selan, Saran (three islands so named to us by an Indian), the middle whereof stands in 3°. We passed the last

[1] Supposed to be Serangan and Candigar, or the Saddle Islands, South of the southernmost point of Mindanao. Other narrators name the islands of "Tagulada, Zelon, and Zewarra," as passed by the voyagers on their way to the Moluccas, the first producing much cinnamon, and the inhabitants of all being friendly with the Portuguese.

[2] The Tulour Islands, about half way between Mindinao and Gilolo.

save one of these, and the first day of the following month in like manner we passed the isle Suaro, in 1° 30′, and the 3d of November we came in sight of the Islands of the Moluccas, as we desired. These are four high-peaked islands: their names, Ternate, Tidore, Matchan, Batchan, all of them very fruitful and yielding abundance of cloves, whereof we furnished ourselves of as much as we desired at a very cheap rate. At the east of them lies a very great island called Gilolo.

We directed our course to have gone to Tidore, but in coasting along a little island [1] belonging to the King of Ternate, November 4th, his deputy or viceroy with all expedition came off to our ship in a canoe, and without any fear or doubting of our good meaning came presently aboard. Who, after some conference with our General, entreated him by any means to run with Ternate, not with Tidore: assuring him that his King would be wondrous glad of his coming, and be ready to do for him what he could, and what our General in reason should require. For which purpose he himself would that night be with his King to carry him the news; with whom if he once dealt, he should find that as he was a King, so his word should stand: whereas if he dealt with the Portuguese, who had the command of Tidore,[2] he should find in them nothing but deceit and treachery. And besides that if he went to Tidore before he came to Ternate, then would his King have nothing to do with us, for he held the Portuguese as an enemy. On these persuasions our General resolved to run with Ternate, where the next day, very early in the morning, we came to anchor: and presently our General sent a messenger to the King with a velvet cloak, for a present and token that his coming should be in peace, and that he required no other thing at his hands, but that (his victuals being spent in so long a voyage) he might have supply from him by way of traffic and exchange of merchandise (whereof he had store of divers sorts) of such things as he wanted. Which he thought he might be the bolder to require at his hands, both for that the thing was lawful, and that he offered him no prejudice or wrong therein; as also because he was entreated to repair to the place by his Viceroy at Motir, who assured him of necessary provision in such manner as now he required the same.

Before this the Viceroy, according to his promise, had been with the King, signifying unto him what a mighty Prince and Kingdom we belonged to; what good things the King might receive from us, not only now, but for hereafter by way of traffic: Yea what honour and benefit it might be to him, to be in league and in friendship with so noble and famous a Prince as we served; and farther, what a discouragement it would be to the Portuguese his enemies to hear and see it. In hearing whereof the King was so presently moved to the well liking of the matter, that before our messenger could come half the way, he had sent the Viceroy, with divers others of his nobles and councillors, to our General, with special message that he should not only have what things he needed, or would require, with peace and friendship, but that he would willingly entertain amity with so famous and renowned a Princess as was ours; and that if it seemed good in her eyes to accept of it, he would sequester the commodities and traffic of his whole island from others (especially from his enemies the Portuguese, from whom he had nothing but by the sword), and reserve it to the intercourse of our nation, if we would embrace it. In token whereof he had now sent to our General his signet, and would within short time after come in his own person, with his brethren and nobles, with boats or canoes, into our ship,

[1] The island of Motir.

[2] They had been expelled from their settlements at Ternate by the warlike monarch whose friendship was offered to Drake, and had established themselves at Tidore.

and be a means of bringing her into a safer harbour. While they were delivering their message to us, our messenger was come unto the Court, who being met by the way by certain noble personages, was with great solemnity conveyed into the King's presence; at whose hands he was most friendly and graciously entertained; and having delivered his errand, together with his present unto the King, the King seemed to him to judge himself blameworthy that he had not sooner hastened in person to present himself to our General, who came so far and from so great a Prince; and presently, with all expedition, he made ready himself, with the chief of all his States and Councillors, to make repair unto us. The manner of his coming, as it was princely, so truly it seemed to us very strange and marvellous: serving at the present not so much to set out his own royal and Kingly state (which was great) as to do honour to Her Highness, to whom we belonged; wherein how willingly he employed himself, the sequel will make manifest.

First, therefore, before his coming, did he send off three great and large canoes, in each whereof were certain of the greatest personages that were about him, attired all of them in white lawn, or cloth of Calicut, having over their heads, from one end of the canoe to the other, a covering of thin and fine mats, borne up by a frame made of reeds, under which every man sat in order according to his dignity; the hoary heads of many of them set forth the greater reverence due to their persons, and manifestly showed that the King used the advice of a grave and prudent Council in his affairs. Besides these were divers others, young and comely men, a great number attired in white, as were the other, but with manifest differences: having their places also under the same covering, but in inferior order, as their calling required. The rest of the men were soldiers, who stood in comely order round about on both sides; on the outside of whom, again, did sit the rowers, in certain galleries, which being three on each side all alongst the canoe, did lie off from the side thereof some three or four yards, one being orderly builded lower than the other: in every of which galleries was an equal number of banks, whereon did sit the rowers, about the number of fourscore in one canoe. In the forepart of each canoe sat two men, the one holding a tabrel,[1] the other a piece of brass, whereon they both at once struck; and observing a due time and reasonable space between each stroke, by the sound thereof directed the rowers to keep their stroke with their oars: as, on the contrary, the rowers ending their stroke with a song, gave warning to the others to strike again; and so continued they their way with marvellous swiftness. Neither were their canoes naked or unfurnished of warlike munition; they had each of them at least one small cast piece, of about a yard in length, mounted upon a stock which was set upright; besides, every man except the rowers had his sword, dagger, and target, and some of them some other weapons, as lances, calivers,[2] bows, arrows, and many darts.

These canoes, coming near our ship in order, rowed round about us one after another; and the men, as they passed by us, did us a kind of homage with great solemnity, the greatest personages beginning first, with reverent countenance and behaviour, to bow their bodies even to the ground: which done, they put their own messenger aboard us again, and signified to us that their King, who himself was coming, had sent them before him to conduct our ship into a better road, desiring a hawser to be given them forth, that they might employ their service, as their King commanded, in towing our ship therewith to the place assigned. The King himself was not far behind, but he also with six grave and ancient fathers in his canoe approaching, did at once, together with them,

[1] A small drum. [2] Guns.

yield us a reverent kind of obeisance, in far more humble manner than was to be expected. He was of a tall stature,[1] very corpulent and well set together, of a very princely and gracious countenance: his respect amongst his own was such, that neither his Viceroy of Motir aforenamed, nor any other of his councillors, durst speak unto him but upon their knees, not rising again till they were licensed. Whose coming, as it was to our General no small cause of good liking, so was he received in the best manner we could, answerable unto his state; our ordnance thundered, which we mixed with great store of small shot, among which sounding our trumpets and other instruments of music, both of still and loud noise; wherewith he was so much delighted, that requesting our music to come into the boat, he joined his canoe to the same, and was towed at least a whole hour together, with the boat at the stern of our ship. Besides this, our General sent him such presents as he thought might both requite his courtesy already received, and work a further confirmation of that good liking and friendship already begun. The King being thus in musical paradise, and enjoying that wherewith he was so highly pleased, his brother, named Moro, with no less bravery[2] than any of the rest, accompanied also with a great number of gallant followers, made the like repair,[3] and gave us like respect; and, his homage done, he fell astern of us till we came to anchor: neither did our General leave his courtesy unrewarded, but bountifully pleased him also before we parted.

The King, as soon as we were come to anchor, craved pardon to be gone, and so took leave, promising us that the next day he would come aboard, and in the mean time would prepare and send such victuals as were requisite and necessary for our provision. Accordingly the same night, and the morrow following, we received what was there to be had in the way of traffic, to wit, rice in pretty quantity, hens, sugar-canes, imperfect and liquid sugar, a fruit which they call Figo (Magellan calls it a fig of a span long, but it is no other than that which the Spaniards and Portuguese have named Plantains), cocoas, and a kind of meal which they call sago, made of the tops of certain trees, tasting in the mouth like sour curds, but melts away like sugar; whereof they make a kind of cake which will keep good at least ten years. Of this last we made the greatest quantity of our provision: for a few cloves we did also traffic, whereof, for a small matter, we might have had greater store than we could well tell where to bestow: but our General's care was, that the ship should not be too much pestered or annoyed therewith.

At the time appointed, our General, having set all things in order to receive him, looked for the King's return; who, failing both in time and promise, sent his brother to make his excuse, and to entreat our General to come on shore, his brother being the while to remain on board, as a pawn for his safe restoring. Our General could willingly have consented, if the King himself had not first broken his word: the consideration whereof bred an utter disliking in the whole company, who by no means would give consent he should hazard himself, especially for that the King's brother had uttered certain words, in secret confidence with our General aboard his cabin, which bred no small suspicion of ill intent. Our General being thus resolved not to go ashore at this time, reserved the Viceroy for a pledge, and so sent certain of his gentlemen to the Court, both to accompany the King's brother, and also with special message to the King himself. They, being come somewhat near unto the castle, were received by another brother of the King's, and certain others of the greatest states, and conducted with great honour towards the castle, where being brought into a large and fair house, they saw gathered

[1] Fuller—"Holy State," page 127—calls him "a true gentleman Pagan."
[2] Magnificence, splendid show.
[3] Paid a similar visit.

together a great multitude of people, by supposition at least 1000, the chief whereof were placed round about the house, according, as it seemed, to their degrees and calling: the rest remained without. The house was in form four-square, covered all over with cloth of divers colours, not much unlike our usual pentadoes,[1] borne upon a frame of reeds, the sides being open from the groundsill to the covering, and furnished with seats round about: it seems it was their Council-house, and not commonly employed to any other use. At the side of this house, next unto the Castle, was seated the chair of state, having directly over it, and extending very largely every way, a very fair and rich canopy, as the ground also, for some ten or twelve paces' compass, was covered with cloth of Arras. Whilst our gentlemen awaited in this place the coming of the King, which was about the space of half-an-hour, they had the better opportunity to observe these things; as also that before the King's coming there were already set threescore noble, grave, and ancient personages, all of them reported to be of the King's privy council. At the nether end of the house were placed a great company of young men, of comely personage and attire. Without the house, on the right side, stood four ancient, comely, hoar-headed men, clothed all in red down to the ground, but attired on their heads not much unlike the Turks. These they called Romans, or strangers, who lay as lidgiers,[2] there to keep perpetual traffic with the people: there were also two Turks, one Italian, as lidgiers, and last of all one Spaniard, who being freed by the King out of the hands of the Portuguese, in the recovering of the island, served him now instead of a soldier.

The King at last coming from the castle, with eight or ten grave Senators following him, had a very rich canopy, adorned in the midst with embossings of gold, borne over him, and was guarded with twelve lances, the points turned downwards. Our men, accompanied with Moro the King's brother, arose to meet him, and he very graciously did welcome and entertain them. He was for person such as we have before described him, of low voice, temperate in speech, of kingly demeanour, and a Moor by nation. His attire was after the fashion of the rest of his country, but far more sumptuous, as his condition and state required: from the waist to the ground was all cloth of gold, and that very rich; his legs bare, but on his feet a pair of shoes of cordovan, dyed red; in the attire of his head were finely wreathed-in divers rings of plated gold, of an inch or an inch and a-half in breadth, which made a fair and princely show, somewhat resembling a crown in form; about his neck he had a chain of perfect gold, the links very great and one fold double. On his left hand were a diamond, an emerald, a ruby, and a turquoise, four very fair and perfect jewels; on his right hand, in one ring, a big and perfect turquoise, and in another ring many diamonds of a smaller size, very artificially set and couched together. As thus he sat in his chair of state, at his right side there stood a page with a very costly fan, richly embroidered and beset with sapphires, breathing and gathering the air to refresh the King, the place being very hot, both by reason of the sun, and the assembly of so great a multitude. After a while, our gentlemen, having delivered their message, and received answer, were licensed to depart, and were safely conducted back again, by one of the

---

[1] Canopies, tents.

[2] Resident or permanent ambassadors; the word is spelled in various other ways, as "leger," "ligier," "legier;" it comes from the Anglo-Saxon "leigan," to lie or remain; and the word "ledger," a book that lies to receive entries, is from the same source. In "Measure for Measure," Isabella, informing her brother of his impending death, says:

"Lord Angelo, having affairs to heaven,
Intends you for his swift ambassador,
Where you shall be an everlasting leiger."

Chiefs of the King's Council, who had charge from the King himself to perform the same. .

Our gentlemen, observing the castle as well as they could, could not conceive it to be a place of any great force; two cannons only there they saw, and those at that present untraversable, because unmounted. These, with all other furniture of like sort which they have, they have gotten them from the Portuguese, by whom the castle itself was also builded, while they inhabited that place and island. Who seeking to settle tyrannous government (as in other places so) over this people, and not contenting themselves with a better estate than they deserved (except they might, as they thought, make sure work by leaving none of the Royal blood alive, who should make challenge to the kingdom), cruelly murdered the King himself—father to him who now reigns—and intended the like to all his sons. Which cruelty, instead of establishing brought such a shaking on their usurped estate, that they were fain without covenanting to carry away goods, munition, or anything else, to quit the place and the whole island, to save their lives. For the present King, with his brethren, in revenge of their father's murder, so bestirred themselves, that the Portuguese were wholly driven from the island, and glad that he yet keeps footing in Tidore. These four years this King hath been increasing, and was (as was affirmed) at that present, Lord of an Hundred Islands thereabout, and was even now preparing his forces to hazard a chance with the Portuguese for Tidore itself. The people are Moors, whose religion consists much in certain superstitious observations of new moons, and certain seasons, with a rigid and strict kind of fasting. We had experience hereof in the Viceroy and his retinue, who lay aboard us all the time for the most part during our abode in this place; who during their prescribed time would neither eat nor drink, not so much as a cup of cold water in the day (so zealous are they in their self-devised worship), but yet in the night would eat three times, and that very largely. This Ternate stands in 27′ North latitude.

While we rode at anchor in the harbour at Ternate, besides the natives there came aboard us another, a goodly gentleman, very well accompanied, with his interpreter, to view our ship and to confer with our General. He was apparelled much after our manner, most neat and court-like, his carriage the most respective and full of discreet behaviour that ever we had seen. He told us that he was himself but a stranger in those islands, being a natural of the province of Paghia in China; his name Pausaos, of the family of Hombu; of which family there had eleven reigned in continual succession these 200 years, and King Bonog, by the death of his elder brother—who died by a fall from his horse—the rightful heir of all China, is the twelfth of this race. He is twenty-two years of age; his mother yet living; he hath a wife, and by her one son; he is well-beloved and highly-honoured of all his subjects, and lives in great peace from any fear of foreign invasion. But it was not this man's fortune to enjoy his part of this happiness, both of his King and country, as he most desired. For being accused of a capital crime, whereof though free,[1] yet he could not evidently make his innocence appear, and knowing the peremptory justice of China to be irrevocable, if he should expect[2] the sentence of the Judges; he beforehand made suit to his King, that it would please him to commit his trial to God's providence and judgment, and to that end to permit him to travel, on this condition, that if he brought not home some worthy intelligence, such as His Majesty had never had before, and were most fit to be known, and most honourable for China, he should for ever live an exile, or else die for daring to set foot again in his own country; for he was assured that the God of heaven had care of innocency.

[1] Guiltless. [2] Await.

The King granted his suit, and now he had been three years abroad; and at this present came from Tidore (where he had remained two months), to see the English General, of whom he heard such strange things, and from him (if it pleased God to afford it) to learn some such intelligence as might make way for his return into his country: and therefore he earnestly entreated our General to make relation to him of the occasion, way, and manner of his coming so far from England thither, with the manifold occurrences that had happened to him by the way. Our General gave ample satisfaction to each part of his request; the stranger hearkened with great attention and delight to his discourse, and as he naturally excelled in memory, besides his help of art to better the same, so he firmly printed it in his mind, and with great reverence thanked God, who had so unexpectedly brought him to the notice of such admirable things. Then fell he to entreat our General with many most earnest and vehement persuasions, that he would be content to see his country before his departure any farther Westward; that it should be a most pleasant, most honourable, and most profitable thing for him; that he should gain thereby the notice, and carry home the description, of one of the most ancient, mightiest, and richest kingdoms in the world. Hereupon he took occasion to relate the number and greatness of the provinces, with the rare commodities and good things they yielded: the number, stateliness, and riches of their cities; with what abundance of men, victuals, munition, and all manner of necessaries and delightful things they were stored with; in particular touching ordnance and great guns—the late invention of a scab-skinned Friar amongst us in Europe[1]—he related that in Suntien, by some called Quinzai, which is the chief city of all China, they had brass ordnance of all sorts (much easier to be traversed than ours were, and so perfectly made that they would hit a shilling) above 2000 years ago. With many other worthy things which our General's own experience, if it would please him to make trial, would better than his relation assure him of. The breeze would shortly serve very fitly to carry him thither, and he himself would accompany him all the way. He accounted himself a happy man that he had but seen and spoken with us; the relation of it might perhaps serve him to recover favour in the country; but if he could prevail with our General himself to go thither, he doubted not but it would be a means of his great advancement, and increase of honour with his King. Notwithstanding, our General could not on such persuasions be induced, and so the stranger parted, sorry that he could not prevail in his request, but yet exceeding glad of the intelligence he had learned.

By the 9th of November, having gotten what provision the place could afford us, we then set sail: and considering that our ship for want of trimming was now grown foul, that our casks and vessels for water were much decayed, and that divers other things stood in need of reparation, our next care was, how we might fall with such a place where with safety we might awhile stay for the redressing of these inconveniences. The calmness of the winds, which are almost continual before the coming of the breeze (which was not yet expected) persuaded us it was the fittest time that we could take. With this resolution we sailed along till November 14th, at what time we arrived at a little island to the southward of Celebes, standing in 1° 40′ towards the Pole Antarctic: which being without inhabitants, gave us the better hope of quiet abode. We anchored, and finding the place con-

[1] This scarcely complimentary description may apply either to the English Monk Roger Bacon, or, more appropriately, to the German Monk Berthold Schwartz, whom accident enlightened as to the projectile force of "villainous saltpetre."

venient for our purposes (there wanting nothing here which we stood in need of, but only water, which we were fain to fetch from another island somewhat farther to the south), made our abode here for twenty-six whole days together. The first thing we did, we pitched our tents and entrenched ourselves as strongly as we could upon the shore, lest at any time perhaps we might have been disturbed by the inhabitants of the greater island, which lay not far to the westward of us. After we had provided thus for our security, we landed our goods, and had a smith's forge set up, both for the making of some necessary shipwork, and for the repairing of some iron-hooped casks, without which they could not long have served our use. And for that our smith's coals were all spent long before this time, there was order given and followed for the burning of charcoal, by which that want might be supplied.

We trimmed our ship, and performed our other businesses to our content. The place affording us not only all necessaries (which we had not of our own before) thereunto, but also wonderful refreshing to our wearied bodies, by the comfortable relief and excellent provision that here we found, whereby of[1] sickly, weak, and decayed, as many of us seemed to be before our coming hither, we in short space grew all of us to be strong, lusty, and healthful persons. Besides this, we had rare experience of God's wonderful wisdom in many rare and admirable creatures which here we saw. The whole island is a through[2] grown wood, the trees for the most part are of large and high stature, very straight and clean, without boughs, save only in the very top; the leaves whereof are not much unlike our brooms in England. Among these trees, night by night, did show themselves an infinite swarm of fiery-seeming worms flying in the air, whose bodies, no bigger than an ordinary fly, did make a show and give such light as if every twig on every tree had been a lighted candle, or as if that place had been the starry sphere. To these we may add the relation of another, almost as strange a creature, which here we saw, and that was an innumerable multitude of huge bats or reremice, equalling or rather exceeding a good hen in bigness. They fly with marvellous swiftness, but their flight is very short; and when they light, they hang only by the boughs, with their backs downward. Neither may we without ingratitude, by reason of the special use we made of them, omit to speak of the huge multitude of a certain kind of crayfish, of such a size, that one was sufficient to satisfy four hungry men at a dinner, being a very good and restorative meat; the special means (as we conceived it) of our increase of health. They are, as far as we could perceive, utter strangers to the sea, living always on the land, where they work themselves earths as do the conies, or rather they dig great and huge caves under the roots of the most huge and monstrous trees, where they lodge themselves by companies together. Of the same sort and kind we found, in other places about the Island Celebes, some that, for want of other refuge, when we came to take them did climb up into trees to hide themselves, whither we were enforced to climb after them if we would have them, which we would not stick to do rather than to be without them. This island we called Crab Island.

All necessary causes of our staying longer in this place being at last finished, our General prepared to be in a readiness to take the first advantage of the coming of the breeze or wind which we expected; and having the day before furnished ourselves with fresh water from the other island, and taken in provision of wood and the like, December 12th we put to sea, directing our course toward the West. The 16th day we had sight of the Island of Celebes or Silébis, but having a bad wind and being

[1] From being.
[2] Thoroughly.

entangled among many islands, encumbered also with many other difficulties, and some dangers, and at last meeting with a deep bay out of which we could not in three days turn out again, we could not by any means recover the North of Celebes, or continue on our course farther west, but were enforced to alter the same towards the South; finding that course also to be both difficult and very dangerous by reason of many shoals, which lay far off, here and there among the islands; insomuch that in all our passages from England hitherto, we had never more care to keep ourselves afloat, and from sticking on them. Thus were we forced to beat up and down with extraordinary care and circumspection, till January 9th, at which time we supposed that we had at last attained a free passage, the lands turning evidently in our sight about to westward, and the wind being enlarged, followed us as we desired with a reasonable gale.

When lo! on a sudden, when we least suspected, no show or suspicion of danger appearing to us, and we were now sailing onward with full sails, in the beginning of the first watch of the said day at night, even in a moment, our ship was laid up fast upon a desperate shoal, with no other likelihood in appearance but that we with her must there presently perish; there being no probability how anything could be saved, or any person escape alive. The unexpectedness of so extreme a danger presently roused us up to look about us, but the more we looked the less hope we had of getting clear of it again, so that nothing now presenting itself to our minds, but the ghastly appearance of instant death, affording no respite or time of pausing, called upon us to deny ourselves, and to commend ourselves into the merciful hands of our most gracious God. To this purpose we presently fell prostrate, and with joined prayers sent up unto the throne of grace, humbly besought Almighty God to extend his mercy unto us in his Son Christ Jesus, and so preparing as it were our necks unto the block, we every minute expected the final stroke to be given unto us. Notwithstanding that we expected nothing but imminent death, yet—that we might not seem to tempt God, by leaving any second means unattempted which he afforded—presently, as soon as prayers were ended, our General (exhorting us to have the especialest care of the better part, to wit, the soul, and adding many comfortable speeches, of the joys of that other life which we now alone looked for) encouraged us all to bestir ourselves, shewing us the way thereto by his own example. And first of all the pump being well plied, and the ship freed of water, we found our leaks to be nothing increased; which though it gave us no hope of deliverance, yet it gave us some hope of respite, insomuch as it assured us that the bulk[1] was sound; which truly we acknowledged to be an immediate providence of God alone, insomuch as no strength of wood and iron could have possibly borne so hard and violent a shock as our ship did, dashing herself under full sail against the rocks, except the extraordinary hand of God had supported the same.

Our next essay was for good ground and anchor-hold to seaward of us, whereon to haul; by which means, if by any, our General put us in comfort, that there was yet left some hope to clear ourselves. In his own person he therefore undertook the charge of sounding, and but even a boat's length from the ship he found that the bottom could not by any length of line be reached unto; so that the beginning of hope, which we were willing to have conceived before, were by this means quite dashed again; yea, our misery seemed to be increased, for whereas at first we could look for nothing but a present end, that expectation was now turned into the awaiting for a lingering death, of the two the far more dreadful to be chosen. One thing fell out

[1] The hull.

happily for us, that the most of our men did not conceive this thing; which had they done, they would in all likelihood have been so much discouraged, that their sorrow would the more disable them to have sought the remedy: our General, with those few others that would judge of the event wisely, dissembling the same, and giving, in the mean time, cheerful speeches and good encouragements unto the rest. For whilst it seemed to be a clear case that our ship was so fast moored that she could not stir, it necessarily followed that either we were there to remain on the place with her, or else, leaving her, to commit ourselves in a most poor and helpless state to seek some other place of stay and refuge, the better of which two choices did carry with it the appearance of worse than a thousand deaths. As touching the ship, this was the comfort that she could give us, that she herself lying there confined already upon the hard and pinching rocks, did tell us plain that she continually expected her speedy despatch, as soon as the sea and winds should come, to be the severe executioners of that heavy judgment by the appointment of the Eternal Judge already given upon her, who had committed her there to Adamantine bonds in a most narrow prison, against their coming for that purpose: so that if we could stay with her, we must peril with her; or if any, by any yet unperceivable means, should chance to be delivered, his escape must needs be a perpetual misery, it being far better to have perished together, than with the loss and absence of his friends to live in a strange land: whether a solitary life (the better choice) among wild beasts, as a bird on the mountains without all comfort, or among the barbarous people of the heathen, in intolerable bondage both of body and mind. And put the case that her day of destruction should be deferred longer than either reason could persuade us, or in any likelihood could seem possible (it being not in the power of earthly things to endure what she had suffered already), yet could our abode there profit us nothing, but increase our wretchedness and enlarge our sorrows; for as her store and victuals were not much—sufficient to sustain us only some few days, without hope of having any increase, no not so much as of a cup of cold water—so must it inevitably come to pass, that we, as children in the mother's womb, should be driven even to eat the flesh from off our own arms, she being no longer able to sustain us; and how horrible a thing this would have proved, is easy by any one to be perceived. And whither, had we departed from her, should we have received any comfort? nay, the very impossibility of going appeared to be no less than those other before mentioned. Our boat was by no means able at once to carry above twenty persons with any safety, and we were fifty-eight in all; the nearest land was six leagues from us, and the wind from the shore directly bent against us; or should we have thought of setting some ashore, and after that to have fetched the rest, there being no place thereabout without inhabitants, the first that had landed must first have fallen into the hands of the enemy, and so the rest in order; and though perhaps we might escape the sword, yet would our life have been worse than death, not alone in respect of our woeful captivity and bodily miseries, but most of all in respect of our Christian liberty, being to be deprived of all public means of serving the true God, and continually grieved with the horrible impieties and devilish idolatries of the heathen. Our misery being thus manifest, the very consideration whereof must needs have shaken flesh and blood, if faith in God's promises had not mightily sustained us, we passed the night with earnest longings that the day would once appear; the mean time we spent in often prayers and other godly exercises, thereby comforting ourselves, and refreshing our hearts, striving to bring ourselves to an humble submission under the hand of God, and to a referring of ourselves wholly to his good will and pleasure.

The day therefore at length appearing, and it being almost full sea about that time, after we had given thanks to God for his forbearing of us hitherto, and had with tears called upon him to bless our labours; we again renewed our travail to see if we could now possibly find any anchor-hold, which we had formerly sought in vain. But this second attempt proved as fruitless as the former, and left us nothing to trust to but prayers and tears; seeing it appeared impossible that ever the forecast, counsel, policy, or power of man could ever effect the delivery of our ship, except the Lord only miraculously should do the same. It was therefore presently motioned, and by general voice determined, to commend our case to God alone, leaving ourselves wholly in his hand to spill[1] or save us, as [might] seem best to his gracious wisdom. And that our faith might be the better strengthened, and the comfortable apprehension of God's mercy in Christ be more clearly felt, we had a sermon and the Sacrament of the body and blood of our Saviour celebrated. After this sweet repast was thus received, and other holy exercises adjoined were ended, lest we should seem guilty in any respect for not using all lawful means we could invent, we fell to another practice yet unessayed, to wit, to unloading of our ship by casting some of her goods into the sea; which thing, as it was attempted most willingly, so it was despatched in very short time. So that even those things which we before this time, nor any other in our case could be without, did now seem as things only worthy to be despised; yea, we were herein so forward, that neither our munition for defence, nor the very meal for sustentation of our lives, could find favour with us, but every thing as it first came to hand went overboard: assuring ourselves of this, that if it pleased God once to deliver us out of that most desperate strait wherein we were, he would fight for us against our enemies, neither would he suffer us to perish for want of bread. But, when all was done, it was not any of our endeavours, but God's only hand, that wrought our delivery; 'twas he alone that brought us even under the very stroke of death; 'twas he alone that said unto us, "Return again, ye sons of men!" 'twas he alone that set us at liberty again, that made us safe and free, after that we had remained in the former miserable condition the full space of twenty hours; to his glorious name be the everlasting praise. The manner of our delivery (for the relation of it will especially be expected) was only this: The place whereon we sat so fast was a firm rock, in a cleft whereof it was we stuck on the larboard side. At low water there was not above six feet of depth in all on the starboard; within little distance, as you have heard, no bottom to be found; the breeze during the whole time that we stayed blew somewhat stiff directly against our broadside, and so perforce kept the ship upright. It pleased God in the beginning of the tide, while the water was yet almost at lowest, to slack the stiffness of the wind; and now our ship, which required thirteen feet of water to make her float, and had not at that time on the one side above seven at most, wanting her prop on the other side, which had too long already kept her up, fell a-heeling towards the deep water, and by that means freed her keel and made us glad men. This shoal is at least three or four leagues in length; it lies in 2°, lacking three or four minutes, South latitude. The day of this deliverance was the 10th of January.

Of all the dangers that in our whole voyage we met with, this was the greatest; but it was not the last, as may appear by what ensueth. Neither could we indeed for a long season free ourselves from the continual care and fear of them; nor could we ever come to any convenient anchoring, but were continually for the most part tossed amongst the many islands and shoals which lie in infinite number round about on the South part

[1] Destroy.

of Celebes, till the 8th day of the following month. January 12th, not being able to bear our sails, by reason of the tempest, and fearing of the dangers, we let fall our anchors upon a shoal in 3° 30′. January 14th, we were gotten a little farther South, where, at an island in 4° 6′, we again cast anchor, and spent a day in watering and wooding. After this we met with foul weather, Westerly winds, and dangerous shoals, for many days together; insomuch that we were utterly weary of this coast of Celebes, and thought best to bear with Timor. The Southernmost cape of Celebes stands in 5° that side[1] the Line. But of this coast of Celebes we could not so easily clear ourselves. The 20th of January we were forced to run with a small island not far from thence; where having sent our boat a good distance from us to search out a place where we might anchor, we were suddenly environed with no small extremities. For there arose a most violent, yea an intolerable flaw and storm out of the South-west against us, making us (who were on a lee-shore amongst most dangerous and hidden shoals) to fear extremely not only the loss of our boat and men, but the present loss of ourselves, our ship, and goods, or the casting of those men, whom God should spare, into the hands of Infidels. Which misery could not by any power or industry of ours have been avoided, if the merciful goodness of God had not, by staying the outrageous extremities wherewith we were set upon, wrought our present delivery; by whose unspeakable mercy our men and boats also were unexpectedly, yet safely, restored unto us. We got off from this place as well as we could, and continued on our course till the 26th day [of January], when the wind took us, very strong against us, W. and WSW., so as that we could bear no more sail till the end of that month was full expired. February 1st, we saw very high land, and as it seemed well inhabited, we would fain have borne with it, to have got some succour, but the weather was so ill that we could find no harbour, but we were very fearful of adventuring ourselves too far amongst the many dangers which were near the shore. The third day also we saw a little island, but being unable to bear any sail, but only to lie at hull,[2] we were by the storm carried away and could not fetch it. February 6th, we saw five islands, one of them towards the East, and four towards the West of us, one bigger than another; at the biggest of which we cast anchor, and the next day watered and wooded.

After we had gone on thence, on February 8th, we descried two canoes, who having descried us, as it seems, before, came willingly unto us, and talked with us, alluring and conducting us to their town not far off, named Barativa; it stands in 7° 13′ South the Line. The people are Gentiles, of handsome body and comely stature, of civil demeanour, very just in dealing, and courteous to strangers; of all which we had evident proof, they showing themselves most glad of our coming, and cheerfully ready to relieve our wants with whatsoever their country could afford. The men all go naked, save their heads and secret parts, every one having one thing or other hanging at his ears. Their women are covered from the middle to the foot, wearing upon their naked arms bracelets, and that in no small number, some having nine at least upon each arm, made for the most part of horn or brass, whereof the lightest, by our estimation, would weigh two ounces. With this people linen cloth, whereof they make rolls for their heads and girdles to wear about their loins, is the best merchandise, and of greatest estimation. They are also much delighted with margarites,[3] which in their language they call "Saleta," and such other like

[1] That is, to the South side.

[2] A ship lies at hull, or a hull, when either in a dead calm or in a storm all her sails are taken in, and she shows only bare masts and rigging.

[3] Beads.

trifles. Their island is both rich and fruitful; rich in gold, silver, copper, tin, sulphur, &c. Neither are they only expert to try those metals, but very skilful also in working of them artificially into divers forms and shapes, as pleaseth them best. Their fruits are diverse likewise and plentiful, as nutmegs, ginger, long pepper, lemons, cucumbers, cocoas, figs, sago, with divers other sorts, whereof we had one in reasonable quantity, in bigness, form, and husk, much like a bay-berry, hard in substance, but pleasant in taste, which being sodden becomes soft, and is a most profitable and nourishing meat. Of each of these we received of them whatsoever we desired for our need, insomuch that (such was God's gracious goodness to us) the old proverb was verified with us, "After a storm cometh a calm, after war peace, after scarcity followeth plenty:" so that in all our voyage, Ternate only excepted, from our departure out of our own country, hitherto we found not anywhere greater comfort and refreshing than we did at this time in this place. In refreshing and furnishing ourselves here we spent two days, and departed hence February 10th. When we were come into the height of 8° 4′, February 12th, in the morning we espied a green island to the Southward; not long after, two other islands on the same side, and a great one more towards the North: they seemed all to be well inhabited, but we had neither need nor desire to go to visit them, and so we passed by them. The 14th day we saw some other reasonably big islands; and February 16th we passed between four or five big islands more, which lay in the height[1] 9° 40′. The 18th, we cast anchor under a little island, whence we departed again the day following; we wooded here, but other relief, except two turtles, we received none. The 22d, we lost sight of three islands on our starboard side, which lay in 10° and some odd minutes. After this we passed on to the Westward without stay or anything to be taken notice of till the 9th of March, when in the morning we espied land, some part thereof very high, in 8° 20′ South latitude. Here we anchored that night, and the next day weighed again, and bearing further North and nearer shore, we came to anchor the second time. The 11th of March we first took in water, and after sent our boat again to shore, where we had traffic with the people of the country; whereupon, the same day, we brought our ship more near the town, and having settled ourselves there that night, the next day our General sent his man ashore to present the King with certain cloth, both linen and woollen, besides some silks; which he gladly and thankfully received, and returned rice, cocoas, hens, and other victuals in way of recompense. This island we found to be the Island of Java, the middle whereof stands in 7° 30′ beyond the Equator. The 13th of March our General himself, with many of his gentlemen and others, went to shore, and presented the King (of whom he was joyfully and lovingly received) with his music, and shewed him the manner of our use of arms, by training his men with their pikes and other weapons which they had, before him. For the present, we were entertained as we desired, and at last dismissed with a promise of more victuals to be shortly sent us.

In this island there is one chief, but many under-governors, or petty kings, whom they call Rajahs, who live in great familiarity and friendship one with another. The 14th day we received victuals from two of them; and the day after that, to wit the 15th, three of these kings in their own persons came aboard to see our General, and to view our ship and warlike munition. They were well pleased with what they saw, and with the entertainment which we gave them. And after these had been with us, and on their return had, as it seems, related what they found, Rajah Donan, the chief King of the whole land, bringing victuals with him for our relief, he also the next

[1] Latitude (South of the Line).

day after came aboard us. Few were the days that one or more of these kings did miss to visit us, insomuch that we grew acquainted with the names of many of them, as of Rajah Pataiára, Rajah Cabocapálla, Rajah Manghángo, Rajah Boccabarra, Rajah Timbánton: whom our General always entertained with the best cheer that we could make, and shewed them all the commodities of our ship, with our ordnance and other arms and weapons, and the several furnitures belonging to each, and the uses for which they served. His music also, and all things else whereby he might do them pleasure, wherein they took exceeding great delight with admiration. One day, amongst the rest, March 21st, Rajah Donan coming aboard us, in requital of our music which was made to him, presented our General with his country music, which though it were of a very strange kind, yet the sound was pleasant and delightful. The same day he caused an ox also to be brought to the water's side and delivered to us, for which he was to his content rewarded by our General with divers sorts of very costly silks, which he held in great esteem. Though our often giving entertainment in this manner did hinder us much in the speedy despatching of our businesses, and made us spend the more days about them, yet there we found all such convenient helps, that to our contents we at last ended them. The matter of greatest importance which we did, besides victualling, was the new trimming and washing of our ship, which by reason of our long voyage was so overgrown with a kind of shellfish sticking fast unto her, that it hindered her exceedingly, and was a great trouble to her sailing. The people, as are their kings, are a loving, a very true, and a just-dealing people. We trafficked with them for hens, goats, cocoas, plantains, and other kinds of victuals, which they offered us in such plenty, that we might have laden our ship if we had needed.[1]

We took our leaves and departed from them the 26th of March, and set our course WSW., directly towards the Cape of Good Hope, or Bon Esperance, and continued without touch of aught but air and water until the 21st of May, when we espied land—to wit, a part of the main of Africa—in some places very high, under the latitude of thirty-one and a half degrees. We coasted along till June 15th, on which day, having very fair weather, and the wind at South-east, we passed the Cape itself so near in sight, that we had been able with our pieces to have shot to land.[2] July 15th we fell with the land again about Rio de Sesto, where we saw many negroes in their boats a-fishing, whereof two came very near us, but we cared not to stay, nor had any talk or dealing with them. The 22d of the same month we came to Sierra Leone, and spent two days for watering in the mouth of Tagoine, and then put to sea again; here also we had oysters,[3]

---

[1] In every village, other narratives inform us, was a house of assembly or public hall, where the people met twice daily to partake of a common meal and enjoy the pleasures of conversation. "To this festival every one contributed, at his pleasure or convenience, fruits, boiled rice, roasted fowls, and sago. The viands were spread on a table raised three feet, and the party gathered round, one rejoicing in the company of another."

[2] The Cape is described by another chronicler as "a most stately thing, and the fairest cape we saw in the whole circumference of the earth." They passed it in perfectly calm and clear weather; making them affirm, that the Portuguese had not less falsely alleged the extreme peril of the passage from continual tempests, than the Spaniards, to deter voyagers of other nations, had exaggerated the dangers of the course round the southern extremity of America.

[3] The voyagers came here upon a kind of oysters which "was found on trees, spawning and increasing infinitely; the oyster suffering no bud to grow."

and plenty of lemons, which gave us good refreshing. We found ourselves under the Tropic of Cancer, August 15th, having the wind at North-east, and we fifty leagues off from the nearest land. The 22d day we were in the height of the Canaries.

And the 26th of September (which was Monday in the just and ordinary reckoning of those that had stayed at home in one place or country, but in our computation was the Lord's Day or Sunday[1]) we safely, with joyful minds and thankful hearts to God, arrived at Plymouth, the place of our first setting forth, after we had spent two years, ten months, and some few odd days besides, in seeing the wonders of the Lord in the deep, in discovering so many admirable things, in going through with so many strange adventures, in escaping out of so many dangers, and overcoming so many difficulties, in this our encompassing of this nether globe, and passing round about the world, which we have related.

Soli rerum maximarum Effectori,
Soli totius mundi Gubernatori,
Soli suorum Conservatori,
Soli Deo sit semper Gloria.

[1] The same circumstance, which "every schoolboy" can now explain, had also astonished the companions of Magellan, who, on their return from their circumnavigation to San Lucar in 1522, discovered that they had "lost a day." Dampier notes the same thing at the commencement of his Fourteenth Chapter. See page 223.

END OF DRAKE'S VOYAGE ROUND THE WORLD.

DRAKE'S LAST VOYAGE.

1595.

[An account of Drake's unfortunate expedition to the West Indies in 1595, written by Thomas Maynarde, one of his companions on the occasion, is still preserved, and is given here—though a little apart from the main purpose of the present volume—as an appropriate sequel to Mr Fletcher's narrative of his most brilliant achievement.]

IT appears by the attempts and known purposes of the Spaniard—as by his greedy desire to be our neighbour in Britain, his fortifying upon the river of Brest, to gain so near us a quiet and safe road for his fleet, his carelessness in losing the strongholds and towns which he possessed in the Low Countries, not following those wars in that heat which he wonted, the rebellious rising of the Earl of Tyrone (wrought or drawn thereto undoubtedly by his wicked practices)—that he leaveth no means unattempted which he judged might be a furtherance to turn our tranquillity into accursed thraldom; so robbing us of that quiet peace which we, from the hands of Her Majesty (next under God), abundantly enjoy. This his bloodthirsty desire foreseen by the wisdom of our Queen and Council, they held no better means to curb his unjust pretences, than by sending forces to invade him in that kingdom from whence he hath feathers to fly to the top of his high desires; they knowing that if for two or three years a blow were given him there that might hinder the coming into Spain of his treasure, his poverty, by reason of his huge daily payments, would be so great, and his men of war, most of them mercenaries, that assuredly would fall from him, so would he have more need of means to keep his own territories, than he now hath of superfluity to thrust into others' rights.

This invasion was spoken of in June 1594, a long time before it was put in execution; and it being partly resolved on, Sir Francis Drake was named General in November following: a man of great spirit and fit to undertake matters: in my poor opinion, better able to conduct forces and discreetly to govern in conducting them to places where service was to be done, than to command in the execution thereof. But, assuredly, his very name was a great terror to the enemy in all those parts, having heretofore done many things in those countries to his honourable fame and profit. But entering into them as the child of fortune, it may be his self-willed and peremptory command was doubted, and that caused Her Majesty, as should seem, to join Sir John Hawkins in equal commission: a man old and wary, entering into matters with so laden a foot, that the other's meat would be eaten before his spit could come to the fire; men of so different natures and dispositions, that what the one desireth the other would commonly oppose against; and though their wary carriages sequestered it from meaner wits, yet was it apparently seen to better judgments before our going from Plymouth, that whom the one loved, the other smally esteemed. Agreeing best, for what I could conjecture, in giving out a glorious title to their intended journey, and in not so well victualling the navy as, I deem, was Her Majesty's pleasure it should be, both of them served them to good purpose; for, from this having the distributing of so great sums, their miserable providing for us would free them from incurring any great loss, whatsoever befell of the journey. And the former[1] drew unto them so great repair of voluntaries,[2] that they had choice to discharge such few as they had pressed, and to enforce the stay of others who gladly would be partakers of their voyage. But notwithstanding matters were very forward, and that they had drawn together three thousand men, and had ready furnished twenty-seven ships, whereof six were Her Majesty's, yet many times was it very doubtful whether the journey should proceed;[3] and had not the news of a galleon of the King of Spain, which was driven into Saint John de Puerto Rico with two millions and a half of treasure, come unto them by the report of certain prisoners, whereof they advertised Her Majesty, it is very likely it had been broken, but Her Majesty, persuaded by them of the easy taking thereof, commanded them to hasten their departure.

So on Thursday, being the 28th of August, in the year 1595, having stayed two months in Plymouth, we went thence twenty-seven sail, and

---

[1] That is, the "giving out a glorious title" to their intended expedition.

[2] Resort of volunteers.

[3] It was detained, among other causes, by artfully propagated rumours that another great Armada was being prepared for the invasion of England—the Spaniards thus gaining time to put their colonies in good defence against the formidable attack now menaced.

were two thousand five hundred men of all sorts. This fleet was divided into two squadrons; not that it was so appointed by Her Majesty, for from her was granted as powerful authority unto either of them over the whole as any part, but Sir Francis victualling the one half and Sir John the other, it made them, as men affecting what they had done,[1] to challenge a greater prerogative over them than the whole; wherein they wronged themselves and the action,[2] for we had not run sixty or seventy leagues in our course, before a flag of council was put out in the Garland, unto which all commanders with the chief masters and gentlemen repaired. Sir Francis complained that he had three hundred men more in his squadron than were in the other, and that he was much pestered in his own ship, whereof he would gladly be eased. Sir John gave no other hearing to this motion, but seemed to dislike that he should bring more than was concluded betwixt them; and this drew them to some choleric speeches. But Sir John would not receive any unless he were entreated; to this Sir Francis' stout[3] heart could never be driven. This was on the 2d of September, and after they were somewhat qualified,[4] they acquainted us that Sir Thomas Baskerville, our Colonel-general, was of their council by virtue of the broad seal, and that they would take unto them Sir Nicholas Clifford and the other captains appointed by Her Majesty, who were, eleven for the land, four for the ships in which they themselves went not. They gave us instructions for directing our course, if, by foul weather or mischance, any should be severed, and orders what allowances we should put our men into for preservation of victuals, with other necessary instructions. In the end, Sir John revealed the places whither we were bound, in hearing of the basest mariner; observing therein no warlike or provident advice, nor was it ever amended to the time of their deaths, but so he named Saint John de Puerto Rico, where the treasure before spoken of was to be taken, even without blows; from whence we should go direct to Nombre de Dios, and so over land to Panama. What other things should fall out by the way, he esteemed them not worth the naming, this being sufficient to make a far greater army rich to their content.

Some seven or eight days after this, we were called aboard the Defiance, where, Sir Francis Drake propounding unto us whether we should give upon the Canaries or Madeiras (for he was resolved to put for one of them by the way), we seeing his bent and the earnestness of the Colonel-general, together with the apparent likelihood of profit, might soon have been drawn thereto, but for considering the weighty matters we had undertaken, and how needful it was to hasten us thither. But General Hawkins utterly misliking this notion—it being a matter, as he said, never before thought of—knew no cause why the fleet should stay in any place till they came to the Indies, unless it should be by his[5] taking in of so great numbers to consume his waters and other provision; the which, if Sir Francis would acknowledge, he would rid him and relieve him the best he could. Now the fire which lay hid in their stomachs began to break forth, and had not the Colonel pacified them, it would have grown farther; but their heat somewhat abated, and they concluded to dine next day aboard the Garland with Sir John, when it was resolved that we should put for the Grand Canaries, though, in my conscience, whatsoever his tongue said, Sir John's heart was against it. These matters were well qualified, and for that place we shaped our course; in which

[1] Taking a greater interest in what had engaged their own attention and touched their own pocket.

[2] Enterprise.

[3] Proud, stubborn.

[4] After their passion had somewhat abated.

[5] Drake's.

we met with a small Fleming bound for the Straits, and a small man-of-war of Weymouth, who kept us company to the Canaries. On Wednesday, the 24th day, we had sight of Lancerotta and Forteventura. The 25th, at night we descried the Canaries, it being a month after our departure from Plymouth. On Friday, being the 26th, we came to anchor some saker-shot from a fort which stands to the WNW. of the harbour. Sir Francis spent much time in seeking out the fittest place to land; the enemy thereby gaining time to draw their forces in readiness to impeach[1] our approach. At length we, putting for the shore in our boats and pinnaces, found a great siege[2] and such power of men to encounter us, that it was then thought it would hazard the whole action if we should give further upon it, whereupon we returned without receiving or doing any harm worthy the writing; but, undoubtedly, had we launched under the fort at our first coming to anchor, we had put fair to be possessors of the town, for the delays gave the enemy great stomachs[3] and daunted our own; and it being the first service our new men were brought into, it was to be doubted they would prove the worse the whole journey following.

We presently weighed hence and came to anchor the 27th at the WSW. part of this island, where we watered. Here Captain Grimstone, one of the twelve captains for land, was slain by the mountaineers, with his boy and a surgeon. Hence we departed the 28th, holding our course SW. three weeks, then we ran WSW. and W. by S. until the 27th of October, on which day we had sight of Maten, an island lying south-east from Dominica. Our Generals meant to water at Guadaloupe, for Dominica being inhabited by Indians, our men straggling soon would have their throats cut. General Drake lying ahead the fleet, ran in by the north of Dominica, Sir John by south. The 29th we anchored under Guadaloupe; Sir Francis being there a day before us. On the 30th, Josias, captain of the Delight, brought news to the Generals, that the Francis, a small ship of company, was taken by nine frigates; whereupon Sir Francis would presently have followed them, either with the whole fleet or some part, for that he knew our intentions were discovered by reason they were so openly made known, as I afore have set down, by Sir John Hawkins. Sir John would in no wise agree to either of these motions, and he was assisted in his opinion by Sir Nicholas Clifford, all others furthering his desires, which might be a means to stay them for going into Puerto Rico before us; but Sir John prevailed, for that he was sickly, Sir Francis being loth to breed his further disquiet. The reason of his stay was, to trim his ships, mount his ordnance, take in water, set by some new pinnaces, and to make things in that readiness, that he cared not to meet with the King's whole fleet. Here we stayed doing these necessaries three days. This is a desert, and was without inhabitants.

On the 4th of November we departed, and being becalmed under the lee of the land, Sir Francis caused the Richard, one of the victuallers, to be unladen and sunk. The 8th we anchored among the Virgins, other west islands: here we drew our company on shore, that every man might know his colours, and we found our company short of the one thousand two hundred promised for land service, few of the captains having above ninety, most not eighty, some not fifty; which fell out partly for that the Generals had selected to them a company for their guard, of many of the gallantest men of the army. Sir John's sickness increased. Sir Francis appointed captains to the merchants' ships; this consumed time till the 11th, when we passed a

[1] Prevent; French, "empêcher." The word is still used in Ireland in the sense of hindering or obstructing.

[2] Fortification.

[3] Courage.

sound, though, by[1] our mariners, never passed by fleet afore, and we came to anchor before Puerto Rico on the 12th about three of the clock in the afternoon, at which time Sir John Hawkins died. I made my men ready presently to have landed, knowing that our sudden resolution would greatly have daunted the enemy, and have held ours in opinion of assured victory; but I was countermanded by authority, and during the time of our deliberation the enemy laboured by all means to cause us to disanchor, so working, that within an hour he had planted three or four pieces of artillery upon the shore next to us, and playing upon the Defiance, knowing her to be the Admiral, whilst our Generals sat at supper with Sir Nicholas Clifford and divers others, a shot came amongst them, wherewith Sir Nicholas, Brute Brown, Captain Strafford, who had Grimstone's company, and some standers-by, were hurt. Sir Nicholas died that night, so seconding Sir John Hawkins in his death, as he did in his opinion at Guadaloupe. My brother Brown lived five or six days after, and died much bewailed. This shot made our General weigh and fall farther to the westward, where we rode safely. The frigates before spoken of rode within their forts: we had no place now to land our men but within them, in the face of the town, which was dangerous, for that both forts and ships could play on us; it was therefore concluded that boats should fire them where they rode. Captain Poore and myself had the command of this service; for the regiments, Captain Salisbury commanding; the grand captain company was sent by the Generals; divers sea commanders were also sent; and on the 13th at night, passing in hard under the fort, we set three of them on fire; only one of which, it was my chance to undertake, was burnt; on the others the fire held not, by reason that being once out they were not maintained with new. The burnt ship gave a great light, the enemy thereby playing upon us with their ordnance and small shot as if it had been fair day, and sinking some of our boats: a man could hardly command his mariners to row, they foolishly thinking every place more dangerous than where they were, when, indeed, none was sure. Thus doing no harm, we returned with two or three prisoners, when, indeed, in my poor opinion, it had been an easier matter to bring them out of the harbour than fire them as we did, for our men aboard the ships numbered five thousand one hundred and sixty pieces of artillery[2] that played on us during this service; and it had been less dangerous to have abidden them close in the frigates and in the dark than as we did. But great commanders many times fail in their judgment, being crossed by a co-partner. But I had cause of more grief than the Indies could yield me of joy, losing my alferez,[3] Davis Pursell; Mr Vaughan, a brother-in-law of Sir John Hawkins, with three others; Thomas Powton, with five or six more, hurt and maimed; and was somewhat discomfited, for the General feigned here to set up his rest; but examining the prisoners, by whom he understood that these frigates were sent for his treasure, and that they would have fallen among us at Guadaloupe had they not taken the Francis, his mind altered: calling to council, he commanded us to give our opinions what we thought of the strength of the place. Most thought it would hazard the whole action.[4] But one Rush, a captain, more to me alleged, that without better putting for it[5] than bare looking upon the outside of the forts, we could hardly give such judg-

[1] According to the report of.

[2] A number wholly incredible; the Spanish accounts say that there were only seventy pieces.

[3] Standard-bearer; a word borrowed from the Arabic, as its prefix plainly enough shows.

[4] That to attack it would bring the whole expedition into jeopardy.

[5] Without some further effort.

ment; and I set it plainly under my hand, that if we resolutely attempted it, all was ours; and that I persuaded myself no town in the Indies could yield us more honour or profit. The General presently said: "I will bring thee to twenty places far more wealthy and easier to be gotten." Such like speeches I think had bewitched the Colonel, for he most desired him to hasten him hence. The enemy, the day after we had fired the frigates, sunk together four to save us labour, but chiefly to strengthen their forts: two other great ships they sunk and fired in the mouth of the harbour, to give them light to play on us from their forts, as we entered the first night. And hence we went the 15th. Here I left all hope of good success.

On the 19th we came to anchor in a fair bay, the Bay of Sta Jermana,[1] at the westernmost part of the island, where we stayed till the 24th, setting up more new pinnaces and unloading the other new victualler, the General taking the most part into his own ship, as he did of the former. Captain Torke, in the Hope, was made Vice-Admiral. This is a very pleasant and fertile island, having upon it good store of cattle, fruits, and fish, with all things necessary to man's sustenance; and were it well manured, no place could yield it in greater abundance or better. Departing hence, we had our course for Curaçoa. The second day after our putting off, the Exchange, a small ship, sprung her mast, and was sunk; the men and part of the victuals were saved by other ships. Upon Curaçoa there is great store of cattle and goats, and we fell with it upon Saturday the 29th; but our General, deceived by the current and westerly course, made it for Aruba,[2] an island lying ten or twelve leagues to the westward, and so made no stay; when, next morning descrying whether he found his error, we bore with Cape de la Vela, and from thence our Colonel, with all the companies in the pinnaces and boats, were sent to the city of Rio de la Hacha, and with small resistance we took it the 1st of December at night. The General came unto us the next morning with the fleet. This town was left bare of goods; the inhabitants, having intelligence of our coming, had carried all in the woods, and hid their treasures in *caches;* but, staying here seventeen days, we made so good search, that little remained unfound within four leagues of the town. We took many prisoners, Spaniards and Negroes, some slaves repairing to us voluntarily. The General with two hundred men went in boats to Lancheria,[3] which is a place where they fish for pearl, standing ten leagues to the eastward of their town, from whence they brought good store of pearl, and took a carvel, in which was some money, wine, and myrrh. During our stay here, the Governor once, divers others often, repaired unto us to redeem their town, Lancheria, their boats, and slaves. They did this to gain time to convey away the King's treasure, and to advertise their neighbour towns to convey their treasure in more safety than themselves had done; for the whole (except the slaves who voluntarily repaired unto us) was yielded unto them for twenty-four thousand pesos, five shillings and sixpence a piece, to be paid in pearls; bringing these to their town at the day, and valuing in double the price they were worth. Our General delivered the hostages and set their town Lancheria and boats on fire, carrying their slaves with us. The wealth we had here was given to countervail the charge of the journey; but I fear it will not so prove in the end. Our Vice-Admiral, Captain Torke, died here of sickness. This is an exceedingly good country, champaign and well inhabited; great store of cattle, horses, sheep, goats, fish, and fowl, whereon we fed, but small store of grain or fruit near the town, rich only in pearl and cattle.

[1] San Germau.

[2] Mistook it for Oruba.

[3] La Rancheria.

The 20th, being Saturday, we came to Santa Martha. We sunk two ketches, before we came to Rio de la Hacha, which we brought out of England. Presently, upon our coming to anchor, we landed and gave upon[1] the town. We found small resistance more than a few shot playing out of the woods as we marched towards the town. Companies were presently sent abroad to discover and search the country. The inhabitants had too long forewarning to carry their goods out of our possibility to find them in so short time; little or nothing of value was gotten, only the Lieutenant-Governor and some others were taken prisoners; and firing the town the 21st, we departed. Captain Worrell, our trenchmaster,[2] died at this town of sickness. This was a very pretty town, and six leagues off there was a gold mine. If part of our company had been sent thither upon our first arrival at Rio de la Hacha, doubtless we had done much good; but now they had scrubbed it very bare. In this place was great store of fruit and much fernandobuck;[3] for that the wind blew so extremely, and the road wild, we could not ship it. Before we departed hence, it was concluded that we should pass Carthagena and go directly for Nombre de Dios. We anchored in the road on Sunday following, being the 28th; and landing presently, receiving some small shot from the town, we found small resistance more than a little fort at the east end of their town, in which they had left one piece of ordnance, which brake at the first shot. They gave upon us as we gave upon them: certain prisoners were taken in the flying, who made it known, that having intelligence long before of our coming, their treasure was conveyed to places of more safety, either to Panama, or secretly hidden; and it might very well be, for the town was left very bare;[4] wherefore it was resolved that we should hasten with speed to Panama. Nombre de Dios standeth on the North Sea, Panama upon the South,[5] distant some eighteen or nineteen leagues. There were only two ways to get thither; one by the River Chagre, which lies to the westward twenty leagues; upon this it is passable within five leagues of Panama: the other through deserts and over mountains void of inhabitants: this was troublesome and hard, as well for want of means to carry our provision of meat and munition, as for the ill passage with an army through these deserts and unknown places. That by the river our General held more dangerous, feigning there was no place for our fleet to wade safely. This made our Colonel yield to the way by the mountain, though he and others foresaw the danger before our setting hence; but he resolved to make trial of what could be done.

So on Monday the 29th we began our journey, taking with us the strongest and lustiest of our army, to the number of fifty[6] men and seven colours. Before our setting hence, we buried Captain Arnold Baskerville, our Sergeant-major-general, a gallant gentleman. The first day we marched three leagues; the next, six leagues, where we came to a great house which the enemy had set on fire, it being a place where the King's mules do use to lodge coming from Panama to Nombre de Dios with his treasure: it is the midway betwixt both places. The house would receive five hundred horses. We had not marched fully a league on Wednesday morning, when we came to a place fortified upon the top of a hill, which the enemy defended. We had no other way to pass nor no

[1] Attacked; French, "donner sur."

[2] Engineer.

[3] Wood of Pernambuco or Fernandobuco; Brasil wood.

[4] They found, however, at the top of an adjacent watchtower, more than 2000 lbs. of silver, with some gold and other valuables.

[5] The Atlantic and Pacific respectively.

[6] The number was really 750.

means to make our approach but a very deep lane, where but one could pass at once, unless it were by clambering upon the banks and creeping up the hill through the brakes, which some of our men did, and came to the trees which they had plashed[1] to make their palisado, over which they could not pass, the many boughs so hindered them. It was my chance, clambering up the banks to repair to three musketeers whom I had helped up, to fall directly between two of their places fortified, coming unto two paths by which they fetched their water, and giving presently upon them, the place being open, my small number found too good resistance, and I was driven to retire with the loss of these few. Here was the only place to beat them from their hold, whereof I sent the Colonel word, Captain Poore and Bartlett and others repairing to me. I shewed them the path; we heard the enemy plashing and felling of trees far before us. The Colonel sent for us to come unto him: he debated with us what he foresaw before our coming from Nombre de Dios, and though he thought, in his opinion, we should fear the enemy hence, yet, having retreats upon retreats, they would kill our best men without taking little or any hurt themselves; and our men began to drop apace; our powder and match were spoiled by much rain and waters which we had passed, unless it were such as some of our soldiers had with more care preserved. The provision for meat at our coming from Nombre de Dios was seven or eight cakes of biscuit or rusk for a man, which was either by wet spoiled, or their greediness had devoured; so there remained to few one day's bread, to most none at all. Our hurt men, as Captain Nicholas Baskerville and some others of account, we should be driven to leave to the mercy of the enemy, unless they could hold company. Before our coming to Panama, had we beaten them from all these holds, which I think would have been too dangerous for us to have attempted, considering the estate we were in, we must have fought with them at a bridge where they had entrenched themselves in a far greater number than we were; and it is manifest, if we had not within three days gotten some relief, we had been overthrown, though no enemy had fought against us. But our stomachs calling these, with other dangers, to his careful consideration, he resolved to retire, and so commanded us to cause the slain to be thrown out of sight, the hurt to be sent to the quarters from whence we came that morning, and the rest to be drawn away. Here were slain Captain Marchant, our Quarter-master, with some other officers, gentlemen, and soldiers. Upon our coming to the quarters, the Colonel took view of the hurt, and for such as could ride he procured all the horses of the army; for the other, he entreated the enemy to treat them kindly, as they expected the like from us towards theirs, of which we had a far greater number. On the 2d of January we returned to Nombre de Dios; our men so wearied with the illness of the way, surbatted[2] for want of shoes, and weak with their diet, that it would have been a poor day's service that we should have done upon an enemy had they been there to resist us. I am persuaded that never army, great or small, undertook a march through so unknown places so weakly provided and with so small means to help themselves, unless it might be some few going covertly to do some sudden exploit before it were thought of by the enemy, and so return unspied; for, undoubtedly, two hundred men foreknowing their intentions and provided with all things necessary, are able to break or weaken the greatest force that any prince in Christendom can bring thither, if he had place to find more than we had. This march had made many swear that he will never venture to buy gold at such a price again. I confess noble spirits,

[1] Pleached, or plaited, like a hedge.

[2] Bruised, wearied, footsore.

desirous to do service to their Prince and country, may soon be persuaded to all hardness and danger, but having once made trial thereof, would be very loth, as I suppose, to carry any force that way again; for beholding it in many places, a man would judge it dangerous for one man to pass alone, almost impossible for horses and an army.

The day that our General had news of our return, he meant to weigh and fall nearer to the River Chagre with the fleet, leaving some few to bring us if we were enforced to retire, whereby he little doubted. But being beaten from the place where it appeared all his hopes rested for gaining to himself and others this mass of treasure which he so confidently promised before, it was high time for him to devise of some other course. Wherefore, on the 4th of January, he called us to council, and debated with us what was now to be done. All these parts had notice long before of all our intentions, as it appeared by letters written from the Governor of Lima to the Governor of Panama and Nombre de Dios, giving them advice to be careful and to look well to themselves, for that Drake and Hawkins were making ready in England to come upon them. Lima is distant from these places more than three hundred leagues, all overlaid with snakes. It appears that they had good intelligence. This made them to convey their treasure to places which they resolved to defend with better force than we were able to attempt. Like as upon the coming of the sun, dews and mists begin to vanish, so our blinded eyes began now to open, and we found that the glorious speeches, of a hundred places that they[1] knew in the Indies to make us rich, was but a bait to draw Her Majesty to give them honourable employments, and us to adventure our lives for their glory; for now charts and maps must be our chiefest directors, he[2] being in these parts at the furthest limit of his knowledge. There he found out a lake called Laguna de Nicaragua, upon which stand certain towns, as Granada, Leon, and others; also the Bay of Honduras, a place known to be of small wealth by itself, unless it be brought thither to be embarked for Spain. He demanded which of these we would attempt; our Colonel said, "Both, one after the other, and all too little to content us if we took them." It was then resolved that we should first for the river, and as matters fell out, for the other. Nombre de Dios, together with their Negro town, was fired; and we sunk and fired fourteen small frigates which we found in the road. We got here twenty bars of silver, with some gold and certain plate; more would have been found had it been well sought: but our General thought it folly to gather our harvest grain by grain, being so likely at Panama to thrust our hands into the whole heaps; and after our return, being troubled in mind, he seemed little to regard any counsel that should be given him to that purpose, but to hasten thence as fast as he might. This is a most wealthy place, being settled upon a ground full of camphire, environed with hilly woods and mountains, the bottom a dampish fen. Hence we departed the 5th, and held our course for Nicaragua.

On the 9th we found a very deep and dangerous bay, playing it here up and down; all men weary of the place. The 10th we descried a small island called Escudes,[3] where we came to anchor; and here we took a frigate which was an advice[4] of the King's. By this we learned that the towns standing upon this Lake[5] were of small wealth and very dangerous, by reason of many shoals and great roughs our mariners should have, it being a hundred leagues: yet if the wind would have permitted, we had

[1] The promoters of the expedition.

[2] Drake.

[3] Escudo Island, near the bottom of Mosquito Bay.

[4] An "aviso," or despatch-boat.

[5] Of Nicaragua.

assuredly put for them, and never returned to one half again. Here we stayed, at a waste island where there was no relief but a few tortoises for such as could catch them, twelve days. This is counted the sickliest place of the Indies; and here died many of our men, victuals beginning to grow scarce with us. In the end, finding the wind to continue contrary, he resolved to depart, and to take the wind as God sent it.

So on the 22d we went hence, having there buried Captain Plott, Egerton, and divers others. I questioned with our General, being often private with him whilst we stayed here, to see whether he would reveal unto me any of his purposes; and I demanded of him, why he so often conjured me, being in England, to stay with him in these parts as long as himself, and where the place was? He answered me with grief, protesting that he was as ignorant of the Indies as myself, and that he never thought any place could be so changed, as it were from a delicious and pleasant arbour into a waste and desert wilderness; besides the variableness of the wind and weather, so stormy and blusterous as he never saw it before. But he most wondered that since his coming out of England he never saw sail worth giving chase unto: yet in the greatness of his mind, he would in the end, conclude with these words: "It matters not, man; God hath many things in store for us; and I know many means to do Her Majesty good service and to make us rich, for we must have gold before we see England;" when, good gentleman, in my conceit, it fared with him as with some careless-living man who prodigally consumes his time, fondly persuading himself that the nurse that fed him in his childhood will likewise nourish him in his old age, and, finding the dug dried and withered, enforced then to behold his folly, tormented in mind, dieth with a starved body. He had, besides his own adventure, gaged his own reputation greatly, in promising Her Majesty to do her honourable service, and to return her a very profitable adventure; and having sufficiently experienced, for seven or eight years together, how hard it was to regain favour once thought ill of, the mistress of his fortune now leaving him to yield to a discontented mind.[1] And since our return from Panama he never carried mirth nor joy in his face; yet no man he loved must conjecture that he took thought thereof. But here he began to grow sickly. At this island we sunk a carvel which we brought out of England, putting her men and victuals into a last-taken frigate. From hence a great current sets towards the eastward; by reason whereof, with the scant of wind we had, on Wednesday, being the 28th, we came to Portobello, which is within eight or nine leagues of Nombre de Dios. It was the best harbour we came into since we left Plymouth.

This morning, about seven of the clock, Sir Francis died. The next day Sir Thomas Baskerville carried him a league off, and buried him in the sea. In this place, the inhabitants of Nombre de Dios meant to build a town, it being far more healthy than where they dwell. Here they began a fort which already cost the King seven thousand purses, and a few houses towards their town, which they called Civitas Sti Philippi. Them we fired, razing the fortification to the ground. Here we found, as in other places, all abandoned; their ordnance cast into the sea, some of which we found, and carried aboard the Garland.

Our Generals being dead, most men's hearts were bent to hasten for England as soon as they might; but Sir Thomas Baskerville, having the command of the army by virtue of

[1] Referring, doubtless, to the failure of the expedition to Portugal, for the restoration of Dom Antonio, which Drake undertook in 1589, with Sir John Novis as commander of the land forces. Though the Admiral was acquitted honourably of all blame, his reputation seems to have for the time lain under a cloud.

Her Majesty's broad seal, endeavoured to prevent the dissevering of the fleet, and to that end talked with such as he heard intended to quit company before they were disembogued,[1] and drew all companies to subscribe to certain articles signifying our purposes: viz., that putting hence, we should turn it back to Santa Martha, if the wind would suffer us, otherwise to run over for Jamaica, where it was thought we should be refreshed with some victuals. Matters thus concluded, the Delight, the Elizabeth, and our late taken frigates were sunk. Many of the Negro men and base[2] prisoners were here put on shore; and here we weighed on Sunday the 8th of February. Our victuals began to shorten apace, yet we had lain a long time at very hard allowance—four men each morning one quart of beer and cake of biscuit for dinner, and for supper one quart of beer and two cakes of biscuit and two cans of water, with a pint of pease, or half a pint of rice, or somewhat more of oatmeal. This was our allowance being at Portobello, and six weeks before, but that we had sometimes stock-fish. From thence there is a current that sets to the eastward, by the help of which, on the 14th, we had sight of an island short of Carthagena fifteen or sixteen leagues; further than this we could not go to the eastward, for that the current had left us. The 15th at night, it being fair weather, we lost sight of our fleet. Here as I grew discontented, knowing it touched my poor regulation so to leave the army; and I had many things to persuade me that it was done of purpose by the captain and master, thereby gaining an excuse to depart; I showed the captain the danger he would run into by leaving so honourable forces when they had need of our company; and God knoweth that had I but had judgment which way to have cast for them, I would rather have lost my life than so forsake the like. He deposed on the Bible, and Christianity made me believe him. But playing it up and down about twelve of the clock, and discovering none of them, the wind blew so contrary that the seamen affirmed by holding this course we should be cast back in the bay, and they were persuaded that our fleet could not attain Santa Martha, but were gone over for Jamaica, whither they would follow them. I plainly foresaw that if we missed them there, it was like that we should no more meet till we came to England, which would have made me to persuade a longer search upon the main; but my hope of their being there, together with the weakness of our men and the small means we had to retain them, fearing lest my delay might endanger Her Majesty's ships and the whole company, I yielded to their persuasions. We were in ten degrees and a half when we put from hence, and we came till the 22d, when we had sight of a very dangerous shoal which our seamen thought they had passed near two days ago. If we had fallen in with it in the night we had been all lost. The shoal is named Secrana.

On Shrove Wednesday, being the 24th, we fell with Jamaica, and by means of a Mulatto and an Indian we had, this night, forty bundles of dried beef, which served our whole company so many days. We came to anchor at the westernmost part of the island, in a fair sandy bay, where we watered, and stayed, in hope to have some news of our fleet, seven days. This our stay brought no intelligence, wherefore, our seamen thought that our fleet, not able to recover this place, were fallen either with Cape Corrientes or Cape Saint Antonio;[3] these places we meant to touch in our course; and hence we went the 1st of March. On the 6th we saw a ship on the leeward of us, and the next morning we made her to be the Pegasine, one of our fleet, who, as they said, lost the Admiral near the time as we did, being by the Colonel

[1] Before they passed the Boccas or narrow seas, and entered the wider Ocean.

[2] Commoner sort.

[3] At the west of Cuba.

sent to the Susan Bonaventure, whom they left in great distress, by reason of a leak they had taken, and I greatly feared, by their report, they are perished. There were in her one hundred and thirty or one hundred and forty persons, many gallant gentlemen and good men. If they perish this ship shall repent it. Holding our course for these places, we descried five sails astern of us. We stayed for them, and soon made them out to be none of our fleet; and we had good reason to persuade us they were enemies. They had the wind of us, but we soon regained it upon them, which made them, upon a piece of ordnance shot off by the greatest ship, tack about; we tacked with them; when the captain of this ship faithfully protested unto me not to shoot a piece of ordnance till we came board and board, and then I promised him, with our small shot, to win the greatest or lose our persons. This we might have done without endangering Her Majesty's ships; but our enemy, playing upon us with their ordnance, made our gunners fall to it ere we were at musket shot, and no nearer could I bring them, though I had no hope to take any of them but by boarding. Here we popped away powder and shot to no purpose, for most of our gunners would hardly have stricken Paul's steeple had it stood there. I am a young seaman, yet my small judgment and knowledge make me avow, that never ship of Her Majesty's went so vilely manned out of her kingdom; not twenty of them worthy to come into her ships; and I know not what had possessed the captain, but his mind was clean altered, telling me that he had no authority to lay any ship aboard, whereby he might endanger this, Her Majesty's; and they being, as he said, the King's men of war, they would rather fire with us than be taken. Had I been a merchant of her burthen (God favouring me) they would have been mine, as many as stood to the trial of their fortune; but the paltry Pegasie we lately met withal never came near us by a league, which was some colour to our men to give them over. So after I had endeavoured, by myself, my lieutenant, and other gentlemen, by persuasion, to work the captain resolutely to attempt them, and finding no disposition in him but to consume powder and shot to no purpose but firing it in the air, I yielded to give them over, persuading myself that God had even ordained that we should not, with any nature, attempt where we were resisted with never so weak forces. Thus away we went, and the wind chopping us southerly, our seamen held that our fleet could neither ride at Corrientes nor at Saint Antonio, which made me condescend to leave the Indies, with all their treasure, and to ply the next course to disembogue, for little hope was left me that we should do Her Majesty any service, or good to ourselves, when, upon the feigned excuse of endangering her ships which she sent forth to fight if occasion were offered; and to persuade myself that Her Majesty prizeth not her ships dearer than the lives of so many faithful subjects, who gladly would have ventured their lives, and upon no brain-sick humour, but from a true desire to do Her Highness some service for the charge and adventure she had been at in this glorious spoken-of journey. Fortune's child was dead, things would not fall into our mouths, nor riches be our portions, how dearly soever we ventured for them. Thus avoiding Scylla (after the proverb) we fell into Charybdis, and indeed we were not now far from it.

Our master, a careful old man, but not experienced upon these coasts, rather following the advice of others than relying on his own judgment, brought us, on the 12th three hours before day, into a very shallow water, upon a dangerous bank, which some held to be the Meltilettes, others the Tortugas, either like enough to have swallowed us, had not God blessed us with fair weather. Freeing ourselves of this danger, upon Monday the 15th of March we entered the Gulf, and by ten of the clock we brought the Cape of Florida west of us. On the 17th (the Lord be thanked) we were disembogued. After this we ran with most foul weather and contrary winds till the 1st of May, when we had soundings in ninety fathoms, being in the Channel, and on the 3d we had sight of Scilly; the which day, ere night we came to anchor (the Lord be therefore praised) 1596.

To give mine opinion of the Indies, I verily think that filching men-of-war shall do more good, than such a fleet if they have any forewarning of their coming. And unless Her Majesty will undertake so royally as to dispossess him of the lands of Puerto Rico, Hispaniola, and Cuba, her charge will be greater in sending thither, than the profit such a fleet can return; for having but a few days' warning, it is easy for them to convey their goods into assured safety, as experience hath taught us. Their towns they dare not redeem, being enjoined the contrary by the King's commandment. These places will be taken and possessed by two thousand men; and by this Her Majesty might debar the King of Spain of his whole profit of the Indies; and the first gaining them will return her a sufficient requital for her adventure. God grant I may live to see such an enterprise put in practice; and the King of Spain will speedily fly to what conditions of peace Her Majesty will require.

Thus I have truly set down the whole discourse of our voyage, using therein many idle words and ill-compared sentences. It was done on the sea, which I think can alter any disposition. Your loves, I think, can pardon these faults, and secrete them from the view of others.

The 1st of March the fleet fell in with the Island of Pinos, on the land of Cuba, which day they had sight of the Spanish fleet by eleven of the clock; where Sir Thomas Baskerville gave directions for the fleet as thus: the Garland, being Admiral, with one half of the fleet, to have the vanguard; the Hope, being Vice-admiral, with the other half, the rearward. The fight continued fiercely three hours within musket-shot. That night they saw the Spanish Vice-admiral, a ship of seven hundred tons, burnt, with other six lost and sunk by the next morning, when they departed. The Hope received a leak and was forced to go from the fleet to an island, called Saint Crusado, inhabited by cannibals, where they had store of hens and Indian wheat for nine weeks. March 8th, the fleet shot the Gulf and came for England, leaving Florida on the starboard side; and when they came to the Enchanted Islands[1] they were dispersed, and came home one by one.

THOMAS MAYNARDE.

[1] The Azores.

END OF DRAKE'S LAST VOYAGE.

DAMPIER'S

VOYAGE ROUND THE WORLD.

1679—1691.

THE AUTHOR'S ACCOUNT OF HIMSELF.

[The Second Edition of Dampier's "Voyage Round the Terrestrial Globe" appeared in two volumes; the first containing the Circumnavigation proper, the second occupied by three Appendices, to which frequent references were made in the chief recital. These Appendices bore the following titles: I. "A Supplement of the Voyage round the World," being a fuller account of the Author's voyagings and observations during the time he spent in the East Indies between his arrival there in 1686 and his departure for England in 1691. II. "Two Voyages to Campeachy," narrating Dampier's experiences among the logwood cutters in the Bay of Campeachy between 1675 and 1678, and describing the western and south-western coast of the Caribbean Sea. III. "A Discourse of Trade Winds, Breezes, Storms, Seasons of the Year, Tides, and Currents of the Torrid Zone," entirely meteorological and professional. In the second Appendix the Author gives an account of himself fuller than any that we have from other sources; and, both from their autobiographical interest, and from the direct way in which they lead up to the greater subject, the main personal incidents of the Campeachy Voyages are here prefixed to the "Voyage Round the World."]

MY friends did not originally design me for the sea, but bred me at school till I came to years fit for a trade.[1] But upon the death of my father and mother, they who had the disposal of me took other measures; and having removed me from the Latin school to learn writing and arithmetic, they soon after placed me with a master of a ship at Weymouth,[2] complying with the inclinations I had very early of seeing the world. With him I made a short voyage to France; and, returning thence, went to Newfoundland, being then about eighteen years of age. In this voyage I spent one summer, but [was] so pinched with the rigour of that cold climate, that upon my return I was absolutely against going to those parts of the world, but went home again to my friends. Yet going up, a while after to London, the offer of a warm voyage and a long one, both which I always desired, soon carried me to sea again. For hearing of an outward-bound East Indiaman, the John and Martha of London, I entered myself aboard, and was employed before the mast, for which my two former voyages had some way qualified me. We went directly for Bantam in the Isle of Java, and staying there about two months, came home again in little more than a year; touching at Santiago of the Cape Verd Islands at our going out, and at Ascension in our return. In this voyage I gained more experience in navigation, but kept no journal. We arrived at Plymouth about two months before Sir Robert Holms went out to fall upon the Dutch Smyrna fleet; and the second Dutch War breaking out upon this, I forebore going to sea that summer, retiring to my brother in Somersetshire. But growing weary of staying ashore, I listed myself on board the Royal Prince, commanded by Sir Edward Spragge, and served under him in the year 1673, being the last of the Dutch War. We had three engagements that summer; I was in two of them, but falling very sick, I

[1] Dampier was born in 1652.

[2] About 1669.

was put on board an hospital ship, a day or two before the third engagement, seeing it at a distance only; and in this Sir Edward Spragge was killed. Soon after I was sent to Harwich, with the rest of the sick and wounded; and having languished a great while, I went home to my brother to recover my health. By this time the war with the Dutch was concluded; and with my health I recovered my old inclination for the sea. A neighbouring gentleman, Colonel Hellier of East Coker in Somersetshire, my native parish, made me a seasonable offer to go and manage a plantation of his in Jamaica, under one Mr Whalley: for which place I set out with Captain Kent in the Content of London. I was then about twenty-two years old, and had never been in the West Indies; and therefore, lest I might be trepanned and sold as a servant after my arrival in Jamaica, I agreed with Captain Kent to work as a seaman for my passage, and had it under his hand to be cleared at our first arrival. We sailed out of the River Thames in the beginning of the year 1674, and, meeting with favourable winds, in a short time got into the trade-wind and went merrily along, steering for the Island of Barbadoes. When we came in sight of it Captain Kent told his passengers, if they would pay his port charges he would anchor in the road, and stop whilst they got refreshment; but the merchants not caring to part with their money, he bore away, directing his course towards Jamaica, . . . where we arrived, . . . bringing with us the first news they had of the peace with the Dutch. Here, according to my contract, I was immediately discharged; and the next day I went to the Spanish Town, called Santiago de la Vega; where meeting with Mr Whalley, we went together to Colonel Hellier's plantation in Sixteen-Mile Walk. . . . I lived with Mr Whalley at Sixteen-Mile Walk for almost six months, and then entered myself into the service of one Captain Heming, to manage his plantation at St Ann's, on the north side of the Island, and accordingly rode from Santiago de la Vega towards St Ann's. This road has but sorry accommodation for travellers. The first night I lay at a poor hunter's hut, at the foot of Mount Diabolo [Devil's Mountain], on the south side of it, where for want of clothes to cover me in the night I was very cold when the land-wind sprang up. . . . The next day, crossing Mount Diabolo, I got a hard lodging at the foot of it on the north side; and the third day after arrived at Captain Heming's plantation. I was clearly out of my element there; and therefore, as soon as Captain Heming came thither, I disengaged myself from him, and took my passage on board a sloop to Port Royal, with one Mr Statham, who used to trade round the Island, and touched there at that time. From Port Royal I sailed with one Mr Fishook, who traded to the north side of the Island, and sometimes round it; and by these coasting voyages I came acquainted with all the ports and bays about Jamaica, as also with the benefit of the land and sea winds. For our business was to bring goods to, or carry them from planters to Port Royal; and we were always entertained civilly by them, both in their houses and plantations, having liberty to walk about and view them. They gave us also plantains, yams, potatoes, &c., to carry aboard with us; on which we fed commonly all our voyage. But after six or seven months I left that employ also, and shipped myself aboard one Captain Hudswell, who was bound to the Bay of Campeachy to load logwood. We sailed from Port Royal about the beginning of August, in 1675, in company with Captain Wren in a small Jamaica bark, and Captain Johnson, commander of a ketch belonging to New England. This voyage is all the way before the wind, and therefore ships commonly sail it in twelve or fourteen days: neither were we longer in our passage; for we had very fair weather, and touched nowhere till we came to

Trist Island, in the Bay of Campeachy, which is the only place they go to. . . . Trist is the road only for big ships. Smaller vessels that draw but a little water run three leagues farther, by crossing over a great lagoon that runs from the island up into the mainland; where they anchor at a place called One Bush Key. We stayed at Trist three days to fill our water, and then with our two consorts sailed thence with the tide of flood; and the same tide arrived there. This Key is not above forty paces long, and five or six broad, having only a little crooked tree growing on it, and for that reason it is called One Bush Key. . . . [It] is about a mile from the shore; and just against the island is a small creek that runs a mile farther, and then opens into another wide lagoon; and through this creek the logwood is brought to the ships riding at the Key. . . . Here we lay to take in our lading. Our cargo to purchase logwood was rum and sugar; a very good commodity for the logwood cutters, who were then about 250 men, most English, that had settled themselves in several places hereabouts. Neither was it long before we had these merchants come aboard to visit us. We were but six men and a boy in the ship, and all little enough to entertain them; for besides what rum we sold by the gallon or firkin, we sold it made into punch, wherewith they grew frolicsome. We had none but small arms to fire at their drinking healths, and therefore the noise was not very great at a distance; but on board the vessels we were loud enough till all our liquor was spent. We took no money for it, nor expected any; for logwood was what we came hither for, and we had of that in lieu of our commodities after the rate of £5 per ton, to be paid at the place where they cut it; and we went with our long boat to fetch small quantities. But because it would have taken up a long time to load our vessel with our own boat only, we hired a periago of the logwood cutters, to bring it on board, and by that means made the quicker despatch. I made two or three trips to their huts, where I and those with me were always very kindly entertained by them with pork and pease, or beef and dough-boys. Their beef they got by hunting in the savannahs. As long as the liquor lasted which they bought of us, we were treated with it, either in drams or punch. . . . It was the latter end of September 1675, when we sailed from One Bush Key with the tide of ebb, and anchored again at Trist that same tide; where we watered our vessel in order to sail. This we accomplished in two days, and the third day sailed from Trist towards Jamaica. A voyage which proved very tedious and hazardous to us, by reason of our ship's being so sluggish a sailer that she would not ply to windward, whereby we were necessarily driven upon several shoals that otherwise we might have avoided, and forced to spend thirteen weeks in our passage, [which] is usually accomplished in half that time.

[Dampier gives a long and particular account of the voyage to Jamaica, with descriptions of the Alacranes Islands or Reefs on which the ship struck, and the Island of Pines, near Cuba, on which the crew landed in pursuit of food. After narrowly escaping capture by the Spaniards, shipwreck, and death by starvation, the mountains of Jamaica were sighted, and the ship anchored at Negril.]

As soon as we came to anchor, we sent our boat ashore to buy provisions to regale ourselves, after our long fatigue and fasting, and were very busy going to drink a bowl of punch; when unexpectedly Captain Rawlings, commander of a small New England vessel that we left at Trist, and Mr John Hooker, who had been in the Bay [of Campeachy] a twelvemonth cutting logwood, and was now coming up to Jamaica to sell it, came aboard, and were invited into the cabin to drink with us. The bowl had not yet been touched (I think there might be six quarts in it), but Mr Hooker being drunk to by Captain Rawlings, who pledged Captain Hudswell, and having the bowl in his hand, said, that he was under an oath to drink but three draughts of strong liquor a day; and putting the bowl to his head, turned it off at one draught, and so making himself drunk, disappointed us of our expectations, till we made another bowl. The next day, having a brisk NW. wind, . . . we arrived at Port Royal, and so ended this troublesome voyage. It was not long after our arrival at Port Royal, before we were paid off, and discharged. Now, Captain Johnson of New England being bound again into the Bay of Campeachy, I took the opportunity of going a passenger with him, being resolved to spend some time at the logwood trade; and accordingly provided such necessaries as were required about it, viz., hatchets, axes, macheats[1] (*i.e.*, long knives), saws, wedges, &c., a pavilion to sleep in, a gun, with powder and shot, &c. And leaving a letter of attorney with Mr Fleming, a merchant of Port Royal, as well to dispose of anything that I should send up to him, as to remit to me what I should order, I took leave of my friends, and embarked. About the middle of February 1675-6 we sailed from Jamaica, and with a fair wind and weather soon got as far as Cape Catoche and there met a pretty strong north, which lasted two days. After that the trade [wind] settled again at ENE., which speedily carried us to Trist Island. In a little time I settled myself in the west creek of the west lagoon with some old logwood cutters, to follow the employment with them.

[Dampier here suspends "the relation of his own affairs," to give a long description of the coast and country bordering on the Bay of Campeachy, with its natural products by land and sea; and an account of the life and habits of the logwood cutters.]

. . . The logwood trade was grown very common before I came hither, there being, as I said before, about 260 or 270 men living in all the lagoon and at Beef Island. This trade had its rise from the decay of privateering; for after Jamaica was well settled by the English, and a peace established with Spain, the Privateers, who had hitherto lived upon plundering the Spaniards, were put to their shifts; for they had prodigally spent whatever they got, and now wanting subsistence, were forced either to go to Petit Goave,[2] where the Privateer trade still continued, or into the Bay for logwood. The more industrious sort of them came hither; yet even these, though they could work well enough if they pleased, yet thought it a dry business to toil at cutting wood. They were good marksmen, and so took more delight in hunting; but neither of these employments affected them[3] so much as privateering; therefore they often made sallies out in small parties among the nearest Indian towns, where they plundered, and brought away the Indian women to serve them at their huts, and sent their husbands to be sold at Jamaica. Besides, they had not their old drinking-bouts forgot, and would still spend £30 or £40 at a sitting aboard the ships that came hither from Jamaica, carousing and firing off guns three or four days together. And though afterwards many sober men came into the Bay to cut wood, yet by degrees the old standers so debauched them, that they could never settle themselves under any civil government, but continued in their wickedness till the Spaniards, encouraged by their careless rioting, fell upon them, and took most of them singly at their own huts, and carried them away prisoners to Campeachy or La Vera Cruz; from whence they were sent to Mexico, and sold to several tradesmen in that city; and from thence, after two or three years, when they could speak Spanish, many of them made their escapes,

[1] Spanish, "machete," a long knife, or cutlass.

[2] See page 166.

[3] They affected, or relished, neither of these employments.

and marched in by paths back to La Vera Cruz, and [were] by the Flota[1] conveyed to Spain, and so to England. I have spoken with many of them since, who told me that none of them were sent to the silver mines to work, but kept in or near the city, and never suffered to go with their caravans to New Mexico or that way. I relate this, because it is generally suggested that the Spaniards commonly send their prisoners thither, and use them very barbarously; but I could never learn that any European has been thus served; whether for fear of discovering their weakness, or for any other reason, I know not. But to proceed: it is most certain that the logwood cutters that were in the Bay when I was there were all routed or taken; a thing I ever feared; and that was the reason that moved me at last to come away, although at a place where a man might have gotten an estate.[2] . . . Though I was a stranger to their employment and manner of living, as being known but to those few only of whom we bought our wood in my former voyage hither, yet that little acquaintance I then got encouraged me to visit them after my second arrival here, being in hopes to strike in to work with them. There were six in company, who had a hundred tons ready cut, logged, and chipped, but not brought to the creek's side; and they expected a ship from New England in a month or two, to fetch it away. When I came thither they were beginning to bring it to the creek; and because the carriage is the hardest work, they hired me to help them, at the rate of a ton of wood per month; promising me that after this carriage was over I should strike in to work with them, for they were all obliged in bonds to procure this 100 tons jointly together, but for no more. This wood lay all in the circumference of 500 or 600 yards, and about 300 from the creek side, in the middle of a very thick wood, impassable with burthens. The first thing we did was to bring it all to one place in the middle; and from thence we cut a very large[3] path to carry it to the creek's side. We laboured hard at this work five days in the week, and on Saturdays went into the savannahs and killed beeves. . . . When my month's service was up, in which time we brought down all the wood to the creek's side, I was presently paid my ton of logwood; with which, and some more that I borrowed, I bought a little provision, and was afterwards entertained as a companion at work with some of my former masters; for they presently broke up consortships, letting the wood lie till either Mr West came to fetch it, according to his contract, or else till they should otherwise dispose of it. Some of them immediately went to Beef Island, to kill bullocks for their hides, which they preserve. . . . I was yet a stranger to this work, therefore remained with three of the old crew to cut more logwood. . . .

[By and by, two of the company, Scotsmen, get tired of the work and go away, the third—a Welshman, Price Morris by name, though the author calls him a Scotsman also—proves lazy and self-indulgent; and Dampier "keeps to his work by himself." He is hindered, however, by a growth of worms in his leg; afterward a great storm makes the region so uninhabitable, that with some other cutters he takes his departure for One Bush Key, and finding little aid from the ships there, themselves sufficiently distressed, goes to Beef Island, to hunt cattle for the sake of their hides. Dampier describes very minutely the features, peoples, and products of the southern and western coasts of the Bay of Campeachy; incidentally, in his mention of Vera Cruz, giving the following account of the Spanish West Indian squadrons:]

[1] An explanation of the terms Armada and Flota will be found in Chapter VII. of the Voyage: see page 161; and the Flota is described on next page.

[2] Enriched himself.

[3] Broad.

The Flota comes hither every three years from Old Spain; and besides goods of the product of the country, and what is brought from the East Indies [across New Spain from the port of Neapulco] and shipped aboard them, the King's plate that is gathered in this kingdom, together with what belongs to the merchants, amounts to a vast sum. Here also comes every year the Barralaventa Fleet in October or November, and stays till March. This is a small squadron, consisting of six or seven sail of stout ships, from 20 to 50 guns. These are ordered to visit all the Spanish seaport towns once every year, chiefly to hinder foreigners from trading, and to suppress Privateers. . . . If they meet with any English or Dutch trading-sloops, they chase and take them, if they are not too nimble for them; the Privateers keep out of their way, having always intelligence where they are.

[The personal narrative is resumed and concluded thus:]

The account I have given of the Campeachy rivers, &c., was the result of the particular observations I made in cruising about that coast, in which I spent eleven or twelve months. For when the violent storm before-mentioned took us, I was but just settling to work; and not having a stock of wood to purchase such provision as was sent from Jamaica, as the old standards had, I, with many more in my circumstances, was forced to range about to seek a subsistence in company of some Privateers then in the Bay. In which rambles we visited all the rivers, from Trist to Alvarado; and made many descents into the country among the villages there, where we got Indian corn to eat with the beef and other flesh that we got by the way, or manatee,[1] and turtle, which was also a great support to us. Alvarado was the westernmost place I was at. Thither we went in two barks with thirty men in each, and had ten or eleven men killed and desperately wounded in taking the fort; being four or five hours engaged in that service, in which time the inhabitants, having plenty of boats and canoes, carried all their riches and best movables away. It was after sunset before the fort yielded; and growing dark, we could not pursue them, but rested quietly that night; the next day we killed, salted, and sent aboard twenty or thirty beeves, and a good quantity of salt fish, and Indian corn, as much as we could stow away. Here were but few hogs, and those ate very fishy; therefore we did not much esteem them, but of cocks, hens, and ducks were sent aboard in abundance. . . . So that, with provision chests, hencoops, and parrot-cages, our ships were full of lumber, with which we intended to sail; but the second day after we took the fort, having had a westerly wind all the morning, with rain, seven armadilloes that were sent from La Vera Cruz appeared in sight, within a mile of the bar, coming in with full sail. But they could scarce stem the current of the river; which was very well for us, for we were not a little surprised. Yet we got under sail, in order to meet them; and clearing our decks by heaving all the lumber overboard, we drove out over the bar, before they reached it: but they being to windward, forced us to exchange a few shot with them. Their admiral was called the Toro; she had 10 guns and 100 men; another had 4 guns and 80 men: the rest, having no great guns, had only 60 or 70 men a piece, armed with muskets, and the vessels barricaded round with bull-hides breast high. We had not above 50 men in both ships, 6 guns in one and 2 in the other. As soon as we were over the bar, we got our larboard tacks aboard and stood to the eastward, as nigh the wind as we could lie. The Spaniards came away quartering on us; and, our ship being the headmost, the Toro came directly towards us, designing to board us. We kept firing at her, in hopes to have lamed either mast or yard; but failing, just as she was sheering aboard, we gave her a good volley, and pre-

[1] Described in Voyage, Chapter III. page 111.

sently clapped the helm a-weather, wore our ship, and got our starboard tacks aboard, and stood to the westward: and so left the Toro, but were saluted by all the small craft as we passed by them, who stood to the eastward, after the Toro, that was now in pursuit [of] and close by our consort. We stood to the westward till we were against the river's mouth; then we tacked, and by the help of a current that came out of the river, we were near a mile to windward of them all: then we made sail to assist our consort, who was hard put to it; but on our approach the Toro edged away towards the shore, as did all the rest, and stood away for Alvarado; and we, glad of the deliverance, went away to the eastward, and visited all the rivers in our return again to Trist. . . . And now the effects of the late storm being almost forgot, the lagoon men settled again to their employments; and I amongst the rest fell to work in the east lagoon, where I remained till my departure for Jamaica. . . . After I had spent about ten or twelve months at the logwood trade, and was grown pretty well acquainted with the way of traffic here, I left the employment; yet with a design to return hither after I had been in England; and accordingly went from hence with Captain Chambers of London, bound to Jamaica. We sailed from Trist the beginning of April 1678, and arrived at Jamaica in May, where I remained a small time, and then returned for England with Captain Loader of London. I arrived there the beginning of August the same year; and at the beginning of the following year I set out again for Jamaica, in order to have gone thence to Campeachy: but it proved to be a Voyage round the World. . . .

MR WILLIAM DAMPIER'S

VOYAGE ROUND THE TERRESTRIAL GLOBE.

THE EPISTLE DEDICATORY.

To the Right Honourable

CHARLES MOUNTAGUE, Esq.,

President of the Royal Society, One of the Lords Commissioners of the Treasury, &c.

Sir,—May it please you to pardon the boldness of a stranger to your person, if upon the encouragement of common fame he presumes so much upon your candour, as to lay before you this Account of his Travels. As the scene of them is not only remote, but for the most part little frequented also, so there may be some things in them new even to you, and some, possibly, not altogether unuseful to the public: And that just veneration which the world pays, as to your general worth, so especially to that zeal for the advancement of knowledge and the interest of your country which you express upon all occasions, gives you a particular right to whatever may any way tend to the promoting these interests, as an offering due to your merit. I have not so much of the vanity of a traveller, as to be fond of telling stories especially of this kind; nor can I think this plain piece of mine deserves a place among your more curious collections, much less have I the arrogance to use your name by way of patronage for the too obvious faults both of the Author and the Work. Yet dare I avow, according to my narrow sphere and poor abilities, a heavy zeal for the promoting of useful knowledge, and of anything that may never so remotely tend to my country's advantage; and I must own an ambition of transmitting to the public through your hands these essays I have made toward those great ends of which you are so deservedly esteemed the patron. This hath been my design in this publication, being desirous to bring in my gleanings here and there in remote regions, to that general magazine of the knowledge of foreign parts which the Royal Society thought you most worthy the custody of when they chose you for their President; and if in perusing these papers your goodness shall so far distinguish the experience of the Author from his faults, as to judge him capable of serving his country either immediately, or by serving you, he will endeavour by some real proofs to show himself,

Sir, Your most Faithful,

Devoted, Humble Servant,

W. Dampier.

THE PREFACE.

Before the Reader proceeds any further in the perusal of this Work, I must bespeak a little of his patience here, to take along with him this short account of it. It is composed of a mixed relation of places and actions, in the same order of time in which they occurred; for which end I kept a Journal of every day's observations.

In the description of places, their produce, &c., I have endeavoured to give what satisfaction I could to my countrymen; though possibly to the describing several things that may have been much better accounted for by others; choosing to be more particular than might be needful with respect to the intelligent Reader, rather than to omit what I thought might tend to the information of persons no less sensible and inquisitive, though not so learned or experienced. For which reason my chief care hath been to be as particular as was consistent with my intended brevity in setting down such observables[1] as I met with, nor have I given myself any great trouble since my return to compare my discoveries with those of others; the rather, because, should it so happen that I have described some places or things which others have done before me, yet in different accounts, even of the same things, it can hardly be but there will be some new light afforded by each of them. But after all, considering that the main of this voyage hath its scene laid in long tracts of the remoter parts both of the East and West Indies, some of which are very seldom visited by Englishmen, and others as rarely by any Europeans, I may without vanity encourage the Reader to expect many things wholly new to him, and many others more fully described than he may have seen elsewhere; for which not only this Voyage, though itself of many years' continuance, but also several former long and distant voyages, have qualified me.

As for the actions of the company among whom I made the greatest part of this Voyage, a thread of which I have carried on through it, it is not to divert the Reader with them that I mention them, much less that I take any pleasure in relating them, but for method's sake, and for the Reader's satisfaction, who could not so well acquiesce in my description of places, &c., without

[1] Notable things or incidents.

knowing the particular traverses I made among them; nor in these, without an account of the concomitant circumstances; besides that I would not prejudice the truth and sincerity of my relation, though by omissions only. And as for the traverses themselves, they make for the Reader's advantage, how little soever for mine, since thereby I have been the better enabled to gratify his curiosity; as one who rambles about a country can give usually a better account of it, than a carrier who jogs on to his inn without ever going out of his road.

As to my style, it cannot be expected that a seaman should affect politeness; for were I able to do it, yet I think I should be little solicitous about it in a work of this nature. I have frequently indeed divested myself of sea-phrases to gratify the land Reader; for which the seamen will hardly forgive me; and yet possibly I shall not seem complacent enough to the other; because I still retain the use of so many sea-terms. I confess I have not been at all scrupulous in this matter, either as to the one or the other of these; for I am persuaded, that if what I say be intelligible, it matters not greatly in what words it is expressed.

For the same reason I have not been curious as to the spelling of the names of places, plants, fruits, animals, &c., which in many of these remoter parts are given at the pleasure of travellers, and vary according to their different humours: neither have I confined myself to such names as are given by learned authors, or so much as inquired after many of them. I write for my countrymen; and have therefore for the most part used such names as are familiar to our English seamen, and those of our colonies abroad, yet without neglecting others that occurred. And it may suffice me to have given such names and descriptions as I could: I shall leave to those of more leisure and opportunity the trouble of comparing these with those which other authors have designed. . . .[1]

I have nothing more to add, but that there are here and there some mistakes made, as to expression and the like, which will need a favourable correction as they occur upon reading. For instance, the log of wood lying out at some distance from [the] sides of the boats described at Guam,[2] and parallel to their keel, which for distinction's sake I have called the little boat, might more clearly and properly have been called the side-log, or by some such name; for though [it is] fashioned at the bottom and ends boatwise, yet [it] is not hollow at top, but solid throughout. In other places also I may not have expressed myself so fully as I ought: but any considerable omission that I shall recollect, or be informed of, I shall endeavour to make up in these accounts I have yet to publish; and for any faults I leave the Reader to the joint use of his judgment and candour.

THE INTRODUCTION.

I FIRST set out of England on this voyage at the beginning of the year 1679, in the Loyal Merchant of London, bound for Jamaica, Captain Knapman, commander. I went a passenger, designing when I came thither to go from thence to the Bay of Campeachy, in the Gulf of Mexico, to cut logwood; where in a former voyage I had spent about three years in that employ, and so was well acquainted with the place and the work. We sailed with a prosperous gale, without any impediment or remarkable passage in our voyage: unless that, when we came in sight of the Island of Hispaniola, and were coasting along on the south side of it. by the little Isles of Vacca, or Ash,[3] I

[1] Two paragraphs are omitted here, which refer to the Appendices already noticed, and to the "maps and draughts" that illustrated the earlier editions of the work.

[2] Chapter X.

[3] La Vache is a small island at the south-west end of Hayti; in Dam-

observed Captain Knapman was more vigilant than ordinary, keeping at a good distance off shore, for fear of coming too near those small low islands; as he did once, in a voyage from England, about the year 1673, losing his ship there by the carelessness of his mates. But we succeeded better, and arrived safe at Port Royal in Jamaica some time in April 1679, and went immediately ashore. I had brought some goods with me from England which I intended to sell here, and stock myself with rum and sugar, saws, axes, hats, stockings, shoes, and such other commodities as I knew would sell among the Campeachy logwood cutters. Accordingly I sold my English cargo at Port Royal; but upon some maturer considerations of my intended voyage to Campeachy, I changed my thoughts of that design, and continued at Jamaica all that year, in expectation of some other business.

I shall not trouble the Reader with my observations at that isle, so well known to Englishmen; nor with the particulars of my own affairs during my stay there. But in short, having there made a purchase of a small estate in Dorsetshire, near my native country of Somerset, of one whose title to it I was well assured of, I was just embarking myself for England, about Christmas 1679, when one Mr Hobby invited me to go first a short trading voyage to the country of the Mosquitoes. I was willing to get up some money before my return, having laid out what I had at Jamaica; so I sent the writing of my new purchase along with the same friends whom I should have accompanied to England, and went on board Mr Hobby. Soon after our setting out, we came to an anchor again in Negril Bay, at the west end of Jamaica; but finding there Captains Coxon, Sawkins, Sharpe, and other Privateers, Mr Hobby's men all left him to go with them upon an expedition they had contrived, leaving not one with him besides myself; and being thus left alone, after three or four days' stay with Mr Hobby, I was the more easily persuaded to go with them too.

It was shortly after Christmas 1679 when we set out. The first expedition was to Portobello, which being accomplished, it was resolved to march by land over the Isthmus of Darien, upon some new adventures in the South Seas. Accordingly, on the 5th of April 1680, we went ashore on the Isthmus, near Golden Island, one of the Sambaloes,[1] to the number of between 300 and 400 men, carrying with us such provisions as were necessary, and toys wherewith to gratify the wild Indians through whose country we were to pass. In about nine days' march we arrived at Santa Maria, and took it; and after a stay there of about three days, we went on to the South Sea coast, and there embarked ourselves in such canoes and peringoes,[2] as our Indian friends furnished us withal. We were in sight of Panama by the 23d of April, and having in vain attempted Pueblo Nuevo, before which Sawkins, then commander-in-chief, and others, were killed, we made some stay at the neighbouring Isles of Quibo. Here we resolved to change our course and stand away to the southward for the coast of Peru. Accordingly we left the Keys or Isles of Quibo the 6th of June, and spent the rest of the year in that southern course; for, touching at the isles of Gorgona and Plata, we came to Ylo, a small town on the coast of Peru, and took it. This was in October, and in November we went thence to Coquimbo on the same coast, and about Christmas were got as far as the Isle of Juan Fernan-

pier's time called "Ash" by English seamen.

[1] Probably corresponding with what is now called the Muletas Archipelago, a number of small islands and rocks extending along the north-eastern coast of the Isthmus of Darien, from Point San Blas.

[2] Piroques; large canoes made square at one of the ends; called also "piraguas:" Italian, "piroga;" Spanish, "piragua."

dez, which was the farthest of our course to the southward. After Christmas, we went back again to the northward, having a design upon Arica, a strong town advantageously situated in the hollow of the elbow or bending of the Peruvian coast. But being there repulsed with great loss, we continued our course northward, till by the middle of April we were come in sight of the Isle of Plata, a little to the southward of the Equinoctial Line. . . .

While we lay at the Isle of Juan Fernandez, Captain Sharpe[1] was by general consent displaced from being commander, the company being not satisfied either with his courage or behaviour. In his stead, Captain Watling was advanced; but, he being killed shortly after before Arica, we were without a commander during all the rest of our return towards Plata. Now, Watling being killed, a great number of the meaner sort began to be as earnest for choosing Captain Sharpe again into the vacancy, as before they had been as forward as any to turn him out; and, on the other side, the abler and more experienced men, being altogether dissatisfied with Sharpe's former conduct, would by no means consent to have him chosen. In short, by the time we were come in sight of the Island of Plata, the difference between the contending parties was grown so high, that they resolved to part companies, having first made an agreement, that which party soever should, upon polling, appear to have the majority, they should keep the ship, and the other should content themselves with the launch or longboat, and canoes, and return back over the Isthmus, or go to seek their fortune other ways, as they would. Accordingly, we put it to the vote, and, upon dividing, Captain Sharpe's party carried it. I, who had never been pleased with his management, though I had hitherto kept my mind to myself, now declared myself on the side of those that were outvoted; and, according to our agreement, we took our shares of such necessaries as were fit to carry overland with us (for that was our resolution), and so prepared for our departure.

[1] Who had been made chief in command after Sawkins was killed at Pueblo Nuevo.

CHAPTER I.

APRIL the 17th, 1681, about 10 o'clock in the morning, being twelve leagues NW. from the Island of Plata, we left Captain Sharpe and those who were willing to go with him in the ship, and embarked into our launch and canoes, designing for the River of Santa Maria,[2] in the Gulf of San Miguel, which is about 200 leagues from the Isle of Plata. We were in number forty-four white men, who bore arms; a Spanish Indian, who bore arms also; and two Mosquito Indians, who always bare arms amongst the Privateers, and are much valued by them for striking fish and turtle, or tortoise, and manatee or sea-cow; and five slaves taken in the South Seas, who fell to our share. The craft which carried us was a launch or longboat, one canoe, and another canoe which had been sawn asunder in the middle, in order to have made bumkins, or vessels for carrying water, if we had not separated from our ship. This we joined together again and made it tight, providing sails to help us along; and for three days before we parted, we sifted as much flour as we could well carry, and rubbed up 20 or 30 lbs. of chocolate, with sugar to sweeten it; these things and a kettle the slaves carried also on their backs after we landed. And because there were some who designed to go with us that we knew were not well able to march, we gave out, that if any man faltered in the journey overland, he must expect to be shot to death;

[2] Now, apparently, the Tuyra, which flows into the south-east corner of the Gulf.

for we knew that the Spaniards would soon be after us, and one man falling into their hands might be the ruin of us all, by giving an account of our strength and condition; yet, this would not deter them from going with us. We had but little wind when we parted from the ship, but before 12 o'clock the sea breeze came in strong, which was like to founder us before we got in with the shore. For our security, therefore, we cut up an old dry hide that we brought with us, and barricaded the launch all round with it, to keep the water out. About 10 o'clock at night we got in about seven leagues to windward of Cape Pasado, under the Line, and then it proved calm, and we lay and drove all night, being fatigued the preceding day. The 18th we had little wind till the afternoon, and then we made sail, standing along the shore to the northward, having the wind at SSW., and fair weather. At 7 o'clock we came abreast of Cape Pasado, and found a small bark at anchor in a small bay to leeward of the Cape, which we took, our own boats being too small to transport us. We took her just under the Equinoctial Line. She was not only a help to us, but in taking her we were safe from being described. We did not design to have meddled with any when we parted with our consorts, nor to have seen any if we could have helped it. The bark came from Gallo, laden with timber, and was bound for Guayaquil. The 19th, in the morning, we came to an anchor about twelve leagues to the southward of Cape San Francisco, to put our new bark into a better trim. In three or four hours' time we finished our business, and came to sail again, and steered along the coast with the wind at SSW., intending to touch at Gorgona.

Being to the northward of Cape San Francisco, we met with very wet weather; but, the wind continuing, we arrived at Gorgona the 24th, in the morning, before it was light: we were afraid to approach it in the day-time, for fear the Spaniards should lie there for us, it being the place where we careened lately, and where they might expect us. When we came ashore we found the Spaniards had been there to seek after us, by a house they had built which would entertain 100 men, and by a great cross before the doors. This was token enough that the Spaniards did expect us this way again, therefore we examined our prisoners if they knew anything of it, who confessed they had heard of a periago, that rowed with fourteen oars, which was kept in a river on the main, and once in two or three days came over to Gorgona purposely to seek for us; and that, having discovered us, she was to make all speed to Panama with the news, where they had three ships ready to send after us. We lay here all the day, and scrubbed our new bark, that if ever we should be chased we might the better escape; we filled our water, and in the evening went from thence, having the wind at SW., a brisk gale. The 25th we had much wind and rain, and we lost the canoe that had been cut and was joined together; we would have kept all our canoes to carry us up the river, the bark not being so convenient. The 27th we went from thence with a moderate gale of wind at SW. In the afternoon we had excessive showers of rain.

The 28th was very wet all the morning; betwixt ten and eleven it cleared up, and we saw two great ships about a league and a half to the westward of us, we being then two leagues from the shore, and about ten leagues to the southward of Point Garachina. These ships had been cruising between Gorgona and the Gulf six months; but whether our prisoners did know it, I cannot tell. We presently furled our sails, and rowed in close under the shore, knowing that they were cruisers. The glare did not continue long before it rained again, and kept us from the sight of each other; but if they had seen and chased us, we were resolved to run our bark and canoes ashore, and take ourselves to the mountains and travel overland, for we knew that the Indians which lived in these parts never had any commerce with the Spaniards, so we might have had a chance for our lives. The 29th, at 9 o'clock in the morning, we came to an anchor at Point Garachina, about seven leagues from the Gulf of San Miguel, which was the place where we first came into the South Seas, and the way by which we designed to return. Here we lay all the day, and went ashore and dried our clothes, cleaned our guns, dried our ammunition, and fixed ourselves[1] against our enemies if we should be attacked; for we did expect to find some opposition at landing; we likewise kept a good lookout all the day, for fear of those two ships that we saw the day before. The 30th, in the morning at 8 o'clock, we came into the Gulf of San Miguel's mouth; for we put from Point Garachina in the evening, designing to have reached the islands in the Gulf before day, that we might the better work our escape from our enemies, if we should find any of them waiting to stop our passage. About 9 o'clock we came to an anchor a mile without a large island, which lies four miles from the mouth of the river; we had other small islands without us, and might have gone up into the river, having a strong tide of flood, but would not adventure farther till we had looked well about us. We immediately sent a canoe ashore on the island, where we saw (what we always feared) a ship at the mouth of the river, lying close by the shore, and a large tent by it, by which we found it would be a hard task for us to escape them. When the canoe came aboard with this news, some of our men were a little disheartened; but it was no more than I ever expected.

Our care was now to get safe overland, seeing we could not land here according to our desire; therefore, before the tide of flood was spent, we manned our canoe and rowed again to the island, to see if the enemy was yet in motion. When we came ashore we dispersed ourselves all over the island, to prevent our enemies from coming any way to view us; and presently after high water, we saw a small canoe coming over from the ship to the island that we were on, which made us all get into our canoe and wait their coming; and we lay close till they came within pistol shot of us, and then, being ready, we started out and took them. There were in her one white man and two Indians, who, being examined, told us that the ship which we saw at the river's mouth had lain there six months guarding the river, waiting for our coming; that she had 12 guns, and 150 seamen and soldiers; that the seamen all lay aboard, but the soldiers lay ashore in their tent; that there were 300 men at the mines, who had all small arms, and would be aboard in two tides' time. They likewise told us, that there were two ships cruising in the bay, between this place and Gorgona; the biggest had 20 guns and 200 men; the other 10 guns and 150 men. Besides all this, they told us that the Indians on this side the country were our enemies, which was the worst news of all. However, we presently brought these prisoners aboard, and got under sail, turning out with the tide of ebb, for it was not convenient to stay longer there. We did not long consider what to do, but intended to land that night or the next day betimes; for we did not question but we should either get a good commerce with the Indians by such toys as we had purposely brought with us, or else force our way through their country in spite of all their opposition; and we did not fear what these Spaniards could do against us in case they should land and come after us. We had a strong southerly wind, which blew right in; and the tide of ebb being far spent, we could not

[1] Prepared ourselves, put ourselves in trim: the so-called Americanism "to fix," like other words now specially used in the United States, having really its origin in a — possibly technical or local—English use of the word.

turn out. I persuaded[1] them to run into the River of Congo, which is a large river, about three leagues from the islands where we lay; which, with a southerly wind, we could have done; and when we were got as high as the tide flows, then we might have landed. But all the arguments I could use were not of force sufficient to convince them that there was a large river so near us; but they would land somewhere, they neither did know how, where, nor when. When we had rowed and towed against the wind all night, we just got about Cape San Lorenzo in the morning, and sailed about four miles farther to the westward, and ran into a small creek within two keys[2] or little islands, and rowed up to the head of the creek, being about a mile up, and there we landed, May 1st, 1681. We got out all our provision and clothes, and then sunk our vessel. While we were landing and fixing our snapsacks[3] to march, our Mosquito Indians struck a plentiful dish of fish, which we immediately dressed, and therewith satisfied our hunger.

Having made mention of the Mosquito Indians, it may not be amiss to conclude this Chapter with a short account of them. They are tall, well made, rawboned, lusty, strong, and nimble of foot; long-visaged, lank black hair, look stern, hard-favoured, and of a dark copper-coloured complexion. They are but a small nation or family, and not 100 men of them in number, inhabiting on the main, on the north side, near Cape Gracias Dios, between Cape Honduras and Nicaragua. They are very ingenious at throwing the lance, fisgig,[4] harpoon, or any manner of dart, being bred to it from their infancy, for the children, imitating their parents, never go abroad without a lance in their hands, which they throw at any object, till use has made them masters of the art. Then they learn to put by[5] a lance, arrow, or dart. The manner is thus:—Two boys stand at a small distance, and dart a blunt stick at one another, each of them holding a small stick in his right hand, with which he strikes away that which was darted at him. As they grow in years they become more dexterous and courageous, and then they will stand a fair mark to any one that will shoot arrows at them, which they will put by with a very small stick no bigger than the rod of a fowling-piece; and when they are grown to be men, they will guard themselves from arrows though they come very thick at them, provided two do not happen to come at once. They have extraordinary good eyes, and will descry a sail at sea farther, and see anything better, than we. Their chief employment in their own country is to strike fish, turtle, or manatee, the manner of which I describe elsewhere (Chapter III.). For this they are esteemed and coveted by all Privateers, for one or two of them in a ship will maintain 100 men; so that when we careen our ships, we choose commonly such places where there is plenty of turtle or manatee for these Mosquito men to strike; and it is very rare to find Privateers destitute of one or more of them when the commander or most of the men are English; but they do not love the French, and the Spaniards they hate mortally. When they come among Privateers they get the use of guns, and prove very good marksmen. They behave themselves very boldly in

[1] Advised.

[2] A key or cay (Latin, "cautes," a cliff; Spanish, "cayo;" French, "cayes") is a low island or ledge of rocks rising above the water; it is generally of coralline formation, and differs from a reef inasmuch as the latter is either below water, or washed by the waves. Keys are numerous among the West Indian Islands, and in the Gulf of Mexico, &c.

[3] (Swedish, "snappäck") soldiers' bags, knapsacks.

[4] A kind of harpoon or spear, with several barbed prongs, and a line attached; it is used for striking fish at sea, and is also called a "fishgig" or "fisgy."

[5] Parry.

fight, and never seem to flinch nor hang back, for they think that the White men with whom they are know better than they do when it is best to fight; and, let the disadvantage of their party be never so great, they will never yield nor give back while any of their party stand. I could never perceive any religion, nor any ceremonies, or superstitious observations[1] among them, being ready to imitate us in whatsoever they saw us do at any time. Only they seem to fear the devil, whom they call *Wallesaw;* and they say he often appears to some among them, whom our men commonly call their priests, when they desire to speak with him on urgent business, but the rest know not anything of him, nor how he appears, otherwise than as these priests tell them. Yet they all say they must not anger him, for then he will beat them, and that sometimes he carries away these their priests. Thus much I have heard from some of them who speak good English.

They marry but one wife, with whom they live till death separates them. At their first coming together the man makes a very small plantation, for there is land enough, and they may choose what spot they please. They delight to settle near the sea, or by some river, for the sake of striking fish, their beloved employment. Far within land there are other Indians with whom they are always at war. After the man has cleared a spot of land, and has planted it, he seldom minds it afterwards but leaves the managing of it to his wife, and goes out a-striking. Sometimes he seeks only for fish, at other times for turtle or manatee; and whatever he gets he brings home to his wife, and never stirs out to seek for more till it is all eaten. When hunger begins to bite, he either takes his canoe and seeks for more game at sea, or walks out into the woods and hunts about for peccary,[2] warree—each a sort of wild hogs—or deer, and seldom returns empty-handed, nor seeks for any more so long as any of it lasts. Their plantations are so small that they cannot subsist with what they produce, for their largest plantations have not above twenty or thirty plantain-trees, a bed of yams and potatoes, a bush of Indian pepper, and a small spot of pine-apples, which last fruit is a main thing they delight in, for with these they make a sort of drink which our men call pine-drink, much esteemed by these Mosquitoes, and to which they invite each other to be merry, providing fish and flesh also. Whoever of them makes of this liquor treats his neighbours, making a little canoe full at a time, and so enough to make them all drunk; and it is seldom that such feasts are made but the party that makes them hath some design, either to be revenged for some injury done him, or to debate of such differences as have happened between him and his neighbours, and to examine into the truth of such matters. Yet before they are warmed with drink they never speak one word of their grievances; and the women, who commonly know their husbands' designs, prevent them from doing any injury to each other by hiding their lances, harpoons, bows and arrows, or any other weapon that they have.

While they are among the English they wear good clothes, and take delight to go neat and tight; but when they return again to their own country they put by all their clothes, and go after their own country fashion, wearing only a small piece of linen tied about their waists hanging down to their knees.

CHAPTER II.

Being landed, May the 1st, we began our march about 3 o'clock in the afternoon, directing our course by our pocket compasses NE.; and having gone about two miles we came to the foot of a hill, where we built small

---

[1] Observances.

[2] The Mexican hog, or tajaçu—*Dicotyles tajaçu.*

huts and lay all night, having excessive rains till 12 o'clock. The 2d, in the morning, having fair weather, we ascended the hill, and found a small Indian path, which we followed till we found it ran too much easterly, and then, doubting[1] it would carry us out of our way, we climbed some of the highest trees on the hill, which was not meanly furnished with as large and tall trees as ever I saw. At length we discovered some houses in a valley on the north side of the hill, but it being steep [we] could not descend on that side, but followed the small path, which led us down the hill on the east side, where we presently found several other Indian houses. The first that we came to at the foot of the hill had none but women at home, who could not speak Spanish, but gave each of us a good calabash or shell full of corn-drink. The other houses had some men at home, but none that spoke Spanish; yet we made a shift to buy such food as their houses or plantations afforded, which we dressed and ate all together, having all sorts of our provision in common, because none should live better than others, or pay dearer for anything than it was worth. This day we had marched six miles. In the evening the husbands of those women came home, and told us in broken Spanish that they had been on board the guard-ship which we fled from two days before; that we were now not above three miles from the mouth of the River of Congo, and that they could go from thence aboard the guard-ship in half a tide's time. This evening we supped plentifully on fowls and peccary which we bought of the Indians; yams, potatoes, and plantains served us for bread, whereof we had enough. After supper we agreed with one of these Indians to guide us a day's march into the country towards the north side; he was to have for his pains a hatchet, and his bargain was to bring us to a certain Indian's habitation who could speak Spanish, from whom we were in hopes to be better satisfied of our journey.

The 3d, having fair weather, we began to stir betimes, and set out betwixt 6 and 7 o'clock, marching through several old ruined plantations. This morning one of our men, being tired, gave us the slip. By 12 o'clock we had gone eight miles, and arrived at the Indian's house, who lived on the bank of the River Congo, and spoke very good Spanish; to whom we declared the reason of this visit. At first he seemed to be very dubious of entertaining any discourse with us, and gave very impertinent answers to the questions that we demanded of him; he told us he knew no way to the north side of the country, but could carry us to Chepo, or to Santa Maria, which we knew to be Spanish garrisons, the one lying to the eastward of us, the other to the westward: either of them at least twenty miles out of our way. We could get no other answer from him, and all his discourse was in such an angry tone as plainly declared he was not our friend. However, we were forced to make a virtue of necessity and humour him; for it was neither time nor place to be angry with the Indians, all our lives lying in their hand. We were now at a great loss, not knowing what course to take, for we tempted him with beads, money, hatchets, macheats or long knives, but nothing would work on him, till one of our men took a sky-coloured petticoat out of his bag, and put it on his wife; who was so much pleased with the present, that she immediately began to chatter to her husband, and soon brought him into a better humour. He could then tell us that he knew the way to the north side, and would have gone with us, but that he had cut his foot two days before, which made him incapable of serving us himself: but he would take care that we should not want a guide; and therefore he hired the same Indian who brought us hither, to conduct us two days' march farther for another hatchet. The old man would have stayed us here all the day, be-

[1] Suspecting, apprehending.

cause it rained very hard; but our business required more haste, our enemies lying so near us, for he told us that he could go from his house aboard the guard-ship in a tide's time; and this was the fourth day since they saw us. So we marched three miles farther, and then built huts, where we stayed all night; it rained all the afternoon, and the greatest part of the night. The 4th, we began our march betimes, for the forenoons were commonly fair, but much rain after noon; though whether it rained or shined it was much at one with us, for I verily believed we crossed the rivers thirty times this day: the Indians having no paths to travel from one part of the country to another, and therefore, guiding themselves by the rivers. We marched this day twelve miles, and then built our huts and lay down to sleep; but we always kept two men on the watch, otherwise our own slaves might have knocked us on the head while we slept. It rained violently all the afternoon and most part of the night. We had much ado to kindle a fire this evening: our huts were but very mean or ordinary, and our fire small, so that we could not dry our clothes, scarce warm ourselves, and no sort of food for the belly; all which made it very hard with us. I confess these hardships quite expelled the thoughts of an enemy; for now, having been four days in the country, we began to have but few other cares than how to get guides and food: the Spaniards were seldom in our thoughts. The 5th, we set out in the morning betimes, and having travelled seven miles in those wild pathless woods, by 10 o'clock in the morning we arrived at a young Spanish Indian's house who had formerly lived with the Bishop of Panama. The young Indian was very brisk, spoke very good Spanish, and received us very kindly. This plantation afforded us store of provision, yams, and potatoes, but nothing of any flesh besides two fat monkeys we shot, part whereof we distributed to some of our company who were weak and sickly; for others we got eggs, and such refreshments as the Indians had; for we still provided for the sick and weak. We had a Spanish Indian in our company, who first took up arms with Captain Sawkins, and had been with us ever since his death. He was persuaded to live here by the master of the house, who promised him his sister in marriage, and to be assistant to him in clearing a plantation; but we would not consent to part with him here for fear of some treachery, but promised to release him in two or three days, when we were certainly out of danger of our enemies. We stayed here all the afternoon, dried our clothes and ammunition, cleared our guns, and provided ourselves for a march the next morning. Our Surgeon, Mr Wafer, came to a sad disaster here. Being drying his powder, a careless fellow passed by with his pipe lighted, and set fire to his powder, which blew up and scorched his knee, and reduced him to that condition that he was not able to march; wherefore we allowed him a slave to carry his things, being all of us the more concerned at the accident, because liable ourselves every moment to misfortune, and none to look after us but him. This Indian plantation was seated on the bank of the River Congo, in a very fat soil; and thus far we might have come in our canoe, if I could have persuaded them to it.

The 6th, we set out again, having hired another guide. Here we first crossed the River Congo in a canoe, having been from our first landing on the west side of the river; and being over, we marched to the eastwards two miles, and came to another river, which we forded several times, though it was very deep. Two of our men were not able to keep company with us, but came after us as they were able. The last time we forded the river, it was so deep, that our tallest men stood in the deepest place and handed the sick, weak, and short men; by which means we all got over safe, except those two who were behind. Foreseeing a necessity of wading through rivers frequently in our land march, I took care, before I left the ship, to provide myself a large joint of bamboo, which I stopped at both ends, closing it with wax, so as to keep out any water. In this I preserved my journal and other writings from being wet, though I was often forced to swim. When we were over this river, we sat down to wait the coming of our consorts who were left behind, and in half an hour they came. But the river by that time was so high, that they could not get over it; neither could we help them over, but bid them be of good comfort and stay till the river did fall: but we marched two miles farther by the side of the river, and there built our huts, having gone this day six miles. We had scarce finished our huts before the river rose much higher, and, overflowing the banks, obliged us to remove into higher ground: but the night came on before we could build more huts, so we lay straggling in the woods, some under one tree, some under another, as we could find conveniency; which might have been indifferent comfortable if the weather had been fair, but the greatest part of the night we had extraordinary hard rain, with much lightning and terrible claps of thunder. These hardships and inconveniences made us all careless, and there was no watch kept (though I believe nobody did sleep); so our slaves, taking opportunity, went away in the night, all but one who was hid in some hole and knew nothing of their design, or else fell asleep. Those that went away carried with them our Surgeon's gun and all his money. The next morning, being the 8th, we went to the river's side and found it much fallen; and here our guide would have us ford it again, which, being deep, and the current running swift, we could not. Then we contrived[1] to swim over; those that could not swim we were resolved to help over as well as we could; but this was not so feasible, for we should not be able to get all our things over. At length we concluded to send one man over with a line, who should haul over all our things first, and then get the men over. This being agreed on, one George Gayny took the end of a line, and made it fast about his neck, and left the other end ashore; and one man stood by the line, to clear it away to him. But when Gayny was in the midst of the water, the line in drawing after him chanced to kink, or grow entangled; and he that stood by to clear it away stopped the line, which turned Gayny on his back, and he that had the line in his hand threw it all into the river after him, thinking he might recover himself; but the stream running very swift, and the man having three hundred dollars at his back, [he] was carried down, and never seen more by us. Those two men whom we left behind the day before told us afterwards that they found him lying dead in a creek, where the eddy had driven him ashore, and the money on his back; but they meddled not with any of it, being only in care how to work their way through a wild unknown country. This put a period to that contrivance. This was the fourth man that we lost in this land journey; for those two men that we left the day before did not come to us till we were in the North Seas, so we yielded them also for lost. Being frustrated of getting over the river this way, we looked about for a tree to fell across the river. At length we found one, which we cut down, and it reached clear over; on this we passed to the other side, where we found a small plantain walk, which we soon ransacked. While we were busy getting plantains, our guide was gone; but in less than two hours came to us again, and brought with him an old Indian, to whom he delivered up his charge; and we gave him a hatchet and dismissed him, and entered ourselves under the conduct of our new guide: who immediately led us away, and crossed another river, and entered into a large valley of the fattest land I did ever take notice of; the trees were not very thick, but the largest that I saw in all my travels. We

[1] Planned, sought to devise means.

saw great tracks which were made by the peccaries, but saw none of them. We marched in this pleasant country till 3 o'clock in the afternoon, in all about four miles, and then arrived at the old man's country-house, which was only a habitation for hunting; there was a small plantain walk, some yams and potatoes. Here we took up our quarters for this day, and refreshed ourselves with such food as the place afforded, and dried our clothes and ammunition. At this place our young Spanish Indian provided to leave us, for now we thought ourselves past danger. This was he that was persuaded to stay at the last house we came from, to marry the young man's sister; and we dismissed him according to our promise.

The 9th the old man conducted us towards his own habitation. We marched about five miles in this valley, and then ascended a hill, and travelled about five miles farther over two or three small hills before we came to any settlement. Half a mile before we came to the plantations we light of[1] a path, which carried us to the Indians' habitations. We saw many wooden crosses erected in the way, which created some jealousy[2] in us that here were some Spaniards; therefore we new-primed all our guns, and provided ourselves for an enemy; but coming into the town [we] found none but Indians, who were all got together in a large house to receive us: for the old man had a little boy with him that he sent before. They made us welcome to such as they had, which was very mean; for these were new plantations, the corn being not eared. Potatoes, yams, and plantains they had none but what they brought from their old plantations. There were none of them spoke good Spanish; two young men could speak a little; it caused us to take more notice of them. To these we made a present and desired them to get us a guide to conduct us to the north side, or part of the way; which they promised to do themselves if we would reward them for it; but told us we must lie still the next day. But we thought ourselves nearer the North Sea than we were, and proposed to go without a guide rather than stay here a whole day. However some of our men who were tired, resolved to stay behind; and Mr Wafer, our Surgeon, who marched in great pain ever since his knee was burned with powder, was resolved to stay with them. The 10th we got up betimes, resolving to march, but the Indians opposed it as much as they could; but seeing they could not persuade us to stay, they came with us; and having taken leave of our friends we set out. Here therefore we left the Surgeon and two more, as we said, and marched away to the eastward, following our guides. But we often looked on our pocket compasses, and showed them to the guides, pointing at the way that we would go; which made them shake their heads, and say they were pretty things, but not convenient[3] for us. After we had ascended the hill on which the town stood, we came down into a valley, and guided ourselves by a river which we crossed thirty-two times; and having marched nine miles, we built huts and lay there all night. This evening I killed a quam,[4] a large bird as big as a turkey, wherewith we treated our guides; for we brought no provision with us. This night our last slave ran away. The 11th we marched ten miles farther and built huts at night, but went supperless to bed. The 12th, in the morning we crossed a deep river, passing over it on a tree, and marched seven miles in a low swampy ground, and came to the side of a great deep river, but could not get over. We built huts upon its banks, and lay there all night, upon our

[1] Came upon, lighted upon.

[2] Suspicion; to "jalouse" is still used in Scotland for to suspect.

[3] Of no advantage.

[4] Or Guan; *Penelope cristata*, a bird resembling the curassow, thirty inches long, of a dusky black above, glossed with green and olive, the neck and breast spotted with white.

barbecues, or frames of sticks, raised about three feet from the ground. The 13th, when we turned out, the river had overflowed its banks, and was two feet deep in our huts, and our guides went from us, not telling us their intent, which made us think they were returned home again. Now we began to repent our haste in coming from the last settlements, for we had no food since we came from thence. Indeed we got macaw-berries in this place, wherewith we satisfied ourselves this day, though coarsely. The 14th, in the morning betimes, our guides came to us again, and, the waters being fallen within their bounds, they carried us to a tree that stood on the bank of the river, and told us if we could fell that tree across it, we might pass; if not, we could pass no farther. Therefore we set two of the best axemen that we had, who felled it exactly across the river, and the boughs just reached over; on this we passed very safe. We afterwards crossed another river three times, with much difficulty; and at 3 o'clock in the afternoon we came to an Indian settlement, where we met a drove of monkeys, and killed four of them, and stayed here all night; having marched this day six miles. Here we got plantains enough, and a kind reception of the Indian that lived here all alone, except one boy to wait on him. The 15th, when we set out, the kind Indian and his boy went with us in a canoe, and set us over such places as we could not ford, and being past those great rivers he returned back again, having helped us at least two miles. We marched afterwards five miles, and came to large plantain walks, where we took up our quarters that night; we there fed plentifully on plantains, both ripe and green, and had fair weather all the day and night. I think these were the largest plantain walks, and the biggest plantains that ever I saw; but no house [was] near them. We gathered what we pleased by our guides' orders. The 16th, we marched three miles, and came to a large settlement, where we abode all day. Not a man of us but wished the journey at an end: our feet being blistered, and our thighs stripped with wading through so many rivers; the way being almost continually through rivers or pathless woods. In the afternoon five of us went to seek for game, and killed three monkeys, which we dressed for supper. Here we first began to have fair weather, which continued with us till we came to the North Seas. The 18th, we set out at 10 o'clock; and the Indians with five canoes carried us a league up a river, and when we landed, the kind Indians went with us and carried our burthens. We marched three miles farther, and then built our huts, having travelled from the last settlements six miles. The 19th, our guides lost their way, and we did not march above two miles. The 20th, by 12 o'clock, we came to Chepo River. The rivers we crossed hitherto, ran all into the South Seas; and this of Chepo was the last we met that ran that way. Here an old man who came from the last settlements distributed his burthen of plantains amongst us, and taking his leave returned home. Afterwards we forded the river and marched to the foot of a very high mountain, where we lay all night. This day we marched about nine miles. The 21st, some of the Indians returned back, and we marched up a very high mountain; being on the top, we went some miles on a ridge, and steep on both sides; then descended a little, and came to a fine spring, where we lay all night, having gone this day about nine miles; the weather still very fair and clear. The 22d, we marched over another very high mountain, keeping on the ridge, five miles. When we came to the north end, we to our great comfort saw the sea; then we descended and parted ourselves into three companies, and lay by the side of a river, which was the first we met that runs into the North Sea. The 23d, we came through several large plantain walks, and at 10 o'clock came to an Indian habitation not far from the

North Sea. Here we got canoes to carry us down the River Concepcion to the seaside; having gone this day about seven miles. We found a great many Indians at the mouth of this river. They had settled themselves here for the benefit of trade with the Privateers, and their commodities were yams, potatoes, plantains, sugar-canes, fowls, and eggs. These Indians told us that there had been a great many English and French ships here, which were all gone but one barcolongo,[1] a French Privateer, that lay at La Sound's Key or Island. This island is about three leagues from the mouth of the River Concepcion, and is one of the Sambaloes, a range of islands reaching for about twenty leagues from Point Samballas[2] to Golden Island eastward. These islands or keys, as we call them, were first made the rendezvous of Privateers in the year 1679, being very convenient for careening, and had names given to some of them by the Captains of the Privateers; as this La Sound's Key particularly. Thus we finished our journey from the South Sea to the North in twenty-three days; in which time, by my account, we travelled 110 miles, crossing some very high mountains; but our common march was in the valleys among deep and dangerous rivers.

On the 24th of May, having lain one night at the river's mouth, we all went on board the Privateer who lay at La Sound's Key. It was a French vessel; Captain Tristian, commander. The first thing we did was to get such things as we could to gratify our Indian guides, for we were resolved to reward them to their hearts' content. This we did by giving them beads, knives, scissors, and looking-glasses, which we bought of the Privateer's crew; and half-a-dollar a man from each of us, which we would have bestowed in goods also, but could not get any, the Privateer having no more toys. They were so well satisfied with these, that they returned with joy to their friends, and were very kind to our consorts whom we left behind; as Mr Wafer our Surgeon, and the rest of them told us, when they came to us some months afterwards, as shall be said hereafter.

[1] A small, low, long, sharp-built vessel without a deck, going with oars and sails; Spanish, "barcalonga."

[2] Point San Blas, from which Dampier's title for the islands seems to be corrupted.

CHAPTER III.

THE Privateer on board which we went being now cleaned, and our Indian guides satisfied and set ashore, we set sail in two days for Springer's Key, another of the Sambaloes Isles, about seven or eight leagues from La Sound's Key. Here lay eight sail of Privateers more, viz.:—

English commanders and English men.

Captain Coxon, 10 guns, 100 men.
Captain Payne, 10 guns, 100 men.
Captain Wright, a barcolongo, 4 guns, 40 men.

Captain Williams, a small barcolongo.

Captain Yanky, a barcolongo, 4 guns, about 60 men, English, Dutch, and French; himself a Dutchman.

French commanders and men.

Captain Archembo, 8 guns, 40 men.
Captain Tucker, 6 guns, 70 men.
Captain Rose, a barcolongo.

An hour before we came to the fleet, Captain Wright, who had been sent to the Chagres River, arrived at Springer's Key, with a large canoe or periago laden with flour, which he took there. Some of the prisoners belonging to the periago came from Panama not above six days before he took her, and told the news of our coming overland, and likewise related the condition and strength of Panama, which was the main thing they inquired after; for Captain Wright was sent thither purposely to get a prisoner that was able to inform them of the strength of that city, because these Privateers designed to join all their force, and by the assistance of the Indians (who had promised to be their guides) to march overland to Panama; and there is no other way of getting prisoners for that purpose but by absconding[1] between Chagres and Porto Bello, because there are much goods brought that way from Panama, especially when the Armada lies at Porto Bello. All the commanders were aboard of Captain Wright when we came into the fleet, and were mighty inquisitive of the prisoners to know the truth of what they related concerning us. But as soon as they knew we were come, they immediately came aboard of Captain Tristian, being all overjoyed to see us; for Captain Coxon and many others had left us in the South Seas about twelve months since, and had never heard what became of us since that time. They inquired of us what we did there? how we lived? how far we had been? and what discoveries we made in those seas? After we had answered these general questions, they began to be more particular in examining us concerning our passage through the country from the South Seas. We related the whole matter, giving them an account of the fatigues of our march, and the inconveniences we suffered by the rains, and disheartened them quite from that design. Then they proposed several other places where such a party of men as were now got together might make a voyage; but the objections of some or other still hindered any proceeding. For the Privateers have an account of most towns within twenty leagues of the sea on all the coast from Trinidad down to La Vera Cruz, and are able to give a near guess of the strength and riches of them, for they make it their business to examine all prisoners that fall into their hands concerning the country, town, or city they belong to: whether born there, or how long they have known it? how many families? whether most Spaniards? or whether the major part are not copper-coloured, as mulattoes, Mustesoes, or Indians? whether rich, and what their riches consist in? and what their chief manufactures? If fortified, how many great guns, and what number of small arms? whether it is possible to come undescried on them? how many look-outs or sentinels? for such the Spaniards always keep; how the look-outs are placed? whether possible to avoid the look-outs or take them? If any river or creek comes near it, or where the best landing? with innumerable other such questions, which their curiosity leads them to demand. And if they have had any former discourse of such places from other prisoners, they compare one with the other; then examine again, and inquire if he or any of them are capable to be guides to conduct a party of men thither; if not, where and how any prisoner may be taken that may do it; and from thence they afterwards lay their schemes to prosecute whatever design they take in hand.

It was seven or eight days before any resolution was taken, yet consultations were held every day. The French seemed very forward to go to any town that the English could or would propose, because the Governor of Petit Goave (from whom the Privateers take commissions) had recommended a gentleman lately come from France to be General of the expedition, and sent word by Captain Tucker, with whom this gentleman came, that they should if possible make an attempt on some town before he returned again. The English, when they were in company with the French, seemed to approve of what the French said, but never looked on that General to be fit for the service in hand. At length it was concluded to go to a town called Coretaga,[2]

[1] Lying in ambush or concealment.

[2] Cartago, near San Jose, the present capital of Costa Rica; the "Carpenter's River" afterwards mentioned would be the Matina.

which lies a great way in the country, but not such a tedious march as it would be from hence to Panama. Our way to it lay up Carpenter's River, which is about sixty leagues to the westward of Porto Bello. Our greatest obstruction in this design was our want of boats, therefore it was concluded to go with all our fleet to St Andreas,[1] a small uninhabited island lying near the Isle of Providence, to the W. of it, in 13° 15′ N. Lat., and from Porto Bello NNW. about seventy leagues, where we should be but a little way from Carpenter's River. And besides, at this island we might build canoes, it being plentifully stored with large cedars for such a purpose; and for this reason the Jamaica-men come hither frequently to build sloops, cedar being very fit for building, and being to be had here at free cost, besides other wood. Jamaica is well stored with cedars of its own, chiefly among the Rocky Mountains; these also of St Andreas grow in stony ground, and are the largest that ever I knew or heard of, the bodies alone being ordinarily forty or fifty feet long, many sixty or seventy and upwards, and of a proportionable bigness. The Bermudas Isles are well stored with them; so is Virginia, which is generally a sandy soil. I saw none in the East Indies, nor in the South Sea coast, except on the Isthmus as I came over it.[2] We reckon the periagoes and canoes that are made of cedar to be the best of any; they are nothing but the tree itself made hollow boat-wise, with a flat bottom, and the canoe generally sharp at both ends, the periago at one only, with the other end flat. But what is commonly said of cedar, that the worm will not touch it, is a mistake; for I have seen it very much worm-eaten.

All things being thus concluded on, we sailed from hence, directing our course toward St Andreas. We kept company the first day, but at night it blew a hard gale at NE., and some of our ships bore away. The next day others were forced to leave us, and the second night we lost all our company. I was now belonging to Captain Archembo, for all the rest of the fleet were over-manned. Captain Archembo wanting men, we that came out of the South Seas must either sail with him or remain among the Indians. Indeed, we found no cause to dislike the Captain; but his French seamen were the saddest creatures that ever I was among; for though we had bad weather that required many hands aloft, yet the biggest part of them never stirred out of their hammocks but to eat or ease themselves. We made a shift to find the Island the fourth day, where we met Captain Wright, who came thither the day before, and had taken a Spanish tartane,[3] wherein were thirty men, all well armed. She had four petereroes,[4] and some long guns placed in a swivel on the gunwale. They fought an hour before they yielded. The news they related was, that they came from Carthagena in company of eleven armadilloes (which are small vessels of war) to seek for the fleet of Privateers lying in the Sambaloes; that they parted from the armadilloes two days before; that they were ordered to search the Sambaloes for us, and if they did not find us, then they were ordered to go to Porto Bello, and lie there till they had further intelligence of us; and he supposed these armadilloes to be now there. We that came overland out of the South Seas, being weary of living among the French, desired Captain Wright to fit up his prize the tartane, and make a man-of-war of

[1] St Andrew's Island.

[2] The Author afterwards (Chapter IX., page 193) tells us that he found large cedars at the Three Marias Islands, off the coast of Mexico.

[3] A small coasting vessel used in the Mediterranean, with one mast and a large lateen sail; Spanish, "tartana;" French, "tartane."

[4] Or pedereroes (Spanish, "pedrero," from "piedra," a stone); a sort of swivel-gun which, before the invention of iron balls, were loaded with stone shot.

her for us, which he at first seemed to decline, because he was settled among the French on Hispaniola, and was very well beloved both by the Governor of Petit Goave, and all the gentry; and they would resent it ill that Captain Wright, who had no occasion of men, should be so unkind to Captain Archembo as to seduce his men from him; he being so meanly manned that he could hardly sail his ship with his Frenchmen. We told him we would no longer remain with Captain Archembo, but would go ashore there and build canoes to transport ourselves down to the Mosquitoes if he would not entertain us; for Privateers are not obliged to any ship, but free to go ashore where they please, or to go into any other ship that will entertain them, only paying for their provision. When Captain Wright saw our resolution, he agreed with us on condition we should be under his command as one ship's company, to which we unanimously consented.

We stayed here about ten days to see if any more of our fleet would come to us; but there came no more of us to the island but three, Captain Wright, Captain Archembo, and Captain Tucker. Therefore we concluded the rest were bore away either for Boca del Toro or Blewfields River on the main; and we designed to seek them. We had fine weather while we lay here, only some tornadoes or thunder-showers. But in this Isle of St Andreas there being neither fish, fowl, nor deer, and it being therefore but an ordinary place for us who had but little provision, we sailed from hence again in quest of our scattered fleet, directing our course for some islands lying near the main, called by the Privateers the Corn Islands; being in hopes to get corn there. These islands I take to be the same which are generally called in the maps the Pearl Islands, lying about the lat. of 12° 10′ N. Here we arrived the next day, and went ashore on one of them, but found none of the inhabitants, for there are but a few poor naked Indians that live here, who have been so often plundered by the Privateers, that they have but little provision; and when they see a sail they hide themselves, otherwise ships that come here would take them, and make slaves of them; and I have seen some of them that have been slaves. They are people of a mean stature, yet strong limbs; they are of a dark copper-colour, black hair, full round faces, small black eyes, their eye-brows hanging over their eyes, low foreheads, short thick noses, not high but flattish; full lips, and short chins. They have a fashion to cut holes in the lips of the boys when they are young, close to their chin, which they keep open with little pegs till they are fourteen or fifteen years old: then they wear beards in them made of turtle or tortoise shell. A little notch at the upper end they put in through the lip, where it remains between the teeth and the lip; the under part hangs down over their chin. This they commonly wear all day, and when they sleep they take it out. They have likewise holes bored in their ears, both men and women, when young; and by continual stretching them with great pegs, they grow to be as big as a milled five-shilling piece: herein they wear pieces of wood cut very round and smooth, so that their ear seems to be all wood, with a little skin about it. Another ornament the women use is about their legs, which they are very curious in; for from the infancy of the girls their mothers make fast a piece of cotton cloth about the small of their leg, from the ankle to the calf, very hard, which makes them have a very full calf: this the women wear to their dying day. Both men and women go naked, only a clout about their waists; yet they have but little feet though they go barefoot. Finding no provision here, we sailed towards Blewfields River, where we careened our tartane; and there Captain Archembo and Captain Tucker left us, and went towards Boca del Toro. This Blewfields River [1] comes

[1] Marked on some modern maps as the River Escondido or Segovia; it

out between the Rivers of Nicaragua and Veragua. It had this name from Captain Blewfield, a famous Privateer living on Providence Island long before Jamaica was taken: which Island of Providence was settled by the English and belonged to the Earls of Warwick. In this river we found a canoe coming down the stream; and though we went with our canoes to seek for inhabitants, yet we found none, but saw in two or three places signs that Indians had made on the side of the river. The canoe which we found was but meanly made for want of tools; therefore we concluded these Indians have no commerce with the Spaniards nor with other Indians that have.

While we lay here, our Mosquito-men went in their canoe and struck us some manatee or sea-cow.[1] Besides this Blewfields River I have seen the manatee in the Bay of Campeachy, on the coasts of Boca del Drago and Boca del Toro, in the River of Darien, and among the south keys or little islands of Cuba. I have heard of their being found on the north of Jamaica, a few; and in the rivers of Surinam in great multitudes, which is a very low land. I have seen them also at Mindanao, one of the Philippine Islands, and on the coast of New Holland. This creature is about the bigness of a horse, and ten or twelve feet long. The mouth of it is much like the mouth of a cow, having great thick lips. The eyes are no bigger than a small pea. The ears are only two small holes on each side of the head. The neck is short and thick, bigger than the head. The biggest part of this creature is at the shoulders, where it has two large fins, one on each side of its belly. Under each of these fins the female has a small dug to suckle her young. From the shoulders towards the tail it retains its bigness for about two feet, then grows smaller and smaller to the very tail, which is flat, and about fourteen inches broad and twenty inches long, and in the middle four or five inches thick, but about the edges not above two inches thick. From the head to the tail it is round and smooth, without any fin but those two before-mentioned. I have heard that some have weighed above 1200 lbs., but I never saw any so large. The manatee delights to live in brackish water; and they are commonly in creeks and rivers near the sea. They live on grass seven or eight inches long, and of a narrow blade, which grows in the sea in many places, especially among islands near the main; this grass grows likewise in creeks or in great rivers near the sides of them, in such places where there is but little tide or current. They never come ashore, nor into shallower water than where they can swim. Their flesh is white, both the fat and the lean, and extraordinary sweet wholesome meat. The tail of a young cow is most esteemed; but if old both head and tail are very tough. A calf that sucks is the most delicate meat; Privateers commonly roast them, as they do also great pieces cut out of the bellies of the old ones. The skin of the manatee is of great use to Privateers; for they cut them out into straps, which they make fast on the sides of their canoes, through which they put their oars in rowing, instead of tholes[2] or pegs. The skin of the bull or of the back of the cow is too thick for this use; but of it they make horse-whips, cutting them two or three feet long; at the handle they leave the full substance of the skin, and from thence cut it away tapering, but very even and square [on] all the four sides. While the thongs are green they twist them and hang them to dry; which in a week's time become as hard as wood. The Mosquito-men have always a small canoe for their use, to

---

runs south-eastward to the sea through the Mosquito country, and at its mouth are Blewfields town and lagoon.

[1] The description generally applies to some variety of the *Phoca*, or seal—known in different seas and to different navigators as the sea-dog, the sea-calf, the sea-cow, and the sea-lion.

[2] Pins in the gunwale to support the oar in rowing; also called "thowls."

strike fish, tortoise, or manatee: which they keep usually to themselves, and very neat and clean. They use no oars, but paddles, the broad part of which does not go tapering towards the staff, pole, or handle of it as in the oar; nor do they use it in the same manner, by laying it on the side of the vessel, but hold it perpendicularly, gripping the staff hard with both hands, and putting back the water by main strength and very quick strokes. One of the Mosquitoes (for there go but two in a canoe) sits in the stern, the other kneels down in the head; and both paddle till they come to the place where they expect their game. Then they lie still, or paddle very softly, looking well about them; and he that is in the head of the canoe lays down his paddle and stands up with his striking staff in his hand. This staff is about eight feet long, almost as big as a man's arm at the great end, in which there is a hole to place his harpoon in. At the other end of his staff there is a piece of light wood called bob-wood, with a hole in it through which the small end of the staff comes; and on this piece of bob-wood there is a line of ten or twelve fathoms wound neatly about, and the end of the line made fast to it. The other end of the line is made fast to the harpoon, which is at the great end of the staff; and the Mosquito-man keeps about a fathom of it loose in his hand. When he strikes, the harpoon presently comes out of the staff, and as the manatee swims away the line runs off from the bob; and although at first both staff and bob may be carried under water, yet as the line runs off it will rise again. Then the Mosquito-men paddle with all their might to get hold of the bob again, and spend usually a quarter of an hour before they get it. When the manatee begins to be tired, it lies still; and then the Mosquito-men paddle to the bob and take it up, and begin to haul in the line. When the manatee feels them he swims away again with the canoe after him; then he that steers must be nimble to turn the head of the canoe that way that his consort points, who being in the head of his canoe, and holding the line, both sees and feels which way the manatee is swimming. Thus the canoe is towed with a violent motion till the manatee's strength decays. Then they gather in the line, which they are often forced to let all go to the very end. At length, when the creature's strength is spent, they haul it up to the canoe's side, and knock it on the head, and tow it to the nearest shore, where they make it fast, and seek for another; which having taken, they go ashore with it to put it into their canoe. For it is so heavy that they cannot lift it in; but they haul it up in shoal water as near the shore as they can, and then overset the canoe, laying one side close to the manatee. Then they roll it in, which brings the canoe upright again; and when they have heaved out the water they fasten a line to the other manatee that lies afloat, and tow it after them.[1]

When we had cleaned our tartane we sailed from hence, bound for Boca del Toro, which is an opening between two islands about 10° 10′ N., between the Rivers of Veragua and Chagres. Here we met with Captain Yanky, who told us that there had been a fleet of Spanish armadilloes to seek us: that Captain Tristian having fallen to leeward, was coming to Boca del Toro, and fell in amongst them, supposing them to be our fleet: that they fired and chased him, but he rowed and towed, and they supposed he got away; that Captain Payne was likewise chased by them,

[1] The manner of striking the tortoise is also given at length, but being much the same as the method employed in striking the sea-cow, it has been omitted. Instead of a harpoon, a four-square sharp iron peg on the end of a striking staff is used, with a line attached. When the tortoise is struck it flies off, but the iron with the end of the line attached being buried beneath the shell, there is no possibility of its escape.

and Captain Williams, and that they had not seen them since; that they lay within the islands; that the Spaniards never came in to him; and that Captain Coxon was in at the careening place. This Boca del Toro is a place that the Privateers use to resort to as much as any place in all the coast, because here is plenty of green tortoise and a good careening place. The Indians here have no commerce with the Spaniards, but are very barbarous, and will not be dealt with. They have destroyed many Privateers, as they did not long after this some of Captain Payne's men; who having built a tent ashore to put his goods in while he careened his ship, and some men lying there with their arms, in the night the Indians crept softly into the tent and cut off the heads of three or four men, and made their escape; nor was this the first time they had served the Privateers so. There grow on this coast vinelloes in great quantity, with which chocolate is perfumed; these I shall describe elsewhere.[1]

Our fleet being thus scattered, there were now no hopes of getting together again; therefore every one did what they thought most conducing to obtain their ends. Captain Wright, with whom I now was, was resolved to cruise on the coast of Carthagena; and it being now almost the westerly wind season, we sailed from hence, and Captain Yanky with us; and we consorted, because Captain Yanky had no commission, and was afraid the French would take away his bark. We passed by Scuda,[2] a small island where it is said Sir Francis Drake's bowels were buried, and came to a small river to westward of Chagres, where we took two new canoes, and carried them with us into the Sambaloes. We had the wind at W. with much rain; which brought us to Point Samballas.[3] Here Captain Wright and Captain Yanky left us in the tartane to fix the canoes, while they went on the coast of Carthagena to seek for provision. We cruised in among the islands, and kept our Mosquito-men or strikers out, who brought aboard some half-grown tortoise; and some of us went ashore every day to hunt for what we could find in the woods. Sometimes we got peccary, warree, or deer; at other times we lighted on a drove of large fat monkeys or quams, corrosoes[4] (each a large sort of fowl), pigeons, parrots, or turtle-doves. We lived very well on what we got, not staying long in one place: but sometimes we would go on the islands, where there grow great groves of sappodillas,[5] which is a sort of fruit much like a pear, but more juicy; and under those trees we found plenty of soldiers,[6] a little kind of animals that live in shells, and have two great claws like a crab, and are good food. One time our men found a great many large ones, and, being sharp-set had them dressed, but most of them were very sick afterwards, being poisoned by them: for on this island were many manchineel trees,[7] whose fruit

[1] See Chapter VIII.

[2] Escudo de Veragua, off the Lagoon of Chiriqui. In Maynarde's narrative, however (*ante*, page 81), it is stated that Drake was buried at sea.

[3] San Blas.

[4] For Quam or Guan see Note 4, page 105. The Corroso, or Curassow, is described elsewhere by Dampier as "a larger fowl than the quam: the cock is black, the hen is of a dark brown. The cock has a crown of black feathers on his head, and appears very stately. These live also on berries, and are very good to eat; but their bones are said to be poisonous; therefore we do either burn or bury them, or throw them into the water for fear our dogs should eat them."

[5] Sappodilla is the name applied to plants of the genus *Achras*, natives of the West Indies and some parts of South America; the plum, or fruit, according to Lindley, is esteemed as an article of the dessert; the bark is employed in medicine as an astringent.

[6] Soldier-crab, or hermit-crab.

[7] Spanish, "Manzanilla;" a West Indian tree, used for furniture, and

is like a small crab,[1] and smells very well, but they are not wholesome; and we commonly take care of meddling with any animals that eat them. And this we take for a general rule: when we find any fruits that we have not seen before, if we see them pecked by birds we may freely eat; but if we see no such sign we may let them alone; for of this fruit no birds will taste. Many of these islands have these manchineel trees growing on them. Thus cruising in among these islands at length we came again to La Sound's Key; and the day before, having met with a Jamaica sloop that was come over on the coast to trade, she went with us. It was in the evening when we came to an anchor, and the next morning we fired two guns for the Indians that lived on the main to come aboard; for by this time we concluded we should hear from our five men that we left in the heart of the country among the Indians, this being about the latter end of August, and it was the beginning of May when we parted from them. According to our expectation the Indians came aboard and brought our friends with them. Mr Wafer wore a clout about him, and was painted like an Indian; and he was some time aboard before I knew him. One of them, named Richard Cobson, died within three or four days after, and was buried on La Sound's Key. After this we went to other keys to the eastward of these, to meet Captain Wright and Captain Yanky, who met with a fleet of periagoes laden with Indian corn, hog, and fowls, going to Carthagena; being convoyed by a small armadillo of two guns and six petereroes. Her they chased ashore, and most of the periagoes; but they got two of them off and brought them away. Here Captain Wright's and Captain Yanky's barks were cleaned; and we stocked ourselves with corn and then went towards the coast of Carthagena. In our way thither we passed by the River of Darien; which is very broad at the mouth, but not above six feet [of] water on a spring-tide; for the tide rises but little here. Captain Coxon, about six months before we came out of the South Seas, went up this river with a party of men: every man carried a small strong bag to put his gold in, expecting great riches there, though they got little or none. They rowed up about 100 leagues before they came to any settlement, and then found some Spaniards who lived there to truck[2] with the Indians for gold; there being gold scales in every house. The Spaniards admired[3] how they came so far from the mouth of the river; because there are a sort of Indians living between that place and the sea who are very dreadful to the Spaniards, and will not have any commerce with them, nor with any white people.

To return therefore to the prosecution of our voyage. Meeting with nothing of note, we passed by Carthagena, which is a city so well known that I shall say nothing of it. We sailed by in sight of it, for it lies open to the sea; and had a fair view of Madre de Popa, or Nuestra Señora de Popa, a monastery of the Virgin Mary, standing on the top of a very steep hill just behind Carthagena. It is a place of incredible wealth, by reason of the offerings made there continually; and for this reason often in danger of being visited by the Privateers, did not the neighbourhood of Carthagena keep them in awe. It is, in short, the very Loretto of the West Indies: it has innumerable miracles related of it. Any misfortune that befalls the Privateers is attributed to this Lady's doing; and the Spaniards report that she was abroad that night the Oxford man-of-war was blown up at the Isle of Vacca near Hispaniola, and that she came home all wet; as, belike, she often returns with her clothes dirty and torn with passing

well known for its poisonous white juice; the *Hippomane mancinella.*

[1] Apple.

[2] Barter, traffic by exchange; Spanish, "trocar," French, "troquer."

[3] Wondered.

through woods and bad ways when she has been out upon any expedition; deserving doubtless a new suit for such eminent pieces of service. From hence we passed on to the Rio Grande,[1] where we took up fresh water at sea, a league off the mouth of that river. From thence we sailed eastward, passing by Santa Marta, a large town and good harbour belonging to the Spaniards: yet hath it within these few years been twice taken by the Privateers. It stands close upon the sea, and the hill within land is a very large one, towering up a great height from a vast body of land.[2] I am of opinion that it is higher than the Peak of Teneriffe; others also that have seen both, think the same; though its bigness makes its height less sensible. I have seen it in passing by, thirty leagues off, at sea; others, as they told me, above sixty. Its head is generally hid in the clouds; but in clear weather, when the top appears, it looks white, supposed to be covered with snow. Santa Marta lies in Lat. 12°. Being advanced five or six leagues to the eastward of Santa Marta, we left our ships at anchor, and returned back in our canoes to the Rio Grande, entering it by a mouth of it that disembogues itself near Santa Marta: purposing to attempt some towns that lie a pretty way up that river. But this design meeting with discouragements, we returned to our ships and set sail to Rio la Hacha. This hath been a strong Spanish town, and is well built; but being often taken by the Privateers, the Spaniards deserted it some time before our arrival. It lies to the westward of a river; and right against the town is a good road for ships, the bottom clean and sandy. The Jamaica sloops used often to come over to trade here: and I am informed that the Spaniards have again settled themselves in it and made it very strong. We entered the fort and brought two small guns aboard. From thence we went to the Rancherias, one or two small Indian villages where the Spaniards keep two barks to fish for pearl.

When we had spent some time here, we returned again towards the coast of Carthagena; and being between Rio Grande and that place, we met with westerly winds, which kept us still to the eastward of Carthagena three or four days; then in the morning we descried a sail off at sea, and we chased her at noon. Captain Wright, who sailed best, came up with her and engaged her; and in half an hour after, Captain Yanky, who sailed better than the tartane (the vessel that I was in), came up with her likewise and laid her aboard, then Captain Wright also; and they took her before we came up. They lost two or three men, and had seven or eight wounded. The prize was a ship of 12 guns and forty men, who had all good small arms; she was laden with sugar and tobacco, and had eight or ten tons of marmalade on board: she came from Santiago de Cuba, and was bound to Carthagena. We went back with her to Rio Grande to fix our rigging, which was shattered in the fight, and to consider what to do with her; for these were commodities of little use to us, and not worth going into a port with. At the Rio Grande, Captain Wright demanded the prize as his due by virtue of his commission; Captain Yanky said it was his due by the law of Privateers. Indeed Captain Wright had the most right to her, having by his commission protected Captain Yanky from the French, who would have turned him out because he had no commission; and he likewise began to engage her first. But the company were all afraid that Captain Wright would presently carry her into a port; therefore most of Captain Wright's men stuck to Captain Yanky, and Captain Wright losing[3] his prize, burned his own bark, and had Captain

[1] Now Rio Magdalena.

[2] The Sierra de Santa Marta, the highest point of which, about thirty miles from the town, is 19,000 feet in elevation.

[3] Yielding up.

Yanky's, it being bigger than his own; the tartane was sold to a Jamaica trader, and Captain Yanky commanded the prize ship. We went again from hence to Rio la Hacha, and set the prisoners ashore: and it being now the beginning of November, we concluded to go to Curaçoa to sell our sugar, if favoured by westerly winds, which were now come in. We sailed from thence, having fair weather and winds to our mind, which brought us to Curaçoa, a Dutch island. Captain Wright went ashore to the Governor, and offered him the sale of the sugar: but the Governor told him he had a great trade with the Spaniards, therefore he could not admit us in there; but if we would go to St Thomas, which is an island and free port belonging to the Danes, and a sanctuary for Privateers, he would send a sloop with such goods as we wanted, and money to buy the sugar, which he would take at a certain rate; but it was not agreed to.

Curaçoa is the only island of importance that the Dutch have in the West Indies. It is about five leagues in length, and may be nine or ten in circumference: the northernmost point is laid down in N. Lat. 12° 40′, and it is about seven or eight leagues from the main, near Cape San Roman. On the south side of the east end is a good harbour called Santa Barbara; but the chief harbour is about three leagues from the SE. end, on the south side of it, where the Dutch have a very good town and a very strong fort. At the east end are two hills, one of them much higher than the other, and steepest towards the north side. The rest of the island is in different level; where of late some rich men have made sugar-works; which formerly was all pasture for cattle. There are also some small plantations of potatoes and yams, and they have still a great many cattle on the island: but it is not so much esteemed for its produce, as for its situation for the trade with the Spaniards. Formerly the harbour was never without ships from Carthagena and Porto Bello, that did use to buy of the Dutch here, 1000 or 1500 Negroes at once, besides great quantities of European commodities; but of late that trade is fallen into the hands of the English at Jamaica: yet still the Dutch have a vast trade over all the West Indies, sending from Holland ships of good force laden with European goods, whereby they make very profitable returns. The Dutch have two other islands there, but of little moment in comparison of Curaçoa; the one lies seven or eight leagues to the westward of Curaçoa, called Oruba; the other nine or ten leagues to the eastward of it, called Buen Ayre. From these islands the Dutch fetch in sloops, provision for Curaçoa, to maintain their garrison and Negroes. I was never at Oruba, therefore cannot say anything of it as to my own knowledge; but by report it is much like Buen Ayre, which I shall describe, only not so big. Between Curaçoa and Buen Ayre is a small island called Little Curaçoa; it is not above a league from Great Curaçoa. The King of France has long had an eye on Curaçoa, and made some attempts to take it, but never yet succeeded. I have heard that about twenty-three or twenty-four years since the Governor had sold it to the French, but died a small time before the fleet came to demand it; and by his death that design failed. Afterwards, in the year 1678, the Count D'Estrées, who a year before had taken the Isle of Tobago from the Dutch, was sent hither also with a squadron of stout ships, very well manned, and fitted with bombs and carcasses,[1] intending to take it by storm. This fleet first came to Martinico; where, while they stayed, orders were sent to Petit Goave for all Privateers to repair thither and assist the Count in his design. There were but two Privateers' ships that went thither to him, which were

[1] Carcasses (Italian, "carcassa," Spanish, "carcax" or "carcaza") are hollow cases made of ribs of iron, filled with inflammable matter, and thrown into besieged places with incendiary intent.

manned partly with French, partly with Englishmen. These set out with the Count; but in their way to Curaçoa, the whole fleet was lost on a reef or ridge of rocks that runs off from the Isle of Aves; not above two ships escaping, one of which was one of the Privateers: and so that design perished.

Wherefore not driving a bargain for our sugar with the Governor of Curaçoa, we went from thence to Buen Ayre, another Dutch island, where we met a Dutch sloop come from Europe laden with Irish beef which we bought in exchange for some of our sugar. Buen Ayre is the easternmost of the Dutch islands, and the largest of the three, though not the most considerable. The middle of the island is laid down in Lat. 12° 16′ [N.]. It is about twenty leagues from the main, and nine or ten from Curaçoa, and is accounted sixteen or seventeen leagues round. The road is on the SW. side, near the middle of the island, where there is a pretty deep bay runs in. The houses are about half-a-mile within land, right in the road. There is a Governor lives here, a deputy to the Governor of Curaçoa, and seven or eight soldiers, with five or six families of Indians. There is no fort; and the soldiers in peaceable times have little to do but to eat and sleep, for they never watch but in time of war. The Indians are husbandmen, and plant maize and Guinea corn, and some yams and potatoes: but their chief business is about cattle; for this island is plentifully stocked with goats; and they send great quantities every year in salt to Curaçoa. There are some horses, and bulls and cows; but I never saw any sheep, though I have been all over the island. The south side is plain low land, and there are several sorts of trees, but none very large. There is a small spring of water by the houses, which serves the inhabitants, though it is brackish. At the west end of the island there is a good spring of fresh water, and three or four Indian families live there; but no water nor houses at any other place. On the south side near the east end is a good salt-pond, where Dutch sloops come for salt. From Buen Ayre we went to the Isle of Aves, or birds; so called from its great plenty of birds, as man-of-war and boobies, but especially boobies. The booby is a water-fowl, somewhat less than a hen, of a light greyish colour. I observed the boobies of this island to be whiter than others. This bird has a strong bill, longer and bigger than a crow's, and broader at the end; its feet are flat like a duck's feet. The man-of-war (as it is called by the English) is about the bigness of a kite, and in shape like it, but black; and the neck is red. It lives on fish, yet never lights on the water, but soars aloft like a kite, and when it sees its prey, it flies down headforemost to the water's edge, very swiftly takes its prey out of the sea with its bill, and immediately mounts again as swiftly, never touching the water with its body. His wings are very long; his feet are like other land-fowl; and he builds on trees, where he finds any; but where they are wanting, on the ground. This Island of Aves lies about eight or nine leagues to the eastward of the island Buen Ayre, about fourteen or fifteen leagues from the main, and about the Lat. of 11° 45′ north. It is but small, not above four miles in length, and towards the east end not half-a-mile broad. On the north side it is low land, commonly overflown with the tide; but on the south side there is a great rocky bank of coral thrown up by the sea. The west end is, for near a mile's space, plain even savannah land, without any trees. There are two or three wells dug by Privateers, who often frequent this island, because there is a good harbour about the middle of it on the north side, where they may conveniently careen. The reef or bank of rocks on which the French fleet was lost, runs along from the east end to the northward about three miles, then trends away to the westward, making as it were a half-moon. This reef breaks off all the sea, and there is good riding in even sandy ground to the westward of it. There are two or three small low sandy keys or islands within this reef, about three miles from the main island. The Count D'Estrées lost his fleet here in this manner: Coming from the eastward, he fell in on the back of the reef, and fired guns to give warning to the rest of his fleet. But they, supposing their Admiral was engaged with enemies, hoised up their topsails, and crowded all the sail they could make, and ran full sail ashore after him; all within half-a-mile of each other. For his light being in the maintop was an unhappy beacon for them to follow; and there escaped but one King's ship, and one Privateer. The ships continued whole all day, and the men had time enough, most of them, to get ashore; yet many perished in the wreck: and many of those that got safe on the island, for want of being accustomed to such hardships, died like rotten sheep. But the Privateers who had been used to such accidents, lived merrily; from whom I had this relation: and they told me that if they had gone to Jamaica with £30 a man in their pockets they could not have enjoyed themselves more. For they kept in a gang by themselves, and watched when the ships broke to get the goods that came from them; and though much was staved against the rocks, yet abundance of wine and brandy floated over the reef, where these Privateers waited to take it up. They lived here about three weeks, waiting an opportunity to transport themselves back again to Hispaniola; in all which time they were never without two or three hogsheads of wine and brandy in their tents, and barrels of beef and pork, which they could live on without bread well enough, though the new comers out of France could not. There were about forty Frenchmen on board one of the ships where there was good store of liquor, till the after part of her broke away and floated over the reef, and was carried away to sea, with all the men drinking and singing, who, being in drink, did not mind the danger, but were never heard of afterwards. In a short time after this great shipwreck, Captain Payne, commander of a Privateer of 6 guns, had a pleasant accident befell him at this island. He came hither to careen, intending to fit himself very well; for here lay driven on the island, masts, yards, timbers, and many things that he wanted, therefore he hauled into the harbour, close to the island, and unrigged his ship. Before he had come, a Dutch ship of 20 guns was sent from Curaçoa to take up the guns that were lost on the reef. But seeing a ship in the harbour, and knowing her to be a French Privateer, they thought to take her first, and came within a mile of her, and began to fire at her, intending to warp in the next day; for it is very narrow going in. Captain Payne got ashore some of his guns, and did what he could to refit them, though he did in a manner conclude he must be taken. But, while his men were thus busied, he spied a Dutch sloop turning to get into the road, and saw her at the evening anchor at the west end of the island. This gave him some hope of making his escape, which he did, by sending two canoes in the night aboard the sloop, who took her, and got considerable purchase in her, and he went away in her, making a good reprisal, and leaving his own empty ship to the Dutch man-of-war. There is another island to the eastward of the Isle of Aves about four leagues, called by Privateers the Little Isle of Aves, which is overgrown with mangrove trees. I have seen it, but was never on it. There are no inhabitants that I could learn on either of these islands but boobies, and a few other birds. While we were at the Isle of Aves we careened Captain Wright's bark, and scrubbed the sugar prize, and got two guns out of the wrecks, continuing here till the beginning of February 1681-2. We went from hence to the Isles Roques to careen the sugar prize, which the Isle of Aves was not a place so convenient for. Accordingly we hauled close to one of the small islands, and got our guns ashore the first thing we did, and built a breastwork on the point, and planted all our guns there, to hinder an enemy from coming to us while we lay on the careen; then we made a house, and covered it with our sails, to put our goods and provisions in. While we lay here, a French man-of-war, of 36 guns, came through the keys or little islands, to whom we sold about ten tons of sugar. I was aboard twice or thrice, and very kindly welcomed both by the captain and his lieutenant, who was a Cavalier of Malta; and they both offered me great encouragement in France, if I would go with them; but I ever designed to continue with those of my own nation. The Islands Roques are a parcel of small uninhabited islands, lying about the Lat. of 11° 40′, about fifteen or sixteen leagues from the main, and about twenty leagues NW. by W. from Tortuga, and six or seven leagues W. of Orchillo, another island, lying about the same distance from the main. [Los] Roques stretch themselves E. and W. about five leagues, and their breadth [is] about three leagues. The northernmost of these islands is the most remarkable, by reason of a high white rocky hill at the west end of it, which may be seen a great way; and on it there are abundance of tropic birds, men-of-war, boobies, and noddies, which breed there. The booby and man-of-war I have described already.[1] The middle of this island is low plain land, overgrown with long grass, where there are multitudes of small grey fowls, no bigger than a blackbird; yet [they] lay eggs bigger than a magpie's: and they are therefore by Privateers called egg-birds. The east end of the island is overgrown with black mangrove trees.[2] The other islands are low, and have red mangroves, and other trees on them. Here also ships may ride; but no such place for careening as where we lay, because at that place ships may haul close to the shore; and, if they have but four guns on the point, may secure the channel, and hinder any enemy from coming near them.

After we had filled what water we could from hence, we set out again in April 1682, and came to Salt Tortuga; so called to distinguish it from the shoals of Dry Tortugas, near Cape Florida, and from the Isle of Tortugas by Hispaniola, which was called formerly French Tortugas; though not having heard any mention of that name a great while, I am apt to think it is swallowed up in that of Petit Goave, the chief garrison the French have in those parts. This island we arrived at is pretty large, uninhabited, and abounds with salt. It is in Lat. 11° N., and lies W. and a little N. from Margarita, an island inhabited by the Spaniards, strong and wealthy; it is distant from it about fourteen leagues, and seventeen or eighteen from Cape Blanco on the main.[3] At this isle we

[1] The noddy is described as a small black bird, about the size of an English blackbird, and esteemed good for food by voyagers there. In shape, they are round and plump like a partridge, and all white, save two or three feathers in each wing, which are of a light grey.

On the Roques Islands here described, the water was found to taste "copperish," and after two or three days' use of it, no other water seemed to possess any taste.

[2] The mangrove trees according to Dampier, are of three sorts, black, red, and white. Of these, the black and red form the most serviceable timber. The young saplings were used by the Privateers for making that part of the oar within the boat, called the "loom" or handle.

[3] Some remarks not generally interesting, are here made on a large salt pond at the east end of the Island of Tortuga, and often visited by ships to lade salt. This island had its name from the turtle or tortoise, which came upon the sandy bays to lay their eggs.

thought to have sold our sugar among the English ships that come hither for salt; but failing there we designed for Trinidad, an island near the main inhabited by the Spaniards, tolerably strong and wealthy: but the current and easterly winds hindering us, we passed through between Margarita and the main, and went to Blanco,[1] a pretty large island almost north of Margarita, about thirty leagues from the main, and in 11° 50′ N. Lat. It is a flat, even, low, uninhabited island, dry and healthy, most savannah of long grass, and has some trees of *Lignum-vitæ* growing in spots, with shrubby bushes of other wood about them. It is plentifully stored with guanos,[2] which are an animal like a lizard, but much bigger. The body is as big as the small of a man's leg, and from the hind quarter the tail grows tapering to the end, which is very small. If a man takes hold of the tail, except very near the hind quarter, it will part and break off in one of the joints, and the guano will get away. They lay eggs, as most of those amphibious creatures do, and are very good to eat. Their flesh is much esteemed by Privateers, who commonly dress them for their sick men; for they make very good broth. They are of divers colours, as almost black, dark brown, light brown, dark green, light green, yellow, and speckled; they all live as well in the water as on land, and some of them are constantly in the water, and among rocks: these are commonly black. Others that live in swampy wet ground are commonly on bushes and trees: these are green. But such as live on dry ground, as here at Blanco, are commonly yellow; yet these also will live in the water, and are sometimes on trees. There are sandy bays round the island, where turtle or tortoise come up in great abundance, going ashore in the night. Those that frequent this island are called green turtle; and they are the best of that sort, both for largeness and sweetness, of any in all the West Indies.

We stayed at the Isle of Blanco not above ten days, and then went back to Salt Tortuga again, where Captain Yanky parted with us. And from thence, after about four days, all which time our men were drunk and quarrelling, we in Captain Wright's ship went to the coast of Caracas on the mainland.[3] The cacao tree[4] has a body about a foot and a half thick (the largest sort) and seven or eight feet high to the branches, which are large, and spreading like an oak, with a pretty thick, smooth, dark-green leaf, shaped like that of a plum tree, but larger. The nuts are enclosed in cods as big as both a man's fists put together, at the broad end of which there is a small, tough, limber[5] tree, by which they hang pendulous from the body of the tree in all parts of it from top to bottom, scattered at irregular distances, and from the greater branches a little way up, especially at the joints of them, or partings, where they hang thickest, but never on the smaller boughs. There may be ordinarily about twenty or thirty of these cods upon a well-bearing tree, and they have two crops of them in a year, one in December, but the best in June. The cod itself, or shell, is almost half-an-inch thick; neither spongy nor woody, but of a substance between both, brittle, yet harder than the rind of a lemon, like which, its

[1] Or Blanquilla.

[2] Guana, or iguana, is the designation of several species of lizards, the best known being the *Iguana tuberculatum*, found in many parts of America and the West Indies, and valued for its flesh.

[3] The low-lying lands on the coast of Caracas are here characterised as extremely fertile, well watered, and inhabited by Spaniards and their Negroes, and that the main product of these valleys is the cacao nut, of which the chocolate is made, and of which such a painstaking description follows.

[4] Not to be confused with the cocoa-nut tree.

[5] Supple, flexible.

surface is grained or knobbed, but more coarse and unequal. The cods at first are of a dark green, but the side of them next the sun of a muddy red. As they grow ripe the green turns to a fine bright yellow and the muddy to a more lively beautiful red, very pleasant to the eye. They neither ripen nor are gathered [all] at once; but for three weeks or a month, when the season is, the overseers of the plantations go every day about to see which are turned yellow, cutting at once, it may be, not above one from a tree. The cods thus gathered they lay in several heaps to sweat, and then, bursting the shell with their hands, they pull out the nuts, which are the only substance they contain, having no stalk or pith among them; and (excepting that these nuts lie in regular rows) are placed like the grains of maize, but sticking together, and so closely stowed, that after they have been once separated, it would be hard to place them again in so narrow a compass. There are generally near 100 nuts in a cod, in proportion to the greatness of which, for it varies, the nuts are bigger or less. When taken out they dry them in the sun upon mats spread on the ground, after which they need no more care, having a thin hard skin of their own, and much oil, which preserves them. Salt water will not hurt them, for we had our bags rotten lying in the bottom of our ships, and yet the nuts never the worse. They raise the young trees [from] nuts set with the great end downward in fine black mould, and in the same places where they are to bear, which they do in four or five years' time without the trouble of transplanting. There are ordinarily of these trees from 500 to 2000 and upwards in a plantation, or cacao walk as they call them; and they shelter the young trees from the weather with plantains set about them for two or three years, destroying all the plantains by such time the cacao trees are of a pretty good body and able to endure the heat, which I take to be most pernicious to them of anything; for though these valleys lie open to the north winds, unless a little sheltered here and there by some groves of plantain trees which are purposely set near the shores of the several bays, yet, by all that I could either observe or learn, the cacaos in this country are never blighted, as I have often known them to be in other places. Cacao nuts are used as money in the Bay of Campeachy.

The chief town of this country is called Caracas, a good way within land; it is a large wealthy place, where live most of the owners of these cacao walks that are in the valleys by the shore, the plantations being managed by overseers and Negroes. It is in a large savannah country that abounds with cattle; and a Spaniard of my acquaintance, a very sensible man who hath been there, tells me that it is very populous, and he judges it to be three times as big as Coruña in Galicia. The way to it is very steep and craggy, over that ridge of hills which I said closes up the valleys and partition hills of the cacao coast. In this coast itself the chief place is La Guayra, a good town close by the sea; and though it has but a bad harbour, yet it is much frequented by the Spanish shipping, for the Dutch and English anchor in the sandy bays that lie here and there in the mouth of several valleys, and where there is very good riding. The town is open, but has a strong fort, yet both were taken some years since by Captain Wright and his Privateers. It is seated about four or five leagues to the westward of Cape Blanco, which is the easternmost boundary of this coast of Caracas. Farther eastward, about twenty leagues, is a great lake or branch of the sea, called La Laguna de Venezuela, about which are many rich towns; but the mouth of the lake is [so] shallow that no ships can enter. Near this mouth is a place called Cumana, where the Privateers were once repulsed without daring to attempt it any more, being the only place in the North Seas they attempted in vain for many years; and the Spaniards since throw it in their teeth frequently as a word of reproach or defiance to them. Not far from that place is Varinas, a small village and Spanish plantation famous for its tobacco, reputed the best in the world. But to return to Caracas. All this coast is subject to dry winds, generally north-east, which caused us to have scabby lips; and we always found it thus, and that in different seasons of the year, for I have been on this coast several times. In other respects it is very healthy, and a sweet clear air. The Spaniards have look-outs or scouts on the hills, and breastworks in the valleys, and most of their Negroes are furnished with arms also for defence of the bays. The Dutch have a very profitable trade here almost to themselves. I have known three or four great ships at a time on the coast, each, it may be, of 30 or 40 guns. They carry hither all sorts of European commodities, especially linen, making vast returns, chiefly in silver and cacao. And I have often wondered and regretted that none of my own countrymen find the way thither directly from England, for our Jamaica-men trade thither indeed, and find the sweet[1] of it, though they carry English commodities at second or third hand.

While we lay on this coast we went ashore in some of the bays and took seven or eight tons of cacao; and after that, three barks, one laden with hides, the second with European commodities, the third with earthenware and brandy. With these three barks we went again to the Islands of Roques, where we shared our commodities, and separated, having vessels enough to transport us all whither we thought most convenient. Twenty of us (for we were about sixty) took one of the vessels and our share of the goods, and went directly for Virginia. In our way thither we took several of the sucking-fishes,[2] for when we see them about the ship we cast out a line and hook, and they will take it with any manner of bait, whether fish or flesh. The sucking-fish is about the bigness of a large whiting, and much of the same shape towards the tail, but the head is flatter. From the head to the middle of its back there grows a sort of flesh of a hard gristly substance, like that part of the limpet, a shell-fish tapering up pyramidically, which sticks to the rocks; or like the head or mouth of a shell-snail, but harder. This excrescence is of a flat oval form about seven or eight inches long and five or six broad, and rising about half-an-inch high. It is full of small ridges, with which it will fasten itself to anything that it meets with in the sea, just as a snail does to a wall. When any of them happen to come about a ship, they seldom leave her, for they will feed on such filth as is daily thrown overboard, or on mere excrements. When it is fair weather and but little wind, they will play about the ship; but in blustering weather, or when the ship sails quick, they commonly fasten themselves to the ship's bottom, from whence neither the ship's motion, though never so swift, nor the most tempestuous sea, can remove them. They will likewise fasten themselves to any other bigger fish, for they never swim fast themselves if they meet with anything to carry them. I have found them sticking to a shark after it was hauled in on the deck, though a shark is so strong and boisterous a fish, and throws about him so vehemently for half-an-hour together, it may be, when caught, that did not the sucking-fish stick at no ordinary rate it must needs be cast off by so much violence. It is usual also to see them sticking to turtle, to any old trees, planks, or the like, that lie driving at sea. Any knobs or inequalities at a ship's bottom are a great hindrance to the swiftness of its sailing, and ten or twelve of these sticking to it must needs retard it as much, in a manner, as if its bottom were foul. So that I am inclined to think that this fish is the *Remora*, of which the ancients tell such stories: if it be not, I know no other that is,

[1] Advantage, gratification.

[2] The *Echeneis remora*, or sea lamprey.

and I leave the reader to judge.[1] I have seen these sucking-fishes in great plenty in the Bay of Campeachy, and in all the sea between that and the coast of Caracas, as about those islands particularly I have lately described, Roques, Blanco, Tortuga, &c. They have no scales, and are very good meat.

We met nothing else worth remark in our voyage to Virginia, where we arrived in July 1682. That country is so well known to our nation, that I shall say nothing of it; nor shall I detain the reader with the story of my own affairs, and the troubles that befell me during about thirteen months of my stay there: but in the next Chapter enter immediately upon my Second Voyage into the South Seas and round the Globe.

CHAPTER IV.

Being now entering upon the relation of a new voyage, which makes up the main body of this book, proceeding from Virginia by the way of Tierra del Fuego and the South Seas, the East Indies, and so on, till my return to England by the way of the Cape of Good Hope, I shall give my reader this short account of my first entrance upon it. Among those who accompanied Captain Sharpe into the South Seas in our former expedition, and, leaving him there, returned overland (as is said in the Introduction and in the First and Second Chapters), there was one Mr Cooke, an English native of St Christopher's, a Creole, as we call all born of European parents in the West Indies. He was a sensible man, and had been some years a Privateer. At our joining ourselves with those Privateers we met at our coming again to the North Seas, his lot was to be with Captain Yanky, who kept company for some considerable time with Captain Wright, in whose ship I was, and parted with us at our second anchoring at the Isle of Tortuga. After our parting, this Mr Cooke, being Quarter-master under Captain Yanky, the second place in the ship, according to the law of Privateers, laid claim to a ship they took from the Spaniards; and such of Captain Yanky's men as were so disposed, particularly all those who came with us overland, went aboard this prize ship, under the new Captain Cooke. This distribution was made at the Isle of Vacca, or the Isle of Ash, as we call it; and here they parted also such goods as they had taken. But Captain Cooke having no commission, as Captain Yanky, Captain Tristian, and some other French commanders had, who lay then at that island, and they grudging the English such a vessel, they all joined together, plundered the English of their ship, goods, and arms, and turned them ashore. Yet Captain Tristian took in about eight or ten of these English, and carried them with him to Petit Goave; of which number Captain Cooke was one, and Captain Davis another, who with the rest found means to seize the ship as she lay at anchor in the road, Captain Tristian and many of his men being then ashore. And the English sending ashore such Frenchmen as remained in the ship and were mastered by them, though superior in number, stood away with her immediately for the Isle of Vacca, before any notice of this surprise could reach the French Governor of that Isle; so deceiving him also by a stratagem, they got on board the rest of their countrymen who had been left on

[1] Pliny, in the opening chapter of his 32d book, is very eloquent on the powers of the *echineis*, or *remora*, or delaying-fish. "Let the winds rush," he says, among other grandiose things, "and the storms rage, one little fishling lays commands on their fury, and controls their mighty forces, and compels the ships to stand still: a thing that could be done by no bonds, by no anchor cast with irrevocable weight. It curbs the shocks and tames the madness of the world by no labour of its own, not by holding back, nor in any other way than simply by adhering."

that island; and going thence they took a ship newly come from France laden with wines. They also took a ship of good force, in which they resolved to embark themselves and make a new expedition into the South Seas, to cruise on the coast of Chili and Peru. But first they went for Virginia with their prizes; where they arrived the April after my coming thither. The best of their prizes carried eighteen guns: this they fitted up there with sails and everything necessary for so long a voyage; selling the wines they had taken for such provisions as they wanted. Myself and those of our fellow-travellers over the Isthmus of America who came with me to Virginia the year before this (most of whom had since made a short voyage to Carolina, and were again returned to Virginia), resolved to join ourselves to these new adventurers; and as many more engaged in the same design as made our whole crew consist of about seventy men. So having furnished ourselves with necessary materials, and agreed upon some particular rules, especially of temperance and sobriety, by reason of the length of our intended voyage, we all went on board our ship.

August 23d, 1683, we sailed from Achamack[1] in Virginia, under the command of Captain Cooke, bound for the South Seas. I shall not trouble the reader with an account of every day's run, but hasten to the less known parts of the world, to give a description of them: only relating such memorable accidents as happened to us, and such places as we touched at by the way. We met nothing worth observation till we came to the Islands of Cape Verd, except a terrible storm, which [we] could not escape: this happened in a few days after we left Virginia, with a SSE. wind just in our teeth. The storm lasted above a week: it drenched us all like so many drowned rats, and was one of the worst storms I ever was in. One I met with in the East Indies was more violent for the time, but of not above twenty-four hours' continuance. After that storm we had favourable winds and good weather; and in a short time we arrived at the Island [of] Sal, which is one of the easternmost of the Cape Verd Islands. Of these there are ten in number, so considerable as to bear distinct names; and they lie several degrees off from Cape Verd in Africa, whence they receive that appellation; taking up about 5° of longitude in breadth, and about as many of latitude in their length, viz., from near 14° to 19° North. They are mostly inhabited by Portuguese banditti. This of Sal is an island, lying in Lat. 16°, in Long. 19° 33′ W. from the Lizard in England, stretching from north to south about eight or nine leagues, and not above a league and a half or two leagues wide. It has its name from the abundance of salt that is naturally congealed there, the whole island being full of large salt ponds. The land is very barren, producing no tree that I could see, but some small shrubby bushes by the sea-side; neither could I discern any grass; yet there are some poor goats on it. [The island was also well stocked with wild fowl, especially flamingoes, which build their nests in shallow ponds among the mud. The bird itself is in shape like a heron, but bigger, and of a reddish colour. The flesh of both the young and old birds they found eatable, especially the tongue, "a dish of flamingoes' tongues being fit for a prince's table."]

There were not above five or six men on this Island of Sal, and a poor Governor, as they called him, who came aboard in our boat, and brought three or four poor lean goats for a present to our Captain, telling him they were the best that the island did afford. The Captain, minding more the poverty of the giver than the value of the present, gave him in

---

[1] Accomack is a county in or rather of Virginia, lying in some sort as an *enclave* in the peninsula of Maryland, which runs down towards Cape Charles between the Chesapeake and the Atlantic Ocean.

requital a coat to clothe him; for he had nothing but a few rags on his back, and an old hat not worth three farthings; which yet I believe he wore but seldom, for fear he should want before he might get another, for he told us there had not been a ship in three years before. We bought of him about twenty bushels of salt for a few old clothes; and he begged a little powder and shot. We stayed here three days: in which time one of these Portuguese offered to some of our men a lump of ambergris in exchange for some clothes, desiring them to keep it secret; for he said if the Governor should know it he should be hanged. At length one Mr Coppinger bought it for a small matter; yet I believe he gave more than it was worth. We had not a man in the ship that knew ambergris: but I have since seen it in other places, and therefore am certain it was not right. It was of a dark colour like sheep's dung, and very soft, but of no smell; and possibly it was some of their goats' dung. . . . We went from this Island of Sal to San Nicolas, another of the Cape Verd Islands, lying WSW. from Sal about twenty-two leagues. We arrived there the next day after we left the other, and anchored on the SE. side of the island. This is a pretty large island; it is one of the biggest of all the Cape Verd, and lies in a triangular form. The largest side, which lies to the east, is about thirty leagues long, and the other two above twenty leagues each. It is a mountainous barren island, and rocky all round towards the sea; yet in the heart of it there are valleys where the Portuguese which inhabit here have vineyards and plantations and wood for fuel. Here are many goats, which are but poor in comparison with those in other places, yet much better than those at Sal; there are likewise many asses. The Governor of this island came aboard us, with three or four gentlemen more in his company, who were all indifferently well clothed, and accoutred with swords and pistols; but the rest that accompanied him to the sea-side, which were about twenty or thirty men more, were but in a ragged garb. The Governor brought aboard some wine made in the island, which tasted much like Madeira wine; it was of a pale colour, and looked thick. He told us the chief town was in a valley fourteen miles from the bay where we rode; that he had there under him above one hundred families besides other inhabitants that lived scattering in valleys more remote. They were all very swarthy; the Governor was the clearest of them, yet of a dark tawny complexion. At this island we scrubbed the bottom of our ship; and here also we dug wells ashore on the bay, and filled all our water; and after five or six days' stay we went from hence to Mayo, another of the Cape Verd Islands, lying about forty miles E. and by S. from the other; arriving there the next day, and anchoring on the NW. side of the island. We sent our boat on shore, intending to have purchased some provision, as beef or goats, with which this island is better stocked than the rest of the islands. But the inhabitants would not suffer our men to land; for about a week before our arrival, there came an English ship, the men of which came ashore pretending friendship, and seized on the Governor with some others, and carrying them aboard made them send ashore for cattle to ransom their liberties: and yet after this set sail, and carried them away, and they had not heard of them since. The Englishman that did this, as I was afterwards informed, was one Captain Bond of Bristol. Whether ever he brought back those men again, I know not. He himself and most of his men have since gone over to the Spaniards: and it was he who had like to have burnt our ship after this in the Bay of Panama, as I shall have occasion to relate.[1] This Isle of Mayo is but small and environed with shoals, yet a place much frequented by shipping, for its great plenty of salt; and though there is but bad

---

[1] In Chapter VII.

landing, yet many ships lade here every year. Here are plenty of bulls, cows, and goats; and at a certain season in the year, as May, June, July, and August, a sort of small sea tortoise come hither to lay their eggs: but these turtle are not so sweet as those in the West Indies. The inhabitants plant corn, yams, potatoes, and some plantains, and breed a few fowls; living very poor, yet much better than the inhabitants of any other of these islands, Santiago excepted, which lies four or five leagues to the westward of Mayo, and is the chief, the most fruitful, and best inhabited of all the Islands of Cape Verd; yet mountainous, and much barren land in it.

On the east side of the Isle of Santiago is a good port, which in peaceable times especially is seldom without ships; for this hath long been a place which ships have been wont to touch at for water and refreshments, as those outward bound to the East Indies, English, French, and Dutch; many of the ships bound to the coast of Guinea, and Dutch to Surinam, and their own Portuguese Fleet going for Brazil, which is generally about the latter end of September: but few ships call in here in their return for Europe. When any ships are here the country people bring down their commodities to sell to the seamen and passengers, viz., bullocks, hogs, goats, fowls, eggs, plantains, and cocoa-nuts; which they will give in exchange for shirts, drawers, handkerchiefs, hats, waistcoats, breeches, or in a manner for any sort of cloth, especially linen; for woollen is not much esteemed there. They care not willingly to part with their cattel [1] of any sort but in exchange for money, or linen, or some other valuable commodity. Travellers must have a care of these people, for they are very thievish, and, if they see an opportunity, will snatch anything from you and run away with it. We did not touch at this island in this voyage; but I was there before this in the year 1670, when I saw a fort here lying on the top of an hill, and commanding the harbour. The Governor of this island is chief over all the rest of the islands. I have been told that there are two large towns on this island, some small villages, and a great many inhabitants; and that they make a great deal of wine, such as is that of San Nicolas. I have not been on any other of the Cape Verd Islands, nor near them, but have seen most of them at a distance. They seem to be mountainous and barren, some of these before mentioned being the most fruitful and most frequented by strangers, especially Santiago and Mayo. As to the rest of them, Fogo and Brava are two small islands lying to the westward of Santiago, but of little note; only Fogo is remarkable for its being a volcano. It is all of it one large mountain of a good height, out of the top whereof issue flames of fire, yet only discerned in the night; and then it may be seen a great way at sea. Yet this island is not without inhabitants, who live at the foot of the mountain near the sea. Their subsistence is much the same as in the other islands; they have some goats, fowls, plantains, cocoa-nuts, &c., as I am informed. The remainder of these islands of Cape Verd are San Antonia, Santa Lucia, San Vincente, and Bona Vista: of which I know nothing considerable.

Our entrance among these islands was from the NE.; for in our passage from Virginia we ran pretty far toward the coast of Gualata [2] in Africa, to preserve the trade-wind, lest we should be borne off too much to the westward, and so lose the islands. We anchored at the south of Sal, and passing by the south of San Nicolas anchored again at Mayo, as hath been said; where we made the shorter stay, because we could get no flesh among the inhabitants, by reason of

[1] Goods, chattels.

[2] Apparently the coast north of Cape Blanco, under the Tropic of Cancer; two Arab tribes with the designation of Aoulâd or Walad inhabit the interior.

the regret they had at their Governor and his men being carried away by Captain Bond. So leaving the Isles of Cape Verd we stood away to the southward with the wind at ENE., intending to have touched no more till we came to the Straits of Magellan. But when we came into the Lat. of 10° N., we met the winds at S. by W. and SSW., therefore we altered our resolutions, and steered away for the coast of Guinea, and in few days came to the mouth of the River of Sherboro', which is an English factory lying south of Sierra Leone. We had one of our men who was well acquainted there; and by his direction we went in among the shoals, and came to an anchor. Sherboro' was a good way from us, so I can give no account of the place, or our factory there; save that I have been informed, that there is a considerable trade driven there for a sort of red wood for dyeing, which grows in that country very plentifully; it is called by our people Camwood. A little within the shore where we anchored was a town of Negroes, natives of this coast. It was screened from our sight by a large grove of trees that grew between them and the shore; but we went thither to them several times during the three or four days of our stay here, to refresh ourselves; and they as often came aboard us, bringing with them plantains, sugar-canes, palm-wines, rice, fowls, and honey, which they sold us. They were no way shy of us, being well acquainted with the English, by reason of our Guinea factories and trade. This town seemed pretty large; the houses but low and ordinary; but one great house in the midst of it, where their chief men meet and receive strangers: and here they treated us with palm-wine. As to their persons they are like other Negroes. While we lay here we scrubbed the bottom of our ship, and then filled all our water-casks; and buying up two puncheons of rice for our voyage, we departed from hence about the middle of November 1683, prosecuting our intended course towards the Straits of Magellan.

We had but little wind after we got out, and very hot weather, with some fierce tornadoes, commonly rising out of the NE., which brought thunder, lightning, and rain. These did not last long; sometimes not a quarter of an hour; and then the wind would shuffle about to the southward again, and fall flat calm; for these tornadoes commonly come against the wind that is then blowing, as our thunder-clouds are often observed to do in England. At this time many of our men were taken with fevers: yet we lost but one. While we lay in the calms we caught several great sharks; sometimes two or three in a day, and ate them all, boiling and squeezing them dry, and then stewing them with vinegar, pepper, &c., for we had but little flesh aboard. We took the benefit of every tornado, which came sometimes three or four in a day, and carried what sail we could to get to the southward, for we had but little wind when they were over; and those small winds between the tornadoes were much against us till we passed the Equinoctial Line. In the Lat. of 5° S. we had the wind at ESE., where it stood a considerable time, and blew a fresh topgallant gale. We then made the best use of it, steering on briskly with all the sail we could make; and this wind by the 18th of January carried us into the Lat. of 36° S. In all this time we met with nothing worthy remark; not so much as a fish, except flying fish, which have been so often described, that I think it needless for me to do it. Here we found the sea much changed from its natural greenness, to a white or palish colour, which caused us to sound, supposing we might strike ground; for whenever we find the colour of the sea to change, we know we are not far from land, or shoals which stretch out into the sea, running from some land. But here we found no ground with 100 fathom line. The 20th, one of our Surgeons died, much lamented, because we had but one more for such a dangerous voyage.

January 28th, we made the Sibbel

de Wards,[1] which are three islands lying in the Lat. of 51° 25′ S., and Long. W. from the Lizard in England, by my account, 57° 28′. I had, for a month before we came hither, endeavoured to persuade Captain Cooke and his company to anchor at these islands, where I told them we might probably get water, as I then thought; and in case we should miss it here, yet by being good husbands of what we had, we might reach Juan Fernandez in the South Seas, before our water was spent. This I urged to hinder their designs of going through the Straits of Magellan, which I knew would prove very dangerous to us; the rather, because our men being Privateers, and so more wilful and less under command, would not be so ready to give a watchful attendance in a passage so little known. For although these men were more under command than I had ever seen any Privateers, yet I could not expect to find them at a minute's call in coming to anchor, or weighing anchor: besides, if ever we should have occasion to moor, or cast out two anchors, we had not a boat to carry out or weigh an anchor. These Islands of Sibbel de Wards were so named by the Dutch. They are all three rocky barren islands without any tree, only some dildo bushes growing on them; and I do believe there is no water on any one of them, for there was no appearance of any water.

Leaving therefore the Sibbel de Ward Islands, as having neither good anchorage nor water, we sailed on, directing our course for the Straits of Magellan. But the winds hanging in the wester-board, and blowing hard, oft put us by our topsails; so that we could not fetch it. The 6th of February we fell in with the Straits of Le Maire, which is very high land on both sides, and the Straits very narrow. We had the wind at NNW. a fresh gale; and seeing the opening of the Straits, we ran in with it, till within four miles of the mouth, and then it fell calm, and we found a strong tide setting out of the Straits to the northward, and like to founder our ship; but whether flood or ebb I know not; only it made such a short cockling sea as if we had been in a race, or place where two tides meet. For it ran every way, sometimes breaking in over our waist, sometimes over our poop, sometimes over our bow, and the ship tossed like an egg-shell, so that I never felt such uncertain jerks in a ship. At 8 o'clock in the evening we had a small breeze at WNW., and steered away to the eastward, intending to go round the Staten Island, the east end of which we reached the next day by noon, having a fresh breeze all night. At the east end of Staten Island are three small islands, or rather rocks, pretty high, and white with the dung of fowls. Having observed the sun, we hauled up south, designing to pass round to the southward of Cape Horn, which is the southernmost land of Tierra del Fuego. The winds hung in the western quarter betwixt the NW. and the W., so that we could not get much to the westward, and we never saw Tierra del Fuego after that evening that we made the Straits of Le Maire. I have heard that there have been smokes and fires on Tierra del Fuego, not on the tops of hills, but in plains and valleys, seen by those who have sailed through the Straits of Magellan; supposed to be made by the natives.[2]

The 14th of February, being in Lat. 57°, and to the west of Cape Horn, we had a violent storm, which held us till the 3d day of March, blowing commonly at SW. and SW. by W.

---

[1] The Sebaldine group, lying on the north-west of the Falkland Islands; they were discovered by the Dutch navigator Sebald de Wert in 1600, and, until Commodore Byron rechristened them in 1765, they gave their name to the whole group now called the Falklands.

[2] In the account of Drake's voyage (*ante*, page 56), we find it stated: "The people inhabiting these parts made fires as we passed by in divers places."

and WSW., thick weather all the time, with small drizzling rain, but not hard. We made a shift, however, to save twenty-three barrels of rain-water besides what we dressed our victuals withal. March the 3d, the wind shifted at once, and came about at S., blowing a fierce gale of wind; soon after it came about to the eastward, and we stood into the South Seas. The 9th, having an observation of the sun, not having seen it of late, we found ourselves in Lat. 47° 10′. The wind stood at SE., we had fair weather, and a moderate gale; and the 17th, we were in Lat. 36° by observation. The 19th day, when we looked out in the morning, we saw a ship to the southward of us coming with all the sail she could make after us. We lay muzzled to let her come up with us, for we supposed her to be a Spanish ship come from Valdivia bound to Lima; we being now to the northward of Valdivia, and this being the time of the year when ships that trade thence to Valdivia return home. They had the same opinion of us, and therefore made sure to take us, but coming nearer we both found our mistakes. This proved to be one Captain Eaton, in a ship sent purposely from London for the South Seas. We hailed each other, and the Captain came on board, and told us of his actions on the coast of Brazil and in the river of Plate. He met Captain Swan, one that came from England to trade here, at the east entrance into the Straits of Magellan, and they accompanied each other through the Straits, and were separated after they were through by the storm before mentioned. Both we and Captain Eaton being bound for Juan Fernandez's Isle, we kept company, and we spared him bread and beef and he spared us water, which he took in as he passed through the Straits.

March the 22d, 1684, we came in sight of the island, and the next day got in and anchored in a bay at the south end of the island, in twenty-five fathom water, not two cables' length from the shore. We presently got out our canoe and went ashore to seek for a Mosquito Indian whom we left here when we were chased hence by three Spanish ships in the year 1681, a little before we went to Arica, Captain Watling being then our commander, after Captain Sharpe was turned out. This Indian lived here alone above three years, and although he was several times sought after by the Spaniards, who knew he was left on the island, yet they could never find him. He was in the woods hunting for goats when Captain Watling drew off his men, and the ship was under sail before he came back to shore. He had with him his gun and a knife, with a small horn of powder, and a few shot, which being spent, he contrived a way, by notching his knife, to saw the barrel of his gun into small pieces, wherewith he made harpoons, lances, hooks, and a long knife; heating the pieces first in the fire, which he struck with his gun-flint, and a piece of the barrel of his gun, which he hardened, having learnt to do that among the English. The hot pieces of iron he would hammer out and bend as he pleased with stones, and saw them with his jagged knife, or grind them to an edge by long labour, and harden them to a good temper as there was occasion. All this may seem strange to those that are not acquainted with the sagacity of the Indians; but it is no more than these Mosquito men are accustomed to in their own country, where they make their own fishing and striking instruments without either forge or anvil, though they spend a great deal of time about them. Other wild Indians who have not the use of iron, which the Mosquito men have from the English, make hatchets of a very hard stone, with which they will cut down trees (the cotton tree especially, which is a soft tender wood), to build their houses or make canoes; and though in working their canoes hollow they cannot dig them so neat and thin, yet they will make them fit for their service. This their digging or hatchet-work they help out by fire, whether for the felling of the trees or for the making the inside of their canoes hollow. These contrivances are used particularly by the savage Indians of Blewfields River, whose canoes and stone hatchets I have seen. These stone hatchets are about ten inches long, four broad, and three inches thick in the middle. They are ground away flat and sharp at both ends; right in the midst, and clear round it, they make a notch, so wide and deep that a man might place his finger along it; and taking a stick or withe about four feet long, they bind it round the hatchet-head in that notch, and so twisting it hard, use it as a handle or helve,[1] the head being held by it very fast. Nor are other wild Indians less ingenious. Those of Patagonia, particularly, head their arrows with flint cut or ground, which I have seen and admired.[2]

But to return to our Mosquito man on the Isle of Juan Fernandez. With such instruments as he made in that manner, he got such provision as the island afforded, either goats or fish. He told us that at first he was forced to eat seal, which is very ordinary meat, before he had made hooks; but afterwards he never killed any seals but to make lines, cutting their skins into thongs. He had a little house or hut half-a-mile from the sea, which was lined with goatskin; his couch, or barbecue, of sticks, lying along about two feet distant from the ground, was spread with the same, and was all his bedding. He had no clothes left, having worn out those he brought from Watling's ship, but only a skin about his waist. He saw our ship the day before we came to an anchor, and did believe we were English, and therefore killed three goats in the morning before we came to anchor, and dressed them with cabbage to treat us when we came ashore. He came then to the sea-side to congratulate our safe arrival. And when we landed, a Mosquito Indian, named Robin, first leaped ashore, and running to his brother Mosquito-man, threw himself flat on his face at his feet, who, helping him up and embracing him, fell flat with his face on the ground at Robin's feet, and was by him taken up also. We stood with pleasure to behold the surprise and tenderness and solemnity of this interview, which was exceedingly affectionate on both sides; and when their ceremonies of civility were over, we also that stood gazing at them drew near, each of us embracing him we had found here, who was overjoyed to see so many of his old friends come hither, as he thought, purposely to fetch him. He was named Will, as the other was named Robin. These were names given them by the English, for they have no names among themselves; and they take it as a great favour to be named by any of us, and will complain for want of it if we do not appoint them some name when they are with us, saying, of themselves they are poor men and have no name.

This island is in Lat. 34° 15′, and about 120 leagues from the main. It is about twelve leagues round, full of high hills and small pleasant valleys, which, if manured, would probably produce anything proper for the climate. The sides of the mountains are part savannahs, part woodland. Savannahs are clear pieces of land without woods, not because more barren than the woodland, for they are frequently spots of as good land as any, and often are intermixed with woodland. [The grass in these savannahs is here described as long and flaggy, and the valleys well stocked with wild goats, these having been first left there by Juan Fernandez in his voyage from Lima to Valdivia. The sea about it is described as swarming with fish, "so plentiful that two men in an hour's time will take with hook and line as many as will serve 100 men."]

Seals swarm as thick about this island as if they had no other place to live in, for there is not a bay nor rock that one can get ashore on but is full of them. The seals are a sort of creatures pretty well

[1] From Anglo-Saxon "helf," a haft or handle.

[2] Marvelled at.

known, yet it may not be amiss to describe them. They are as big as calves, the head of them like a dog, therefore called by the Dutch sea-hounds. Under each shoulder grows a long thick fin; these serve them to swim with them in the sea, and are instead of legs to them when on the land, for raising their bodies up on end by the help of these fins or stumps, and so having their tail-parts drawn close under them, theyrebound, as it were, and throw their bodies forward, drawing their hinder parts after them; and then again rising up and springing forward with their fore-parts alternately, they lie tumbling up and down all the while they are moving on land. From their shoulders to their tails they grow tapering like fish, and have two small fins on each side of the rump, which is commonly covered with their fins. These fins serve instead of a tail in the sea, and on land they sit on them when they give suck to their young. Their hair is of divers colours, as black, grey, dun, spotted, looking very sleek and pleasant when they come first out of the sea; for these at Juan Fernandez have fine thick short fur, the like I have not taken notice of anywhere but in these seas. Here are always thousands, I might say possibly millions of them, sitting on the bays or going and coming in the sea round the island, which is covered with them, as they lie at the top of the water playing and sunning themselves for a mile or two from the shore. When they come out of the sea, they bleat like sheep for their young; and though they pass through hundreds of others' young ones before they come to their own, yet they will not suffer any of them to suck. The young ones are like puppies, and lie much ashore; but when beaten by any of us, they, as well as the old ones, will make towards the sea, and swim very swift and nimble, though on shore they lie very sluggishly, and will not go out of our ways unless we beat them, but snap at us. A blow on the nose soon kills them. Large ships might here load themselves with seal-skins and train-oil, for they are extraordinary fat. Seals are found as well in cold as hot climates, and in the cold places they love to get on lumps of ice, where they will lie and sun themselves as here on the land. They are frequent in the northern parts of Europe and America, and in the southward parts of Africa, as about the Cape of Good Hope, and at the Straits of Magellan; and though I never saw any in the West Indies but in the Bay of Campeachy, at certain islands called the Alecranes, and at others called the Desertas, yet they are over all the American coast of the South Seas, from Tierra del Fuego up to the Equinoctial Line; but to the north of the Equinox again in these seas I never saw any till as far as 21° N. Nor did I ever see any in the East Indies. In general they seem to resort where there is plenty of fish, for that is their food; and fish such as they feed on, as cods, groopers, &c., are most plentiful on rocky coasts, and such is mostly the western coast of South America.

The sea-lion[1] is a large creature about twelve or fourteen feet long. The biggest part of his body is as big as a bull: it is shaped like a seal, but six times as big. The head is like a lion's head; it hath a broad face, with many long hairs growing about its lips like a cat. It has a great goggle eye, the teeth three inches long, about the bigness of a man's thumb. In Captain Sharpe's time some of our men made dice with them. They have no hair on their bodies like the seal; they are of a dun colour, and are all extraordinary fat: one of them being cut up and boiled will yield a hogshead of oil, which is very sweet and wholesome to fry meat withal. The lean flesh is black, and of a coarse grain, yet indifferent good food. They will lie a week at a time ashore if not disturbed. Where three or four or more of them come ashore together, they huddle one on

[1] A large species of seal, the male of which has a mane on its neck.

another like swine, and grunt like them, making a hideous noise. They eat fish, which I believe is their common food. The snapper is a fish made much like a roach, but a great deal bigger. It has a large head and mouth, and great gills. The back is of a bright red, the belly of a silver colour. The scales are as broad as a shilling. The snapper is excellent meat. They are in many places in the West Indies and the South Seas. The rock-fish[1] is called by seamen a grooper: the Spaniards call it "baccalao," which is the name for cod, because it is much like it. It is rounder than the snapper, of a dark brown colour, and hath small scales no bigger than a silver penny. This fish is good sweet meat, and is found in great plenty on all the coast of Peru and Chili.

There are only two bays in the whole island where ships may anchor; these are both at the east end, and in both of them is a rivulet of good fresh water. Either of these bays may be fortified, with little charge, to that degree that fifty men in each may be able to keep off 1000; and there is no coming into these bays from the west end but with great difficulty, over the mountains, where if three men are placed they may keep down as many as come against them on any side. This was partly experienced by five Englishmen that Captain Davis left here, who defended themselves against a great body of Spaniards who landed in the bays, and came here to destroy them; and though the second time one of their consorts deserted and fled to the Spaniards, yet the other four kept their ground, and were afterward taken in from hence by Captain Strong of London.

We remained at Juan Fernandez sixteen days. Our sick men were ashore all the time, and one of Captain Eaton's doctors (for he had four in his ship) tending and feeding them with goat, and several herbs, whereof here is plenty growing in the brooks; and their diseases were chiefly scorbutic.

[1] The *Gobius niger*, or black goby.

CHAPTER V.

THE 8th of April 1684, we sailed from the Isle of Juan Fernandez with the wind at SE. We were now two ships in company: Captain Cooke's, whose ship I was in, and who here took the sickness of which he died a while after; and Captain Eaton's. Our passage lay now along the Pacific Sea, properly so called. For though it be usual with our map-makers to give that name to this whole Ocean, calling it Mare Australe, Mare del Zur, or Mare Pacificum; yet, in my opinion, the name of the Pacific Sea ought not to be extended from S. to N. farther than from 30° to about 4° S. Lat., and from the American shore westward indefinitely. In this sea we made the best of our way towards the Line, till in the Lat. of 24° S., where we fell in with the mainland of South America. All this course of the land, both of Chili and Peru, is vastly high; therefore we kept twelve or fourteen leagues off from shore, being unwilling to be seen by the Spaniards dwelling there. The land (especially beyond this, from 24° S. Lat. to 17°, and from 14° to 10°) is of a most prodigious height. It lies generally in ridges parallel to the shore, and three or four ridges one within another, each surpassing other in height; and those that are farthest within land are much higher than the others. They always appear blue when seen at sea: sometimes they are obscured with clouds, but not so often as the high lands in other parts of the world; for here are seldom or never any rains on these hills, any more than in the sea near it; neither are they subject to fogs. These are the highest mountains that ever I saw, far surpassing the Peak of Teneriffe, or Santa Marta, and I believe any mountains in the world. The excessive height of these mountains may possibly be the reason that there are no rivers of note that fall into these seas. Some small rivers indeed there are, but very few of them, for in some places there is not one that comes out into the sea in 150 or 200 leagues; and where they are thickest, they are thirty, forty, or fifty leagues asunder, and too little and shallow to be navigable. Besides, some of these do not constantly run, but are dry at certain seasons of the year, being rather torrents or land-floods caused by their rains at certain seasons far within land than perennial streams.

We kept still along in sight of this coast, but at a good distance from it, encountering nothing of note, till in the Lat. of 9° 40′ S., on the 3d of May, we descried a sail to the northward of us, plying to windward. We chased her, and Captain Eaton being ahead soon took her. She came from Guayaquil about a month before, laden with timber, and was bound to Lima. Three days before we took her she came from Santa, whither she had gone for water, and where they had news of our being in these seas by an express from Valdivia; for, as we afterwards heard, Captain Swan had been at Valdivia to seek a trade there, and he having met Captain Eaton in the Straits of Magellan, the Spaniards of Valdivia were doubtless informed of us by him; suspecting him also to be one of us, though he was not. Upon this news, the Viceroy of Lima sent expresses to all the seaports, that they might provide themselves against our assaults. We immediately steered away for the Island of Lobos, which lies in Lat. 6° 24′ S., and is five leagues from the main: it is called Lobos de la Mar,[1] to distinguish it from another that is not far from it, and extremely like it, called Lobos de la Tierra, for it lies near the main. Lobos, or Lovos, is the Spanish name for a seal, of which there are great plenty about these and several other islands in these seas that go by this name. The 9th of May, we arrived at this Isle of Lobos de la Mar, and came to an anchor with our prize. This Lobos consists indeed of two little islands, each about a mile round, of an indifferent height, a small channel between, fit for boats only; and several rocks lying on the north side of the islands, a little way from shore. Within land they are both of them partly rocky and partly sandy, barren, without any fresh water, tree, shrub, grass, or herbs; or any land animals (for the seals and sea-lions come ashore here) but fowls, of which there are great multitudes; as boobies, but mostly penguins, which I have seen plentifully all over the South Seas, on the coast of Newfoundland, and off the Cape of Good Hope. They are a sea fowl, about as big as a duck, and such feet, but a sharp bill; feeding on fish. They do not fly, but flutter, having rather stumps like a young gosling's than wings; and these are instead of fins to them in the water. Their feathers are downy. Their flesh is but ordinary food; but their eggs are good meat. There is another sort of small black fowl, that make holes in the sand for their night habitations, whose flesh is good sweet meat: I never saw any of them but here, and at Juan Fernandez.

Here we scrubbed our ships, and being in readiness to sail, the prisoners were examined, to know if any of them could conduct us to some town where we might make some attempt; for they had before informed us that we were descried by the Spaniards, and by that we knew that they would send no riches by sea so long as we were here. Many towns were considered on, as Guayaquil, Sana, Truxillo, and others. At last Truxillo was pitched on as the most important, therefore the likeliest to make us a voyage if we could conquer it, which we did not much question, though we knew it to be a very populous city. But the greatest difficulty was in landing; for Huanchaco [to the north of Truxillo], which is the nearest seaport to it, but six miles off, is an ill place to land, since sometimes the very fishermen that live there are not able to go out in three or four days. However, the 17th of May, in the afternoon, our men were

[1] Or Lobos de Afuera.

mustered of both ships' companies, and their arms proved. We were in all 108 men fit for service, besides the sick; and the next day we intended to sail and take the wood prize with us. But the next day one of our men, being ashore betimes on the island, descried three sail bound to the northward; two of them without the island to the westward, the other between it and the continent. We soon got our anchors up and chased; and Captain Eaton, who drew the least draught of water, put through between the westernmost island and the rocks, and went after those two that were without the islands. We in Captain Cooke's ship went after the other, which stood in for the mainland; but we soon fetched her up; and, having taken her, stood in again with her to the island, for we saw that Captain Eaton wanted no help, having taken both those that he went after. He came in with one of his prizes; but the other was so far to leeward, and so deep, that he could not then get her in, but he hoped to get her in the next day; but being deeply laden, as designed to go down before the wind to Panama, she would not bear sail. The 19th, she turned all day, but got nothing nearer the island. Our Mosquito strikers, according to their custom, went out and struck six turtles; for here are indifferent plenty of them. These ships that we took the day before we came from Huanchaco were all three laden with flour, bound for Panama. Two of them were laden as deep as they could swim; the other was not above half laden, but was ordered by the Viceroy of Lima to sail with the other two, or else she should not sail till we were gone out of the seas; for he hoped they might escape us by setting out early. In the biggest ship was a letter to the President of Panama from the Viceroy of Lima, assuring him that there were enemies come into that sea; for which reason he had despatched these three ships with flour, that they might not want (for Panama is supplied from Peru), and desired him to be frugal of it, for he knew not when he should send more. In this ship were likewise seven or eight tons of marmalade of quinces, and a stately mule sent to the President, and a very large image of the Virgin Mary in wood, carved and painted, to adorn a new church at Panama, and sent from Lima by the Viceroy; for this great ship came from thence not long before. She brought also from Lima 800,000 pieces of eight, to carry with her to Panama; but while she lay at Huanchaco, taking in her lading of flour, the merchants, hearing of Captain Swan's being at Valdivia, ordered the money ashore again. These prisoners likewise informed us that the gentlemen, inhabitants of Truxillo, were building a fort at Huanchaco, close by the sea, purposely to hinder the designs of any that should attempt to land there. Upon this news we altered our former resolutions, and resolved to go with our three prizes to the Galapagos, which are a great many large islands, lying some under the Equator, others on each side of it.

The 19th, in the evening, we sailed from the Island of Lobos, with Captain Eaton in our company. We carried the three flour prizes with us, but our first prize, laden with timber, we left here at anchor. We steered away NW. by N., intending to run into the latitude of the Isles of Galapagos, and steer off W., because we did not know the certain distance, and therefore could not shape a direct course to them. When we came within 40′ of the Equator, we steered W., having the wind at S., a very moderate gentle gale. It was the 31st of May when we first had sight of the Islands Galapagos. Some of them appeared on our weatherbow, some on our lee bow, others right ahead. We at first sight trimmed our sails, and steered as nigh the wind as we could, striving to get to the southernmost of them; but our prizes being deep laden, their sails but small and thin, and a very small gale, they could not keep up with us. Therefore we likewise edged away again a point from the wind, to keep near them; and, in the evening, the ship that I was in, and Captain Eaton, anchored on the east side of one of the easternmost islands, a mile from the shore, in sixteen fathoms water, clean, white, hard sand. The Galapagos Islands are a great number of uninhabited islands lying under and on both sides of the Equator. The easternmost of them are about 110 leagues from the main. The Spaniards who first discovered them, and in whose draughts alone they are laid down, report them to be a great number, stretching north-west from the Line as far as 5° N.; but we saw not above fourteen or fifteen. They are some of them seven or eight leagues long, and three or four broad. They are of a good height, most of them flat and even on the top; four or five of the easternmost are rocky, barren, and hilly, producing neither tree, herb, nor grass, but a few dildo trees, except by the sea-side. The dildo tree is a green prickly shrub, that grows about ten or twelve feet high, without either leaf or fruit.. It is as big as a man's leg from the root to the top, and it is full of sharp prickles, growing in thick rows from top to bottom. This shrub is fit for no use, not so much as to burn. Close by the sea there grow in some places bushes of Burton-wood, which is very good firing. This sort of wood grows in many places in the West Indies, especially in the Bay of Campeachy, and in the Sambaloes. I did never see any in these seas but here. There is water on these barren islands, in ponds and holes among the rocks. Some others of these islands are mostly plain and low, and the land more fertile; producing trees of divers sorts unknown to us. Some of the westernmost of these islands are nine or ten leagues long, and six or seven broad; the mould deep and black. These produce trees of great and tall bodies, especially mammee trees,[1] which grow here in great groves. In these large islands there are some pretty big rivers; and on many of the other lesser islands there are brooks of good water. The Spaniards, when they first discovered these islands, found multitudes of guanas, and land-turtle or tortoise, and named them the Galapagos Islands. I do believe there is no place in the world that is so plentifully stored with these animals. The guanas here are as fat and large as any that I ever saw; they are so tame, that a man may knock down twenty in an hour's time with a club. The land-turtle are so numerous, that 500 or 600 men might subsist on them alone for several months, without any other sort of provision; they are extraordinary large and fat, and so sweet, that no pullet eats more pleasantly. One of the largest of these creatures will weigh 150 or 200 lbs., and some of them are two feet or two feet six inches over the gallapee[2] or belly. I did never see any but at this place that will weigh above 30 lbs. I have heard that at the Isle of St Lawrence or Madagascar, and at the English Forest, an island near it, called also Don Mascarin,[3] and now possessed by the French, there are very large ones; but whether so big, fat, and sweet as these, I know not. There are three or four sorts of these creatures in the West Indies. One is called by the Spaniards "hecatee;" these live most in fresh-water ponds, and seldom come on land. They weigh about 10 or 15 lbs.; they have small legs and flat feet, and small

[1] The *Mammeo Americana*, a genus with only one species; it bears a fruit sweet in taste and aromatic in odour.

[2] The callipee is the gelatinous substance, of a light yellowish colour, which forms part of the lower shield of the turtle; callipash is the similar substance, of a dull greenish hue, which belongs to the upper shield.

[3] The general name for the group of islands in the Indian Ocean that comprises Mauritius and Reunion, is the Mascarenhas Islands, so called from the name of their Portuguese discoverer, in 1545.

long necks. Another sort is called terrapin ;[1] these are a great deal less than the hecatee ; the shell on their backs is all carved naturally, finely wrought and well clouded ; the backs of these are rounder than those before mentioned ; they are otherwise much of the same form : these delight to live in wet swampy places, or on the land near such places. Both these sorts are very good meat. They are in great plenty on the Isle of Pines near Cuba . there the Spanish hunters, when they meet them in the woods, bring them home to their huts, and mark them by notching their shells, then let them go ; this they do to have them at hand, for they never ramble far from thence. When these hunters return to Cuba, after about a month or six weeks' stay, they carry with them 300 or 400, or more, of these creatures to sell ; for they are very good meat, and every man knows his own by their marks. These tortoises in the Galapagos are more like the hecatee, except that, as I said before, they are much bigger, and they have very long small necks and little heads. There are some green snakes on these islands, but no other land animal that I did ever see. There are great plenty of turtle doves, so tame, that a man may kill five or six dozen in a forenoon with a stick. They are somewhat less than a pigeon, and are very good meat, and commonly fat.

There are good wide channels between these islands, fit for ships to pass, and in some places shoal water, where there grows plenty of turtle-grass ; therefore these islands are plentifully stored with sea-turtle, of that sort which is called the green turtle. There are four sorts of sea-turtle—viz., the trunk turtle, the loggerhead, the hawksbill, and the green turtle. The trunk turtle is commonly bigger than the others, their backs are higher and rounder, and their flesh rank and not wholesome. The loggerhead is so called because it has a great head, much bigger than the other sorts ; their flesh is likewise very rank and seldom eaten but in case of necessity; they feed on moss that grows about rocks. The hawksbill turtle is the least kind ; they are so called because their mouths are long and small, somewhat resembling the bill of a hawk. Hawksbill turtle are in many places of the West Indies. They have islands and places peculiar to themselves, where they lay their eggs, and seldom come among any other turtle. These, and all other turtle, lay eggs in the sand ; in N. Latitude, their time of laying is in May, June, July ; in S. Latitude, about Christmas ; some begin sooner, some later ; they lay three times in a season, and at each time eighty or ninety eggs. Their eggs are as a big as a hen's egg, and very round, covered only with a white tough skin. There are some bays on the north side of Jamaica, where these hawksbills resort to lay. In the Bay of Honduras are islands which they likewise make their breeding-places, and many places along all the coast on the main of the West Indies, from Trinidad to La Vera Cruz, in the Bay of Nova Hispania. When a sea-turtle turns out of the sea to lay, she is at least an hour before she returns again ; for she is to go above high-water mark, and if it be low-water when she comes ashore, she must rest once or twice, being heavy, before she comes to the place where she lays. When she has found a place for her purpose, she makes a great hole with her fins in the sand, wherein she lays her eggs, then covers them two feet deep with the same sand which she threw out of the hole, and so returns ; sometimes they come up the night before they intend to lay, and take a view of the place ; and so, having made a tour or semicircular march, they return to the sea again, and they never fail to come ashore the next night to lay near that place. All sorts of turtle use the same methods in laying. I knew a man

[1] Otherwise "terrapene," the box-tortoise.

in Jamaica that made £8 sterling of the shell of these hawksbill turtle which he got in one season, and in one small bay not half a mile long. The manner of taking them is to watch the bay by walking from one part to the other all night; making no noise, nor keeping any sort of light. When the turtle come ashore, the man that watches for them turns them on their backs, then hauls them above high-water mark, and leaves them till the morning. A large green turtle, with her weight and struggling, will puzzle two men to turn her. The hawksbill turtle are not only found in the West Indies, but on the coast of Guinea, and in the East Indies; I never saw any in the South Seas.

The green turtle are so called because their shell is greener than any other. It is very thin and clear, and better clouded than the hawksbill; but it is used only for inlays, being extraordinary thin. These turtles are generally larger than the hawksbill; one will weigh 200 or 300 lbs.; their backs are flatter than the hawksbill, their heads round and small. Green turtle are the sweetest of all the kinds; but there are degrees of them, both in respect to their flesh and their bigness. I have observed that at Blanco, in the West Indies, the green turtle (which is the only kind there) are larger than any others in the North Seas; there they commonly will weigh 280 or 300 lbs. Their fat is yellow and the lean white, and their flesh extraordinary sweet. At Boca del Toro, west of Porto Bello, they are not so large, their flesh not so white, nor the fat so yellow. Those in the Bays of Honduras and Campeachy are somewhat smaller still; their fat is green, and the lean of a darker colour than those at Boca del Toro. I heard of a monstrous green turtle once taken at Port Royal, in the Bay of Campeachy, that was four feet deep from the back to the belly, and the belly six feet broad. Captain Rocky's son, of about nine or ten years of age, went in it, as in a boat, on board his father's ship about a quarter of a mile from the shore; the leaves[1] of fat afforded eight gallons of oil. The turtle that live among the keys or small islands on the south side of Cuba are a mixed sort, some bigger, some less; and so their flesh is of a mixed colour, some green, some dark, some yellowish. With these, Port Royal, in Jamaica, is kept constantly supplied by sloops that come hither with nets to take them. They carry them alive to Jamaica, where the turtles have wires made with stakes in the sea to preserve them alive; and the market is every day plentifully stored with turtle, it being the common food there, chiefly for the ordinary sort of people.

There is another sort of green turtle in the South Seas, which are but small, yet pretty sweet; these lie westward, on the coast of Mexico. One thing is very strange and remarkable in these creatures; that, at the breeding time, they leave for two or three months their common haunts where they feed most of the year, and resort to other places, only to lay their eggs. And it is not thought that they eat anything during this season; so that both he's and she's grow very lean, but the he's to that degree that none will eat them. The most remarkable places that I did ever hear of for their breeding is at an island, in the West Indies, called Cayman, and the Isle of Ascencion, in the Western Ocean; and when the breeding time is past there is none remaining. Doubtless they swim some hundreds of leagues to come to those two places. For it has been often observed that at Cayman, at the breeding time, there are found all those sorts of turtle before described. The South Keys of Cuba are above forty leagues from thence, which is the nearest place that these creatures can come from; and it is most certain that there could not live so many there as come here in one season. Those that go to lay at Ascencion must needs travel much

[1] Layers.

farther, for there is no land nearer it than 300 leagues. And it is certain that these creatures live always near the shore. In the South Sea, likewise, the Galapagos is the place where they live the biggest part of the year; yet they go from thence at their season over to the main to lay their eggs; which is 100 leagues, the nearest place. Although multitudes of these turtles go from their common places of feeding and abode to those laying places, yet they do not all go. And at the time when the turtle resort to these places to lay their eggs, they are accompanied with abundance of fish, especially sharks; the places which the turtle then leave being at that time destitute of fish, which follow the turtle. When the she's go thus to their places to lay, the males accompany them, and never leave them till their return. Both male and female are fat [in] the beginning of the season; but, before they return, the males, as I said, are so lean, that they are not fit to eat, but the females are good to the very last, yet not so fat as at beginning of the season. It is reported of these creatures, that they are nine days engendering, and in the water, the male on the female's back. It is observable that the male, while engendering, do not easily forsake their female; for I have gone and taken hold of the male when engendering, and a very bad striker may strike them then; for the male is not shy at all, but the female, seeing a boat when they rise to blow, would make her escape, but that the male grasps her with his two fore fins and holds her fast. When they are thus coupled, it is best to strike the female first, then you are sure of the male also. These creatures are thought to live to a great age; and, it is observed by the Jamaica turtlers, that they are many years before they come to their full growth.

The air of these islands is temperate enough, considering the clime. There is constantly a fresh sea breeze all day, and cooling refreshing winds in the night; therefore the heat is not so violent here as in most places near the Equator. The time of the year for the rains is in November, December, and January. Then there is oftentimes excessive dark tempestuous weather mixed with much thunder and lightning. Sometimes before and after these months there are moderate refreshing showers; but in May, June, July, and August, the weather is always very fair. We stayed at one of these islands, which lies under the Equator, but one night; because our prizes could not get into an anchor. We refreshed ourselves very well, both with land and sea turtle: and the next day we sailed from thence. The next island of the Galapagos that we came to is but two leagues from this: it is rocky and barren like this; it is about five or six leagues long, and four broad. We anchored in the afternoon, at the north side of the island, a quarter of a mile from the shore, in sixteen fathoms water. It is steep all round this island, and no anchoring, only at this place. As soon as we came to an anchor, we made a tent ashore for Captain Cooke who was sick. Here we found the sea-turtle lying ashore on the sands; this is not customary in the West Indies. We turned them on their backs that they might not get away. The next day more came up; when we found it to be their custom to lie in the sun: so we never took care to turn them afterwards, but sent ashore the cook every morning, who killed as many as served for the day. This custom we observed all the time we lay here, feeding sometimes on land-turtle, sometimes on sea-turtle, there being plenty of either sort. Captain Davis came hither again a second time; and then he went to other islands on the west side of these. There he found such plenty of land-turtle, that he and his men ate nothing else for three months that he stayed there. They were so fat, that he saved sixty jars of oil out of those that he spent. This oil served instead of butter to eat with dough-boys and dumplings in his return out of these seas. He found very convenient places to careen, and good channels between the islands; and very good anchoring in many places. There he found also plenty of brooks of good fresh water, and firewood enough; there being plenty of trees fit for many uses. Captain Harris, one that we shall speak of hereafter, came hither likewise, and found some islands that had plenty of mammee trees, and pretty large rivers. The sea about these islands is plentifully stored with fish, such as are at Juan Fernandez. They are both large and fat, and as plentiful here as at Juan Fernandez; here are particularly abundance of sharks. These Isles of the Galapagos have plenty of salt. We stayed here but twelve days: in which time we put ashore 5000 packs of flour, for a reserve, if we should have occasion of any before we left these seas. Here one of our Indian prisoners informed us that he was born at Realejo, and that he would engage to carry us thither. He being examined of the strength and riches of it, satisfied the company so well that they were resolved to go thither.

Having thus concluded, the 12th of June, we sailed from hence, designing to touch at the Island of Cocos, as well to put ashore some flour there, as to see the island, because it was in our way to Realejo. [But] despairing as the winds were, to find the Island of Cocos, we steered over to the main. The Island of Cocos is so named by the Spaniards, because there are abundance of cocoa-nut trees growing on it. They are not only in one or two places, but grow in great groves all round the island, by the sea. This is an uninhabited island; it is seven or eight leagues round, and pretty high in the middle, where it is destitute of trees, but looks very green and pleasant, with an herb called by the Spaniards "gramadel." It is low land by the sea-side. We had very fair weather and small winds in this voyage from the Galapagos, and at the beginning of July we fell in with Cape Blanco, on the main of Mexico. This is so called from two white rocks lying off it. When we are off at sea, right against the cape, they appear as part of the cape; but being near the shore, either to the eastward or westward of the cape, they appear like two ships under sail at first view, but coming nearer they are like two high towers, they being small, high, and steep on all sides, and they are about half-a-mile from the cape. This cape is in Lat. 9° 56′. It is about the height of Beachy Head in England, on the coast of Sussex. It is a full point, with steep rocks to the sea. The top of it is flat and even for about a mile; then it gradually falls away on each side with a gentle descent. It appears very pleasant, being covered with great lofty trees. From the cape on the NW. side, the land runs in NE. for about four leagues, making a small bay called by the Spaniards Caldera. From the bottom of this bay it is but fourteen or fifteen leagues to the Lake of Nicaragua, on the North Sea coast: the way between is somewhat mountainous, but mostly savannah. Captain Cooke, who was taken sick at Juan Fernandez, continued so till we came within two or three leagues of Cape Blanco, and then died of a sudden, though he seemed that morning to be as likely to live as he had been some weeks before; but it is usual with sick men coming from the sea, where they have nothing but the sea air, to die off as soon as ever they come within view of the land. About four hours after, we all came to an anchor (namely, the ship that I was in, Captain Eaton, and the great meal prize), a league within the cape, right against [a] brook of fresh water, in fourteen fathoms, clean hard sand. Presently after we came to an anchor, Captain Cooke was carried ashore to be buried; twelve men carried their arms to guard those that were ordered to dig the grave; for although we saw no appearance of inhabitants, yet we did not know but the country might be thickly inhabited. And before Captain Cooke was interred, three Spanish Indians came to the place where our men were digging the grave, and demanded what they were, and whence they came? To whom our men answered, they came from Lima and were bound to Realejo, but that the captain of one of the ships, dying at sea, obliged them to come into this place to give him Christian burial. The three Spanish Indians, who were very shy at first, began to be more bold, and drawing nearer, asked many silly questions, and our men did not stick to soothe them up with as many falsehoods, purposely to draw them into their clutches. Our men often laughed at their temerity, and asked them if they never saw any Spaniards before? They told them, that they themselves were Spaniards, and that they lived among Spaniards, and that although they were born there, yet they had never seen three ships there before. Our men told them, that neither now might they have seen so many, if it had not been on an urgent occasion. At length they drilled[1] them by discourse so near, that our men laid hold on all three at once; but before Captain Cooke was buried, one of them made his escape; the other two were brought off aboard our ship. Captain Eaton immediately came aboard, and examined them; they confessed they came purposely to view our ship, and if possible to inform themselves what we were; for the President of Panama not long before sent a letter of advice to Nicoya, informing the magistrates thereof that some enemies were come into these seas, and that therefore it behoved them to be careful of themselves. Nicoya is a small Mulatta town about twelve or thirteen leagues east from hence, standing on the banks of a river of that name. It is a place very fit for building ships, therefore most of the inhabitants are carpenters, who are commonly employed in building new or repairing old ships. It was here that Captain Sharpe, just after I left him, in the year 1681, got carpenters to fix his ship before he returned for England; and for that reason it behoved the Spaniards to be careful, according to the Governor of Panama's advice, lest any men at other times wanting such necessaries as that place afforded might again be supplied there. These Spanish Indians told us likewise that they were sent to the place where they were taken, in order to view our ships, as fearing these were those mentioned by the President of Panama. It being demanded of them to give an account of the estate and riches of the country, they said, that the inhabitants were mostly husbandmen, who were employed either in planting and manuring of corn, or chiefly about cattle; they having large savannahs, which were well stored with bulls, cows, and horses: that by the sea-side in some places there grew some red wood useful in dyeing; of this they said there was little profit made, because they were forced to send it to the Lake of Nicaragua, which runs into the North Seas: that they sent thither also great quantities of bull and cow hides, and brought thence in exchange European commodities: as hats, linen, and woollen, wherewith they clothed themselves; that the flesh of the cattle turned to no other profit than sustenance for their families; as for butter and cheese, they made but little in those parts.

After they had given this relation, they told us, that if we wanted provision, there was a beef estantion,[2] or farm of bulls or cows, about three miles off, where we might kill what we pleased. This was welcome news, for we had no sort of flesh since we left the Galapagos; therefore twenty-four of us immediately entered into two boats, taking one of these Spanish Indians with us for a pilot, and went ashore about a league from the ship. There we hauled up our boats dry, and marched all away, following our guide, who soon brought us to some houses, and a large pen for cattle. This pen stood in a large savannah, about two miles from our boats; there were a great many fat bulls and

[1] Enticed.

[2] Spanish, "Estancia," a mansion or farm, or place of store.

cows feeding in the savannahs. Some of us would have killed three or four to carry on board; but others opposed it, and said it was better to stay all night, and in the morning drive the cattle into the pen, and then kill twenty or thirty, or as many as we pleased. I was minded to return aboard, and endeavoured to persuade them all to go with me, but some would not; therefore I returned with twelve, which was half, and left the other twelve behind. At this place I saw three or four tons of the red wood, which I take to be that sort of wood called in Jamaica bloodwood or Nicaragua wood. We who returned aboard met no one to oppose us, and the next day we expected our consorts that we left ashore, but none came; therefore at four o'clock in the afternoon ten men went in our canoe to see what was become of them. When they came to the bay where we landed to go to the estantion, they found our men all on a small rock, half a mile from the shore, standing in the water up to their waists. These men had slept ashore in the house, and turned out betimes in the morning to pen the cattle: two or three went one way, and as many another way, to get the cattle to the pen; and others stood at the pen to drive them in. When they were thus scattered, about forty or fifty armed Spaniards came in among them. Our men immediately called to each other, and drew together in a body before the Spaniards could attack them, and marched to their boat, which was hauled up dry on the sand; but when they came to the sandy bay they found their boat all in flames. This was a very unpleasing sight, for they knew not how to get aboard, unless they marched by land to the place where Captain Cooke was buried, which was near a league. The greatest part of the way was thick woods, where the Spaniards might easily lay in ambush for them, at which they are very expert. On the other side, the Spaniards now thought them secure; and therefore came to them and asked them if they would be pleased to walk to their plantations, with many other such flouts; but our men answered never a word. It was about half ebb when one of our men took notice of a rock a good distance from the shore, just appearing above water; he showed it to his consorts and told them it would be a good castle for them if they could get thither. They all wished themselves there; for the Spaniards, who lay as yet at a good distance from them behind the bushes, as secure of their prey, began to whistle now and then a shot among them. Having therefore well considered the place, together with the danger they were in, they proposed to send one of the tallest men to try if the sea between them and the rock were fordable. This counsel they presently put in execution, and found it according to their desire. So they all marched over to the rock, where they remained till the canoe came to them; which was about seven hours. It was the latter part of the ebb when they first went over, and then the rock was dry; but when the tide of flood returned again the rock was covered, and the water still flowing; so that if our canoe had stayed but one hour longer they might have been in as great danger of their lives from the sea as before from the Spaniards; for the tide rises here about eight feet. The Spaniards remained on the shore, expecting to see them destroyed, but never came from behind the bushes where they first planted themselves; they having not above three or four hand-guns, the rest of them being armed with lances. The Spaniards in these parts are very expert in heaving or darting the lance, with which, upon occasion, they will do great feats, especially in ambuscades; and by their good will they care not for fighting otherwise, but content themselves with standing aloof, threatening and calling names, at which they are as expert as at the other; so that if their tongues be quiet, we always take it for granted they have laid some ambush. Before night our canoe came aboard, and brought our men all safe.

The day before we went from hence, Mr Edward Davis, the company's Quarter-master, was made Captain by consent of all the company; for it was his place by succession. The 20th day of July we sailed from this Bay of Caldera, with Captain Eaton, and our prize which we brought from the Galapagos, in company, directing course for Realejo. The wind was at N., which, although but an ordinary wind, yet carried us in three days abreast of our intended port. Realejo is the most remarkable land on all this coast; for there is a high-peaked burning mountain, called by the Spaniards Volcano Viejo, or the Old Volcano. The volcano may be easily known, because there is not any other so high a mountain near it, neither is there any that appears in the like form all along the coast; besides it smokes all the day, and in the night it sometimes sends forth flames of fire. This mountain may be seen twenty leagues. Being within three leagues of the harbour, the entrance into it may be seen. There is a small flat low island which makes the harbour. This harbour is capable of receiving 200 sail of ships. The best riding is near the main, where there is seven or eight fathoms water; clean hard sand. Realejo town is two leagues from hence, and there are two creeks that run towards it; the westernmost comes near the back-side of the town, the other runs up to the town; but neither ships nor barks can go so far. These creeks are very narrow, and the land on each side drowned, and full of red mangrove-trees. About a mile and a half below the town, on the banks of the east creek, the Spaniards had cast up a strong breastwork; it was likewise reported they had another on the west creek, both so advantageously placed that ten men might with ease keep 200 men from landing.

We were now in sight of the volcano, being, by estimation, seven or eight leagues from the shore; and the mountain bearing NE., we took in our topsails, and hauled up our courses, intending to go with our canoes into the harbour in the night. In the evening we had a very hard tornado out of the NE., with much thunder, lightning, and rain. The violence of the wind did not last long, yet it was 11 o'clock at night before we got out our canoes, and then it was quite calm. We rowed in directly for the shore, and thought to have reached it before day; but it was 9 o'clock in the morning before we got into the harbour. When we came within a league of the Island of Realejo, that makes the harbour, we saw a house on it; and coming nearer we saw two or three men, who stood and looked on us till we came within half-a-mile of the island, then they went into their canoe, which lay on the inside of the island, and rowed towards the main; but we overtook them before they got over, and brought them back again to the island. There was a horseman right against us on the main when we took the canoe, who immediately rode away towards the town as fast as he could. The rest of our canoes rode heavily, and did not come to the island till 12 o'clock; therefore we were forced to stay for them. Before they came, we examined the prisoners, who told us that they were set there to watch, for the Governor of Realejo received a letter about a month before, wherein he was advised of some enemies come into the sea, and therefore admonished to be careful; that immediately thereupon the Governor had caused a house to be built on this island, and ordered four men to be continually there to watch night and day; and if they saw any ship coming thither, they were to give notice of it. They said they did not expect to see boats or canoes, but looked out for a ship. At first they took us in our advanced canoe to be some men that had been cast away and lost our ship; till, seeing three or four canoes more, they began to suspect what we were. They told us likewise, that the horseman we saw did come to them every morning, and that in less than an hour's time he could be at the town. When Captain Eaton and his canoes came ashore, we told them what had happened. It was now three hours since the horseman rode away, and we could not expect to get to the town in less than two hours; in which time the Governor, having notice of our coming, might be provided to receive us at his breastworks; therefore we thought it best to defer this design till another time. Here we stayed till 4 o'clock in the afternoon; then our ships being come within a league of the shore, we all went on board, and steered for the Gulf of Amapalla, intending there to careen our ships.

The 26th of July, Captain Eaton came aboard our ship to consult with Captain Davis how to get some Indians to assist us in careening. It was concluded, that when we came near the Gulf, Captain Davis should take two canoes, well manned, and go before, and Captain Eaton should stay aboard. According to this agreement, Captain Davis went away for the Gulf the next day. The Gulf of Amapalla[1] is a great arm of the sea, running eight or ten leagues into the country. It is bounded on the S. side of its entrance with Point Casivina, and on the NW. side with St Michael's Mount. Both these places are very remarkable. Point Casivina is in Lat. 12° 40′ N. It is a high round point, which at sea appears like an island, because the land within it is very low. St Michael's Mount is a very high peaked hill, not very steep: the land at the foot of it on the SE. side is low and even for at least a mile. From this low land the Gulf of Amapalla enters on that side. Between this low land and Point Casivina are two considerable high islands; the southernmost is called Mangera, the other is called Amapalla; and they are two miles asunder.[2] . . . There are a great many more islands in this Bay, but none inhabited as these. There is one pretty large island, belonging to a nunnery, as the Indians told us; this was stocked with bulls and cows. Three or four Indians lived there to look after the cattle, for the sake of which we often frequented this island while we lay in the bay. They are all low islands, except Amapalla and Mangera. There are two channels to come into this gulf; one between Point Casivina and Mangera, the other between Mangera and Amapalla; the latter is the best.

It was into this gulf that Captain Davis was gone with the two canoes, to endeavour for a prisoner, to gain intelligence, if possible, before our ships came in. He came the first night to Mangera, but for want of a pilot did not know where to look for the town. In the morning he found a great many canoes hauled up on the bay; and from that bay found a path which led him and his company to the town. The Indians saw our ships in the evening coming towards the island, and being before informed of enemies in the sea, they kept scouts out all night for fear; who seeing Captain Davis coming, ran into the town, and alarmed all the people. When Captain Davis came thither, they all ran into the woods. The Friar happened to be there at this time; who, being unable to ramble

[1] Marked in the modern maps as the Gulf of Fonseca. The southern headland is Cape Cosiguina, called Casivina by Dampier; the northern, which he called St Michael's Mount, is Cape Candadillo.

[2] Mangera is described as a high round island, about two leagues in compass, and appearing from the sea like a tall grove. There is mention made of one town, about the middle of the island. Amapalla is much larger than Mangera, with two towns on it. The Indians of both places cultivate maize, a few plantains, and the hog plum. The towns were governed from St Michael's, to which they paid tribute in maize. There was but one friar or padre living amongst them, who exacted a tenth from the natives, and who was the only white man on the island.

into the woods, fell into Captain Davis's hands; there were two Indian boys with him who were likewise taken. Captain Davis went only to get a prisoner, therefore was well satisfied with the Friar, and immediately came down to the sea-side. He went from thence to the Island of Amapalla, carrying the Friar and the two Indian boys with him. These were his pilots to conduct him to the landing-place, where they arrived about noon. They made no stay here, but left three or four men to look after the canoes, and Captain Davis, with the rest, marched to the town, taking the Friar with them. The town, as is before noted, is about a mile from the landing-place, standing in a plain on the top of a hill, having a very steep ascent to go to it. All the Indians stood on the top of the hill, waiting Captain Davis's coming. The Secretary, mentioned before, had no great kindness for the Spaniards. It was he that persuaded the Indians to wait Captain Davis's coming; for they were all running into the woods; but he told them, that if any of the Spaniards' enemies came thither, it was not to hurt them, but the Spaniards, whose slaves they were; and that their poverty would protect them. This man, with the Cacique, stood more forward than the rest, at the bank of the hill, when Captain Davis with his company appeared beneath. They called out therefore in Spanish, demanding of our men what they were, and whence they came? To whom Captain Davis and his men replied, they were Biscayers, and were sent thither by the King of Spain to clear those seas from enemies; that their ships were coming into the gulf to careen, and that they came thither before the ships to seek a convenient place for it, as also to desire the Indians' assistance. The Secretary, who, as I said before, was the only man that could speak Spanish, told them that they were welcome, for he had a great respect for any Old Spain men, especially for the Biscayers, of whom he had heard a very honourable report; therefore he desired them to come up to their town. Captain Davis and his men immediately ascended the hill, the Friar going before; and they were received with a great deal of affection by the Indians. The Cacique and Secretary embraced Captain Davis; and the other Indians received his men with the like ceremony.

These salutations being ended, they all marched towards the church, for that is the place of all public meetings, and all plays and pastimes are acted there also; therefore in the churches belonging to Indian towns they have all sorts of vizards and strange antic dresses both for men and women, and abundance of musical hautboys and strumstrums. The strumstrum is made somewhat like a cittern; most of those that the Indians use are made of a large gourd, cut in the midst, and a thin board laid over the hollow, which is fastened to the sides. This serves for the belly, over which the strings are placed. The nights before any holidays, or the nights ensuing, are the times when they all meet to make merry. Their mirth consists in singing, dancing, and sporting in those antic habits, and using as many antic gestures. If the moon shine they use but few torches; if not, the church is full of light. They meet at these times all sorts of both sexes. All the Indians that I have been acquainted with who are under the Spaniards seem to be more melancholy than other Indians that are free; and at these public meetings, when they are in the greatest of their jollity, their mirth seems to be rather forced than real. Their songs are very melancholy and doleful, so is their music; but whether it be natural to the Indians to be thus melancholy, or the effect of their slavery, I am not certain. But I have always been prone to believe that they are then only condoling their misfortunes, the loss of their country and liberties, which, although those that are now living do not know nor remember what it was to be free, yet there seems to be a deep impression in their thoughts of the slavery which the Spaniards have brought them under, increased probably by some traditions of their ancient freedom. Captain Davis intended, when they were all in the church, to shut the doors and then make a bargain with them, letting them know what he was, and so draw them afterwards by fair means to our assistance, the Friar being with him, who had also promised to engage them to it. But before they were all in the church, one of Captain Davis's men pushed one of the Indians, to hasten him into the church. The Indian immediately ran away, and all the rest, taking the alarm, sprang out of the church like deer; it was hard to say which was first; and Captain Davis, who knew nothing of what happened, was left in the church only with the Friar. When they were all fled, Captain Davis's men fired, and killed the Secretary; and thus our hopes perished by the indiscretion of one foolish fellow.

In the afternoon the ships came into the gulf between Point Casivina and Mangera, and anchored near the Island of Amapalla, on the E. side, in ten fathoms water, clean hard sand. In the evening Captain Davis and his company came aboard, and brought the Friar with them, who told Captain Davis, that if the Secretary had not been killed he could have sent him a letter by one of the Indians that was taken at Mangera, and persuaded him to come to us; but now the only way was to send one of those Indians to seek the Cacique, and [he] himself would instruct him what to say, and did not question but the Cacique would come on his word. The next day we sent ashore one of the Indians, who before night returned with the Cacique and six other Indians, who remained with us all the time that we stayed here. These Indians did us good service, especially in piloting us to an island, where we killed beef whenever we wanted; and for this their service we satisfied them to their hearts' content. It was at this Island of Amapalla that a party of Englishmen and Frenchmen came afterwards and stayed a great while, and at last landed on the main, and marched overland to the Cape River, which disembogues into the North Seas near Cape Gracias a Dios, and is therefore called the Cape River.[1] Near the head of this river they made bark-logs (which I shall describe in the next Chapter), and so went into the North Seas. This was the way that Captain Sharpe had proposed to go if he had been put to it, for this way was partly known to Privateers by the discovery that was made into the country about thirty years since by a party of Englishmen that went up that river in canoes, about as far as the place where these Frenchmen made their bark-logs; there they landed and marched to a town called Segovia in the country. They were near a month getting up the river, for there are many cataracts where they were often forced to leave the river and haul their canoes ashore over the land till they were past the cataracts, and then launch their canoes again into the river. I have discoursed [with] several men that were in that expedition, and if I mistake not, Captain Sharpe was one of them. But to return to our voyage in hand; when both our ships were clean, and our water filled, Captain Davis and Captain Eaton broke off consortships. Captain Eaton took aboard of his ships 400 packs of flour, and sailed out of the gulf the 2d of September.

CHAPTER VI.

THE 3d of September 1684, we sent the Friar ashore, and left the Indians in possession of the prize which we brought in hither, though she was still half-laden with flour; and we sailed out with the land-wind, passing between Amapalla and Mangera. When we were a league out,

[1] Variously called in modern maps the Vanquez, or Yanks, or Tints, or Segovia, or Coco River.

we saw a canoe coming with sail and oars after us, therefore we shortened sail and stayed for her. She was a canoe sent by the Governor of St Michael's town to our Captain, desiring him not to carry away the Friar. The messenger being told that the Friar was set ashore again at Amapalla, he returned with joy, and we made sail again, having the wind at WNW. We steered towards the coast of Peru. We had tornadoes every day till we made Cape San Francisco, which from June to November are very common on these coasts; and we had with the tornadoes very much thunder, lightning, and rain. When the tornadoes were over, the wind, which while they lasted was most from the SE., came about again to the W., and never failed us till we were in sight of Cape San Francisco. This cape is in Lat. 1° N.; it is a high bluff or full point of land, clothed with tall great trees. The land in the country within this cape is very high, and the mountains commonly appear very black. When we came in with this cape we overtook Captain Eaton plying under the shore; he in his passage from Amapalla, while he was on that coast, met with such terrible tornadoes of thunder and lightning that, as he and all his men related, they had never met with the like in any place. They were very much affrighted by them, the air smelling very much of sulphur, and they apprehending themselves [to be] in great danger of being burnt by the lightning. He touched at the island of Cocos, and put ashore 200 packs of flour there, and loaded his boat with cocoa-nuts, and took in fresh water. In the evening we separated again from Captain Eaton, for he stood off to sea, and we plied up under the shore, making our best advantage both of sea and land winds.

The 20th of September we came to the Island of Plata, and anchored in sixteen fathoms. We had very good weather from the time that we fell in with Cape San Francisco, and were now fallen in again with the same places from whence I begin the account of this voyage in the First Chapter, having now compassed in the whole continent of South America. The Island of Plata, as some report, was so named by the Spaniards after Sir Francis Drake took the Cacafuego,[1] a ship chiefly laden with plate, which they say he brought hither and divided it here with his men. It is about four miles long and a mile and a half broad, and of a good height. It is bounded with high steep cliffs clear round, only at one place on the east side. The top of it is flat and even, the soil sandy and dry; the trees it produces are but small-bodied, low, and grow thin; and there are only three or four sorts of trees, all unknown to us. I observed they were much overgrown with long moss. There is good grass, especially in the beginning of the year. There is no water on this island, but at one place on the east side close by the sea; there it drills[2] slowly down from the rocks, where it may be received into vessels. There were plenty of goats, but they are now all destroyed. There is no other sort of land animal that I did ever see; there are plenty of boobies and man-of-war birds. At this island are plenty of those small sea-turtle spoken of in my last Chapter.

The 21st, Captain Eaton came to an anchor by us; he was very willing to have consorted with us again, but Captain Davis's men were so unreasonable that they would not allow Captain Eaton's men an equal share with them in what they got; therefore Captain Eaton stayed there but one night, and the next day sailed from hence, steering away to the southward. We stayed no longer than the day ensuing, and then we sailed toward Point Santa Elena, intending there to land some men purposely to get prisoners for intelligence.

Point Santa Elena bears S. from

---

[1] The capture of this rich prize is narrated in Drake's Voyage. See page 44.

[2] Penetrates, trickles. Bishop Taylor uses the word "drill" to signify a small water-course.

the Island of Plata. It lies in Lat. 2° 15′ S. The point is pretty high, flat, and even at top; overgrown with many great thistles but no sort of tree; at a distance it appears like an island because the land within it is very low. This point strikes out west into the sea, making a pretty large bay on the north side. . . . When we were abreast of this point we sent away our canoes in the night to take the Indian village. They landed in the morning betimes close by the town, and took some prisoners. They took likewise a small bark which the Indians had set on fire, but our men quenched it, and took the Indian that did it, who being asked wherefore he set the bark on fire, said, that there was an order from the Viceroy lately set out commanding all seamen to burn their vessels if attacked by us, and betake themselves to their boats. There was another bark in a small cove a mile from the village; thither our men went, thinking to take her, but the seamen that were aboard set her in flames and fled. In the evening, our men came aboard, and brought the small bark with them, the fire of which they had quenched; and then we returned again towards Plata, where we arrived the 26th of September.

In the evening we sent out some men in our bark lately taken and canoes, to an Indian village called Manta, two or three leagues to the W. of Cape San Lorenzo, hoping there to get other prisoners, for we could not learn from those we took at Point Santa Elena the reason why the Viceroy should give such orders to burn the ships. They had a fresh sea breeze till 12 o'clock at night, and then it proved calm, wherefore they rowed away with their canoes as near to the town as they thought convenient, and lay still till day. Manta is a small Indian village on the main, distant from the Island of Plata seven or eight leagues. It stands so advantageously to be seen, being built on a small ascent, that it makes a very fair prospect to the sea, yet but a few poor scattering Indian houses. There is a very fine church, adorned with a great deal of carved work. It was formerly a habitation of Spaniards, but they are all removed from hence now. The land about it is dry and sandy, bearing only a few shrubby trees. These Indians plant no manner of grain or root, but are supplied from other places, and commonly keep a stock of provision to relieve ships that want, for this is the first settlement that ships can touch at which come from Panama bound to Lima, or any other port in Peru. The land, being dry and sandy, is not fit to produce crops of maize, which is the reason they plant none. There is a spring of good water between the village and the sea. On the back of the town, a pretty way up in the country, there is a very high mountain, towering up like a sugar-loaf, called Monte Christo. It is a very good sea mark, for there is none like it on all the coast. The body of this mountain bears due S. from Manta.[1] From Manta to Cape San Lorenzo the land is plain and even, of an indifferent height.

As soon as ever the day appeared, our men landed, and marched towards the village, which was about a mile and a half from their landing-place. Some of the Indians who were stirring saw them coming, and alarmed their neighbours; so that all that were able got away. They took only two old women, who both said, that it was reported that a great many enemies were come overland through the country of Darien into the South Seas, and that they were at present in canoes and periagoes; and that the Viceroy upon this news, had set out the fore-mentioned order for burning their own ships. Our men found no sort of provision here; the Viceroy having likewise sent orders to all seaports to keep no provision, but just to supply themselves. These

[1] It has been conjectured that Chimberazo is here meant, but that mountain lies east by south, and not south, from Manta, and probably Dampier refers to some smaller eminence nearer the coast.

women also said, that the Manta Indians were sent over to the Island of Plata to destroy all the goats there, which they performed about a month ago. With this news our men returned again, and arrived at Plata the next day. We lay still at the Island of Plata, being not resolved what to do, till the 2d of October; and then Captain Swan, in the Cygnet of London, arrived there. He was fitted out by very eminent merchants of that city, on a design only to trade with the Spaniards or Indians, having a very considerable cargo well sorted for these parts of the world; but meeting with divers disappointments, and being out of hopes to obtain a trade in these seas, his men forced him to entertain a company of Privateers which he met with near Nicoya, a town whither he was going to seek a trade; and these Privateers were bound thither in boats to get a ship. These were the men that we had heard of at Manta; they came overland, under the command of Captain Peter Harris, nephew to that Captain Harris who was killed before Panama. Captain Swan was still commander of his own ship, and Captain Harris commanded a small bark under Captain Swan. There was much joy on all sides when they arrived; and immediately hereupon, Captain Davis and Captain Swan consorted, wishing for Captain Eaton again. Our little bark, which was taken at Santa Elena, was immediately sent out to cruise while the ships were fitting; for Captain Swan's ship, being full of goods, was not fit to entertain his new guests, till the goods were disposed of; therefore he, by the consent of the supercargoes, got up all his goods on deck, and sold to any one that would buy, upon trust. The rest was thrown overboard into the sea, except fine goods, as silks, muslins, stockings, &c., and except the iron, whereof he had a good quantity, both wrought and in bars; this was saved for ballast. The third day after our bark was sent to cruise, she brought in a prize of 400 tons, laden with timber; they took her in the bay of Guayaquil; she came from the town of that name, and was bound to Lima. The commander of this prize said, that it was generally reported and believed at Guayaquil, that the Viceroy was fitting out ten sail of frigates to drive us out of the Seas. This news made our unsettled crew wish that they had been persuaded to accept of Captain Eaton's company on reasonable terms. Captain Davis and Captain Swan had some discourse concerning Captain Eaton; they at last concluded to send our small bark towards the coast of Lima, as far as the Island of Lobos, to seek Captain Eaton. This being approved by all hands, she was cleaned the next day, and sent away, manned with twenty men, ten of Captain Davis's, and ten of Swan's men; and Captain Swan wrote a letter directed to Captain Eaton, desiring his company; and the Isle of Plata was appointed for the general rendezvous. When this bark was gone, we turned another bark which we had into a fireship, having six or seven carpenters, who soon fixed her; and while the carpenters were at work about the fireship, we scrubbed and cleaned our men-of-war, as well as time and place would permit. The 19th of October we finished our business, and the 20th we sailed towards the Island of Lobos, where our bark was ordered to stay for us, or meet us again at Plata. We had but little wind, therefore it was the 23d before we passed by Point Santa Elena. The 25th we crossed over the Bay of Guayaquil. The 30th we doubled Cape Blanco. This cape is in Lat. 3° 45′. It is counted the worst cape in all the South Seas to double, passing to the southward. This cape is of an indifferent height. It is fenced with white rocks to the sea; for which reason, I believe, it has this name.[1] The land in the country seems to be full of high, steep, rugged, and barren rocks.

The 2d of November we got as high as Payta. We lay about six leagues

[1] Cabo Blanco—White Cape.

off shore all the day, that the Spaniards might not see us; and in the evening sent our canoes ashore to take it, manned with 110 men. Payta is a small Spanish seaport town, in Lat. 5° 15′ S. It is built on the sand, close by the sea, in a nook, elbow, or small bay, under a pretty high hill. There are not above seventy-five or eighty houses, and two churches. The houses are but low and ill-built. The building in this country of Peru is much alike on all the sea-coast. The walls are built of bricks made with earth and straw kneaded together; they are about three feet long, two feet broad, and a foot and a half thick; they never burn them, but lay them a long time in the sun to dry before they are used in building. In some places they have no roofs, only poles laid across from the side walls, and covered with mats; and then those walls are carried up to a considerable height. But where they build roofs upon their houses, the walls are not made so high, as I said before. The houses in general all over this kingdom are but meanly built: one chief reason, with the common people especially, is the want of materials to build withal; for, however it be more within land, yet here is neither stone nor timber to build with, nor any materials but such brick as I have described; and even the stone which they have in some places is so brittle that you may rub it into sand with your fingers. Another reason why they build so meanly is, because it never rains; therefore they only endeavour to fence themselves from the sun. Yet their walls, which are built but with an ordinary sort of brick in comparison with what is made in other parts of the world, continue a long time as firm as when first made, having never any winds nor rains to rot, moulder, or shake them. However, the richer sort have timber, which they make use of in building; but it is brought from other places. This dry country commences to the northward, from about Cape Blanco to Coquimbo, in about 30° S., having no rain that I could ever observe or hear of, nor any green thing growing in the mountains, neither yet in the valleys, except where here and there watered with a few small rivers dispersed up and down. So that the northernmost parts of this tract of land are supplied with timber from Guayaquil, Galleo, Tumaco, and other places that are watered with rains, where there is plenty of all sorts of timber. In the south parts as about Huasco and Coquimbo, they fetch their timber from the Island of Chiloe, or other places thereabouts. The walls of churches and rich men's houses are whitened with lime both within and without; and the doors and posts are very large, and adorned with carved work, and the beams also in the churches; the insides of the houses are hung round with rich embroidered or painted cloths. They have likewise abundance of fine pictures, which add no small ornament to their houses. These, I suppose, they have from Old Spain. But the houses of Payta are none of them so richly furnished. The churches were large, and fairly carved. At one end of the town there was a small fort, close by the sea, but no great guns in it. This fort, only with muskets, will command all the bay, so as to hinder any boats from landing. There is another fort on the top of the hill, just over the town, which commands both it and the lower fort. There is neither wood nor water to be had here. They fetch their water from an Indian town called Colan, about two leagues NNE. from Payta; for at Colan there is a small river of fresh water which runs out into the sea, from whence ships that touch at Payta are supplied with water and other refreshments, as fowls, hogs, plantains, yams, and maize: Payta being destitute of all these things, only as they fetch them from Colan as they have occasion.

The Indians of Colan are all fishermen. They go out to sea and fish on bark-logs.[1] Bark-logs are made of

[1] This title has been supposed to

many round logs of wood, in the manner of a raft, and very different, according to the use that they are designed for, or the humour of the people that make them, or the matter that they are made of. If they are made for fishing, then they are only three or four logs of light wood, of seven or eight feet long, placed by the side of each other, pinned fast together with wooden pins, and bound hard with withes. The logs are so placed that the middlemost are longer than those by the sides, especially at the head or fore part, which grows narrower gradually into an angle or point, better to cut through the water. Others are made to carry goods. The bottom of these is made of twenty or thirty great trees, of about twenty, thirty, or forty feet long, fastened as the other, side to side, and so shaped. On the top of these they place another shorter row of trees across them, pinned fast to each other, and then pinned to the undermost row. This double row of planks makes the bottom of the float. . . . They always go before the wind, being unable to ply against it, and therefore are fit only for these seas where the wind is always in a manner the same, not varying above a point or two all the way from Lima till such time as they come into the Bay of Panama; and even there they meet with no great sea, but sometimes northerly winds; and then they lower their sails, and drive before it, waiting a change. All their care then is only to keep off from shore, for they are so made that they cannot sink at sea. These rafts carry sixty or seventy tons of goods and upwards. Their cargo is chiefly wine, oil, flour, sugar, Quito cloth, soap, goat-skins dressed, &c. The float is managed usually by three or four men, who, being unable to return with it against the trade-wind, when they come to Panama dispose of the goods and bottom together, getting a passage back again for themselves in some ship or boat bound to the port they came from; and there they make a new bark-log for their next cargo. The smaller sort of bark-logs, described before, which lie flat on the water, and are used for fishing or carrying water to ships or the like (half a ton or a ton at a time), are more governable than the other, though they have masts and sails too. With these they go out at night by the help of the land-wind, which is seldom wanting on this coast, and return back in the daytime with the sea-wind. This sort of floats are used in many places both in the East and West Indies. On the coast of Coromandel in the East Indies they call them Catamarans. These are but one log, or two sometimes, of a sort of light wood, and are made without sail or rudder, and so small that they carry but one man, whose legs and breech are always in the water; and he manages his log with a paddle, appearing at a distance like a man sitting on a fish's back.

November the 3d, at 6 o'clock in the morning, our men landed about four miles to the south of the town, and took some prisoners that were sent thither to watch for fear of us; and these prisoners said, that the Governor of Piura came with 100 armed men to Payta the night before, purposely to oppose our landing there if we should attempt it. Our men marched directly to the fort on the hill, and took it without the loss of one man. Hereupon the Governor of Piura with all his men, and the inhabitants of the town, ran away as fast as they could. Then our men entered the town, and found it emptied both of money and goods; there was not so much as a meal of victuals left for them. The prisoners told us a ship had been here a little before and burnt a great ship in the road, but did not land their men; and that here they put ashore all their prisoners and pilots. We knew this must be Captain Eaton's ship

---

be a mistranslation of "barcolongo" (see Note 1, p. 107). But the description which follows shows plainly enough that the word means just what it says—that is, barks of log or log-barks.

which had done this; and by these circumstances we supposed he was gone to the East Indies, it being always designed by him. The prisoners told us also, that since Captain Eaton was here, a small bark had been off the harbour and taken a pair of bark-logs a-fishing, and made the fishermen bring aboard twenty or thirty jars of fresh water. This we supposed was our bark that was sent to Lobos to seek Captain Eaton. In the evening we came in with our ships, and anchored before the town in ten fathoms water, near a mile from the shore. Here we stayed till the 6th day, in hopes to get a ransom for the town. Our Captains demanded 300 packs of flour, 3000 lbs. of sugar, twenty-five jars of wine, and 1000 jars of water to be brought off to us; but we got nothing of it. Therefore Captain Swan ordered the town to be fired, which was presently done. Then all our men came aboard, and Captain Swan ordered the bark which Captain Harris commanded, to be burnt, because she did not sail well.

At night, when the land-wind came off, we sailed from hence towards Lobos. The 10th, in the evening we saw a sail, bearing NW. by N., as far as we could well discern her on our deck. We immediately chased, separating ourselves, the better to meet her in the night, but we missed her. Therefore the next morning we again trimmed sharp, and made the best of our way to Lobos de la Mar. The 14th, we had sight of the Island of Lobos de Tierra: it bore E. from us; we stood in towards it, and betwixt 7 and 8 o'clock in the night came to an anchor at the NE. end of the island, in fourteen fathoms water. This island at sea is of an indifferent height, and appears like Lobos de la Mar. About a quarter of a mile from the north end there is a great hollow rock, and a good channel between, where there is seven fathoms water. The 15th, we went ashore, and found abundance of penguins and boobies, and seals in great quantities. We sent aboard of all these to be dressed, for we had not tasted any flesh in a great while before; therefore some of us did eat very heartily. Captain Swan, to encourage his men to eat this coarse flesh, would commend it for extraordinary good food, comparing the seal to roasting pig, the boobies to hens, and the penguins to ducks. This he did to train them to live contentedly on coarse meat, not knowing but we might be forced to make use of such food before we departed out of these seas; for it is generally seen among Privateers that nothing emboldens them sooner to mutiny than want, which we could not well suffer in a place where there are such quantities of these animals to be had, if men could be persuaded to be contented with them.

[Dampier now sailed from *Lobos de Tierra* to *Lobos de la Mar* on the 19th. On the 21st he sent out his Mosquito strikers for turtle, which they brought in, in great abundance. On the evening of the 26th, a suspicious-looking bark was observed about three leagues NNW. from the island. The next morning she stood off to sea, which they allowed her to do without giving chase. On the 28th day the ships' bottoms were scrubbed. On the morning of the 29th they were steering for the Bay of Guayaquil. In the vicinity, the cat-fish are said to be abundant. It is so called from its great wide mouth and the strings pointing out from each side of it like cats' whiskers.]

From the Island Santa Clara to Punta Arenas is seven leagues ENE. This Punta Arenas, or Sandy Point, is the westernmost point of the Island of Puna. Here all ships bound into the River of Guayaquil anchor, and must wait for a pilot, the entrance being very dangerous for strangers. The Island of Puna is a pretty large flat low island, stretching E. and W., about twelve or fourteen leagues long, and about four or five leagues wide. The tide runs very strong all about this island, but so many different ways, by reason of the branches, creeks, and rivers that run into the sea near it, that it casts up many dangerous shoals on all sides of it. There is in the island only one Indian town, on the south side of it, close by the sea, and seven leagues from Point Arenas, which town is also called Puna. The Indians of this town are all seamen, and are the only pilots in these seas, especially for this river. Their chief employment, when they are not at sea, is fishing. These men are obliged by the Spaniards to keep good watch for ships that anchor at Point Arenas. The place where they keep this watch is at a point of land on the Island of Puna that starts out into the sea, from whence they can see all ships that anchor at Point Arenas. The Indians come thither in the morning, and return at night on horseback. From this watching point to Point Arenas it is four leagues, all drowned mangrove-land: and midway between these two points is another small point, where these Indians are obliged to keep another watch, when they fear an enemy. The sentinel goes thither in a canoe in the morning, and returns at night; for there is no coming thither by land through that mangrove marshy ground.[1] . . . There are in the town of Puna about twenty houses, and a small church. The houses stand all on posts, ten or twelve feet high, with ladders on the outside to go up into them. I did never see the like building anywhere but among the Malayans in the East Indies. They are thatched with palmetto leaves, and their chambers well boarded, in which last they exceed the Malayans.

From Puna to Guayaquil is reckoned seven leagues. It is one league before you come to the River of Guayaquil's mouth, where it is above two miles wide; from thence upwards the river lies pretty straight, without any considerable turnings. Both sides of the river are low swampy land, overgrown with red mangroves, so that there is no landing. Four miles before you come to the town of Guayaquil, there is a low island standing in the river; this island divides the river into two parts, making two very fair channels for ships to pass up and down. The SW. channel is the widest; the other is as deep, but narrower and narrower yet, by reason of many trees and bushes which spread over the river both from the main and from the island; and there are also several great stumps of trees standing upright in the water on either side. The island is above a mile long. From the upper part of the island to the town of Guayaquil is almost a league, and near as much from one side of the river to the other. In that spacious place, ships of the greatest burthen may ride afloat; but the best place for ships is nearest to that part of the land where the town stands; and this place is seldom without ships. Guayaquil stands facing the island, close by the river, partly on the side, and partly at the foot of a gentle hill declining towards the river, by which the lower part of it is often overflown. There are two forts, one standing in the low ground, the other on the hill. This town makes a very fine prospect, it being beautified with several churches and other good buildings. Here lives a Governor, who, as I have been informed, has his patent from the King of Spain. Guayaquil may be reckoned one of the chief seaports in the South Seas: the commodities which are exported from hence are cacao, hides, tallow, sarsaparilla, and other drugs, and woollen cloth, commonly called cloth of Quito. The cacao grows on both sides of the river above the town. It is a small nut, like the Campeachy nut, I think the smallest of the two. They produce as much cacao here as serves all the kingdom of Peru; and much of it is sent to Acapulco, and from thence to the Philippine Islands. Sarsaparilla grows in the water by the sides of the river, as I have been informed. The Quito cloth comes from a rich town in

[1] The middle of the island is described as good pasture land, with ridges of woodland, abounding in palmettoes. The Indians cultivated part of these ridges with maize, yams, and potatoes.

the country within land, called Quito.[1] There is a great deal made, both serges and broadcloth. This cloth is not very fine, but is worn by the common sort of people throughout the whole kingdom of Peru. This and all other commodities which come from Quito are shipped off at Guayaquil for other parts; and all imported goods for the city of Quito pass by Guayaquil: by which it may appear that Guayaquil is a place of no mean trade. Quito, as I have been informed, is a very populous city, seated in the heart of the country. It is inhabited partly by Spaniards; but the major part of its inhabitants are Indians under the Spanish Government. It is environed with mountains of a vast height, from whose bowels many great rivers have their rise. These mountains abound in gold, which by violent rains is washed with the sand into the adjacent brooks; where the Indians resort in troops, washing away the sand, and putting up the gold-dust in their calabashes or gourd-shells. Quito is the place in all the kingdom of Peru[2] that abounds most with this rich metal, as I have been often informed. The country is subject to great rains, and very thick fogs, especially the valleys. For that reason it is very unwholesome and sickly. The chief distempers are fevers, violent headache, pains in the bowels and fluxes. I know no place where gold is found but what is very unhealthy. Guayaquil is not so sickly as Quito and other towns farther within land; yet in comparison with the towns that are on the coast of *Mare Pacificum*, south of Cape Blanco, it is very sickly.

It was to this town of Guayaquil that we were bound; therefore we left our ships off Cape Blanco, and ran into the Bay of Guayaquil with our bark and canoes, steering in for the Island of Santa Clara, where we arrived the next day after we left our ships; and from thence we sent away two canoes the next evening to Point Arenas. At this point there are abundance of oysters, and other shell-fish, as cockles and mussels; therefore the Indians of Puna often come hither to get these fish. Our canoes got over before day, and absconded[3] in a creek, to wait for the coming of the Puna Indians. The next morning some of them, according to their custom, came thither on bark-logs, at the latter part of the ebb, and were all taken by our men. The next day, by their advice, the two watchmen of the Indian town of Puna were taken by our men, and all its inhabitants, not one escaping. The next ebb they took a small bark laden with Quito cloth. She came from Guayaquil that tide, and was bound to Lima; they having advice that we were gone off the coast, by the bark which I said we saw while we lay at the Island of Lobos. The master of this cloth-bark informed our men that there were three barks coming from Guayaquil laden with Negroes; he said they would come from thence the next tide. The same tide of ebb that they took the cloth-bark, they sent a canoe to our bark, where the biggest part of the men were, to hasten them away with speed to the Indian town. The bark was now riding at Point Arenas; and the next flood she came with all the men, and the rest of the canoes, to Puna. The tide of flood being now far spent, we lay at this town till the last of the ebb, and then rowed away, leaving five men aboard our bark, who were ordered to lie still till 8 o'clock the next morning, and not to fire at any boat or bark; but after that time they might fire at any object: for it was supposed that before that time we should be masters of Guayaquil. We had not rowed above two miles, before we met and

[1] "Coarse cottons," says M'Culloch, "and woollen cloths, baizes, flannels, ponchos, and stockings are made in Quito."

[2] Quito was annexed to the empire of Peru not long before the Spanish conquest; it is now the capital of the Republic of Ecuador.

[3] Concealed themselves.

took one of the three barks laden with Negroes; the master of her said, that the other two would come from Guayaquil the next tide of ebb. We cut her mainmast down, and left her at anchor. It was now strong flood, and therefore we rowed with all speed towards the town, in hopes to get thither before the flood was down; but we found it further than we did expect it to be; or else our canoes, being very full of men, did not row so fast as we would have them. The day broke when we were two leagues from the town, and then we had not above an hour's flood more; therefore our Captain desired the Indian pilot to direct us to some creek where we might abscond all day, which was immediately done, and one canoe was sent towards Puna to our bark, to order them not to move nor fire till the next day. But she came too late to countermand the first orders: for the two barks before mentioned, laden with Negroes, came from the town the last quarter of the evening tide, and lay in the river, close by the shore on one side, and we rowed up on the other side and missed them; neither did they see nor hear us. As soon as the flood was spent, the two barks weighed and went down with the ebb towards Puna. Our bark, seeing them coming directly towards them, and both full of men, supposed that we by some accident had been destroyed, and that the two barks were manned with Spanish soldiers, sent to take our ships; and therefore they fired three guns at them a league before they came near. The two Spanish barks immediately came to an anchor, and the masters got into their boats and rowed for the shore; but our canoe that was sent from us took them both. The firing of these three guns made a great disorder among our advanced men, for most of them did believe they were heard at Guayaquil, and that therefore it could be no profit to lie still in the creek, but either row away to the town, or back again to our ships. It was now quarter ebb; therefore we could not move upwards, if we had been disposed so to do. At length Captain Davis said he would immediately land in the creek where they lay, and march directly to the town, if but forty men would accompany him; and without saying more words, he landed among the mangroves in the marshes. Those that were so minded followed him, to the number of forty or fifty. Captain Swan lay still with the rest of the party in the creek, for they thought it impossible to do any good that way.

Captain Davis and his men were absent about four hours, and then returned all wet and quite tired, and could not find any passage out into the firm land. He had been so far, that he almost despaired of getting back again; for a man cannot pass through those red mangroves but with very much labour. When Captain Davis was returned, we concluded to be going towards the town the beginning of the next flood; and if we found that the town was alarmed, we purposed to return again without attempting anything there. As soon as it was flood we rowed away, and passed by the island through the NE. channel, which is the narrowest. There are so many stumps in the river, that it is very dangerous passing in the night (and that is the time we always take for such attempts); for the river runs very swift, and one of our canoes stuck on a stump, and had certainly overset if she had not been immediately rescued by others. When we were come almost to the end of the island, there was a musket fired at us out of the bushes on the main. We then had the town open before us, and presently saw lighted torches or candles all the town over, whereas before the gun was fired there was but one light: therefore we now concluded we were discovered. Yet many of our men said that it was a holiday the next day, as it was indeed, and that therefore the Spaniards were making fireworks, which they often do in the night against such times. We rowed therefore a little farther, and found firm land; and Captain Davis pitched his canoe ashore and landed with his men. Captain Swan and most of his men did not think it convenient to attempt anything, seeing the town was alarmed; but at last, being upbraided with cowardice, Captain Swan and his men landed also. The place where we landed was about two miles from the town. It was all overgrown with woods, so thick that we could not march through in the night; and therefore we sat down waiting for the light of the day. We had two Indian pilots with us; one that had been with us a month, who, having received some abuses from a gentleman of Guayaquil, to be revenged, offered his service to us, and we found him very faithful; the other was taken by us not above two or three days before, and he seemed to be as willing as the other to assist us. This latter was led by one of Captain Davis's men, who showed himself very forward to go to the town, and upbraided others with faint-heartedness. Yet this man, as he afterwards confessed, notwithstanding his courage, privately cut the string that the guide was made fast with, and let him go to the town by himself, not caring to follow him; but when he thought the guide was got far enough from us, he cried out that the pilot was gone, and that somebody had cut the cord that tied him. This put every man into a moving posture to seek the Indian, but all in vain; and our consternation was great, being in the dark and among woods; so the design was wholly dashed, for not a man after that had the heart to speak of going farther. Here we stayed till day, and then rowed out into the middle of the river, where we had a fair view of the town; which, as I said before, makes a very pleasant prospect. We lay still about half an hour, being a mile, or something better, from the town. They did not fire one gun at us, nor we at them. Thus our design on Guayaquil failed; yet Captain Townley and Captain François Gronet took it a little while after this. When we had taken a full view of the town, we rowed over the river, where we went ashore to a beef estantion or farm, and killed a cow, which we dressed and ate. We stayed there till the evening tide of ebb, and then rowed down the river, and the 9th December in the morning arrived at Puna. In our way thither we went aboard the three barks laden with Negroes, that lay at anchor in the river, and carried the barks away with us. There were 1000 Negroes in the three barks, all lusty young men and women. When we came to Puna, we sent a canoe to Point Arenas, to see if the ships were come thither. The 12th day she returned again, with tidings that they were both there at anchor. Therefore in the afternoon we all went aboard of our ships, and carried the cloth-bark with us, and about forty of the stoutest Negro men, leaving their three barks with the rest; and out of these also Captain Davis and Captain Swan chose about fourteen or fifteen a-piece, and turned the rest ashore.

There was never a greater opportunity put into the hands of men to enrich themselves than we had, to have gone with these Negroes, and settled ourselves at Santa Maria on the Isthmus of Darien, and employed them in getting gold out of the mines there, which might have been done with ease; for about six months before this, Captain Harris, who was now with us, coming overland from the North Seas with his body of Privateers, had routed the Spaniards away from the town and gold mines of Santa Maria, so that they had never attempted to settle there again since. Add to this, that the Indian neighbourhood, who were mortal enemies to the Spaniards, and had been flushed by their successes against them through the assistance of the Privateers for several years, were our fast friends, and ready to receive and assist us. We had, as I have said, 1000 Negroes to work for us; we had 200 tons of flour that lay at the Galapagos; there was the River of Santa Maria, where we could careen and fit our ships, and might fortify the mouth so, that if all the strength the Spaniards have in Peru had come against us we could have kept them out. If they lay with guard-ships of strength to keep us in, yet we had a great country to live in, and a great nation of Indians that were our friends. Besides, which was the principal thing, we had the North Seas to befriend us; from whence we could export ourselves or effects, or import goods or men to our assistance; for in a short time we should have had assistance from all parts of the West Indies, many thousands of Privateers from Jamaica and the French islands especially would have flocked over to us, and long before this time we might have been masters not only of those mines (the richest gold mines ever yet found in America), but of all the coast as high as Quito; and much more than I say might then probably have been done.

But these may seem to the reader but golden dreams. To leave them, therefore; the 13th day we sailed from Point Arenas towards Plata, to seek our bark that was sent to the Island of Lobos in search of Captain Eaton. We were two ships in company, and two barks; and the 16th day we arrived at Plata, but found no bark there, nor any letter. The next day we went over to the main to fill water, and in our passage met our bark; she had been a second time at the Island of Lobos, and, not finding us, was coming to Plata again. They had been in some want of provision since they left us, and therefore they had been at Santa Elena and taken it; where they got as much maize as served them three or four days; and that, with some fish and turtle which they struck, lasted them till they came to the Island of Lobos de Tierra. They got boobies' and penguins' eggs, of which they laid in a store; and went from thence to Lobos de la Mar, where they replenished their stock of eggs, and salted up a few young seal, for fear they should want; and being thus victualled, they returned again towards Plata. When our water was filled we went over again to the Island of Plata. There we parted the cloths that were taken in the cloth-bark into two lots or shares; Captain Davis and his men had one part, and Captain Swan and his men had the other part. The bark which the cloth was in, Captain Swan kept for a tender. At this time there were at Plata a great many large turtle, which I judge came from the Galapagos; for I had never seen any here before, though I had been here several times: this was their coupling-time, which is much sooner in the year here than in the West Indies properly so called. Our strikers brought aboard every day more than we could eat. Captain Swan had no striker, and therefore had no turtle but what was sent him from Captain Davis; and all his flour too he had from Captain Davis: but since our disappointment at Guayaquil, Captain Davis's men murmured against Captain Swan, and did not willingly give him any provision, because he was not so forward to go thither as Captain Davis. However, at last these differences were made up, and we concluded to go into the Bay of Panama, to a town called La Velia; but because we had not canoes enough to land our men, we were resolved to search some rivers where the Spaniards have no commerce, there to get Indian canoes.

CHAPTER VII.

THE 23d of December 1684, we saile from the Island of Plata towards the Bay of Panama; the wind at SSE., a fine brisk gale, and fair weather. The next morning we passed by Cape Pasado. This cape is in Lat. 0° 28' S. of the Equator. It runs out into the sea with a high round point, which seems to be divided in the midst. It is bald against the sea,[1] but within land, and on both sides, it is full of short trees. The land in the country is very high and mountainous, and it appears to be very woody. Between Cape Pasado and Cape San Francisco, the land by the

[1] Bare on the side facing the sea.

sea is full of small points, making as many little sandy bays between them, and is of an indifferent height, covered with trees of divers sorts.[1] . . .

It was to the River Santiago that we were bound to seek for canoes; therefore the 26th, supposing ourselves to be abreast of it, we went from our ships with four canoes. The 27th in the morning we entered at half flood into the smaller branch of that river, and rowed up six leagues before we met any inhabitants. There we found two small huts thatched with palmetto leaves. The Indians, seeing us rowing towards their houses, got their wives and little ones, with their household-stuff, into their canoes, and paddled away faster than we could row; for we were forced to keep in the middle of the river because of our oars, but they with their paddles kept close under the banks, and so had not the strength of the stream against them as we had. These huts were close by the river on the east side of it, just against the end of the island. We saw a great many other houses a league from us on the other side of the river; but the main stream into which we were now come, seemed to be so swift, that we were afraid to put over for fear we should not be able to get back again. We found only a hog, some fowls, and plantains in the huts; we killed the hog and the fowls, which were dressed presently. Their hogs they got, as I suppose, from the Spaniards by some accident, or from some neighbouring Indians who converse[2] with the Spaniards; for this that we took was of their European kind, which the Spaniards have introduced into America very plentifully, especially into the Islands of Jamaica, Hispaniola, and Cuba above all, this last being very largely stored with them, where they feed in the woods in the daytime, and at night come in at the sounding of a conch shell, and are put up in their crawls[3] or pens. And yet some turn wild, which, nevertheless, are often decoyed in by the others; which being all marked, whenever they[4] see an unmarked hog in the pen, they know it is a wild one, and shoot him presently. These crawls I have not seen on the continent, where the Spaniards keep them tame at home. Among the wild Indians, or in their woods, are no hogs, but peccary and warree, a sort I have mentioned before. After we had refreshed ourselves, we returned towards the mouth of the river. It was the evening when we came from thence, and we got to the river's mouth the next morning before day. Our ships when we left them were ordered to go to Gallo, where they were to stay for us. Gallo is a small uninhabited island, lying in between 2° and 3° N. Lat. It lies in a wide bay about three leagues from the mouth of the River Tumaco, and four leagues and a half from a small Indian village called Tumaco. The Island of Gallo is of an indifferent height; it is clothed with very good timber trees, and is therefore often visited by barks from Guayaquil and other places; for most of the timber carried from Guayaquil and Lima is first fetched from Gallo. Tumaco is a large river that takes its name from the Indian village so called. It is reported to spring from the rich mountains about Quito. It is thickly inhabited with Indians, and there are some Spaniards that live there, who traffic with the Indians for gold. The village Tumaco is but small, and is seated not far from the mouth of the river. It is a place to entertain the

[1] Passing Cape San Francisco, they came to the River Santiago (now supposed to be the River Mira, which, rising N. of Quito, enters the sea S. of the Bay of Tumaco), where it was their intention to search for canoes. The land near the river's mouth is of a deep black mould, producing both the cotton and the cabbage trees in great abundance. The particular description of these trees is now omitted.

[2] Have intercourse

[3] "Crawl," a corruption of the Spanish "corral," is a pen or enclosure of hurdles for fish or turtle.

[4] The Spaniards.

Spanish merchants that come to Gallo to load timber, or to traffic with the Indians for gold. From the branch of the River Santiago, where we now lay, to Tumaco is about five leagues; the land low, and full of creeks, so that canoes may pass within land through those creeks, and from thence into Tumaco River.

[On the 28th they left the River Santiago, and coming to Tumaco town about 12 o'clock at night, they took all the inhabitants of the village, including a Spanish knight called Don Diego de Pinas, who had come in a ship from Lima to lade timber. On the 1st of January 1685, they went from Tumaco towards Gallo. On the way they had news of a Spanish Armada, which they determined to try and intercept among the King's Islands. On the 8th they took a ship laden with about ninety tons of flour.]

We jogged on after this with a gentle gale towards Gorgona, an island lying about twenty-five leagues from the Island of Gallo. The 9th we anchored at Gorgona, on the west side of the island, in thirty-eight fathoms, clean ground, not two cables' length from the shore. Gorgona is an uninhabited island, in Lat. about 3° N. It is a pretty high island, and very remarkable by reason of two saddles, or risings and fallings, on the top. It is about two leagues long, and a league broad, and it is four leagues from the main. At the west end is another small island. The soil or mould of it is black and deep in the low ground, but on the side of the high land it is a kind of a red clay. This island is very well clothed with large trees of several sorts, that are flourishing and green all the year. It is very well watered with small brooks that issue from the high land. Here are a great many little black monkeys, some Indian conies, and a few snakes, which are all the land animals that I know there. Here are pearl oysters in great plenty; they grow to the loose rocks in four, five, or six fathoms water, by beards or little small roots, as a mussel. These oysters are commonly flatter and thinner than other oysters, otherwise much alike in shape. The fish is not sweet, nor very wholesome; it is as slimy as a shell snail. They taste very copperish, if eaten raw, and are best boiled. The Indians, who gather them for the Spaniards, hang the meat of them on strings, like Jews'-ears,[1] and dry them before they eat them. The pearl is found at the head of the oyster, lying between the meat and the shell. Some will have twenty or thirty small seed pearls, some none at all, and some will have one or two pretty large ones. The inside of the shell is more glorious than the pearl itself. I did never see any in the South Seas but here. It is reported there are some at the south end of California. In the West Indies, the Rancho Reys or Rancherias, spoken of in Chapter III.,[2] is the place where they are found most plentifully. It is said there are some at the Island of Margarita, near St Augustine, a town in the Gulf of Florida, &c. In the East Indies, the Island of Ainam,[3] near the south end of China, is said to have plenty of these oysters, more productive of large round pearls than those in other places. They are found also in other parts of the East Indies, and on the Persian coast.

At this Island of Gorgona we rummaged our prize, and found a few boxes of marmalade, and three or four jars of brandy, which were equally shared between Captain Davis, Captain Swan, and their men. Here we filled all our water, and Captain Swan furnished himself with flour; afterwards we turned ashore a great many prisoners, but kept the chief to put them ashore in a better place. The 13th we sailed from hence toward the King's Islands. We were now six sail—two men-of-war, two tenders, a fireship, and the prize. The 16th we passed by Cape Corrientes. This cape is in Lat. 5° 10′ [N.]; it is high

---

[1] A tough, thin, rumpled fungus, like a flat and variously-hollowed cup.

[2] See page 115.

[3] Hai-man, in the Gulf of Tonquin.

bluff land, with three or four small hillocks on the top. It appears at a distance like an island. The day after we passed by the cape, we saw a small white island, which we chased, supposing it had been a sail, till, coming near, we found our error. The 21st we saw Point Garachina. This point is in Lat. 7° 20′ N.;[1] it is pretty high land, rocky, and destitute of trees, yet within land it is woody. It is fenced with rocks against the sea. Within the point by the sea, at low water, you may find store of oysters and mussels. The King's Islands, or Pearl Keys, are about twelve leagues distant from this point. Between Point Garachina and them there is a small, low, flat, barren island called Galera, at which Captain Harris was sharing with his men the gold he took in his pillaging Santa Maria, which I spoke of a little before, when on a sudden five Spanish barks, fitted out on purpose at Panama, came upon him; but he fought them so stoutly with one small bark he had and some few canoes, boarding their admiral particularly, that they were all glad to leave him. By this island we anchored, and sent our boats to the King's Islands for a good careening place.

The King's Islands are a great many low, woody islands lying NW. by N. and SE. by S. They are about seven leagues from the main, and fourteen leagues in length, and from Panama about twelve leagues. Why they are called the King's Islands I know not; they are sometimes, and mostly in maps, called the Pearl Islands.[2] I cannot imagine wherefore they are called so, for I did never see one pearl oyster about them, nor any pearl oyster shells; but on the other oysters I have made many a meal there. The northernmost island of all this range is called Pacheca or Pacheque; this is but a small island, distant from Panama eleven or twelve leagues. The southernmost of them is called St Paul's. Some of these islands are planted with plantains and bananas, and there are fields of rice on others of them. The gentlemen of Panama, to whom they belong, keep Negroes there to plant, weed, and husband the plantations. Many of them, especially the largest, are wholly untilled, yet very good fat land, full of large trees. These unplanted islands shelter many runaway Negroes, who abscond[3] in the woods all day, and in the night boldly pillage the plantain walks. Betwixt these islands and the main is a channel of seven or eight leagues wide; there is good depth of water, and good anchoring all the way. The islands border thick on each other, yet they make many small, narrow, deep channels fit only for boats to pass between most of them. At the SE. end, about a league from St Paul's Island, there is a good place for ships to careen or haul ashore. It is surrounded with the land, and has a good deep channel on the north side to go in at. The tide rises here about ten feet perpendicular. We brought our ships into this place on the 25th, but were forced to tarry for a spring-tide before we could have water enough to clean them; therefore we first cleaned our barks, that they might cruise before Panama while we lay there. The 27th, our barks being clean, we sent them out with twenty men in each. The fourth day after, they returned with a prize laden with maize or Indian corn, salt beef, and fowls. She came from Lavelia, and was bound to Panama. Lavelia is a town we once designed to attempt; it is pretty large, and stands on the banks of a river on the north side of the Bay of Panama, six or seven leagues from the sea. Nata is another such town, standing in a plain near another branch of the same river. In these towns, and some

---

[1] It is really in 8° 10′ N.

[2] The Isola del Rey, or King's Island, is an island of considerable size in the Bay of Panama; and the Pearl Keys are an archipelago of small islets between King's Island and the coast to the north-west.

[3] Hide.

others on the same coast, they breed hogs, fowls, bulls, and cows, and plant maize, purposely for the support of Panama, which is supplied with provision mostly from other towns and the neighbouring islands.[1]

The 14th of February 1685, we made an end of cleaning our ship, filled all our water, and stocked ourselves with firewood. The 15th, we went out from among the islands, and anchored in the channel between them and the main, in twenty-five fathoms water, soft oozy ground. The Plate Fleet was not yet arrived; therefore we intended to cruise before the city of Panama, which is from this place about twenty-five leagues. The next day we sailed towards Panama, passing in the channel between the King's Islands and the main. When we came abreast of Old Panama we anchored, and sent our canoe ashore with our prisoner Don Diego de Pinas, with a letter to the Governor, to treat about an exchange for our man they had spirited away, as I said, and another Captain Harris left in the River of Santa Maria the year before, coming overland. Don Diego was desirous to go on the errand in the name and with the consent of the rest of our Spanish prisoners; but by some accident he was killed before he got ashore, as we heard afterwards.

Old Panama was formerly a famous place; but it was taken by Sir Henry Morgan about the year 1673,[2] and at that time great part of it was burned to ashes, and it was never re-edified since. New Panama is a very fair city, standing close by the sea, about four miles from the ruins of the old town. It gives name to a large bay, which is famous for a great many navigable rivers, some whereof are very rich in gold; it is also very pleasantly sprinkled with islands that are not only profitable to their owners, but very delightful to the passengers and seamen that sail by them. It is encompassed on the back side with a pleasant country, which is full of small hills and valleys, beautified with many groves and spots of trees, that appear in the savannahs like so many little islands. This city is encompassed with a high stone wall; the houses are said to be of brick; their roofs appear higher than the top of the city wall. It is beautified with a great many fair churches and religious houses, besides the President's house, and other eminent buildings; which altogether make one of the finest objects that I did ever see, in America especially. There are a great many guns on the walls, most of which look toward the land. They had none at all against the sea when I first entered those seas with Captain Sawkins, Captain Coxon, Captain Sharpe, and others; for till then they did not fear any enemy by sea: but since then they have planted guns clear round. This is a flourishing city, by reason it is a thoroughfare for all imported or exported goods and treasure to and from all parts of Peru and Chili, whereof their storehouses are never empty. The road also is seldom or never without ships. Besides, once in three years, when the Spanish Armada comes to Porto Bello, then the Plate Fleet also from Lima comes hither with the King's treasure, and abundance of merchant ships full of goods and plate. At that time the city is full of merchants and gentlemen; the seamen are busy in landing the treasure and goods, and the carriers or caravanmasters employed in carrying it overland on mules, in vast droves every day, to Porto Bello, and bringing back European goods from thence. Though the city be then so full, yet during this heat of business there is no hiring of an ordinary slave under a piece of eight a-day; houses

---

[1] The island where they here careened their ships is described as being environed with rocks, on which they gathered abundance of oysters, clams, mussels, and limpets.

[2] Really at the end of January 1671, when Morgan and his men committed atrocities that made the capture of Panama conspicuous even among the brutal records of the Buccaneers.

also, chambers, beds, and victuals, are then extraordinary dear.

Now I am on the subject, I think it will not be amiss to give the reader an account of the progress of the Armada from Old Spain which comes thus every three years into the Indies. Its first arrival is at Carthagena, from whence, as I have been told, an express is immediately sent overland to Lima, through the Southern Continent, and another by sea to Porto Bello, with two packets of letters, one for the Viceroy of Lima, the other for the Viceroy of Mexico. I know not which way that of Mexico goes after its arrival at Porto Bello, whether by land or sea; but I believe by sea to La Vera Cruz. That for Lima is sent by land to Panama, and from thence by sea to Lima. Upon mention of these packets I shall digress yet a little further, and acquaint my reader, that before my first going over into the South Seas with Captain Sharpe (and indeed before any Privateers, at least since Drake and Oxenham, had gone that way which we afterwards went, except La Sound, a French Captain, who by Captain Wright's instructions had ventured as far as Chepo Town with a body of men, but was driven back again), I being then on board Captain Coxon, in company with three or four more Privateers, about four leagues to the east of Porto Bello, we took the packets bound thither from Carthagena. We opened a great quantity of the merchants' letters, and found the contents of many of them to be very surprising; the merchants of several parts of Old Spain thereby informing their correspondents of Panama and elsewhere of a certain prophecy that went about Spain that year, the tenor of which was, that there would be English Privateers that year in the West Indies, who would make such great discoveries as to open a door into the South Seas which they supposed was fastest shut; and the letters were accordingly full of cautions to their friends to be very watchful and careful of their coasts. This door they spoke of we all concluded must be the passage overland through the country of the Indians of Darien, who were a little before this become our friends, and had lately fallen out with the Spaniards, breaking off the intercourse which for some time they had with them: and upon calling also to mind the frequent invitations we had from those Indians a little before this time, to pass through their country and fall upon the Spaniards in the South Seas, we from thenceforward began to entertain such thoughts in earnest, and soon came to a resolution to make those attempts which we afterwards did with Captains Sharpe, Coxon, &c. So that the taking of these letters gave the first life to those bold undertakings: and we took the advantage of the fears the Spaniards were in from that prophecy, or probable conjecture, or whatever it were; for we sealed up most of the letters again, and sent them ashore to Porto Bello.

The occasion of this our late friendship with those Indians was thus: About fifteen years before this time, Captain Wright being cruising near that coast, and going in among the Sambaloes Isles to strike fish and turtle, took there a young Indian lad as he was paddling about in a canoe. He brought him aboard his ship, and gave him the name of John Gret, clothing him, and intending to breed him among the English. But his Mosquito strikers, taking a fancy to the boy, begged him of Captain Wright, and took him with them at their return into their own country, where they taught him their art; and he married a wife among them, and learnt their language, as he had done some broken English while he was with Captain Wright, which he improved among the Mosquitoes, who, corresponding so much with us, do all of them smatter English after a sort; but his own language he had almost forgot. Thus he lived among them for many years; till, about six or eight months before our taking these letters, Captain Wright being again among the Sambaloes, took thence another Indian boy about ten or twelve years old, the son of a man of some account among those Indians; and wanting a striker, he went away to the Mosquito country, where he took in John Gret, who was now very expert at it. John Gret was much pleased to see a lad there of his own country, and it came into his mind to persuade Captain Wright upon this occasion to endeavour a friendship with those Indians; a thing our Privateers had long coveted, but never durst attempt, having such dreadful apprehensions of their numbers and fierceness. But John Gret offered the Captain that he would go ashore and negotiate the matter; who accordingly sent him in his canoe till he was near the shore, which of a sudden was covered with Indians, standing ready with their bows and arrows. John Gret, who had only a clout about his middle, as the fashion of the Indians is, leapt then out of the boat and swam, the boat retiring a little way back; and the Indians ashore, seeing him in that habit, and hearing him call to them in their own tongue (which he had recovered by conversing with the boy lately taken, suffered him quietly to land, and gathered all about to hear how it was with him. He told them particularly, that he was one of their countrymen, and how he had been taken many years ago by the English, who had used him very kindly; that they were mistaken in being so much afraid of that nation, who were not enemies to them, but to the Spaniards. To confirm this, he told them how well the English treated another young lad of theirs they had lately taken, such an one's son; for this he had learnt of the youth; and his father was one of the company that was got together on the shore. He persuaded them, therefore, to make a league with these friendly people, by whose help they might be able to quell the Spaniards; assuring, also, the father of the boy, that if he would but go with him to the ship, which they saw at anchor at an island there (it was Golden Island, the easternmost of the Sambaloes, a place where there is good striking for turtle), he should have his son restored to him, and they might all expect a very kind reception. Upon these assurances, twenty or thirty of them went off presently, in two or three canoes laden with plantains, bananas, fowls, &c.; and, Captain Wright having treated them on board, went ashore with them, and was entertained by them, and presents were made on each side. Captain Wright gave the boy to his father in a very handsome English dress, which he had caused to be made purposely for him; and an agreement was immediately struck up between the English and these Indians, who invited the English through their country into the South Seas. Pursuant to this agreement, the English, when they came upon any such design, or for traffic with them, were to give a certain signal which they pitched upon, whereby they might be known. But it happened that Mr La Sound, the French Captain spoken of a little before, being then one of Captain Wright's men, learnt this signal, and staying ashore at Petit Goave, upon Captain Wright's going thither soon after, who had his commission from thence, he gave the other French there such an account of the agreement before mentioned, and the easiness of entering the South Seas thereupon, that he got at the head of about 120 of them, who made that unsuccessful attempt upon Chepo, as I said; making use of the signal they had learnt for passing the Indians' country, who at that time could not distinguish so well between the several nations of the Europeans as they can since. From such small beginnings arose those great stirs that have been since made all over the South Seas,—viz., from the letters we took, and from the friendship contracted with these Indians by means of John Gret. Yet this friendship had like to have been stifled in its infancy; for, within a few months after, an English trading sloop came on this coast from Jamaica, and John Gret, who by this time had advanced himself to be a grandee among these

Indians, together with five or six more of that quality, went off to the sloop in their long gowns, as the custom is for such to wear among them. Being received aboard, they expected to find everything friendly, and John Gret talked to them in English; but these Englishmen, having no knowledge at all of what had happened, endeavoured to make them slaves, as is commonly done; for upon carrying them to Jamaica they could have sold them for £10 or £12 a-piece. But John Gret and the rest perceiving this, leapt all overboard, and were by the others killed every one of them in the water. The Indians on shore never came to the knowledge of it; if they had, it would have endangered our correspondence. Several times after, upon our conversing with them, they inquired of us what was become of their countrymen; but we told them we knew not, as, indeed, it was a great while after that we heard this story; so they concluded the Spaniards had met with them, and killed or taken them.

But to return to the account of the progress of the Armada, which we left at Carthagena. After an appointed stay there of about sixty days, as I take it, it goes thence to Porto Bello, where it lies thirty days and no longer. Therefore the Viceroy of Lima, on notice of the Armada's arrival at Carthagena, immediately sends away the King's treasure to Panama, where it is landed, and lies ready to be sent to Porto Bello upon the first news of the Armada's arrival there. This is the reason partly of their sending expresses so early to Lima, that upon the Armada's first coming to Porto Bello, the treasure and goods may lie ready at Panama to be sent away upon the mules; and it requires some time for the Lima fleet to unlade, because the ships ride not at Panama, but at Perica, which are three small islands two leagues from thence. The King's treasure is said to amount commonly to about 24,000,000 pieces of eight; besides abundance of merchants' money. All this treasure is carried on mules, and there are large stables at both places to lodge them. Sometimes the merchants, to steal the custom, pack up money among goods, and send it to Venta de Cruces, on the River Chagres; from thence down the river, and afterwards by sea to Porto Bello: in which passage I have known a whole fleet of periagoes and canoes taken. The merchants who are not ready to sail by the thirtieth day after the Armada's arrival are in danger to be left behind; for the ships all weigh the thirtieth day precisely, and go to the harbour's mouth. Yet sometimes, on great importunity, the Admiral may stay a week longer; for it is impossible that all the merchants should get ready, for want of men. When the Armada departs from Porto Bello it returns again to Carthagena, by which time all the King's revenue which comes out of the country is got ready there. Here also meets them again a great ship, called the Patache,[1] one of the Spanish galleons, which, before their first arrival at Carthagena, goes from the rest of the Armada, on purpose to gather the tribute of the coast, touching at the Margaritas and other places in her way thence to Carthagena, as Punta de Guayra, Maracaibo, Rio la Hacha, and Santa Marta, and at all these places takes in treasure for the King. After the set stay at Carthagena, the Armada goes away to the Havana, in the Isle of Cuba, to meet there the Flota, which is a small number of ships that go to La Vera Cruz, and there take in the effects of the city and country of Mexico, and what is brought thither in the ship which comes thither every year from the Philippine Islands; and having joined the rest at the Havana, the whole Armada sets sail for Spain through the Gulf of Florida. The ships in the South Seas lie a great

---

[1] A patache (French, from Italian, "patascia") is a vessel, generally small, used in the conveyance of men, stores, or orders from ship to ship or place to place; a kind of messenger or auxiliary ship.

deal longer at Panama before they return to Lima. The merchants and gentlemen which come from Lima stay as little time as they can at Porto Bello, which is at the best but a sickly place, and at this time is very full of men from all parts. But Panama, as it is not overcharged with men so unreasonably as the other, though very full, enjoys a good air, lying open to the sea-wind, which rises commonly about 10 or 11 o'clock in the morning, and continues till 8 or 9 o'clock at night; then the land-wind comes, and blows till 8 or 9 in the morning. There are no woods nor marshes near Panama, but a brave, dry, champaign land, not subject to fogs nor mists. The wet season begins in the latter end of May and continues till November. The rains are not so excessive about Panama itself as on either side of the bay; yet in the months of June, July, and August they are severe enough. Gentlemen that come from Peru to Panama, especially in these months, cut their hair close to preserve them from fevers; for the place is sickly to them, because they come out of a country which never has any rains or fogs, but enjoys a constant serenity; but I am apt to believe this city is healthy enough to any other people. Thus much for Panama.

The 20th, we went and anchored within a league of the Islands of Perica, which are only three little barren rocky islands, in expectation of the President of Panama's answer to the letter I said we sent him by Don Diego, treating about exchange of prisoners; this being the day on which he had given us his parole to return with an answer. The 21st, we took another bark laden with hogs, fowl, salt beef, and molasses; she came from Lavelia and was going to Panama. In the afternoon we sent another letter ashore by a young Mosteso (a mixed breed of Indians and Europeans), directed to the President; and three or four copies of it to be dispersed abroad among the common people. This letter, which was full of threats, together with the young man's managing the business, wrought so powerfully among the common people, that the city was in an uproar. The President immediately sent a gentleman aboard, who demanded the flour prize that we took off Gallo, and all the prisoners, for the ransom of our two men; but our captains told him they would exchange man for man. The gentleman said he had not orders for that, but if we would stay till the next day he would bring the Governor's answer. The next day he brought aboard our two men, and had about forty prisoners in exchange. [On the 24th, they ran over to the Island of Taboga, about six leagues south from Panama. Its principal products are said to be the plantain, the banana, and the cocoa-nut. A small town, with a church at one end, is described as standing by the sea, the whole having been much destroyed by Privateers.] . . .

While we lay at Taboga we had like to have had a scurvy trick played us by a pretended merchant from Panama, who came as by stealth to traffic with us privately; a thing common enough with the Spanish merchants, both in the North and South Seas, notwithstanding the severe prohibitions of the Governors; who yet sometimes connive at it, and will even trade with the Privateers themselves. Our merchant was by agreement to bring out his bark laden with goods in the night, and we to go and anchor at the south of Perica. Out he came, with a fireship instead of a bark, and approached very near, hailing us with the watchword we had agreed upon. We, suspecting the worst, called to them to come to an anchor, and upon their not doing so, fired at them; when immediately their men going out into the canoes set fire to their ship, which blew up and burnt close by us; so that we were forced to cut our cables in all haste and scamper away as well as we could. The Spaniard was not altogether so politic in appointing to meet us at Perica, for there we had sea-room; whereas had he come thus upon us at Taboga, the land-wind bearing hard upon us as it did, we must either have been burnt by the fireship, or, upon loosing our cables, have been driven ashore. But I suppose they chose Perica rather for the scene of their enterprise, partly because they might there best skulk among the islands, and partly because, if their exploit failed, they could thence escape best from our canoes to Panama, but two leagues off. During this exploit, Captain Swan, whose ship was less than ours, and so not so much aimed at by the Spaniards, lay about a mile off, with a canoe at the buoy of his anchor, as fearing some treachery from our pretended merchant; and a little before the bark blew up, he saw a small float on the water, and, as it appeared, a man on it, making towards his ship; but the man dived, and disappeared of a sudden, as thinking probably that he was discovered. This was supposed to be one coming with some combustible matter to have stuck about the rudder. For such a trick Captain Sharpe was served at Coquimbo, and his ship had like to have been burnt by it, if by mere accident it had not been discovered. I was then aboard Captain Sharpe's ship. Captain Swan, seeing the blaze by us, cut his cables as we did; his bark did the like; so we kept under sail all the night, being more scared than hurt. The bark that was on fire drove burning towards Taboga; but after the first blast she did not burn clear, only made a smother; for she was not well made, though Captain Bond had the framing and management of it.

This Captain Bond was he of whom I made mention in my Fourth Chapter.[1] He, after his being at the Isles of Cape Verd, stood away for the South Seas, at the instigation of one Richard Morton, who had been with Captain Sharpe in the South Seas. In his way he met with Captain Eaton, and they two consorted a day or two; at last Morton went aboard of Captain Eaton, and persuaded him to lose Captain Bond in the night, which Captain Eaton did; Morton continuing aboard of Captain Eaton, as finding his the better ship. Captain Bond thus losing both his consort Eaton, and Morton his pilot, and his ship being but an ordinary sailer, despaired of getting into the South Seas; and he had played such tricks among the Caribbee Isles, as I have been informed, that he did not dare to appear at any of the English Islands. Therefore he persuaded his men to go to the Spaniards, and they consented to do anything that he should propose; so he presently steered away into the West Indies, and the first place where he came to an anchor was at Porto Bello. He presently declared to the Governor that there were English ships coming into the South Seas, and that if they questioned it, he offered to be kept a prisoner till time should discover the truth of what he said; but they believed him, and sent him away to Panama, where he was in great esteem. This several prisoners told us. The Spaniards of Panama could not have fitted out their fireship without this Captain Bond's assistance: for it is strange to say how grossly ignorant the Spaniards in the West Indies, but especially in the South Seas, are of sea affairs. They build indeed good ships; but this is a small matter, for any ship of a good bottom will serve for these seas on the south coast. They rig their ships but untowardly, have no guns but in three or four of the King's ships; and are as meanly furnished with warlike provisions, and as much at a loss for the making any fireships or other less usual machines. Nay, they have not the sense to have their guns run within the sides upon their discharge, but have platforms without for the men to stand on to charge them; so that when we come near we can fetch them down with small shot out of our boats. A main reason of this is, that the native Spaniards are too proud to be seamen, but use the Indians for all those offices; one Spaniard, it may be, going in the ship to command it,

[1] See page 125.

and himself of little more knowledge than those poor ignorant creatures; nor can they gain much experience, seldom going far off to sea, but coasting along the shores.

But to proceed. In the morning when it was light we came again to an anchor close by our buoys, and strove to get our anchors again; but our buoy-ropes, being rotten, broke. While we were puzzling about our anchors, we saw a great many canoes full of men pass between Taboga and the other island. This put us into a new consternation; we lay still some time, till we saw that they came directly towards us, then we weighed and stood towards them; and when we came within hail, we found that they were English and French Privateers come out of the North Seas through the Isthmus of Darien. They were 280 men, in twenty-eight canoes; 200 of them French, the rest English. They were commanded by Captain Gronet and Captain Lequie. We presently came to an anchor again, and all the canoes came aboard. These men told us that there were 180 Englishmen more, under the command of Captain Townley, in the country of Darien, making canoes (as these men had been) to bring them into these seas. All the Englishmen that came over in this party were immediately entertained by Captain Davis and Captain Swan in their own ships; and the Frenchmen were ordered to have our flour prize to carry them, and Captain Gronet, being the eldest commander, was to command them there; and thus they were all disposed of to their hearts' content. Captain Gronet, to retaliate this kindness, offered Captain Davis and Captain Swan each of them a new commission from the Governor of Petit Goave. It has been usual for many years past for the Governor of Petit Goave to send blank commissions to sea by many of his captains, with orders to dispose of them to whom they saw convenient; those of Petit Goave by this means making themselves the sanctuary and asylum of all people of desperate fortunes, and increasing their own wealth and the strength and reputation of their party thereby. Captain Davis accepted of one, having before only an old commission which fell to him by inheritance at the decease of Captain Cooke, who took it from Captain Tristian, together with his bark, as is before mentioned.[1] But Captain Swan refused it, saying he had an order from the Duke of York neither to give offence to the Spaniards nor to receive any affront from them; and that he had been injured by them at Valdivia, where they had killed some of his men, and wounded several more; so that he thought he had a lawful commission of his own to right himself. I never read any of these French commissions while I was in these seas, nor did I then know the import of them; but I have learned since that the tenor of them is, to give a liberty to fish, fowl, and hunt. The occasion of this is, that the Island of Hispaniola, where the garrison of Petit Goave is, belongs partly to the French and partly to the Spaniards, and in time of peace these commissions are given as a warrant to those of each side to protect them from the adverse party; but in effect the French do not restrain them to Hispaniola, but make them a pretence for a general ravage in any part of America, by sea or land.

Having thus disposed of our associates, we intended to sail towards the Gulf of San Miguel, to seek Captain Townley, who by this time we thought might be entering into these seas. Accordingly the 2d of March 1685, we sailed from hence towards the Gulf of San Miguel. This gulf lies near thirty leagues from Panama towards the SE. The way thither from Panama is to pass between the King's Islands and the main. It is a place where many great rivers, having finished their courses, are swallowed up in the sea. It is bounded on the S. by Point Garachina, which lies in N. Lat. 6° 40′, and on the N.

[1] At the opening of Chapter IV.

by Cape San Lorenzo. . . . The chief rivers which run into this gulf are Santa Maria, Sambu, and Congo. The River Congo (which is the river I would have persuaded our men to have gone up as their nearest way in our journey overland, mentioned Chapter I.) comes directly out of the country, and swallows up many small streams that fall into it from both sides; and at last loses itself on the N. side of the gulf, a league within Cape San Lorenzo. It is not very wide, but deep, and navigable some leagues within land. There are sands without it, but a channel for ships. It is not made use of by the Spaniards, because of the neighbourhood of Santa Maria River; where they have most business on account of the mines. The River of Sambu seems to be a great river, for there is a great tide at its mouth; but I can say nothing more of it, having never been in it. This river falls into the sea on the south side of the gulf, near Point Garachina. Between the mouths of these two rivers on either side the gulf runs in towards the land somewhat narrower, and makes five or six small islands, which are clothed with great trees, green and flourishing all the year, and good channels between the islands. Beyond which, farther in still, the shore on each side closes so near, with two points of low mangrove land, as to make a narrow or strait scarce half-a-mile wide. This serves as a mouth or entrance to the inner part of the gulf, which is a deep bay two or three leagues over every way; and about the east end thereof are the mouths of several rivers. The River of Santa Maria is the largest of all the rivers of this gulf; it is navigable eight or nine leagues up, for so high the tide flows. Beyond that place the river is divided into many branches, which are only fit for canoes. [They now sailed towards the Gulf of San Miguel in search of Captain Townley. Near the mouth of the River of Santa Maria which falls into the gulf, the Spaniards, twenty years before, made their first discovery of the gold there, and built the town of Santa Maria. These mines were still being worked by the Spaniards and native Indians, when visited by Dampier. Another town known by its Indian name of Nisperal, also stood at the mouth of the river, described as being more airy and habitable than Santa Maria itself. On the 2d of March they anchored at Pacheque; on leaving and sailing out towards the gulf they met Captain Townley, who had taken two barks bound for Panama, and laden with wine, brandy, and sugar, &c.] The 10th, we took a small bark that came from Guayaquil; she had nothing in her but ballast. The 12th there came an Indian canoe out of the River of Santa Maria, and told us that there were 300 English and French men more coming overland from the North Seas. The 15th we met a bark with five or six Englishmen in her, that belonged to Captain Knight, who had been in the South Seas five or six months and was now on the Mexican coast. There he had spied this bark; but not being able to come up with her in his ship, he detached these five or six men in a canoe, who took her, but when they had done could not recover their own ship again, losing company with her in the night; and therefore they came into the Bay of Panama, intending to go overland back into the North Seas, but that they luckily met with us: for the Isthmus of Darien was now become a common road for Privateers to pass between the North and South Seas at their pleasure. This bark of Captain Knight's had in her forty or fifty jars of brandy: she was now commanded by Mr Henry More; but Captain Swan, intending to promote Captain Harris, caused Mr More to be turned out, alleging that it was very likely these men were run away from their commander. Mr More willingly resigned her, and went aboard of Captain Swan, and became one of his men.

It was now the latter end of the dry season here; and the water at the King's or Pearl Islands, of which there was plenty when we first came hither, was now dried away. Therefore we were forced to go to Point Garachina, thinking to water our ships there. Captain Harris, being now commander of the new bark, was sent into the River of Santa Maria, to see for those men that the Indians told us of, whilst the rest of the ships sailed towards Point Garachina; where we arrived the 21st day, and anchored two miles from the point, and found a strong tide running out of the River Sambu. The next day we ran within the point, and anchored in four fathoms at low water. The Indians that inhabit in the River Sambu came to us in canoes, and brought plantains and bananas. They could not speak nor understand Spanish; therefore I believe they have no commerce with the Spaniards. We found no fresh water here neither; so we went from hence to Puerto Pinas, which is seven leagues S. by W. from hence. Puerto Pinas lies in Lat 7° N. It is so called because there are many pine trees growing there. The land is pretty high, rising gently as it runs into the country. This country near the sea is all covered with pretty high woods; the land that bounds the harbour is low in the middle, but high and rocky at both sides. The 25th we arrived at this harbour of pines, but did not go in with our ships, finding it but an ordinary place to lie at. We sent in our boats to search it, and they found a stream of good water running into the sea: but there were such great swelling surges came into the harbour, that we could not conveniently fill our water there. The 26th day we returned to Point Garachina again. In our way we took a small vessel laden with cacao: she came from Guayaquil. The 29th we arrived at Point Garachina. There we found Captain Harris, who had been in the River of Santa Maria; but he did not meet the men that he went for; yet he was informed again by the Indians that they were making canoes in one of the branches of the River of Santa Maria. Here we shared our cacao lately taken. Because we could not fill our water here, we designed to go to Taboga again, where we were sure to be supplied. Accordingly, on the 30th we set sail, being now nine ships in company; and had a small wind at SSE. The 3d of April in the evening we anchored by Perica, and the next morning went to Taboga, where we found our four canoes. Here we filled our water and cut firewood; and from hence we sent four canoes over to the main, with one of the Indians lately taken, to guide them to a sugar-work; for, now we had cacao, we wanted sugar to make chocolate. But the chief of their business was to get coppers; for each ship having now so many men, our pots would not boil victuals fast enough, though we kept them boiling all the day. About two or three days after, they returned aboard with three coppers.

While we lay here, Captain Davis's bark went to the Island of Otoque. This is another inhabited island in the Bay of Panama, not so big as Taboga, yet there are good plantain walks on it, and some Negroes to look after them. These Negroes rear fowls and hogs for their masters, who live at Panama; as at the King's Islands. It was for some fowls or hogs that our men went thither; but by accident they met also with an express that was sent to Panama with an account that the Lima fleet was at sea. Most of the letters were thrown overboard and lost, yet we found some that said positively that the fleet was coming with all the strength that they could make in the kingdom of Peru; yet were ordered not to fight us, except they were forced to it (though afterwards they choose to fight us, having first landed their treasure at Lavelia); and that the pilots of Lima had been in consultation what course to steer to miss us. [Dampier here inserts translations of two of the captured letters reporting the resolutions taken by the committee of pilots, and laying down the course to be steered by the fleet.] The 10th we sailed from Taboga towards the King's Islands again, because our pilots told us that the

King's ships did always come this way. The 11th we anchored at the place where we careened. Here we found Captain Harris, who had gone a second time into the River of Santa Maria, and fetched the body of men that last came overland, as the Indians told us, but they fell short of the number they told us of. The 19th we sent 250 men in fifteen canoes to the River Chepo to take the town of Chepo. The 21st, all our ships but Captain Harris, who stayed to clean his ship, followed after. The 22d we arrived at the Island of Chepillo, the pleasantest island in the Bay of Panama. It is but seven leagues from the city of Panama, and a league from the main. This island is about two miles long, and almost so broad; it is low on the north side, and rises by a small ascent towards the south side. The soil is yellow, a kind of clay. The high side is stony; the low land is planted with all sorts of delicate fruits, viz., sappodillas, Avocado pears, mammees, mammee-sapotas, star apples, &c. The middle of the island is planted with plantain trees, which are not very large, but the fruit extraordinary sweet.[1]

The River Chepo is very deep, and about a quarter of a mile broad; but the mouth of it is choked up with sands, so that no ships can enter, but barks may. There is a small Spanish town of the same name within six leagues of the sea; it stands on the left hand going from the sea. The land about it is champaign, with many small hills clothed with woods, but the biggest part of the country is savannah. On the south side of the river it is all woodland for many leagues together. It was to this town that our 250 men were sent. The 24th they returned out of the river, having taken the town without any opposition, but they found nothing in it. By the way going thither they took a canoe, but most of the men escaped ashore upon one of the King's Islands; she was sent out well appointed with armed men to watch our motions. The 25th, Captain Harris came to us, having cleaned his ship. The 26th we went again toward Taboga; our fleet now, upon Captain Harris joining us again, consisted of ten sail. We arrived at Taboga the 28th; there our prisoners were examined concerning the strength of Panama; for now we thought ourselves strong enough for such an enterprise, being near 1000 men. Out of these, on occasion, we could have landed 900; but our prisoners gave us small encouragement to it, for they assured us that all the strength of the country was there, and that many men were come from Porto Bello besides its own inhabitants, who of themselves were more in number than we. These reasons, together with the strength of the place, which has a high wall, deterred us from attempting it. While we lay here at Taboga, some of our men burned the town on the island. [From the 4th of May till the 27th, they were cruising among the King's Islands waiting for the Spanish fleet from Lima.] The 28th we had a very wet morning, for the rains were come in, as they do usually in May or June sooner or later; so that May is here a very uncertain month. However about 11 o'clock it cleared up, and we saw the Spanish fleet about three leagues WNW. from the Island of Pacheque, standing close on a wind to the E.; but they could not fetch the island by a league. We were riding a league SE. from the island, between it and the main; only Captain Gronet was about a mile to the northward of us near the island. He weighed so soon as they came in sight, and stood over for the main, and we lay still, expecting when he would tack and come to us; but he took care to keep himself out of harm's way.

Captains Swan and Townley came aboard of Captain Davis to order how to engage the enemy, who we saw came purposely to fight us, they being in all fourteen sail, besides periagoes rowing with twelve and fourteen oars

[1] The particular description of these fruits is now omitted.

a-piece. Six sail of them were ships of good force: first the Admiral, 48 guns, 450 men; the Vice-Admiral, 40 guns, 400 men; the Rear-Admiral, 36 guns, 360 men; a ship of 24 guns, 300 men; one of 18 guns, 250 men; and one of 8 guns, 200 men; two great fireships, six ships only with small arms, having 800 men on board them all; beside two or three hundred men in periagoes. This account of their strength we had afterwards from Captain Knight, who, being to the windward on the coast of Peru, took prisoners, of whom he had this information, being what they brought from Lima.. Besides these men, they had also some hundreds of Old Spain men that came from Porto Bello, and met them at Lavelia, from whence they now came; and their strength of men from Lima was 3000 men, being all the strength they could make in that kingdom, and for greater security they had first landed their treasure at Lavelia. Our fleet consisted of ten sail: first Captain Davis, 36 guns, 156 men, mostly English; Captain Swan, 16 guns, 140 men, all English. These were the only ships of force that we had, the rest having none but small arms. Captain Townley had 110 men, all English; Captain Gronet, 308 men, all French; Captain Harris, 100 men, mostly English; Captain Branley, 36 men, some English, some French; Davis's tender, 8 men; Swan's tender, 8 men; Townley's bark, 80 men; and a small bark of thirty tons made a fireship, with a canoe's crew in her. We had in all 960 men. But Captain Gronet came not to us till all was over, yet we were not discouraged at it, but resolved to fight them; for being to windward of the enemy, we had it at our choice whether we would fight or not. It was 3 o'clock in the afternoon when we weighed, and being all under sail we bore down right afore the wind on our enemies, who kept close on a wind to come to us; but night came on without anything besides the exchanging of a few shot on each side. When it grew dark, the Spanish Admiral put out a light as a signal for his fleet to come to an anchor. We saw the light in the Admiral's top, which continued about half-an-hour, and then it was taken down. In a short time after, we saw the light again; and being to windward, we kept under sail, supposing the light had been in the Admiral's top; but, as it proved, this was only a stratagem of theirs, for this light was put out the second time at one of their barks' topmast-head, and then she went to the leeward, which deceived us, for we thought still the light was in the Admiral's top, and by that means thought ourselves to windward of them. In the morning, therefore, contrary to our expectation, we found they had got the weather-guage of us, and were coming upon us with full sail; so we ran for it, and after a running fight all day, and having taken a turn almost round the Bay of Panama, we came to an anchor again at the Isle of Pacheque, in the very same place from whence we set out in the morning. Thus ended this day's work, and with it all that we had been projecting for five or six months, when, instead of making ourselves masters of the Spanish fleet and treasure, we were glad to escape them, and owed that too, in a great measure, to their want of courage to pursue their advantage.

The 30th day, in the morning, when we looked out, we saw the Spanish fleet all together, three leagues to leeward of us, at anchor. It was but little wind till 10 o'clock, and then sprang up a small breeze at S., and the Spanish fleet went away to Panama. What loss they had I know not; we lost but one man; and having held a consult,[1] we resolved to go to the Keys of Quibo or Cabaya, to seek Captain Harris, who was forced away from us in the fight; that being the place appointed for our rendezvous upon any such accident. As for Gronet, he said his men would not suffer him to join us in the fight; but we were not satisfied with that excuse; so we suffered him to go with

[1] Council, or consultation.

us to the Isles of Quibo, and there cashiered our cowardly companion. Some were for taking from him the ship which we had given him; but at last he was suffered to keep it with his men, and we sent them away in it to some other place.

CHAPTER VIII.

According to the resolutions we had taken, we set out, June the 1st, 1685, passing between Point Garachina and the King's Islands. The 3d we passed by the Island of Chuche, the last remainder of the isles in the Bay of Panama. In our passage to Quibo, Captain Branley lost his mainmast, therefore he and all his men left his bark, and came aboard Captain Davis's ship. Captain Swan also sprung his maintopmast, and got up another; but while he was doing it, and we were making the best of our way, we lost sight of him, and were now on the north side of the bay; for this way all ships must pass from Panama, whether bound towards the coast of Mexico or Peru. The 10th we passed by Morro de Puercos, or the Mountain of Hogs, why so called I know not; it is a high round hill on the coast of Lavelia. This side of the Bay of Panama runs out westerly to the Islands of Quibo; there are on this coast many rivers and creeks, but none so large as those on the south side of the bay. It is a coast that is partly mountainous, partly lowland, and very thick of woods bordering on the sea; but a few leagues within land it consists mostly of savannahs, which are stocked with bulls and cows. The rivers on this side are not wholly destitute of gold, though not so rich as the rivers on the other side of the bay. The coast is but thinly inhabited; for except the rivers that lead up to the towns of Nata and Lavelia I know of no other settlement between Panama and Puebla Nueva. The Spaniards may travel by land from Panama through all the kingdom of Mexico, as being full of savannahs; but towards the coast of Peru they cannot pass farther than the River Chepo, the land there being so full of thick woods, and watered with so many great rivers, besides less rivers and creeks, that the Indians themselves who inhabit there cannot travel far without much trouble.

We met with very wet weather in our voyage to Quibo, and with SSW. and sometimes SW. winds, which retarded our course. It was the 15th of June when we arrived at Quibo, and found there Captain Harris whom we sought. The Island of Quibo or Cabaya is in Lat. 7° 14′ N. of the Equator; it is about six or seven leagues long, and three or four broad. The land is low, except only near the NE. end; it is all over plentifully stored with great tall flourishing trees of many sorts, and there is good water on the E. and NE. sides of the island. Here are some deer, and plenty of pretty large black monkeys, whose flesh is sweet and wholesome; besides a few guanas and some snakes. I know no other sort of land animal on the island. There are many other islands, lying some on the SW. side, others on the N. and NE. sides, of this island; as the Island of Quicara, which is a pretty large island SW. of Quibo, and on the north of it is a small island called Rancheria, on which are plenty of Palma-Maria trees. The Palma-Maria is a tall, straight-bodied tree, with a small head, but very unlike the palm tree, notwithstanding the name. It is greatly esteemed for making masts, being very tough, as well as of a good length; for the grain of the wood runs not straight along it, but twisting gradually about it. These trees grow in many places of the West Indies, and are frequently used both by the English and Spaniards there for that use. The Islands of Canales and Cantarras are small islands lying on the NE. of Rancheria. These have all channels to pass between, and good anchoring about them, and they are as well stored with trees and water as Quibo. Captain Swan gave to several of these islands the names of those English merchants and gentlemen who were owners of his ship. [On June 16th, Captain Swan came to anchor by them, when they held a consultation as to how they might advance their fortunes, as the sea was promising them little. The result was a decision to take the town of Puebla Nueva, which they accordingly did with 150 men. On the 5th of July Captain Knight came to them, having captured two bark-logs laden with flour. After this, each ship's company began the making of canoes.]

Captain Davis made two very large canoes: one was thirty-six feet long, and five or six feet wide; the other thirty-two feet long, and near as wide as the other. In a month's time we finished our business, and were ready to sail. Here Captain Harris went to lay his ship aground to clean her, but she being old and rotten, fell in pieces; and therefore he and all his men went aboard of Captain Davis and Captain Swan. While we lay here we struck turtle every day, for they were now very plentiful; but from August to March there are not many. The 18th of July, John Rose, a Frenchman, and fourteen men more belonging to Captain Gronet, having made a new canoe, came in her to Captain Davis, and desired to serve under him; and Captain Davis accepted of them because they had a canoe of their own. The 20th of July we sailed from Quibo, bending our course for Realejo, which is the port for Leon, the city that we now designed to attempt. We were now 640 men, in eight sail of ships, commanded by Captain Davis, Captain Swan, Captain Townley, and Captain Knight; with a fireship, and three tenders, which last had not a constant crew. We passed out between the River Quibo and Rancheria, leaving Quibo and Quicaro on our larboard side, and Rancheria, with the rest of the islands, and the main, on the starboard side. The wind at first was at SSW. We coasted along shore, passing by the Gulf of Nicoya, the Gulf of Dulce, and the Island Cano. All this coast is low land, overgrown with thick woods; and there are but few inhabitants near the shore. The 8th of August, being in Lat. 11° 20′ by observation, we saw a high hill in the country, towering up like a sugar loaf, which bore NE. by N. We supposed it to be Volcano Viejo by the smoke which ascended from its top; therefore we steered in N., and made it plainer, and then knew it to be that volcano, which is the sea-mark for the harbour for Realejo. When we had brought this mountain to bear NE., we got out all our canoes, and provided to embark into them the next day.

The 9th in the morning, being about eight leagues from the shore, we left our ships under the charge of a few men, and 520 of us went away in thirty-one canoes, rowing towards the harbour of Realejo. We had fair weather and little wind till 2 o'clock in the afternoon; then we had a tornado from the shore, with much thunder, lightning, and rain, and such a gust of wind that we were all like to be foundered. In this extremity we put right afore the wind, every canoe's crew making what shift they could to avoid the threatening danger. The small canoes, being most light and buoyant, mounted nimbly over the surges; but the great heavy canoes lay like logs in the seas, ready to be swallowed by every foaming billow. Some of our canoes were half full of water, yet kept two men constantly heaving it out. The fierceness of the wind continued about half-an-hour, and abated by degrees; and as the wind died away, so the fury of the sea abated; for in all hot countries, as I have observed, the sea is soon raised by the wind, and as soon down again when the wind is gone: and therefore it is a proverb among the seamen, "Up wind, up sea: down wind, down sea." At 7 o'clock in the evening it was quite calm, and the sea as smooth as a millpond. Then we tugged to get into the shore, but finding we could not do it before day, we rowed off again, to keep ourselves out of sight. By the time it was day, we were five leagues from the land, which we thought was far enough off shore. Here we intended to lie till the evening; but at 3 o'clock in the afternoon we had another tornado more fierce than that which we had the day before. This put us in greater peril of our lives, but did not last so long. As soon as the violence of the tornado was over, we rowed in for the shore, and entered the harbour in the night. The creek which leads towards Leon lies on the SE. side of the harbour. Our pilot being very well acquainted here, carried us into the mouth of it, but could carry us no farther till day, because it is but a small creek, and there are other creeks like it. The next morning as soon as it was light, we rowed into the creek, which is very narrow; the land on both sides lying so low, that every tide it is overflown with the sea. This sort of land produces red mangrove trees, which are here so plentiful and thick that there is no passing through them. Beyond these mangroves, on the firm land close by the side of the river, the Spaniards have built a breastwork purposely to hinder an enemy from landing. When we came in sight of the breastwork we rowed as fast as we could to get ashore; the noise of our oars alarmed the Indians who were set to watch; and presently they ran away towards the city of Leon to give notice of our approach. We landed as soon as we could, and marched after them: 470 men were drawn out to march to the town, and I was left with fifty-nine men more to stay and guard the canoes till their return. [The city of Leon, twenty miles up the country, is here described as surrounded with long grassy savannahs, and clumps of high woods. It was thought at the time by some to be the pleasantest place in all America, and the paradise of the Indies. The wealth of the place consisted in the pastures, cattle, and plantations of sugar.]

Our men were now marching to Leon; they went from the canoes about 8 o'clock. Captain Townley with eighty of the briskest men marched before, Captain Swan with 100 men marched next, Captain Davis with 170 men marched next, and Captain Knight brought up the rear. Captain Townley, who was near two miles ahead of the rest, met about seventy horsemen four miles before he came to the city, but they never stood him. About 3 o'clock Captain Townley only with his eighty men entered the town, and was briskly charged in a broad street by 170 or 200 Spanish horsemen; but two or three of their leaders being knocked down, the rest fled. Their foot consisted of about 500 men, which were drawn up in the Parade; for the Spaniards in these parts make a large square in every town, though the town itself be small. This square is called the Parade; commonly the church makes one side of it, and the gentlemen's houses with their galleries about them another. But the foot also, seeing their horse retire, left an empty city to Captain Townley, beginning to save themselves by flight. Captain Swan came in about 4 o'clock, Captain Davis with his men about five, and Captain Knight, with as many men as he could encourage to march, came in about six, but he left many men tired on the road; these, as is usual, came dropping in one or two at a time, as they were able. The next morning the Spaniards killed one of our tired men. He was a stout old grey-headed man, aged about eighty-four, who had served under Oliver in the time of the Irish Rebellion; after which he was at Jamaica, and had followed privateering ever since. He would not accept of the offer our men made him to tarry ashore, but said he would venture as far as the best of them; and when surrounded by the Spaniards he refused to take quarter, but discharged his gun amongst them, keeping a pistol still charged; so they shot him dead at a distance. His name was Swan. He was a very merry, hearty old man, and always used to declare he would never take quarter. But they took Mr Smith, who was tired also; he was a merchant belonging to Captain Swan; and being carried before the Governor of Leon, was known by a Mulatto woman that waited on him. Mr Smith had lived many years in the Canaries, and could speak and write very good Spanish; and it was there this Mulatto woman remembered him. He, being examined how many men we were, said 1000 at the city and 500 at the canoes; which made well for us at the canoes, who straggling about every day might easily have been destroyed. But this so daunted the Governor, that he did never offer to molest our men, although he had with him above 1000 men, as Mr Smith guessed. He sent in a flag of truce about noon, pretending[1] to ransom the town rather than let it be burnt; but our captains demanded 300,000 pieces of eight for its ransom, and as much provision as would victual 1000 men four months, and Mr Smith to be ransomed for some of their prisoners; but the Spaniards did not intend to ransom the town, but only capitulated day after day to prolong time till they had got more men. Our captains therefore, considering the distance that they were from the canoes, resolved to be marching down. The 14th, in the morning, they ordered the city to be set on fire, which was presently done, and then they came away; but they took more time in coming down than in going up. The 15th, in the morning, the Spaniards sent in Mr Smith, and had a gentlewoman in exchange. Then our captains sent a letter to the Governor, to acquaint him that they intended next to visit Realejo, and desired to meet him there; they also released a gentleman on his promise of paying 150 beeves for his ransom, and to deliver them to us at Realejo; and the same day our men came to their canoes, where having stayed all night, the next morning we all entered our canoes, and came to the harbour of Realejo, and in the afternoon our ships came thither to an anchor.

The creek that leads to Realejo lies from the NW. part of the harbour, and runs in northerly. It is about two leagues from the island in the harbour's mouth to the town; two-thirds of the way it is broad, then you enter a narrow deep creek bordered on both sides with red mangrove trees, whose limbs reach almost from one side to the other. A mile from the mouth of the creek it turns away west. There the Spaniards have made a very strong breastwork, fronting towards the mouth of the creek, in which were placed 100 soldiers to hinder us from landing; and twenty yards below that breastwork there was a chain of great trees placed across the creek, so that ten men could have kept off 500 or 1000. When we came in sight of the breastwork we fired but two guns, and they all ran away; and we were afterwards near half-an-hour cutting the boom or chain. Here we landed, and marched to the town of Riolexo or Realejo,[2] which is about a mile from hence. This town stands on a plain by a small river. It is a pretty large town, with three churches and an hospital that hath a fine garden belonging to it, besides many large fair houses; they all stand at a good distance one from another, with yards about them. This is a very sickly place, and I believe hath need enough of an hospital, for it is seated so nigh the creeks and swamps that it is never free from a noisome smell. The land about it is a strong yellow clay, yet where the town stands it seems to be sand. Here are several sorts of

[1] Professing his desire.

[2] The chief trade of Realejo was in pitch, tar, and cordage, with the produce of their sugar-works and estantions or beef-farms. They stayed from the 17th to the 24th helping themselves to the produce of the country; when on leaving, some of the more mischievous of the crew set on fire the town, which they left burning.

fruits, as guavas,[1] pine-apples, melons, and prickly pears.[2] . . .

The 25th, Captain Davis and Captain Swan broke off consortships, for Captain Davis was minded to return again on the coast of Peru, but Captain Swan desired to go farther to the westward. I had till this time been with Captain Davis, but now left him and went aboard of Captain Swan. It was not from any dislike to my old Captain, but to get some knowledge of the northern parts of this continent of Mexico; and I knew that Captain Swan determined to coast it as far north as he thought convenient, and then pass over for the East Indies, which was a way very agreeable to my inclination. Captain Townley, with his two barks, was resolved to keep us company; but Captain Knight and Captain Harris followed Captain Davis. The 27th, in the morning, Captain Davis, with his ships, went out of the harbour, having a fresh land-wind. They were in company, Captain Davis's ship, with Captain Harris in her, Captain Davis's bark and fireship, and Captain Knight in his own ship, in all four sail. Captain Swan took his last farewell of him by firing fifteen guns, and he fired eleven in return of the civility. [While lying here, they were visited by a malignant fever, of which several of the men died. On the 3d September, all their prisoners and pilots were turned ashore, they being unacquainted with the coast farther westward. The same day they steered westward, meeting with a severe storm in the passage.]

We had kept at a good distance off shore, and saw no land till the 14th day; but then being in Lat. 12° 50′, the volcano of Guatemala appeared in sight. This is a very high mountain with two peaks or heads appearing like two sugar-loaves. It often belches forth flames of fire and smoke from between the two heads, and this, as the Spaniards do report, happens chiefly in tempestuous weather. It is called so from the city of Guatemala, which stands near the foot of it, about eight leagues from the South Sea,[3] and by report, forty or fifty leagues from the Gulf of Amatique in the Bay of Honduras in the North Seas. This city is famous for many rich commodities that are produced thereabouts, some almost peculiar to this country and yearly sent into Europe, especially four rich dyes—indigo, otta or anatta, silvester, and cochineal.

Indigo is made of an herb which grows a foot and a half or two feet high, full of small branches, and the branches full of leaves resembling the leaves which grow on flax, but more thick and substantial. They cut this herb or shrub and cast it into a large cistern made in the ground for that purpose, which is half full of water. The indigo stalk or herb remains in the water till all the leaves, and I think the skin, the rind or bark, rot off and in a manner dissolve; but if any of the leaves should stick fast, they force them off by much labour, tossing and tumbling the mass in the water till all the pulpy substance is dissolved. Then the shrub, or woody part, is taken out, and the water, which is like ink, being disturbed no more, settles, and the indigo falls to the bottom of the cistern like mud. When it is thus settled they draw off the water, and take the mud and lay it in the sun to dry, which there becomes hard as you see it brought

[1] The guava grows on a hard scrubbed shrub, is a fruit much like a pear, with a thin rind, and full of small seeds. It is one of the few West Indian fruits which may be eaten while still green.

[2] The prickly pear, according to Dampier, thrives best in barren sandy ground near the sea, the fruit being "as big as a large plum, small near the leaf, and big towards the top, where it opens like a medlar."

[3] The new city of Guatemala stands to the south-east of the old city about twenty-five miles, and only some sixteen miles from the sea. The old city was destroyed by an earthquake in 1776, but has been rebuilt.

home. Otta, or anatta,[1] is a red sort of dye. It is made of a red flower that grows on shrubs seven or eight feet high. It is thrown into a cistern of water as the indigo is, but with this difference, that there is no stalk, nor so much as the head of the flower, but only the flower itself pulled off from the head as you peel rose-leaves from the bud. This remains in the water till it rots, and by much jumbling it dissolves to a liquid substance like the indigo; and being settled, and the water drawn off, the red mud is made up into rolls or cakes and laid in the sun to dry.

Indigo is produced all over the West Indies, on most of the Caribbee Islands, as well as the main; yet no part of the main yields such great quantities, both of indigo and otta, as this country about Guatemala. I know not what quantities either of indigo or otta are made at Cuba or Hispaniola; but the place most used[2] by our Jamaica sloops for these things is the Island of Porto Rica, where our Jamaica traders did use to buy indigo for three reals and otta for four reals the pound, which is but 2s. 3d. of our money; and yet at the same time otta was worth in Jamaica 5s. the pound, and indigo 3s. 6d. the pound; and even this also paid in goods, by which means alone they got 50 or 60 per cent. Our traders had not then found the way of trading with the Spaniards in the Bay of Honduras; but Captain Coxon went thither (as I take it) at the beginning of the year 1679, under pretence to cut logwood, and went into the Gulf of Matique, which is in the bottom of that bay. There he landed with his canoes, and took a whole storehouse full of indigo and otta in chests, piled up in several parcels, and marked with different marks, ready to be shipped off aboard two ships that then lay in the road purposely to take it in; but these ships could not come at him, it being shoal water. He opened some of the chests of indigo, and, supposing the other chests to be all of the same species, ordered his men to carry them away. They immediately set to work, and took the nearest at hand; and having carried out one heap of chests, they seized on another great pile of a different mark from the rest, intending to carry them away next. But a Spanish gentleman, their prisoner, knowing that there was a great deal more than they could carry away, desired them to take only such as belonged to the merchants, whose marks he undertook to show them, and to spare such as had the same mark with those in that great pile they were then entering upon; because, he said, those chests belonged to the ship captains, who following the seas as themselves did, he hoped they would for that reason rather spare their goods than the merchants'. They consented to his request; but upon their opening their chests (which was not before they came to Jamaica, where by connivance they were permitted to sell them) they found that the Don had been too sharp for them; the few chests which they had taken of the same mark with the great pile proving to be otta, of greater value by far than the other; whereas they might as well have loaded the whole ship with otta as with indigo. The cochineal is an insect bred in a sort of fruit much like the prickly pear. The tree or shrub that bears it is like the prickly-pear tree, about five feet high, and so prickly, only the leaves are not quite so big, but the fruit is bigger.[3] . . .

[1] Otherwise called arnotto or annotto; it is obtained from the seeds of the tree *Bixa orellana*, and used, besides dyeing, for colouring cheese, butter, and liquors.

[2] Frequented.

[3] The gathering of the cochineal insect and the red grain called silvester by the Indians, is here minutely described. The cochineal is bred in the fruit of a shrub like a prickly-pear tree, from which it is dislodged, and on being exposed to the heat of the sun falls down dead, and is caught

When we first saw the mountain of Guatemala, we were by judgment twenty-five leagues' distance from it. As we came nearer the land it appeared higher and plainer, yet we saw no fire, but a little smoke proceeding from it. The land by the sea was of a good height, yet but low in comparison with that in the country. The sea for about eight or ten leagues from the shore was full of floating trees, or drift-wood, as it is called (of which I have seen a great deal, but nowhere so much as here), and pumice-stones floating, which probably are thrown out of the burning mountains, and washed down to the shore by the rains, which are very violent and frequent in this country; and on the side of Honduras it is excessively wet. The 24th, we were in Lat. 14° 30′ N., and the weather more settled. Then Captain Townley took with him 106 men in nine canoes, and went away to the westward, where he intended to land and rummage in the country for some refreshment for our sick men; we having at this time near half our men sick, and many were dead since we left Realejo. We in the ships lay still, with our topsails furled and our courses or lower sails hauled up, this day and the next, that Captain Townley might get the start of us. The 26th we made sail again, coasting to the westward, having the wind at N., and fair weather. We ran along by a tract of very high land, which came from the eastward more within land than we could see; after we fell in with it, it bare us company for about ten leagues, and ended with a pretty gentle descent towards the west. The 2d of October, Captain Townley came aboard; he had coasted along shore in his canoes, seeking for an entrance, but found none. At last, being out of hopes to find any bay, creek, or river into which he might safely enter, he put ashore on a sandy bay, but overset all his canoes; he had one man drowned, and several lost their arms, and some of them that had not waxed up their cartridge or cartouche boxes wet all their powder. Captain Townley with much ado got ashore, and dragged the canoes up dry on the bay; then every man searched his cartouche-box, and drew the wet powder out of his gun, and provided to march into the country; but finding it full of great creeks, which they could not ford, they were forced to return again to their canoes. In the night they made good fires to keep themselves warm; the next morning 200 Spaniards and Indians fell on them, but were immediately repulsed, and made greater speed back than they had done forward. Captain Townley followed them, but not far, for fear of his canoes. These men came from Tehuantepec, a town that Captain Townley went chiefly to seek, because the Spanish books made mention of a large river there, but whether it was run away at this time, or rather Captain Townley and his men were short-sighted, I know not; but they could not find it. Upon his return, we presently made sail, coasting still westward. We saw no opening nor sign of any place to land at, so we sailed about twenty leagues farther, and came to a small high island called Tangola, where there is good anchoring. The island is indifferently well furnished with wood and water, and lies about a league from the shore. The main against the island is pretty high champaign savannah land by the sea; but two or three leagues within land it is higher and very woody. We coasted a league farther and came to Huatulco. This port is in Lat. 15° 30′; it is one of the best in all this kingdom of Mexico. Near a mile from the mouth of the harbour, on the east side, there is a little island close by the shore; and on the west side of

in a cloth spread on the ground to receive them. These when dried yield the much-esteemed scarlet dye. The silvester is the seed of the cochineal fruit, which readily falls out of the fruit, on being shaken when ripe. The dye it yields is very similar to the cochineal, but not quite so valuable.

the mouth of the harbour there is a great hollow rock, which by the continual working of the sea in and out makes a great noise which may be heard a great way. Every surge that comes in forces the water out of a little hole on its top, as out of a pipe, from whence it flies out just like the blowing of a whale; to which the Spaniards also liken it. They call this rock and spout the *Buffadore*;[1] upon what account I know not. Even in the calmest seasons the sea beats in there, making the water spout out at the hole; so that this is always a good mark to find the harbour by. The harbour is about three miles deep, and one mile broad; it runs in NW.; and at the bottom of the harbour there is a fine brook of fresh water running into the sea. Here formerly stood a small Spanish town, or village, which was taken by Sir Francis Drake;[2] but now there is nothing remaining of it besides a little chapel standing among the trees about 200 paces from the sea. The land appears in small short ridges parallel to the shore, and to each other; the innermost still gradually higher than that nearer the shore; and they are all clothed with very high flourishing trees, that it is extraordinary pleasant and delightful to behold at a distance. I have nowhere seen anything like it.

At this place, Captain Swan, who had been very sick, came ashore, and all the sick men with him, and the surgeon to tend them. Captain Townley again took a company of men with him, and went into the country to seek for houses or inhabitants. He marched away to the eastward, and came to the River Capalita, which is a swift river, yet deep near the mouth, and is about a league from Huatulco. There two of his men swam over the river, and took three Indians that were placed there as sentinels to watch for our coming. These could none of them speak Spanish; yet our men by signs made them understand that they desired to know if there was any town or village near; who, by the signs which they made, gave our men to understand that they could guide them to a settlement; but there was no understanding by them whether it was a Spanish or Indian settlement, nor how far it was thither. They brought these Indians aboard with them; and the next day, which was the 6th of October, Captain Townley, with 140 men (of whom I was one) went ashore again, taking one of these Indians with us for a guide to conduct us to this settlement. Our men that stayed aboard filled our water, and cut wood, and mended our sails; and our Mosquito men struck three or four turtle every day. They were a small sort of turtle, and not very sweet, yet very well esteemed by us all, because we had eaten no flesh a great while. The 8th we returned out of the country, having been about fourteen miles directly within land before we came to any settlement. There we found a small Indian village, and in it a great quantity of vinelloes drying in the sun. The vinello is a little cod full of small black seeds; it is four or five inches long, about the bigness of the stem of a tobacco leaf, and when dried much resembling it; so that our Privateers at first have often thrown them away when they took any, wondering why the Spaniards should lay up tobacco stems. This cod grows on a small vine, which climbs about and supports itself by the neighbouring trees; it first bears a yellow flower, from whence the cod afterwards proceeds. It is first green, but when ripe it turns yellow; then the Indians, whose manufacture it is, and who sell it cheap to the Spaniards, gather it and lay it in the sun, which makes it soft; then it changes to a chestnut colour. Then they frequently press it between their fingers, which makes it flat. If the Indians

[1] The Roarer, or Snorter; from the Spanish "bufar," signifying to puff and blow with anger, to snort.

[2] It was here that he parted with Nuno da Silva, the Portuguese pilot, captured at the Cape Verd Islands.

do anything to them besides, I know not; but I have seen the Spaniards sleek them with oil. These vines grow plentifully at Boca del Toro, where I have gathered and tried to cure them, but could not, which makes me think that the Indians have some secret, that I know not of, to cure them. I have often asked the Spaniards how they were cured, but I never could meet with any could tell me. One Mr Cree, also, a very curious person, who spoke Spanish well, and had been a Privateer all his life, and seven years a prisoner among the Spaniards at Porto Bello and Carthagena, yet, upon all his inquiry, could not find any of them that understood it. Could we have learnt the art of it, several of us would have gone to Boca del Toro yearly, at the dry season, and cured them, and freighted our vessel. We there might have had turtle enough for food, and store of vinelloes. Mr Cree first showed me those at Boca del Toro. At or near a town, also, called Caihooca,[1] in the Bay of Campeachy, these cods are found. They are commonly sold for threepence a cod among the Spaniards in the West Indies, and are sold by the druggist, for they are much used among chocolate to perfume it. Some will use them among tobacco, for it gives a delicate scent. I never heard of any vinelloes but here in this country, about Caihooca, and at Boca del Toro. The Indians of this village could speak but little Spanish. They seemed to be a poor innocent people; and by them we understood that there are very few Spaniards in these parts; yet all the Indians hereabout are under them. The land from the sea to their houses is black earth, mixed with some stones and rocks; all the way full of very high trees.

The 10th we sent four canoes to the westward, who were ordered to lie for us at Port Angels; where we were in hopes that by some means or other they might get prisoners that might give us a better account of the country than at present we could have; and we followed them with our ships. All our men being now pretty well recovered of the fever, which had raged amongst us ever since we departed from Realejo.

CHAPTER IX.

It was the 12th of October 1685, when we set out of the harbour of Huatulco with our ships. We coasted along to the westward, keeping as near the shore as we could for the benefit of the land-winds, for the sea-winds were right against us; and we found a current setting to the eastward which kept us back, and obliged us to anchor at the Island of Sacrificios,[2] which is a small green island about half-a-mile long. It lies about a league to the W. of Huatulco, and about half-a-mile from the main. [On the 18th they sailed from Point or Island Sacrificios, keeping near the shore till they were abreast of Port Angels (Puerto de los Angelos), where they anchored.]

The 23d we landed about 100 men and marched thither, where we found plenty of fat bulls and cows feeding in the savannahs, and in the house good store of salt and maize, and some hogs, and cocks, and hens; but the owners or overseers were gone. We lay here two or three days, feasting on fresh provisions; but could not contrive to carry any quantity aboard, because the way was so long, and our men but weak, and a great

[1] Such names as Tangola, Capalita River, and Caihooca, will be hard to find on modern maps. Dampier places the first, an island, between Tehuantepec and Point Sacrificios; the second he makes run into the sea a little to the eastward of the point; and the third he lays down, as nearly as may be, at Cape Morillos, not far from the bottom of the Bay of Campeachy.

[2] Apparently off Point Sacrificios, which forms the western limit of the Gulf of Tehuantepec.

wide river to ford. Therefore we returned again from thence the 26th day, and brought every one a little beef or pork for the men that stayed aboard. The two nights that we stayed ashore at this place we heard great droves of jackals, as we supposed them to be, barking all night long not far from us. None of us saw these; but I do verily believe they were jackals, though I did never see those creatures in America, nor hear any but at this time. We could not think that there were less than thirty or forty in a company. We got aboard in the evening, but did not yet hear any news of our two canoes.

The 27th, in the morning, we sailed from hence, and in the evening we anchored in sixteen fathoms water by a small rocky island which lies about half-a-mile from the main and six leagues westward from Port Angels. The 28th, we sailed again with the land-wind; in the afternoon the sea breeze blew hard, and we sprung our maintopmast. This coast is full of hills and valleys, and a great sea falls in upon the shore. In the night we met with the other two of our canoes that went from us at Huatulco. They had been as far as Acapulco to seek Port Angels. Coming back from thence, they went into a river to get water, and were encountered by 150 Spaniards; yet they filled their water in spite of them, but had one man shot through the thigh. Afterwards they went into a lagoon, or lake of salt water, where they found much dried fish, and brought some aboard. We being now abreast of that place, sent in a canoe manned with twelve men for more fish. The mouth of this lagoon is not a pistol-shot wide, and on both sides are pretty high rocks, so conveniently placed by nature that many men may abscond behind; and within the rock the lagoon opens wide on both sides. The Spaniards being alarmed by our two canoes that had been two or three days before, came armed to this place to secure their fish; and seeing our canoe coming, they lay snug behind the rocks, and suffered the canoe to pass in, then they fired their volley and wounded five of our men. Our people were a little surprised at this sudden adventure, yet fired their guns and rowed farther into the lagoon, for they durst not adventure to come out again through the narrow entrance, which was near a quarter of a mile in length. Therefore they rowed into the middle of the lagoon, where they lay out of gun-shot, and looked about to see if there was not another passage to get out at broader than that by which they entered, but could see none. So they lay still two days and three nights in hopes that we should come to seek them, but we lay off at sea, about three leagues distant, waiting for their return, supposing by their long absence that they had made some greater discovery, and were gone farther than the fish range; because it is usual with Privateers, when they enter upon such designs, to search farther than they proposed if they meet any encouragement. But Captain Townley and his bark being near the shore heard some guns fired in the lagoon. So he manned his canoe and went towards the shore, and, beating the Spaniards away from the rocks, made a free passage for our men to come out of their pound, whereelse they must have been starved or knocked on the head by the Spaniards. They came aboard their ships again the 31st of October. From hence we made sail again, coasting to the westward. The 2d of November we passed by a rock called by the Spaniards the Algatross. The land hereabout is of an indifferent height and woody, and more within the country, mountainous. Here are seven or eight white cliffs by the sea, which are very remarkable, because there are none so white and so thick together on all the coast. They are five or six miles to the west of the Algatross Rock. Two leagues to the W. of these cliffs there is a pretty large river which forms a small island at its mouth. The channel on the east side is but shoal and sandy, but the west channel is deep enough for canoes to enter. On the banks of this channel the Spaniards have made a breastwork to hinder an enemy from landing and filling water.

The 3d, we anchored abreast of this river, in fourteen fathoms water, about a mile and a half off shore. The next morning we manned our canoes, and went ashore to the breastwork with little resistance, although there were about 200 men to keep us off. They fired twenty or thirty guns at us, but seeing we were resolved to land, they quitted the place. One chief reason why the Spaniards are so frequently routed by us, although many times much our superiors in numbers, and in many places fortified with breastworks, is their want of small firearms; for they have but few on all the sea-coasts, unless near their larger garrisons. Here we found a great deal of salt, brought hither as I judge to salt fish, which they take in the lagoons. The fish I observed here mostly were what we call snooks,[1] neither a sea-fish nor freshwater-fish, but very numerous in these salt lakes. This fish is about a foot long, and round, and as thick as the small of a man's leg, with a pretty long head; it has scales of a whitish colour, and is good meat. How the Spaniards take them I know not; for we never found any nets, hooks, or lines, neither yet any bark, boat, or canoe among them on all this coast; except the ship I shall mention at Acapulco. We marched two or three leagues into the country, and met with but one house, where we took a Mulatto prisoner, who informed us of a ship that was lately arrived at Acapulco; she came from Lima. Captain Townley wanting a good ship, thought now he had an opportunity of getting one, if he could persuade his men to venture with him into the harbour of Acapulco and fetch this Lima ship out. Therefore he immediately proposed it, and found not only all his own men willing to assist him, but many of Captain Swan's men also. Captain Swan opposed it, because, provision being scarce with us, he thought our time might be much better employed in first providing ourselves with food; and there was plenty of maize in the river where we now were, as we were informed by the same prisoner, who offered to conduct us to the place where it was. But neither the present necessity, nor Captain Swan's persuasion, availed anything, no, nor yet their own interest; for the great design we had then in hand was to lie and wait for a rich ship which comes to Acapulco every year richly laden from the Philippine Islands. But it was necessary we should be well stored with provisions, to enable us to cruise about and wait the time of her coming. However, Townley's party prevailing, we only filled our water here, and made ready to be gone. So the 5th, in the afternoon, we sailed again, coasting to the westward, towards Acapulco. The 7th, in the afternoon, being about twelve leagues from the shore, we saw the high land of Acapulco, which is very remarkable; for there is a round hill standing between other two hills, the westernmost of which is the biggest and highest, and has two hillocks like two paps on its top; the easternmost hill is higher and sharper than the middlemost. From the middle hill the land declines toward the sea, ending in a high round point. There is no land shaped like this on all the coast. In the evening Captain Townley went away from the ships with 140 men in twelve canoes, to try to get the Lima ship out of Acapulco harbour.

Acapulco is a pretty large town 17° N. of the Equator. It is the seaport for the city of Mexico on the west side of the continent; as La Vera Cruz, or San Juan D'Ulloa, in the Bay of Nova Hispania, is on the north side. This town is the only place of trade on all this coast; for there is little or no traffic by sea on all the north-west part of this vast kingdom, there being, as I have said, neither boats, barks, nor ships, that I could ever see, unless only what come hither from other parts, and some boats near the SE. end of California; as I guess by the intercourse between

[1] Or sea-pike; *Centropomus undecimalis.*

that and the main for pearl-fishing. The ships that trade hither are only three; two that constantly go once a-year between this and Manilla in Luzon, one of the Philippine Islands, and one ship more every year to and from Lima. This from Lima commonly arrives a little before Christmas; she brings them quicksilver, cacao, and pieces of eight. Here she stays till the Manilla ship arrives, and takes in a cargo of spices, silks, calicoes, muslins, and other East India commodities, for the use of Peru, and then returns to Lima. This is but a small vessel of twenty guns; but the two Manilla ships are each said to be above 1000 tons. These make their voyages alternately, so that one or other of them is always at Manilla. When either of them sets out from Acapulco, it is at the latter end of March or the beginning of April; she always touches to refresh at Guam, one of the Ladrone Islands, in about sixty days' space after she sets out. There she stays but two or three days, and then prosecutes her voyage to Manilla, where she commonly arrives some time in June. By that time the other is ready to sail from thence laden with East India Commodities. She stretches away to the north as far as 36°, or sometimes 40° N. Lat., before she gets a wind to stand over to the American shore. She falls in first with the coast of California, and then coasts along the shore to the south again, and never misses a wind to bring her away from thence quite to Acapulco. When she gets the length of Cape San Lucas, which is the southernmost point of California, she stretches over to Cape Corrientes, which is in about 20° N. Lat.; from thence she coasts along till she comes to Solagua,[1] and there she sets ashore passengers that are bound to the city of Mexico. From thence she makes her best way, coasting still along shore, till she arrives at Acapulco, which is commonly about Christmas, never more than eight or ten days before or after. Upon the return of this ship to Manilla, the other, which stays there till her arrival takes her turn back to Acapulco. Sir John Narborough therefore was imposed on by the Spaniards who told him that there were six sail or more that used this trade. The port of Acapulco is very commodious for the reception of ships, and so large, that some hundreds may safely ride there without damnifying[2] each other. The harbour runs in north about three miles; then growing very narrow it turns short about to the west, and runs about a mile farther, where it ends. The town stands on the NW. side, at the mouth of this narrow passage, close by the sea; and at the end of the town there is a platform with a great many guns. Opposite to the town, on the east side, stands a high strong castle, said to have forty guns of a very great bore. Ships commonly ride near the bottom of the harbour, under the command both of the castle and the platform. Captain Townley, who, as I said before, with 140 men, left our ships on a design to fetch the Lima ship out of the harbour, had not rowed above three or four leagues before the voyage was like to end with all their lives; for on a sudden they were encountered with a violent tornado from the shore, which had like to have foundered all the canoes: but they escaped that danger, and the second night got safe into Port Marques. Port Marques is a very good harbour a league to the east of Acapulco harbour. Here they stayed all the next day to dry themselves, their clothes, their arms and ammunition; and the next night they rowed softly into Acapulco harbour: and because they would not be heard, they hauled in their oars and paddled as softly as if they had been seeking manatee.

[1] Apparently the Bay of Manzanilla, which is directly west of Mexico, and answers to the position Dampier's map assigns to "Sallagua" or "Solagua." Dampier's subsequent account of the place agrees with the situation and features of Manzanilla.

[2] Incommoding or injuring.

They paddled close to the castle; then struck over to the town, and found the ship riding between the breastwork and the fort, within about 100 yards of each. When they had well viewed her, and considered the danger of the design, they thought it not possible to accomplish it; therefore they paddled softly back again till they were out of command of the forts, and then they went to land, and fell in among a company of Spanish soldiers (for the Spaniards, having seen them the day before, had set guards along the coasts), who immediately fired at them, but did them no damage, only made them retire farther from the shore. They lay afterwards at the mouth of the harbour till it was day, to take a view of the town and castle; and then returned aboard again, being tired, hungry, and sorry for their disappointment. [On the 13th, they made sail farther westward, where the coast is described as low, producing many trees, and the spreading palm in great plenty.]

The land in the country is full of small peaked barren hills, making as many little valleys, which appear flourishing and green. At the W. end of this bay is the hill of Petatlan.[1] We came to an anchor on the NW. side of the hill and went ashore, about 170 men of us, and marched into the country twelve or fourteen miles. There we came to a poor Indian village that did not afford us a meal of victuals. The people all fled, only a Mulatto woman and three or four small children, who were taken and brought aboard. She told us that a carrier (one who drives a caravan of mules) was going to Acapulco laden with flour and other goods, but stopped in the road for fear of us a little to the west of this village (for he had heard of our being on this coast), and she thought he still remained there: and therefore it was we kept the woman to be our guide to carry us to that place. At this place where we now lay, our Mosquito men struck some small turtle and many small Jew-fish.

We went from hence with our ships on the 13th, and steered westward about two leagues farther, to a place called Chequetan.[2] The 14th, in the morning we went with ninety-five men in six canoes to seek for the carrier, taking the Mulatto woman for our guide; but Captain Townley would not go with us. Before day we landed at a place called Istapa, a league to the west of Chequetan. The woman was well acquainted here, having been often at this place for mussels, as she told us, for here are great plenty of them; they seem in all respects like our English mussels. She carried us through the pathless wood by the side of a river for about a league; then we came into a savannah full of bulls and cows; and here the carrier before mentioned was lying at the estantion-house with his mules, not having dared to advance all this while, as not knowing where we lay; so his own fear made him, his mules, and all his goods become a prey to us. He had forty packs of flour, some chocolate, a great many small cheeses, and abundance of earthenware. The eatables we brought away, but the earthen vessels we had no occasion for, and therefore left them. The mules were about sixty; we brought our prize with them to the shore, and so turned them away. Here we also killed some cows, and brought [them] with us to our canoes. In the afternoon our ships came to an anchor half-a-mile from the place where we landed, and then we went aboard. Captain Townley, seeing our good success, went ashore with his men to kill some cows, for there were no inhabitants near to oppose us. The land is very woody, of a good fertile soil, watered with many small rivers, yet it hath but few inhabitants near the sea. Captain Townley killed eighteen beeves, and after he came aboard, our men, con-

[1] Morro de Petatlan.

[2] Probably Siguantanejo, a town of some importance corresponding in position to the indications in the text.

trary to Captain Swan's inclination, gave Captain Townley part of the flour which we took ashore. Afterwards we gave the woman some clothes for her and her children, and put her and two of them ashore; but one of them, a very pretty boy about seven or eight years old, Captain Swan kept. The woman cried and begged hard to have him; but Captain Swan would not, but promised to make much of him, and was as good as his word. He proved afterwards a very fine boy for wit, courage, and dexterity; I have often wondered at his expressions and actions.

The 21st, in the evening, we sailed hence with the land-wind, and coasted along to the westward. The land is high, and full of ragged hills; and west from these ragged hills the land makes many pleasant and fruitful valleys among the mountains. The 25th we were abreast of a very remarkable hill, which, towering above the rest of its fellows, is divided in the top, and makes two small parts. The Spaniards make mention of a town called Thelupan[1] near this hill, which we would have visited if we could have found the way to it. The 26th, Captain Swan and Captain Townley, with 200 men, of whom I was one, went in our canoes to seek for the city of Colima, a rich place by report, but how far within land I could never learn; for, as I said before, here is no trade by sea, and therefore we could never get guides to inform us or conduct us to any town but one or two on this coast; and there is never a town that lies open to the sea but Acapulco; and therefore our search was commonly fruitless, as now, for we rowed above twenty leagues along shore and found it a very bad coast to land; we saw no house nor sign of inhabitants, although we passed by a fine valley called the Valley of Maguella.[2] Only at two places, the one at our first setting out on this expedition, and the other at the end of it, we saw a horseman set, as we supposed, as a sentinel to watch us. At both places we landed with difficulty, and at each place we followed the tracks of the horse on the sandy bay, but where they entered the woods we lost the track; and although we diligently searched for it, yet we could find it no more, so we were perfectly at a loss to find out the houses or town they came from. The 28th, being tired and hopeless to find any town, we went aboard our ships, that were now come abreast of the place where we were; for always when we leave our ships, we either order a certain place of meeting, or else leave them a sign to know where we are by making one or more great smokes. After we came aboard we saw the Volcano of Colima. This is a very high mountain, in about 18° 36′ N., standing five or six leagues from the sea, in the midst of a pleasant valley. It appears with two sharp peaks, from each of which there always issue flames of fire or smoke. The valley in which this volcano stands is called the Valley of Colima, from the town itself, which stands there not far from the volcano. The town is said to be great and rich, the chief of all its neighbourhood; and the valley in which it is seated, by the relation which the Spaniards give of it, is the most pleasant and fruitful valley in all the kingdom of Mexico. This valley is about ten or twelve leagues wide by the sea, where it makes a small bay; but how far the vale runs into the country, I know not. The 29th, our captains went away from our ships with 200 men, intending at the first convenient place to land and search about for a

---

[1] The hills and town appear to correspond to the town of Texupan, and the twin eminences near it called Cabo de Tetas, or the Paps of Texupan, at the mouth of a river of the same name. Telupan is at some distance to the south-eastward.

[2] The valley through which the River Almeria that passes Colima enters the sea has near the coast a little town named Olola; not unlike in sound to the word in the text.

path; for the Spanish books made mention of two or three other towns hereabouts, especially one called Solagua, to the west of this bay. Our canoes rowed along as near the shore as they could; but the sea went so high that they could not land. About ten or eleven o'clock two horsemen came near the shore, and one of them took a bottle out of his pocket, and drank to our men; while he was drinking, one of our men snatched up his gun and let drive at him, and killed his horse; so his consort immediately set spurs to his horse and rode away, leaving the other to come after afoot. But he being booted made but slow haste; therefore two of our men stripped themselves and swam ashore to take him; but he had a macheat, or long knife, wherewith he kept them both from seizing him, they having nothing in their hands wherewith to defend themselves or offend him. The 30th, our men came all aboard again, for they could not find any place to land in.

The 1st of December we passed by the port of Solagua. This port is in Lat. 18° 52′. It is only a pretty deep bay, divided in the middle with a rocky point, which makes, as it were, two harbours. Here we saw a great new thatched house, and a great many Spaniards, both horse and foot, with drums beating and colours flying, in defiance of us as we thought. We took no notice of them till the next morning, and then we landed about 200 men to try their courage; but they presently withdrew. The foot never stayed to exchange one shot; but the horsemen stayed till two or three were knocked down, and then they drew off, our men pursuing them. At last two of our men took two horses that had lost their riders, and mounting them rode after the Spaniards full drive till they came among them, thinking to have taken a prisoner for intelligence, but had like to have been taken themselves; for four Spaniards surrounded them, after they had discharged their pistols, and unhorsed them; and if some of our best footmen had not come to their rescue, they must have yielded or have been killed. They were both cut in two or three places, but their wounds were not mortal. The four Spaniards got away before our men could hurt them, and mounting their horses speeded after their consorts, who were marched away into the country. Our men, finding a broad road leading into the country, followed it about four leagues in a dry stony country, full of short woods; but finding no signs of inhabitants they returned again. In their way back they took two Mulattoes who were not able to march as fast as their consorts, therefore they had skulked in the woods, and by that means thought to have escaped our men. These prisoners informed us that this great road did lead to a great city called Oarrha,[1] from whence many of those horsemen before spoken of came; that this city was distant from hence as far as a horse will go in four days, and that there is no place of consequence nearer; that the country is very poor and thinly inhabited. They said, also, that these men came to assist the Philippine ship, that was every day expected here, to put ashore passengers for Mexico.

We now intended to cruise off Cape Corrientes to wait for the Philippine ship; so the 6th of December we set sail, coasting to the westwards, towards Cape Corrientes. We had fair weather, and but little wind. Here I was taken sick of a fever and ague, that afterwards turned to a dropsy, which I laboured under a long time after; and many of our men died of this distemper, though our surgeons used their greatest skill to preserve their lives. The dropsy is a general distemper on this coast, and the natives say, that the best remedy they can find for it is the stone or cod of an alligator (of which they have four, one near each leg, within the flesh) pulverised and drunk in

[1] Guadalajara, about 160 miles inland.

water. This receipt we also found mentioned in an almanac made at Mexico; I would have tried it, but we found no alligators here, though there are several. There are many good harbours between Solagua and Cape Corrientes; but we passed by them all. As we drew near the Cape, the land by the sea appeared of an indifferent height, full of white cliffs; but in the country, the land is high and barren, and full of sharp peaked hills, unpleasant to the sight. To the west of this ragged land is a chain of mountains running parallel with the shore; they end on the west with a gentle descent, but on the east side they keep their height, ending with a high steep mountain which hath three small sharp peaked tops, somewhat resembling a crown, and therefore called by the Spaniards Coronada, the Crown land. The 11th we were fair in sight of Cape Corrientes; it bore N. by W., and the Crown land bore N. The cape is of an indifferent height, with steep rocks to the sea. It is flat and even on the top, clothed with woods; the land in the country is high and doubled. This cape lies in 20° 28′ N. Here we had resolved to cruise for the Philippine ship, because she always makes this cape in her voyage homeward. We were, as I have said, four ships in company: Captain Swan and his tender, Captain Townley and his tender. It was so ordered that Captain Swan should lie eight or ten leagues off shore, and the rest about a league distant from each other, between him and the cape, that so we might not miss the Philippine ship; but we wanted provision, and therefore we sent Captain Townley's bark, with fifty or sixty men to the west of the cape, to search about for some town or plantations where we might get provision of any sort; the rest of us in the meantime cruising in our stations. The 17th the bark came to us again, but had got nothing; for they could not get about the cape, because the wind on this coast is commonly between the NW. and SW., which makes it very difficult getting to the westward; but they left four canoes with forty-six men at the cape, who resolved to row to the westward. The 18th we sailed to the Keys of Chametly[1] to fill our water. These keys or islands of Chametly are about sixteen or eighteen leagues to the eastward of Cape Corrientes. They are small, low, and woody, environed with rocks; there are five of them, lying in the form of a half moon, not a mile from the shore; and between them and the main is very good riding, secure from any wind. The Spaniards report that here live fishermen to fish for the inhabitants of the city of Purificacion. This is said to be a large town, the best hereabouts; but it is fourteen leagues up in the country. The 20th we entered within these islands, passing in on the SE. side, and anchored between the islands and the main in five fathoms clean sand. Here we found good fresh water and wood, and caught plenty of rock-fish with hook and line, a sort of fish I described at the Isle of Juan Fernandez; but we saw no sign of inhabitants besides three or four old huts, therefore I believe that the Spanish or Indian fishermen come hither only at Lent, or some other such season, but that they do not live here constantly. The 21st, Captain Townley went away with about sixty men to take an Indian village seven or eight leagues from hence to the westward, more towards the cape; and the next day we went to cruise off the cape, where Captain Townley was to meet us.

The 24th, as we were cruising off the cape, the four canoes before mentioned, which Captain Townley's bark left at the Cape, came off to us. They, after the bark left them, passed to the west of the cape, and rowed into the valley of Valderas,[2] or perhaps Val d'Iris; for it signifies the

---

[1] The locality of these islands corresponds with that of the Puerto de Tamatlan, a small bay due west from the city of La Purificacion.

[2] The Bay of Banderas, to the

Valley of Flags. This valley lies in the bottom of a pretty deep bay, that runs in between Cape Corrientes on the SE., and the Point of Pontique[1] on the NW., which two places are about ten leagues asunder. The valley is about three leagues wide; there is a level sandy bay against the sea, and good smooth landing. In the midst of the bay is a fine river, whereinto boats may enter. When our canoes came to this pleasant valley, they landed thirty-seven men, and marched into the country seeking for some houses. They had not gone past three miles before they were attacked by 150 Spaniards, horse and foot. There was a small thin wood close by them, into which our men retreated to secure themselves from the fury of the horse; yet the Spaniards rode in among them and attacked them very furiously, till the Spanish captain and seventeen more tumbled dead off their horses; then the rest retreated, being many of them wounded. We lost four men, and had two desperately wounded. In this action the foot, who were armed with lances and swords, and were the greatest number, never made any attack; the horsemen had each a brace of pistols, and some short guns. If the foot had came in, they had certainly destroyed all our men. When the skirmish was over, our men placed the two wounded men on horses, and came to their canoes. There they killed one of the horses, and dressed it, being afraid to venture into the savannah to kill a bullock, of which there was store. When they had eaten and satisfied themselves, they returned aboard. The 25th, being Christmas, we cruised in pretty near the cape, and sent in three canoes with the strikers to get fish, being desirous to have a Christmas dinner. In the afternoon they returned aboard with three great Jew-fish, which feasted us all; and the next day we sent ashore our canoes again and got three or four more. Captain Townley, who went from us at Chametly, came aboard the 28th, and brought about forty bushels of maize. He had landed to the eastward of Cape Corrientes, and marched to an Indian village that is four or five leagues in the country. The Indians, seeing him coming, set two houses on fire that were full of maize and ran away. Yet he and his men got in other houses as much as they could bring down on their backs, which he brought aboard.

We cruised off the Cape till the 1st of January 1686, and then made towards the valley of Valderas to hunt for beef; and before night we anchored in the bottom of the bay in sixty fathoms water, a mile from the shore. Here we stayed hunting till the 7th, and Captain Swan and Captain Townley went ashore every morning with about 240 men, and marched to a small hill, where they remained with fifty or sixty men to watch the Spaniards, who appeared in great companies on other hills not far distant, but did never attempt anything against our men. Here we killed and salted above two months' meat, besides what we spent fresh; and might have killed as much more if we had been better stored with salt. Our hopes of meeting the Philippine ship were now over, for we did all conclude that while we were necessitated to hunt here for provisions she was past by to the eastwards, as indeed she was, as we did understand afterwards by prisoners. So this design failed through Captain Townley's eagerness after the Lima ship, which he attempted in Acapulco harbour, as I have related. For though we took a little flour hard by, yet the same guide which told us of that ship would have conducted us where we might have had store of beef and maize; but instead thereof, we lost both our time and the opportunity of providing ourselves, and so were forced to be victualling when we should have been cruising off Cape Corrientes in expectation of the Manilla ship. Hitherto

north of Cape Corrientes, seems to be here intended.

[1] Now called Point of Mita.

we had coasted along here with two different designs; the one was to get the Manilla ship, which would have enriched us beyond measure, and this Captain Townley was most for. Sir Thomas Cavendish formerly took the Manilla ship off Cape San Lucas in California (where we also would have waited for her had we been early enough stored with provisions to have met her there), and threw much rich goods overboard. The other design, which Captain Swan and our crew were most for, was to search along the coast for rich towns, and mines chiefly of gold and silver, which we were assured were in this country, and, we hoped, near the shore; not knowing (as we afterwards found) that it was in effect an inland country, its wealth remote from the South Sea coast, and having little or no commerce with it, its trade being driven eastward with Europe by La Vera Cruz. Yet we had still some expectation of mines, and so resolved to steer on farther northward. But Captain Townley, who had no other design in coming on this coast but to meet this ship, resolved to return again towards the coast of Peru. So here we parted, he to the eastward and we to the westward, intending to search as far to the westward as the Spaniards were settled.

It was the 7th of January in the morning when we sailed from this pleasant valley. Before night we passed by Point Pontique; it is high, round, rocky, and barren; at a distance it appears like an island. A league to the W. of this point are two small barren islands called the Islands of Pontique.[1] There are several high, sharp, white rocks that lie scattering about them; we passed between these rocky islands on the left and the main on the right, for there is no danger. The 14th we had sight of a small white rock which appears very much like a ship under sail. This rock is in Lat. 21° 15'; it is three leagues from the main. At night we anchored in six fathoms water, near a league from the main, in good oozy ground. We caught a great many cat-fish here, and at several places on this coast both before and after this. From this island the land runs more northerly, making a fair sandy bay; but the sea falls in with such violence on the shore that there is no landing. We came to an anchor every evening, and in the mornings we sailed off with the land-wind.

The 20th we anchored about three miles on the east side of the Islands of Chametly,[2] different from those of that name before mentioned; for these are six small islands in Lat. 23° 11', a little to the south of the Tropic of Cancer and about three leagues from the main, where a salt lake has its outlet into the sea. These isles are of an indifferent height; some of them have a few shrubby bushes, the rest are bare of any sort of wood. There is a sort of fruit growing on these islands, called penguins, and it is all the fruit they have. The penguin fruit is of two sorts, the yellow and the red. The yellow penguin grows on a green stem as big as a man's arm above a foot high from the ground. The leaves of this stalk are half a foot long and an inch broad; the edges full of sharp prickles. The fruit grows at the head of the stalk in two or three great clusters, sixteen or twenty in a cluster. The fruit is as big as a pullet's egg, of a round form, and in colour yellow. It has a thick skin or rind, and the inside is full of small black seeds mixed among the fruit. The red penguin is of the bigness and colour of a small dry onion, and in shape much like a ninepin, for it grows not on a stalk or stem as the other, but one end on the ground, the other standing upright. There are some guanas on these islands, but no

[1] Two or three small islands, including Corvetena and Marieta, are marked in modern maps in a corresponding situation off Point Mita.

[2] Properly the Islands of Mazatlan, the name in the text being erroneously taken from a town called Chamatla, about forty miles south of Mazatlan.

other sort of land animal. The bays about the islands are sometimes visited by seal.

Captain Swan went away from hence with 100 men in our canoes to the northward to seek for the River Culiacan, possibly the same with the River of Piastla,[1] which some maps lay down in the province or region of Culiacan. This river lies in about 24° N. Lat. We were informed that there is a fair and rich Spanish town seated on the east side of it, with savannahs about it full of bulls and cows, and that the inhabitants of this town pass over in boats to the Island of California, where they fish for pearl. I have been told since by a Spaniard that said he had been at the Island California,[2] that there are great plenty of pearl oysters there, and that the native Indians of California near the pearl fishery are mortal enemies to the Spaniards. Our canoes were absent three or four days, and said they had been above thirty leagues but found no river; that the land by the sea was low and all sandy bay, but such a great sea that there was no landing. They met us in their return in Lat. 23° 30′, coasting along shore after them towards Culiacan, so we returned again to the eastward. This was the farthest that I was to the north on this coast. Six or seven leagues NNW. from the Isles of Chametly there is a small narrow entrance into a lake which runs about twelve leagues easterly, parallel with the shore, making many small low mangrove islands. The mouth of this lake is in Lat. about 23° 30′. It is called by the Spaniards Rio de Sal, for it is a salt lake. There is water enough for boats and canoes to enter, and smooth landing after you are in. On the west side of it there is a house and an estantion or farm of large cattle. Our men went into the lake and landed, and coming to the house, found seven or eight bushels of maize, but the cattle were driven away by the Spaniards; yet there our men took the owner of the estantion and brought him aboard. He said that the beeves were driven a great way into the country for fear we should kill them. While we lay here, Captain Swan went into this lake again, and landed 150 men on the NE. side, and marched into the country. About a mile from the landing-place, as they were entering a dry Salina or salt-pond, they fired at two Indians that crossed the way before them. One of them, being wounded in the thigh, fell down; and being examined, he told our men that there was an Indian town four or five leagues off, and that the way which they were going would bring them thither. While they were in discourse with the Indian, they were attacked by 100 Spanish horsemen, who came with a design to scare them back, but wanted both arms and hearts to do it. Our men passed on from thence, and in their way marched through a savannah of long dry grass. This the Spaniards set on fire, thinking to burn them; but that did not hinder our men from marching forward, though it did trouble them a little. They rambled for want of guides all this day and part of the next before they came to the town the Indian spoke of. There they found a company of Spaniards and Indians, who made head against them, but were driven out of the town after a short dispute. Here our Surgeon and one man more were wounded with arrows, but none of the rest were hurt. When they came into the town they found two or three Indians wounded, who told them that the name of the town was Mazatlan; that there were a few Spaniards living in it, and the rest were Indians; that

[1] The Culiacan and the Piastla are two distinct streams, the former being the more northerly.

[2] In the maps accompanying the second edition of Dampier's Voyage (London, 1697), from which the text is printed, California is more than once laid down as an island, though the author, near the end of the present Chapter, mentions some later Spanish maps in which it is made "to join to the main."

five leagues from this town there were two rich gold mines where the Spaniards of Compostella,[1] which is the chief town in these parts, kept many slaves and Indians at work for gold. Here our men lay that night, and the next morning packed up all the maize that they could find, and brought it on their backs to the canoes, and came aboard.

We lay here till the 2d of February, and then Captain Swan went away with about eighty men to the River Rosario,[2] where they landed and marched to an Indian town of the same name. They found it about nine miles from the sea; the way to it fair and even. This was a fine little town, of about sixty or seventy houses, with a fair church; and it was chiefly inhabited by Indians. They took prisoners there who told them that the River Rosario is rich in gold, and that the mines are not above two leagues from the town. Captain Swan did not think it convenient to go to the mines, but made haste aboard with the maize which he took there, to the quantity of about eighty or ninety bushels, which to us, in the scarcity we were in of provisions, was at that time more valuable than all the gold in the world; and had he gone to the mines the Spaniards would probably have destroyed the corn before his return. The 3d of February we went with our ships also towards the River Rosario, and anchored the next day against the river's mouth. The 7th, Captain Swan came aboard with the maize which he got. This was but a small quantity for so many men as we were, especially considering the place we were in, being strangers and having no pilots to direct or guide us into any river; and we being without all sort of provisions but what we were forced to get in this manner from the shore. The 8th, Captain Swan sent about forty men to seek for the River Olita, which is to the eastward of the River Rosario. The next day we followed after with the ships. In the afternoon our canoes came again to us, for they could not find the River Olita; therefore we designed next for the River Santiago, to the eastward still. The 11th, in the evening, we anchored against the mouth of the river, about two miles from the shore. It is one of the principal rivers on this coast. The mouth of this river is near half-a-mile broad, and very smooth entering. Within the mouth it is broader; for three or four rivers more meet there and issue all out together. The 11th, Captain Swan sent seventy men in four canoes into this river, to seek a town; for although we had no intelligence of any, yet, the country appearing very promising, we did not question but they would find inhabitants before they returned. They spent two days in rowing up and down the creeks and rivers; at last they came to a large field of maize, which was almost ripe: they immediately fell to gathering as fast as they could, and intended to lade the canoes; but seeing an Indian that was set to watch the corn, they quitted that troublesome and tedious work and seized him and brought him aboard, in hopes by his information to have some more easy and expeditious way of a supply by finding corn ready cut and dried. He being examined said that there was a town called Santa Pecaque[3] four leagues from the place where he was taken; and that if we designed to go thither he would undertake to be our guide. Captain Swan immediately ordered his men to make ready, and the same evening went away with eight canoes and 140 men, taking the Indian for their guide.

[1] Nueva Compostella, a city built by Nunez de Guzman, once the see of a bishop, now removed to Guadalaxara, along with the importance, if not the very existence, of the place.

[2] At no great distance east of Mazatlan; several miles up the river is the town of Asilo de Rosario.

[3] Marked on Dampier's map a little way up the left or south bank of the River Santiago, but not traceable in modern maps.

He rowed about five leagues up the river, and landed the next morning. The river at this place was not above pistol-shot wide, the banks pretty high on each side, and the land plain and even. He left twenty-three men to guard the canoes, and marched with the rest to the town. He set out from the canoes at 6 o'clock in the morning, and reached the town by ten. The way through which he passed was very plain, part of it woodland, part savannahs. The savannahs were full of horses, bulls, and cows. The Spaniards seeing him coming ran all away; so he entered the town without the least opposition. This town of Santa Pecaque stands on a plain in a savannah by the side of a wood, with many fruit trees about it. It is but a small town, but very regular, after the Spanish mode, with a parade in the midst. The houses fronting the parade had all balconies; there were two churches, one against the parade, the other at the end of the town. It is inhabited mostly by Spaniards. Their chief occupation is husbandry. There are also some carriers, who are employed by the merchants of Compostella to trade for them to and from the mines. Compostella is a rich town about twenty-one leagues from hence. It is the chief in all this part of the kingdom, and is reported to have seventy White families; which is a great matter in these parts, for it may be that such a town has no less than 500 families of copper-coloured people besides the Whites. The silver mines are about five or six leagues from Santa Pecaque; there, as we were told, the inhabitants of Compostella had some hundreds of slaves at work. The silver here and all over the kingdom of Mexico is said to be finer and richer in proportion than that of Potosi or Peru, though the ore be not so abundant; and the carriers of this town of Santa Pecaque carry the ore to Compostella, where it is refined. These carriers or sutlers also furnish the slaves at the mines with maize, whereof there was great plenty now in the town, designed for that use; there was also sugar, salt, and salt-fish.

Captain Swan's only business at Santa Pecaque was to get provision; therefore he ordered his men to divide themselves into two parts, and by turns carry down the provision to the canoes; one half remaining in the town to secure what they had taken, while the other half were going and coming. In the afternoon they caught some horses; and the next morning, being the 17th, fifty-seven men and some horses went laden with maize to the canoes. They found them, and the men left to guard them, in good order, though the Spaniards had given them a small diversion, and wounded one man; but our men of the canoes landed and drove them away. These that came loaded to the canoes left seven men more there, so that now there were thirty men to guard the canoes. At night the others returned; and the 18th, in the morning, that half which stayed the day before at the town took their turn of going, with every man his burthen, and twenty-four horses laden. Before they returned, Captain Swan and his other men at the town caught a prisoner, who said that there were near 1000 men of all colours, Spaniards and Indians, Negroes and Mulattoes, in arms at a place called Santiago, but three leagues off, the chief town on this river; that the Spaniards were armed with guns and pistols, and the copper-coloured with swords and lances. Captain Swan, fearing the ill consequence of separating his small company, was resolved the next day to march away with the whole party; and therefore he ordered his men to catch as many horses as they could, that they might carry the more provision with them. Accordingly, the next day, being the 19th of February 1686, Captain Swan called out his men betimes to be gone; but they refused to go, and said that they would not leave the town till all the provision was in the canoes; therefore he was forced to yield to them, and suffered half the company to go as before. They had now fifty-four horses laden, which Captain Swan ordered to be tied one to another, and the men to go in two bodies, twenty-five before and as many behind; but the men would go at their own rate, every man leading his horse. The Spaniards observed their manner of marching, and laid an ambush about a mile from the town, which they managed with such success, that falling on our body of men who were guarding the corn to the canoes, they killed them every one. Captain Swan hearing the report of their guns, ordered his men who were then in the town with him to march out to their assistance; but some opposed him, despising their enemies; till two of the Spaniards' horses that had lost their riders came galloping into the town in a great fright, both bridled and saddled, with each a pair of holsters by their sides, and one had a carbine newly discharged; which was an apparent token that our men had been engaged, and that by men better armed than they imagined they should meet with. Therefore Captain Swan immediately marched out of the town, and his men all followed him; and when he came to the place where the engagement had been, he saw all his men that went out in the morning lying dead. They were stripped, and so cut and mangled that he scarce knew one man. Captain Swan had not more men then with him than those were who lay dead before him; yet the Spaniards never came to oppose him, but kept at a great distance; for it is probable the Spaniards had not cut off so many men of ours, but with the loss of a great many of their own. So he marched down to the canoes, and came aboard the ship with the maize that was already in the canoes. We had about fifty men killed.

This loss discouraged us from attempting anything more hereabouts. Therefore Captain Swan proposed to go to Cape San Lucas, on California, to careen. He had two reasons for this: first, that he thought he could lie there secure from the Spaniards; and next, that if he could get a commerce with the Indians there, he might make a discovery in the Lake of California, and by their assistance try for some of the plate of New Mexico. This Lake of California (for so the sea, channel, or strait, between that and the continent is called) is but little known to the Spaniards, by what I could ever learn; for their draughts do not agree about it. Some of them make California an island. Some of their draughts newly made make California to join to the main. I believe that the Spaniards do not care to have this lake discovered, for fear less other European nations should get knowledge of it, and by that means visit the mines of New Mexico. New Mexico, by report of several English prisoners there, and Spaniards I have met with, lies NW. from Old Mexico between 400 and 500 leagues, and the biggest part of the treasure which is found in this kingdom is in that province; but without doubt there are plenty of mines in other parts, as well as in this part of the kingdom where we now were, as in other places; and probably on the main bordering on the Lake of California, although not yet discovered by the Spaniards, who have mines enough, and therefore as yet have no reason to discover more. In my opinion, here might be very advantageous discoveries made by any that would attempt it, for the Spaniards have more than they can well manage. I know yet they would lie like the dog in the manger; although not able to eat themselves, yet they would endeavour to hinder others. But the voyage thither being so far, I take that to be one reason that has hindered the discoveries of these parts; yet it is possible that a man may find a nearer way hither than we came; I mean by the north-west. I know there have been divers attempts made about a north-west passage, and all unsuccessful; yet I am of opinion that such a passage may be found. All our countrymen that have gone to discover the NW. passage, have endeavoured to pass to the westward, beginning their search along Davis's or Hudson's Bay. But if I were to go on this discovery, I would go first into the South Seas, bend my course from thence along by California, and that way seek a passage back into the West Seas. For as others have spent the summer in first searching on this more known side nearer home, and so before they got through, the time of the year obliged them to give over their search and provide for a long course back again, for fear of being left in the winter; on the contrary, I would search first on the less known coasts of the South Sea side, and then as the year passed away I should need no retreat, for I should come farther into my knowledge[1] if I succeeded in my attempt, and should be without that dread and fear which the others must have in passing from the known to the unknown; who, for aught I know, gave over their search just as they were on the point of accomplishing their desires. I would take the same method if I were to go to discover the north-east passage. I would winter about Japan, Corea, or the north-east part of China; and taking the spring and summer before me, I would make my first trial on the coast of Tartary; wherein, if I succeeded, I should come into some known parts, and have a great deal of time before me to reach Archangel or some other port. Captain Wood indeed says this north-east passage is not to be found for ice; but how often do we see that sometimes designs have been given over as impossible, and at another time and by other ways those very things have been accomplished? But enough of this.

The next day after that fatal skirmish near Santa Pecaque, Captain Swan ordered all our water to be filled, and to get ready to sail. The 21st we sailed from thence, directing our course towards California. We passed by three islands, called the [Three] Marias.[2] We beat till the 6th of February, but it was against a brisk wind, and proved labour in vain.

Finding, therefore, that we got nothing, but rather lost ground, being then in 21° 5′ N., we steered away more to the eastward again for the Islands Marias, and the 7th we came to an anchor at the east end of the middle island. The Marias are three uninhabited islands in Lat. 21° 40′; they are distant from Cape San Lucas on California forty leagues, bearing ESE., and from Cape Corrientes twenty leagues, bearing upon the same points of the compass with Cape San Lucas. They stretch NW. and SE. about fourteen leagues. There are two or three small high rocks near them; the westernmost of them is the biggest island of the three, and they are all three of an indifferent height. The soil is stony and dry; the land, in most places, is covered with a shrubby sort of wood, very thick and troublesome to pass through. In some places there is plenty of straight, large cedars. [These islands are described as uninhabitable, but guanas, racoons, turtle, tortoise, and seal were to be had in abundance. Captain Swan named the middle island Prince George's Island. Dampier was here sick of a dropsy, but having been buried half-an-hour in the hot sand to induce perspiration, he got well shortly afterwards.] . . .

We stayed here till the 20th; and then both vessels being clean, we sailed to the valley of Valderas to water. The 28th we anchored in the bottom of the bay of the valley of Valderas, right against the river, where we watered before; but this river was brackish now in the dry season, and therefore we went two or three leagues nearer Cape Corrientes, and anchored by a small round island not half a-mile from the shore. Here our strikers struck nine or ten Jew-

[1] Into the regions of which I had knowledge.

[2] There are really four islands in the group; the fourth, lying farthest to the north-west, is called Santa Juanic. . . . .

fish: some we did eat, and the rest we salted; and the 29th we filled thirty-two tons of very good water.

Having thus provided ourselves, we had nothing more to do but to put in execution our intended expedition to the East Indies, in hopes of some better success there than we had met with on this little frequented coast. We came on it full of expectations; for besides the richness of the country, and the probability of finding some seaports worth visiting, we persuaded ourselves that there must needs be shipping and trade here, and that Acapulco and La Vera Cruz were to the kingdom of Mexico what Panama and Porto Bello are to that of Peru, viz., marts for carrying on a constant commerce between the South and North Seas, as indeed they are. But whereas we expected that this commerce should be managed by sea from the places along the west coast, we found ourselves mistaken; that of Mexico being almost wholly a land trade, and managed more by mules than by ships; so that instead of profit, we met with little on this coast besides fatigues, hardships, and losses, and so were the more easily induced to try what better fortune we might have in the East Indies. But, to do right to Captain Swan, he had no intention to be as a privateer in the East Indies; but, as he has often assured me with his own mouth, he resolved to take the first opportunity of returning to England; so that he feigned a compliance with some of his men who were bent upon going to cruise at Manilla, that he might have leisure to take some favourable opportunity of quitting the privateer trade.

CHAPTER X.

I HAVE given an account in the last Chapter of the resolutions we took of going over to the East Indies. But having more calmly considered on the length of our voyage from hence to Guam, one of the Ladrone Islands, which is the first place that we could touch at, and there also being not certain to find provisions, most of our men were almost daunted at the thoughts of it, for we had not sixty days' provision, at a little more than half a pint of maize a-day for each man, and no other provision except three meals of salted Jew-fish; and we had a great many rats aboard which we could not hinder from eating part of our maize; besides the great distance between Cape Corrientes and Guam, which is variously set down. The Spaniards, who have the greatest reason to know best, make it to be between 2300 and 2400 leagues: our books also reckon it differently—between 90 and 100 degrees, which all comes short indeed of 2000 leagues; but even that was a voyage enough to frighten us, considering our scanty provisions. Captain Swan, to encourage his men to go with him, persuaded them that the English books did give the best account of the distance; his reasons were many, although but weak. He urged, among the rest, that Sir Thomas Cavendish and Sir Francis Drake did run it in less than fifty days, and that he did not question but that our ships were better sailers than those which were built in that age; and that he did not doubt to get there in little more than forty days, this being the best time in the year for breezes, which undoubtedly is the reason that the Spaniards set out from Acapulco about this time; and that although they are sixty days in their voyage, it is because they are great ships, deep-laden, and very heavy sailers; besides, they, wanting nothing, are in no great haste in their way, but sail with a great deal of their usual caution, and when they come near the Island of Guam, they lie by in the night for a week before they make land. In prudence we also should have contrived to lie by in the night when we came near land; for otherwise we might have run ashore, or have outsailed the islands and lost sight of them before morning. But our bold adventurers seldom proceed with such wariness when in any straits. But of all Captain Swan's arguments, that which prevailed most with them was his promising them, as I have said, to cruise off Manilla. So he and his men being now agreed, and they encouraged with the hope of gain, which works its way through all difficulties, we set out from Cape Corrientes, March the 31st, 1686. We were two ships in company, Captain Swan's ship and a bark commanded under Captain Swan by Captain Tait, and we were 150 men—100 aboard of the ship, and 50 aboard the bark, besides slaves, as I said.

The next morning, about 10 o'clock, we had the sea breeze at NNE., so that at noon we were thirty leagues from the cape. It blew a fresh gale of wind, which carried us off into the true trade-wind. At first we had it at NNE., so it came about easterly, and then to the east as we ran off. At 250 leagues' distance from the shore we had it at ENE., and there it stood till we came within forty leagues of Guam. When we had eaten up our three meals of salted Jew-fish in so many days' time, we had nothing but our small allowance of maize. After the 1st of May we made great runs every day, having very fair clear weather and a fresh trade-wind, which we made use of with all our sails, and we made many good observations of the sun. At our first setting out we steered into the Lat. of 13°, which is near the Latitude of Guam; then we steered west, keeping in that Latitude. By the time we had sailed twenty days, our men, seeing we made such great runs, and the wind like to continue, repined because they were kept at such short allowance. Captain Swan endeavoured to persuade them to have a little patience, yet nothing but an augmentation of their daily allowance would appease them. Captain Swan, though with much reluctance, gave way to a small enlargement of our commons, for now we had not above ten spoonfuls of boiled maize a-man once a-day, whereas before we had eight. I do believe that this short allowance did me a great deal of good, though others were weakened by it, for I found that my strength increased and my dropsy wore off. Yet I drank three times every twenty-four hours; but many of our men did not drink in nine or ten days' time, and some not in twelve days; one of our men did not drink in seventeen days' time, and said he was not a-dry when he did drink; yet he made water every day, more or less. One of our men in the midst of these hardships was found guilty of theft, and condemned for the same to have three blows from each man in the ship with a two-inch and a half rope on his bare back. Captain Swan began first, and struck with a good will, whose example was followed by all of us. It was very strange that in all this voyage we did not see one fish, not so much as a flying fish, nor any sort of fowl; but at one time, when we were by my account 4975 miles west from Cape Corrientes; then we saw a great number of boobies, which we supposed came from some rocks not far from us, which were mentioned in some of our sea-charts, but we did not see them.

After we had run the 1900 leagues by our reckoning, which made the English account to Guam, the men began to murmur against Captain Swan for persuading them to come this voyage; but he gave them fair words, and told them that the Spanish account might probably be the truest, and seeing the gale was likely to continue, a short time longer would end our troubles. As we drew nigh the island, we met with some small rain, and the clouds settling in the west were an apparent token that we were not far from land; for in these climates between or near the Tropics, where the trade-wind blows constantly, the clouds, which fly swift overhead, yet seem near the limb[1] of the horizon to hang without much motion or alteration where the land is near. I

[1] The utmost edge or border; an astronomical term applied to the border of the disc of the sun, the moon, or any planet.

have often taken notice of it, especially if it is high land, for you shall then have the clouds hang about it without any visible motion. The 20th day of May our bark, being about three leagues ahead of our ship, sailed over a rocky shoal on which there was but four fathom water, and abundance of fish swimming about the rocks. They imagined by this that the land was not far off; so they clapped on a wind with the bark's head to the north, and being past the shoal, lay by for us. When we came up with them, Captain Tait came aboard us and related what he had seen. We were then in Lat. 12° 55′, steering W. The Island of Guam is laid down in Lat. 13°. N. by the Spaniards, who are masters of it, keeping it as a baiting-place[1] as they go to the Philippine Islands. Therefore we clapped on a wind and stood to the N., being somewhat troubled and doubtful whether we were right, because there is no shoal laid down in the Spanish draughts about the Island of Guam. At four o'clock, to our great joy, we saw the Island Guam at about eight leagues' distance. It was well for Captain Swan that we got sight of it before our provision was spent, of which we had but enough for three days more; for, as I was afterwards informed, the men had contrived[2] first to kill Captain Swan and eat him when the victuals were gone, and after him, all of us who were accessory in promoting the undertaking this voyage. This made Captain Swan say to me after our arrival at Guam, "Ah! Dampier, you would have made them but a poor meal," for I was as lean as the Captain was lusty and fleshy. The wind was at ENE. and the land bore NNE.; therefore we stood to the northward till we brought the island to bear east, and then we turned to get in to an anchor. [Dampier here occupies several pages with a detailed table, showing every day's run during the voyage, with the course steered, the direction of the wind, and the observations made; the result being, by his computation, a total westing of 7323 miles, or 125° 11′ of longitude, "allowing fifty-eight or fifty-nine Italian miles to a degree in these latitudes." And upon the ground of this calculation he disputes the ordinary reckoning of hydrographers, who make the breadth of the South Sea "only about 100 degrees, more or less." The tables and argument are omitted, being purely technical and practically obsolete.]

The Island of Guam, or Guahan (as the native Indians pronounce it), is one of the Ladrone Islands, and belongs to the Spaniards, who have a small fort with six guns in it, with a Governor and twenty or thirty soldiers. They keep it for the relief and refreshment of their Philippine ships that touch here in their way from Acapulco to Manilla, but the winds will not so easily let them take this way back again. The Spaniards of late have named Guam the Island Maria; it is about twelve leagues long and four broad, lying N. and S. It is a pretty high champaign land. The 21st of May 1686, at eleven o'clock in the evening, we anchored near the middle of the Island of Guam, on the west side, a mile from the shore. At a distance it appears flat and even, but coming near it you will find it stands shelving; and the east side, which is much the highest, is fenced with steep rocks that oppose the violence of the sea which continually rages against it, being driven with the constant trade-wind, and on that side there is no anchoring. The west side is pretty low and full of small sandy bays, divided with as many rocky points. The soil of the island is reddish, dry, and indifferent fruitful. The fruits are chiefly rice, pineapples, water-melons, musk-melons, oranges and limes, cocoa-nuts, and a sort of fruit called by us bread-fruit.

The cocoa-nut trees grow by the sea on the western side in great groves, three or four miles in length, and a mile or two broad. This tree

[1] A place of provisioning or refreshment.

[2] Plotted, arranged.

is in shape like the cabbage tree, and at a distance they are not to be known each from other, only the cocoa-nut tree is fuller of branches; but the cabbage tree generally is much higher, though the cocoa-nut trees in some places are very high. . . .

The natives of this island are strong-bodied, large-limbed, and well-shaped. They are copper-coloured like other Indians; their hair is black and long, their eyes meanly proportioned; they have pretty high noses; their lips are pretty full, and their teeth indifferent white. They are long-visaged, and stern of countenance; yet we found them to be affable and courteous. They are many of them troubled with a kind of leprosy.

The natives are very ingenious beyond any people in making boats, or proas as they are called in the East Indies, and therein they take great delight. These are built sharp at both ends. The bottom is of one piece, made like the bottom of a little canoe, very neatly dug, and left of a good substance. This bottom part is instead of a keel; it is about twenty-six or twenty-eight feet long; the under part of this keel is made round, but inclining to a wedge, and smooth; and the upper part is almost flat, having a very gentle hollow, and is about a foot broad. From hence both sides of the boat are carried up to about five feet high with narrow plank, not above four or five inches broad, and each end of the boat turns up round very prettily. But, what is very singular, one side of the boat is made perpendicular, like a wall, while the other side is rounding, made as other vessels are, with a pretty full belly. Just in the middle it is about four or five feet broad aloft, or more, according to the length of the boat. The mast stands exactly in the middle, with a long yard that peaks up and down like a mizzen-yard. One end of it reaches down to the end or head of the boat, where it is placed in a notch that is made there purposely to receive it and keep it fast; the other end hangs over the stern. To this yard the sail is fastened. At the foot of the sail there is another small yard, to keep the sail out square, and to roll up the sail on when it blows hard; for it serves instead of a reef to take up the sail to what degree they please, according to the strength of the wind. Along the belly-side of the boat, parallel with it, at about six or seven feet distance, lies another small boat or canoe, being a log of very light wood, almost as long as the great boat, but not so wide, being not above a foot and a half wide at the upper part, and very sharp like a wedge at each end. And there are two bamboos of about eight or ten feet long, and as big as one's leg, placed over the great boat's side, one near each end of it, and reaching about six or seven feet from the side of the boat; by the help of which the little boat is made firm and contiguous to the other. . . . I have been the more particular in describing these boats, because I believe they sail the best of any boats in the world. I did here for my own satisfaction try the swiftness of one of them; sailing by our log, we had twelve knots on our reel, and she ran it all out before the half-minute glass was half out, which, if it had been no more, is after the rate of twelve miles an hour; but I do believe she would have run twenty-four miles an hour. . . .

The Indians of Guam have neat little houses, very handsomely thatched with palmetto thatch. They inhabit together in villages built by the sea on the west side, and have Spanish priests to instruct them in the Christian religion. The Spaniards have a small fort on the west side, near the south end, with six guns in it. There is a Governor, and twenty or thirty Spanish soldiers. There are no more Spaniards on the island, besides two or three priests. Not long before we arrived here, the natives rose on the Spaniards to destroy them, and did kill many; but the Governor with his soldiers at length prevailed, and drove them out of the fort. So, when they found themselves disappointed of their intent, they destroyed the plantations and stock, and then went away to other islands. There were then 300 or 400 Indians on this island; but now there are not above 100, for all that were in this conspiracy went away. As for those who yet remain, if they were not actually concerned in that broil, yet their hearts also are bent against the Spaniards; for they offered to carry us to the fort, and assist us in the conquest of the island; but Captain Swan was not for molesting the Spaniards here. Before we came to an anchor here, one of the priests came aboard in the night with three Indians. They first hailed us to know from whence we came and what we were; to whom answer was made in Spanish that we were Spaniards, and that we came from Acapulco. It being dark, they could not see the make of our ship, nor very well discern what we were. Therefore they came aboard; but perceiving the mistake they were in in taking us for a Spanish ship, they endeavoured to get from us again; but we held their boat fast, and made them come in. Captain Swan received the priest with much civility, and, conducting him into the great cabin, declared that the reason of our coming to this island was want of provision, and that he came not in any hostile manner, but as a friend, to purchase with his money what he wanted; and therefore desired the priest to write a letter to the Governor to inform him what we were and on what account we came. For having him now aboard, the Captain was willing to detain him as a hostage till we had provision. The Padre told Captain Swan that provision was now scarce on the island, but he would engage that the Governor would do his utmost to furnish us.

In the morning, the Indians, in whose boat or proa the Friar came aboard, were sent to the Governor with two letters, one from the Friar, and another very obliging one from Captain Swan, and a present of four yards of scarlet cloth, and a piece of broad silver and gold lace. The Governor lives near the south end of the island, on the west side, which was about five leagues from the place where we were; therefore we did not expect an answer till the evening, not knowing then how nimble they were. Therefore, when the Indian canoe was despatched away to the Governor, we hoisted out two of our canoes, and sent one a-fishing and the other ashore for cocoa-nuts. Our fishing canoe got nothing, but the men that went ashore for cocoa-nuts came off laden. About 11 o'clock that same morning, the Governor of the island sent a letter to Captain Swan, complimenting him for his present, and promising to support us with as much provision as he could possibly spare; and as a token of his gratitude he sent a present of six hogs of a small sort, most excellent meat, the best, I think, that ever I ate; they are fed with cocoa-nuts, and their flesh is hard as brisket beef. They were doubtless of that breed in America which came originally from Spain. He sent also twelve musk-melons, larger than ours in England, and as many water-melons, both sorts here being a very excellent fruit; and sent an order to the Indians that lived in a village not far from our ship to bake every day as much of the bread-fruit as we desired, and to assist us in getting as many dry cocoa-nuts as we would have, which they accordingly did, and brought of the bread-fruit every day hot, as much as we could eat. After this the Governor sent every day a canoe or two with hogs and fruit, and desired for the same powder, shot, and arms, which were sent according to his request. We had a delicate[1] large English dog, which the Governor did desire, and had it given him very freely by the Captain, though much against the grain of many of his men, who had a great value for that dog. Captain Swan endeavoured to get this Governor's letter of recommendation to some merchants at Manilla, for he had then a design to go to Fort St George,[2]

[1] Handsome, or favourite.
[2] Madras.

and from thence intended to trade at Manilla: but this his design was concealed from the company. While we lay here, the Acapulco ship arrived in sight of the island, but did not come in sight of us; for the Governor sent an Indian proa with advice of our being here. Therefore she stood off to the southward of the island, and coming foul of the same shoal that our bark had run over before, was in great danger of being lost there; for she struck off her rudder, and with much ado got clear, but not till after three days' labour. This we heard afterwards, when we were on the coast of Manilla; but these Indians of Guam did speak of her being in sight of the island while we lay there, which put our men in a great heat to go out after her; but Captain Swan persuaded them out of that humour, for he was now wholly averse to any hostile action.

The 30th of May the Governor sent his last present, which was some hogs, a jar of pickled mangoes, a jar of excellent pickled fish, and a jar of fine rusk, or bread of fine wheat flour, baked like biscuit, but not so hard. He sent besides six or seven packs of rice, desiring to be excused from sending any more provision to us, saying he had no more on the island that he could spare. He sent word also that the west monsoon was at hand; that therefore it behoved us to be jogging from hence, unless we were resolved to return back to America again. Captain Swan returned him thanks for his kindness and advice, and took his leave; and the same day sent the Friar ashore that was seized on at our first arrival, and gave him a large brass clock, an astralobe, and a large telescope; for which present the Friar sent us aboard six hogs and a roasting pig, three or four bushels of potatoes, and fifty pounds of Manilla tobacco. Then we prepared to be gone, being pretty well furnished with provision to carry us to Mindanao, where we designed next to touch. We took aboard as many cocoa-nuts as we could well stow; and we had a good stock of rice, and about fifty hogs in salt.

CHAPTER XI.

While we lay at Guam, we took up a resolution of going to Mindanao, one of the Philippine Islands, being told by the Friar and others that it was exceedingly well stored with provisions; that the natives were Mahometans, and that they had formerly a commerce with the Spaniards, but that now they were at war with them. This island was therefore thought to be a convenient place for us to go to; for besides that it was in our way to the East Indies, which we had resolved to visit; and that the westerly monsoon was at hand which would oblige us to shelter somewhere in a short time; and that we could not expect good harbours in a better place than in so large an island as Mindanao; besides all this, I say, the inhabitants of Mindanao being then, as we were told (though falsely), at war with the Spaniards, our men, who it should seem were very squeamish of plundering without license, derived hopes from thence of getting a commission there from the Prince of the island to plunder the Spanish ships about Manilla, and so to make Mindanao their common rendezvous. And if Captain Swan was minded to go to an English port, yet his men, who thought he intended to leave them, hoped to get vessels and pilots at Mindanao fit for their turn to cruise on the coast of Manilla. As for Captain Swan, he was willing enough to go thither, as best suiting his own design; and therefore this voyage was concluded on by general consent. Accordingly, June 2d, 1686, we left Guam, bound for Mindanao.

The 21st of June, we arrived at the Island St John,[1] which is one of the

[1] It would seem that Dampier was misled by the deep indentation of the coast on the south of Mindanao to fancy two islands when there was

Philippine Islands. The Philippines are a great company of large islands, taking up about 13° of Lat. in length, reaching near upon from 5° N. Lat. to 19°, and in breadth about 6°. of Longitude. They derive this name from Philip the second king of Spain; and even now they do most of them belong to that crown. The chief island in this range is Luconia,[1] which lies on the north of them all. At this island Magellan died in the voyage that he was making round the world. For after he had passed those straits between the south end of America and Tierra del Fuego which now bear his name, and had ranged down in the South Seas on the back of America, from thence stretching over to the East Indies he fell in with the Ladrone Islands, and from thence steering east still he fell in with these Philippine Islands, and anchored at Luconia, where he warred with the native Indians, to bring them in obedience to his master the King of Spain, and was by them killed with a poisoned arrow. It is now wholly under the Spaniards, who have several towns there. The chief is Manilla, which is a large seaport town near the SE. end, opposite to the Island Mindoro. It is a place of great strength and trade; the two great Acapulco ships before mentioned fetching from hence all sorts of East Indian commodities, which are brought hither by foreigners, especially by the Chinese and Portuguese. Sometimes the English merchants of Fort St George send their ships thither as it were by stealth, under the charge of Portuguese pilots and mariners; for as yet we cannot get the Spaniards there to a commerce with us or the Dutch, although they have but few ships of their own. This seems to arise from a jealousy or fear of discovering the riches of these islands; for most if not all the Philippine Islands are rich in gold; and the Spaniards have no place of much strength in all these islands that I could ever hear of, besides Manilla itself. Yet they have villages and towns on several of the islands, and Padres or priests to instruct the native Indians, from whom they get their gold. The Spanish inhabitants, of the smaller islands especially, would willingly trade with us if the government was not so severe against it; for they have no goods but what are brought from Manilla at an extraordinary dear rate. I am of the opinion that if any of our nations would seek a trade with them they would not lose their labour, for the Spaniards can and will smuggle as well as any nation that I know; and our Jamaicans are to their profit sensible enough of it. And I have been informed that Captain Goodlud of London, in a voyage which he made from Mindanao to China, touched at some of these islands, and was civilly treated by the Spaniards, who bought some of his commodities, giving him a very good price for the same. There are about twelve or fourteen more large islands lying to the southwards of Luconia, most of which, as I said before, are inhabited by the Spaniards. Besides these there are an infinite number of small islands of no account; and even the great islands, many of them, are without names, or at least so variously set down, that I find the same islands named by divers names.

The Islands of St John and Mindanao are the southernmost of all these islands, and are the only islands in all this range that are not subject to the Spaniards. St John's Island is on the east side of Mindanao, and distant from it three or four leagues, in Lat. about 7° or 8° N. This island is in length about thirty-eight leagues, stretching NNW. and SSE., and in breadth about twenty-four leagues in the middle of the island; the northernmost end is broader, and the southernmost is narrower.[2] This island is of a good

only one; unless, indeed, he really touched first at Samar, to the north, not the east, of Mindanao.

[1] Luçonia, or Luzon.

[2] This answers fairly enough the description of the eastern part of

height, and is full of many small hills. The land at the SE. end, where I was ashore, is of a black fat mould; and the whole island seems to partake of the same fatness, by the vast number of large trees that it produces, for it looks all over like one great grove. As we were passing by the SE. end we saw a canoe of the natives under the shore; therefore one of our canoes went after to have spoken with her, but she ran away from us, seeing themselves chased, put their canoe ashore, leaving her, fled into the woods, nor would be allured to come to us, although we did what we could to entice them. Besides these men we saw no more here, nor sign of any inhabitants at this end. When we came aboard our ship again, we steered away for the Island Mindanao, which was now fair in sight of us; it being about ten leagues distant from this part of St John's. The 22d we came within a league of the east side of the island, and steered toward the north end, keeping on the east side, till we came into the Lat. of 7° 40′, and there we anchored. Some of our books gave us an account that Mindanao city and isle lie in 7° 40′. We guessed that the middle of the island might lie in this latitude, but we were at a great loss where to find the city, whether on the east or west side. Indeed, had it been a small island, lying open to the eastern wind, we might probably have searched first on the west side; for commonly the islands within the Tropics, or within the bounds of the trade-winds, have their harbours on the west side, as best sheltered; but Mindanao being guarded on the east side by St John's Island, we might as reasonably expect to find the harbour and city on this side as anywhere else. But coming into the Latitude in which we judged the city might be, we found no canoes or people that might give us any umbrage[1] of a city or place of trade near at hand, though we coasted within a league of the shore.

The Island Mindanao is the biggest of all the Philippine Islands except Luconia. It is about sixty leagues long, and forty or fifty broad. The south end is in about 5° N., and the NW. end reaches almost to 8°. It is a very mountainous island, full of hills and valleys. The mould in general is deep and black, and extraordinary fat and fruitful. The sides of the hills are stony, yet productive enough of very large tall trees. In the heart of the country there are some mountains that yield good gold. The valleys are well moistened with pleasant brooks and small rivers of delicate water, and have trees of divers sorts flourishing and green all the year. The trees in general are very large, and most of them are of kinds unknown to us. There is one sort which deserves particular notice, called by the natives libby trees.[2] These grow wild in great groves of five or six miles long by the sides of the rivers. Of these trees sago is made, which the poor country people eat instead of bread three or four months in the year. This tree, for its body and shape, is much like the palmetto tree or the cabbage tree, but not so tall as the latter. The bark and wood are hard and thin like a shell, and full of white pith like the pith of an elder. This tree they cut down and split it in the middle, and scrape out all the pith, which they beat lustily with a wooden pestle in a great mortar or trough, and then put into a cloth or strainer held over a trough, and pouring water in among the pith, they stir it about in the cloth. So the water carries all the substance of the pith through the cloth down into the trough, leaving nothing in the cloth but a light sort of husk which they throw away; but that which falls

Mindanao, which, with a broken but fairly continuous coast line on the north and east, is deeply penetrated on the south-east by Davao Bay, which might easily have misled Dampier into supposing the existence of two islands.

[1] Hint, foreshadowing.

[2] The sago palm; *Sagus Rumphii.*

into the trough settles in a short time to the bottom like mud, and then they draw off the water and take up the muddy substance, wherewith they make cakes, which being baked proves very good bread. The Mindanao people live three or four months of the year on this food for their bread kind. The native Indians of Ternate and Tidore,[1] and all the Spice Islands, have plenty of these trees, and use them for food in the same manner.

The plantain I take to be the king of all fruit, not excepting the cocoa itself. The tree that bears this fruit is about three feet or three feet and a half round, and about ten or twelve feet high. These trees are not raised from seed (for they seem not to have any), but from the roots of other old trees. If these young suckers are taken out of the ground and planted in another place, it will be fifteen months before they bear; but if let stand in their own native soil, they will bear in twelve months. As soon as the fruit is ripe, the tree decays, but then there are many young ones growing up to supply its place. . . . As the fruit of this tree is of great use for food, so is the body no less serviceable to make cloths; but this I never knew till I came to this island. The ordinary people of Mindanao do wear no other cloth. The tree never bearing but once, and so being felled when the fruit is ripe, they cut it down close by the ground if they intend to make cloth with it. One blow with a macheat, or long knife, will strike it asunder: then they cut off the top, leaving the trunk eight or ten feet long, stripping off the outer rind, which is thickest towards the lower end. Having stripped two or three of these rinds, the trunk becomes in a manner all of one bigness, and of a whitish colour: then they split the trunk in the middle, which being done, they split the two halves again as near the middle as they can. This they leave in the sun two or three days, in which time part of the juicy substance of the tree dries away, and then the ends will appear full of small threads. The women, whose employment it is to make the cloth, take hold of those threads one by one, which rend away easily from one end of the trunk to the other, in bigness like whited brown threads; for the threads are naturally of a determinate bigness, as I observed their cloth to be all of one substance and equal fineness; but it is stubborn when new, wears out soon, and when wet feels a little slimy. They make their pieces seven or eight yards long, their warp and woof all one thickness and substance.

The banana tree is exactly like the plantain for shape and bigness, nor easily distinguishable from it but by its fruit, which is a great deal smaller, and not above half so long as a plantain, being also more mellow and soft, less luscious, yet of a more delicate taste. They use this for the making drink oftener than plantains, and it is best when used for drink or eaten as fruit; but it is not so good for bread, nor does it eat well at all when roasted or boiled; for it is only necessity that makes any use it this way. They grow generally where plantains do, being set intermixed with them purposely in their plantain walks. I have not seen the nutmeg trees anywhere; but the nutmegs this island produces are fair and large, yet they have no great store of them, being unwilling to propagate them or the cloves, for fear that should invite the Dutch to visit them and bring them into subjection, as they have done the rest of the neighbouring islands where they grow. For the Dutch being seated among the Spice Islands have monopolised all the trade into their own hands, and will not suffer any of the natives to dispose of it but to themselves alone. Nay, they are so careful to preserve it in their own hands, that they will not suffer the spice to grow in the uninhabited islands, but send soldiers to

[1] Two small islands between Celebes and Gilolo, in the Molucca Passage, south of Mindanao, mentioned by Drake.

cut the trees down. . . . The free merchants are not suffered to trade to the Spice Islands, nor to many other places where the Dutch have factories; but, on the other hand, they are suffered to trade to some places where the Dutch Company themselves may not trade, as to Achin particularly; for there are some princes in the Indies who will not trade with the Company for fear of them. The seamen that go to the Spice Islands are obliged to bring no spice from thence for themselves except a small matter for their own use, about a pound or two. Yet the masters of those ships do commonly so order their business, that they often secure a good quantity, and send it ashore to some place near Batavia before they come into that harbour (for it is always brought thither first before it is sent to Europe); and if they meet any vessel at sea that will buy their cloves, they will sell ten or fifteen tons out of a hundred, and yet seemingly carry their complement to Batavia; for they will pour water among the remaining part of their cargo, which will swell them to that degree that the ship's hold will be as full again as it was before any were sold. This trick they use whenever they dispose of any clandestinely, for the cloves when they first take them in are extraordinary dry, and so will imbibe a great deal of moisture. This is but one instance of many hundreds of little deceitful arts the Dutch seamen in these parts have among them, of which I have both seen and heard several. I believe there are nowhere greater thieves, and nothing will persuade them to discover one another; for should any do it, the rest would certainly knock him on the head. But to return to the products of Mindanao.

The betel nut is much esteemed here, as it is in most places of the East Indies. The betel tree grows like the cabbage tree, but it is not so big nor so high. The body grows straight, about twelve or fourteen feet high, without leaf or branch, except at the head; there it spreads forth long branches, like other trees of the like nature, as the cabbage tree, the cocoa-nut tree, and the palm. These branches are about ten or twelve feet long, and their stems near the head of the tree as big as a man's arm. On the top of the tree, among the branches, the betel nut grows on a tough stem as big as a man's finger, in clusters much as the cocoa nuts do, and they grow forty or fifty in a cluster. This fruit is bigger than a nutmeg, and is much like it, but rounder. It is much used all over the East Indies. This island produces also durians and jacks. The trees that bear the durians are as big as apple trees, full of boughs. The rind is thick and rough; the fruit is so large that they grow only about the bodies, or on the limbs near the body, like the cacao. The fruit is about the bigness of a large pumpkin, covered with a thick, green, rough rind. When it is ripe, the rind begins to turn yellow, but it is not fit to eat till it opens at the top. Then the fruit in the inside is ripe, and sends forth an excellent scent. When the rind is opened, the fruit may be split into four quarters; each quarter has several small cells that enclose a certain quantity of the fruit according to the bigness of the cell, for some are larger than others. The largest of the fruit may be as big as a pullet's egg; it is as white as milk and as soft as cream, and the taste very delicious to those that are accustomed to them; but those who have not been used to eat them will dislike them at first, because they smell like roasted onions. The jack or jaca is much like the durian, both in bigness and shape. The trees that bear them also are much alike, and so is their manner of the fruits growing; but the inside is different, for the fruit of the durian is white, that of the jack is yellow and fuller of stones. The durian is most esteemed, yet the jack is very pleasant fruit, and the stones or kernels are good roasted. There are many other sorts of grain, roots, and fruits in this island, which, to give a particular description of, would fill up a large volume. In this island are also many sorts of beasts both wild and tame, as horses, bulls and cows, buffaloes, goats, wild hogs, deer, monkeys, guanas, lizards, snakes, &c. Of the venomous kind of creatures here are scorpions, whose sting is in their tail; and centipedes, called by the English fork-legs, both which are also common in the West Indies, in Jamaica, and elsewhere. The fowls of this country are ducks and hens: other tame fowl I have not seen, nor heard of any. The wild fowl are pigeons, parrots, paroquets, turtle-doves, and abundance of small fowls. There are bats as big as a kite.

There are a great many harbours, creeks, and good bays for ships to ride in; and rivers navigable for canoes, proas, or barks, which are all plentifully stored with fish of divers sorts; so is also the adjacent sea. The chief fish are bonetas, snooks, cavallies, breams, mullets, ten-pounders, &c. Here are also plenty of sea-turtle and small manatee.

The weather at Mindanao is temperate enough as to heat, for all it lies so near the Equator; and especially on the borders near the sea. There they commonly enjoy the breezes by day, and cooling land-winds at night. The winds are easterly one part of the year and westerly the other. The easterly winds begin to blow in October, and it is the middle of November before they are settled. These winds bring fair weather. The westerly winds begin to blow in May, but are not settled till a month afterwards. The west winds always bring rain, tornadoes, and very tempestuous weather. At the first coming in of these winds they blow but faintly; but then the tornadoes rise, one in a day, sometimes two. These are thunder-showers which commonly come against the wind, bringing with them a contrary wind to what did blow before. After the tornadoes are over, the wind shifts about again, and the sky becomes clear; yet then in the valleys, and the sides of the mountains, there rises a thick fog, which covers the land. The tornadoes continue thus for a week or more; then they come thicker, two or three in a day, bringing violent gusts of wind and terrible claps of thunder. At last they come so fast, that the wind remains in the quarter from whence these tornadoes do rise, which is out of the west, and there it settles till October or November. When these westward winds are thus settled, the sky is all in mourning, being covered with black clouds, pouring down excessive rains, sometimes mixed with thunder and lightning, that nothing can be more dismal; the winds raging to that degree, that the biggest trees are torn up by the roots, and the rivers swell and overflow their banks, and drown the low land, carrying great trees into the sea. Thus it continues sometimes a week together, before the sun or stars appear. The fiercest of this weather is in the latter end of July and in August; for then the towns seem to stand in a great pond, and they go from one house to another in canoes. At this time the water carries away all the filth and nastiness from under their houses. Whilst this tempestuous season lasts, the weather is cold and chilly. In September the weather is more moderate, and the winds are not so fierce, nor the rain so violent. The air thenceforward begins to be more clear and delightsome; but then in the morning there are thick fogs, continuing till 10 or 11 o'clock, before the sun shines out, especially when it has rained in the night. In October the easterly winds begin to blow again, and bring fair weather till April. Thus much concerning the natural state of Mindanao.

CHAPTER XII.

THIS Island is not subject to one prince, neither is the language one and the same; but the people are much alike in colour, strength, and stature. They are all or most of them of one religion, which is Mahometanism, and their customs and manner of living are alike. The Mindanao people, more particularly so called, are the greatest nation in the island; and trading by sea with other nations, they are therefore the more civil.[1] I shall say but little of the rest, being less known to me; but so much as has come to my knowledge take as follows. There are, besides the Mindanayans, the Hilanoons (as they call them), or the Mountaineers, the Sologus, and Alfoores. The Hilanoons live in the heart of the country; they have little or no commerce by sea, yet they have proas that row with twelve or fourteen oars a-piece. They enjoy the benefit of the gold mines, and with their gold buy foreign commodities of the Mindanao people. They have also plenty of bees-wax, which they exchange for other commodities. The Sologus inhabit the NW. end of the island. They are the least nation of all; they trade to Manilla in proas, and to some of the neighbouring islands, but have no commerce with the Mindanao people. The Alfoores are the same with the Mindanayans, and were formerly under the subjection of the Sultan of Mindanao, but were divided between the Sultan's children, and have of late had a Sultan of their own; but having by marriage contracted an alliance with the Sultan of Mindanao, this has occasioned that prince to claim them again as his subjects; and he made war with them a little after we went away, as I afterwards understood.

The Mindanayans, properly so called, are men of mean statures, small limbs, straight bodies, and little heads. Their faces are oval, their foreheads flat, with black small eyes, short low noses, pretty large mouths; their lips thin and red, their teeth black yet very sound, their hair black and straight, the colour of their skin tawny, but inclining to a brighter yellow than some other Indians, especially the women. They have a custom to wear their thumbnails very long, especially that on their left thumb, for they do never cut it, but scrape it often. They are endowed with good natural wits, are ingenious, nimble, and active when they are minded; but generally very lazy and thievish, and will not work except forced by hunger. This laziness is natural to most Indians; but these people's laziness seems rather to proceed not so much from their natural inclinations, as from the severity of their prince, of whom they stand in great awe: for he dealing with them very arbitrarily, and taking from them what they get, this damps their industry, so they never strive to have anything but from hand to mouth. They are generally proud, and walk very stately. They are civil enough to strangers, and will easily be acquainted with them, and entertain them with great freedom; but they are implacable to their enemies, and very revengeful if they are injured, frequently poisoning secretly those that have affronted them. They wear but few clothes; their heads are circled with a short turban, fringed or laced at both ends; it goes once about the head, and is tied in a knot, the laced ends hanging down. They wear frocks and breeches, but no stockings nor shoes.

The women are fairer than the men, and their hair is black and long; which they tie in a knot, that hangs back in their polls.[2] They are more round-visaged than the men, and generally well featured; only their noses are very small, and so low between their eyes, that in some of the female children the rising that should be between the eyes is scarce discernible; neither is there any sensible rising in their foreheads. At a distance they appear very well, but being nigh these impediments are very obvious. They have very small limbs. They wear but two garments; a frock, and a sort of petticoat: the petticoat is only a piece of cloth sewed both ends together: but it is made two feet too big for their waists, so that they may wear either end uppermost: that part

[1] The better civilised.

[2] Behind their heads.

that comes up to their waists, because it is so much too big, they gather in their hands and twist it till it sits close to their waists, tucking in the twisted part between the waist and the edge of the petticoat, which keeps it close. The frock sits loose about them, and reaches down a little below the waist. The sleeves are a great deal longer than their arms, and so small at the end, that their hands will scarce go through. Being on, the sleeve sits in folds about the wrist; wherein they take great pride. The better sort of people have their garments made of long-cloth; but the ordinary sort wear cloth made of plantain-tree, which they call *saggen;* by which name they call the plantain. They have neither stocking nor shoe; and the women have very small feet. The women are very desirous of the company of strangers, especially of white men; and doubtless would be very familiar, if the custom of the country did not debar them from that freedom which seems coveted by them. Yet from the highest to the lowest they are allowed liberty to converse with or treat strangers in the sight of their husbands. There is a kind of begging custom at Mindanao that I have not met elsewhere with in all my travels, and which I believe is owing to the little trade they have; which is thus: when strangers arrive here, the Mindanao men will come aboard, and invite them to their houses, and inquire who has a comrade (which word I believe they have from the Spaniards) or a pagally, and who has not. A comrade is a familiar male friend; a pagally is an innocent platonic friend of the other sex. All strangers are in a manner obliged to accept of this acquaintance and familiarity, which must be first purchased with a small present, and afterwards confirmed with some gift or other to continue the acquaintance: and as often as the stranger goes ashore, he is welcome to his comrade's or pagally's house, where he may be entertained for his money, to eat, drink, or sleep; and complimented with tobacco and betel-nut, which is all the entertainment he must expect gratis. The richest men's wives are allowed the freedom to converse with her pagally in public, and may give or receive presents from him. Even the Sultan's and the General's wives, who are always cooped up, will yet look out of their cages when a stranger passes by, and demand of him if he wants a pagally: and, to invite him to their friendship, will send a present of tobacco and betel-nut to him by their servants.

The chief city on this island is called by the same name of Mindanao. It is seated on the south side of the island in Lat. 7° 20′ N. on the banks of a small river about two miles from the sea. The manner of building is somewhat strange, yet generally used in this part of the East Indies. Their houses are all built on posts about 14, 16, 18, or 20 feet high. These posts are bigger or less, according to the intended magnificence of the superstructure. They have but one floor, but many partitions or rooms, and a ladder or stairs to go up out of the streets. The roof is large, and covered with palmetto or palm leaves. So there is a clear passage like a piazza (but a filthy one) under the house. Some of the poorer people that keep ducks or hens have a fence made round the posts of their houses, with a door to go in and out; and this under-room serves for no other use. Some use this place for the common draught[1] of their houses; but, building mostly close by the river in all parts of the Indies, they make the river receive all the filth of their houses; and at the time of the land-floods all is washed very clean. The Sultan's house is much bigger than any of the rest. It stands on about 180 great posts or trees, a great deal higher than the common building, with great broad stairs made to go up. In the first room he has about twenty iron guns, all saker and minion,[2]

---

[1] Closet.

[2] That is, all of small calibre; the "saker extraordinary," with a charge of 5 lbs. of powder, carried a 7-lb.

placed on field-carriages. The General and other great men have some guns also in their houses. About twenty paces from the Sultan's house there is a small low house built purposely for the reception of ambassadors or merchant strangers. This also stands on posts, but the floor is not raised above three or four feet above the ground, and is neatly matted purposely for the Sultan and his Council to sit on, for they use no chairs, but sit cross-legged like tailors on the floor. The common food at Mindanao is rice or sago, and a small fish or two. The better sort eat buffalo, or fowls, ill dressed, and abundance of rice with it. They use no spoons to eat their rice, but every man takes a handful out of the platter, and by wetting his hand in water that it may not stick to his hand, squeezes it into a lump as hard as possibly he can make it, and then crams it into his mouth. They all strive to make these lumps as big as their mouths can receive them, and seem to vie with each other and glory in taking in the biggest lump, so that sometimes they almost choke themselves. They always wash after meals, or if they touch anything that is unclean; for which reason they spend abundance of water in their houses. This water, with the washing of their dishes, and what other filth they make, they pour down near their fireplace, for their chambers are not boarded but floored with split bamboos like laths, so that the water presently falls underneath their dwelling-rooms, where it breeds maggots and makes a prodigious stink. Besides this filthiness, the sick people ease themselves and make water in their chambers, there being a small hole made purposely in the floor to let it drop through; but healthy sound people commonly ease themselves and make water in the river. For that reason you shall always see abundance of people of both sexes in the river from morning till night—some easing themselves, others washing their bodies or clothes. If they come into the river purposely to wash their clothes, they strip and stand naked till they have done, then put them on and march out again. Both men and women take great delight in swimming and washing themselves, being bred to it from their infancy.

In the city of Mindanao they spoke two languages indifferently, their own Mindanao language and the Malay; but in other parts of the island they speak only their proper language, having little commerce abroad. They have schools, and instruct the children to read and write, and bring them up in the Mahometan religion. Therefore many of the words, especially their prayers, are in Arabic, and many of the words of civility the same as in Turkey; and especially when they meet in the morning, or take leave of each other, they express themselves in that language. Many of the old people, both men and women, can speak Spanish, for the Spaniards were formerly settled among them, and had several forts on this island; and then they sent two friars to this city to convert the Sultan of Mindanao and his people. At that time these people began to learn Spanish, and the Spaniards encroached on them and endeavoured to bring them into subjection; and probably before this time had brought them all under their yoke if they themselves had not been drawn off from this island to Manilla to resist the Chinese, who threatened to invade them there. When the Spaniards were gone, the old Sultan of Mindanao, father to the present, in whose time it was, razed and demolished their forts, brought away their guns, and sent away the friars; and since that time [they] will not suffer the Spaniards to settle on the islands. They are now most afraid of the Dutch, being sensible how they have enslaved many of the neighbouring islands. For that reason they have a long time desired the English to settle among them, and have offered them any convenient

---

ball; the smallest saker, with a 3 lb. charge, a 4¾-lb. ball. The minion was still a smaller piece.

place to build a fort in, as the General himself told us; giving this reason, that they do not find the English so encroaching as the Dutch or Spanish. The Dutch are no less jealous of their admitting the English, for they are sensible what detriment it would be to them if the English should settle here.

There are but few tradesmen at the city of Mindanao. The chief trades are goldsmiths, blacksmiths, and carpenters. There are but two or three goldsmiths; these will work in gold or silver, and make anything that you desire; but they have no shop furnished with ware ready for sale. Here are several blacksmiths who work very well considering the tools that they work with.[1] . . .

The Mindanao men have many wives, but what ceremonies are used when they marry I know not. There is commonly a great feast made by the bridegroom to entertain his friends, and the most part of the night is spent in mirth.

The Sultan is absolute in his power over all his subjects. He is but a poor prince; for, as I mentioned before, they have but little trade, and therefore cannot be rich. If the Sultan understands that any man has money, if it be but twenty dollars, which is a great matter among them, he will send to borrow so much money, pretending urgent occasions for it, and they dare not deny him. Sometimes he will send to sell one thing or another that he has to dispose of to such whom he knows to have money, and they must buy it and give him his price; and if afterwards he has occasion for the same thing he must have it if he sends for it. He is but a little man, between fifty and sixty years old, and by relation very good-natured, but overruled by those about him. He has a queen, and keeps about twenty women, or wives, more, in whose company he spends most of his time. He has one daughter by his Sultaness or queen, and a great many sons and daughters by the rest. These walk about the streets, and would be always begging things of us; but it is reported that the young Princess is kept in a room and never stirs out, and that she did never see any man but her father and Raja Laut her uncle, being then about fourteen years old. When the Sultan visits his friends, he is carried in a small couch on four men's shoulders, with eight or ten armed men to guard him; but he never goes far this way, for the country is very woody, and they have but little paths, which renders it the less commodious. When he takes his pleasure by water, he carries some of his wives along with him. The proas that are built for this purpose are large enough to entertain fifty or sixty persons or more. The hull is neatly built, with a round head and stern, and over the hull there is a small slight house built with bamboos; the sides are made up with split bamboos about four feet high, with little windows in them of the same to open and shut at their pleasure. The roof is almost flat, neatly thatched with palmetto leaves. This house is divided into two or three small partitions or chambers, one particularly for himself. This is neatly matted underneath and round the sides, and there is a carpet and pillows for him to sleep on. The second room is for his women, much like the former. The third is for the servants, who tend them with tobacco and betel-nut, for they are always chewing or smoking.

The Sultan has a brother called Raja Laut, a brave man. He is the second man in the kingdom. All strangers that come hither to trade must make their address to him, for all sea affairs belong to him. He licenses strangers to import or export any commodity, and it is by his permission that the natives themselves

[1] The men there are described as accustomed to the use of the axe and adze. They also built serviceable ships, their principal article of export being gold, bees-wax, and tobacco. The natives were much subject to a kind of leprosy, which showed itself in a dry scurf all over their bodies.

are suffered to trade; nay, the very fishermen must take a permit from him; so that there is no man can come into the river or go out but by his leave. He is two or three years younger than the Sultan, and a little man like him. He has eight women, by some of whom he has issue. He has only one son, about twelve or fourteen years old, who was circumcised while we were there. His eldest son died a little before we came thither, for whom he was still in great heaviness. If he had lived a little longer he should have married the young Princess; but whether this second son must have her I know not, for I did never hear any discourse about it. Raja Laut is a very sharp man; he speaks and writes Spanish, which he learned in his youth. He has, by often conversing with strangers, got a great insight into the customs of other nations, and by Spanish books has some knowledge of Europe. He is General of the Mindanayans, and is accounted an expert soldier and a very stout man; and the women in their dances sing many songs in his praise. The Sultan of Mindanao sometimes makes war with his neighbours the Mountaineers or Alfoores. Their weapons are swords, lances, and some hand cressets.[1] The cresset is a small thing like a bayonet, which they always wear in war or peace, at work or play, from the greatest of them to the poorest and meanest persons. They never meet each other so as to have a pitched battle, but they build small works or forts of timber, wherein they plant little guns, and lie in sight of each other two or three months, skirmishing every day in small parties, and sometimes surprising a breastwork; and whatever side is like to be worsted, if they have no probability to escape by flight, they sell their lives as dear as they can; for there is seldom any quarter given, but the conqueror cuts and hacks his enemies to pieces.

[1] Creeses; the Malay dagger, with zig-zag blade, often poisoned at the point.

The religion of these people is Mahometanism. Friday is their Sabbath; but I did never see any difference that they make between this day and any other day, only the Sultan himself goes then to his mosque twice. Raja Laut never goes to the mosque, but prays at certain hours, eight or ten times in a day; wherever he is, he is very punctual to his canonical hours, and if he be aboard will go ashore on purpose to pray. For no business nor company hinders him from his duty. Whether he is at home or abroad, in a house or in a field, he leaves all his company, and goes about 100 yards off, and there kneels down to his devotion. He first kisses the ground, then prays aloud, and divers times in his prayers he kisses the ground, and does the same when he leaves off. His servants, and his wives and children talk and sing, or play how they please, all the time, but himself is very serious. The meaner sort of people have little devotion; I did never see any of them at their prayers, or go into a mosque. In the Sultan's mosque there is a great drum with but one head, called a gong, which is instead of a clock. This gong is beaten at 12 o'clock, at three, six, and nine; a man being appointed for that service. He has a stick as big as a man's arm, with a great knob at the end, bigger than a man's fist, made with cotton, bound fast with small cords; with this he strikes the gong as hard as he can about twenty strokes, beginning to strike leisurely the first five or six strokes; then he strikes faster, and at last strikes as fast as he can, and then he strikes again slower and slower so many more strokes; thus he rises and falls three times, and then leaves off till three hours after. This is done night and day.

They circumcise the males at eleven or twelve years of age or older; and many are circumcised at once. This ceremony is performed with a great deal of solemnity. There had been no circumcision for some years before our being here, and then there was one for Raja Lant's son. They choose to have a general circumcision when the Sultan or General or some other great person has a son fit to be circumcised; for with him a great many more are circumcised. There is notice given about eight or ten days before, for all men to appear in arms, and great preparation is made against the solemn day. In the morning, before the boys are circumcised, presents are sent to the father of the child that keeps the feast, which, as I said before, is either the Sultan or some great person; and, about 10 or 11 o'clock, the Mahometan priest does his office. After this, most of the men, both in city and country, being in arms before the house, begin to act as if they were engaged with an enemy, having such arms as I described. Only one acts at a time, the rest make a great ring of 200 or 300 yards round about him. He that is to exercise comes into the ring with a great shriek or two, and a horrid look; then he fetches two or three large stately strides, and falls to work. He holds his broadsword in one hand, and his lance in the other, and traverses his ground, leaping from one side of the ring to the other, and in a menacing posture and look, bids defiance to the enemy whom his fancy frames to him, for there is nothing but air to oppose him. Then he stamps and shakes his head, and grinning with his teeth, makes many rueful faces. Then he throws his lance, and nimbly snatches out his cresset, with which he hacks and hews the air like a madman, often shrieking. At last, being almost tired with motion, he flies to the middle of the ring, where he seems to have his enemy at his mercy; and with two or three blows cuts on the ground as if he was cutting off his enemy's head. By this time he is all of a sweat, and withdraws triumphantly out of the ring, and presently another enters with the like shrieks and gestures. Thus they continue combating their imaginary enemy all the rest of the day; towards the conclusion of which the richest men act, and at last the General, and then the Sultan concludes this ceremony. He and the General with some other great men, are in armour, but the rest have none. After this the Sultan returns home, accompanied with abundance of people, who wait on him there till they are dismissed.

But at the time when we were there there was an after-game to be played; for the General's son being then circumcised, the Sultan intended to give him a second visit in the night; so they all waited to attend him thither. The General also provided to meet him in the best manner, and therefore desired Captain Swan with his men to attend him. Accordingly Captain Swan ordered us to get our guns, and wait at the General's house till further orders. So about forty of us waited till 8 o'clock in the evening, when the General, with Captain Swan, and about 1000 men, went to meet the Sultan, with abundance of torches that made it as light as day. The manner of the march was thus: first of all there was a pageant,[1] and upon it two dancing-women gorgeously apparelled, with coronets on their heads full of glittering spangles, and pendants of the same hanging down over their breasts and shoulders. These are women bred up purposely for dancing; their feet and legs are but little employed, except sometimes to turn round very gently; but their hands, arms, head, and body are in continual motion, especially their arms, which they turn and twist so strangely, that you would think them to be made without bones. Besides the two dancing-women, there were two old women in the pageant, holding each a lighted torch in their hands close by the dancing-women, by which light the glittering spangles appeared very gloriously. This pageant was carried by six lusty men. Then came six or seven torches, lighting the General and Captain Swan, who marched side by side next; and

[1] A decorated or triumphal chariot.

we that attended Captain Swan followed close after, marching in order six and six abreast, with each man his gun on his shoulder, and torches on each side. After us came twelve of the General's men, with old Spanish matchlocks, marching four in a row; after them about forty lances, and behind them as many with great swords, marching all in order. After them came abundance only with cressets by their sides, who marched up close without any order. When we came near the Sultan's house the Sultan and his men met us, and we wheeled off to let them pass. The Sultan had three pageants went before him. In the first pageant were four of his sons, who were about ten or eleven years old; they had gotten abundance of small stones, which they roguishly threw about on the people's heads. In the next were four young maidens, nieces to the Sultan, being his sisters' daughters; and in the third there were three of the Sultan's children, not above six years old. The Sultan himself followed next, being carried in his couch, which was not like your Indian palanquins, but open, and very little and ordinary. A multitude of people came after, without any order; but as soon as he was past by, the General and Captain Swan and all our men closed in just behind the Sultan, and so all marched together to the General's house. We came thither between ten and eleven o'clock, where the biggest part of the company were immediately dismissed; but the Sultan and his children and his nieces, and some other persons of quality, entered the General's house. They were met at the head of the stairs by the General's women, who with a great deal of respect conducted them into the house. Captain Swan and we that were with him followed after. It was not long before the General caused his dancing-women to enter the room and divert the company with that pastime. I had forgot to tell you that they have none but vocal music here, by what I could learn, except only a row of kind of bells without clappers; sixteen in number, and their weight increasing gradually from about three to ten pounds' weight. These were set in a row on a table in the General's house, where for seven or eight days together before the circumcision day they were struck each with a little stick for the biggest part of the day, making a great noise; and they ceased that morning. So these dancing-women sung themselves, and danced to their own music. After this the General's women and the Sultan's sons and his nieces danced. Two of the Sultan's nieces were about eighteen or nineteen years old, the other two were three or four years younger. These young ladies were very richly dressed with loose garments of silk, and small coronets on their heads. They were much fairer than any women that I did ever see there, and very well featured; and their noses, though but small, yet higher than the other women's, and very well proportioned. When the ladies had very well diverted themselves and the company with dancing, the General caused us to fire some sky-rockets that were made by his and Captain Swan's orders purposely for this night's solemnity; and after that the Sultan and his retinue went away with a few attendants, and we all broke up; and thus ended this day's solemnity.

They are not, as I said before, very curious or strict in observing any days or times of particular devotion, except it be the Ramdam time,[1] as we call it. The Ramdam time was then in August, as I take it, for it was shortly after our arrival here. In this time they fast all day, and about 7 o'clock in the evening they spend near an hour in prayer. Towards the latter end of their prayer they loudly invoke their Prophet for about a quarter of an hour, both old and young bawling out very strangely, as if they intended to fright him out of his sleepiness or neglect of them. After their prayer is ended, they

---

[1] The Fast of Ramadan, the Mahometan Lent.

spend some time in feasting before they take their repose. Thus they do every day for a whole month at least, for sometimes it is two or three days longer before the Ramdam ends; for it begins at the new moon, and lasts till they see the next new moon, which sometimes in thick, hazy weather is not till three or four days after the change, as it happened while I was at Achin, where they continued the Ramdam till the new moon's appearance. The next day after they have seen the new moon, the guns are all discharged about noon, and then the time ends. A main part of their religion consists in washing often, to keep themselves from being defiled, or after they are defiled to cleanse themselves again. They also take great care to keep themselves from being polluted by tasting or touching anything that is accounted unclean; therefore swine's flesh is very abominable to them; nay, any one that has either tasted of swine's flesh, or touched those creatures, is not permitted to come into their houses in many days after; and there is nothing will scare them more than a swine. Yet there are wild hogs in the island, and those so plentiful that they will come in troops out of the woods in the night into the very city, and come under their houses, to rummage up and down the filth that they find there. The natives therefore would even desire us to lie in wait for the hogs to destroy them, which we did frequently by shooting them and carrying them presently on board; but were prohibited their houses afterwards. And now I am on this subject, I cannot omit a story concerning the General. He once desired to have a pair of shoes made after the English fashion, though he did very seldom wear any; so one of our men made him a pair, which the General liked very well. Afterwards somebody told him, that the threads wherewith the shoes were sewed were pointed with hog's bristles. This put him into a great passion; so he sent the shoes to the man that made them, and sent him withal more leather to make another pair, with threads pointed with some other hair, which was immediately done, and then he was well pleased.

CHAPTER XIII.

HAVING in the two last Chapters given some account of the natural, civil, and religious state of Mindanao, I shall now go on with the prosecution of our affairs during our stay there. It was in a bay on the NE. side of the island that we came to an anchor, as has been said. We lay in this bay but one night and part of the next day. Yet there we got speech with some of the natives, who by signs made us understand that the city Mindanao was on the west side of the island. We endeavoured to persuade one of them to go with us to be our pilot, but he would not; therefore in the afternoon we loosed from hence, steering again to the SE., having the wind at SW. When we came to the SE. end of the Island Mindanao, we saw two small islands about three leagues distant from it.[1] We might have passed between them and the main island, as we learned since; but not knowing them nor what dangers we might encounter there, we chose rather to sail to the eastward of them. But meeting very strong westerly winds, we got nothing forward in many days. In this time we first saw the Islands Meangis, which are about sixteen leagues distant from the Mindanao, bearing SE. The 4th of July we got into a deep bay, four leagues NW. from the two small islands before mentioned. But the night before, in a violent tornado, our bark, being unable to beat any longer, bore away; which put us in some pain for fear she was overset, as we had like to have been ourselves. We anchored on the SW. side of the bay, in fifteen fathoms water, about a cable's length from the shore. Here

---

[1] The Scrangani Islands, off the southernmost point of Mindanao.

we were forced to shelter ourselves from the violence of the weather, which was so boisterous with rains and tornadoes and a strong westerly wind, that we were very glad to find this place to anchor in, being the only shelter on this side from the west winds. On the west side of the bay, the land is of a mean height, with a large savannah bordering on the sea, and stretching from the mouth of the bay a great way to the westward. This savannah abounds with long grass, and it is plentifully stocked with deer. The adjacent woods are a covert for them in the heat of the day; but mornings and evenings they feed in the open plains, as thick as in our parks in England. I never saw anywhere such plenty of wild deer, though I have met with them in several parts of America, both in the North and South Seas. The deer live here pretty peaceably and unmolested, for there are no inhabitants on that side of the bay. We visited this savannah every morning, and killed as many deer as we pleased, sometimes sixteen or eighteen in a day; and we did eat nothing but venison all the time we stayed here. We saw a great many plantations by the sides of the mountains on the east side of the bay, and we went to one of them, in hopes to learn of the inhabitants whereabouts the city was, that we might not oversail it in the night, but they fled from us.

We lay here till the 12th before the winds abated of their fury, and then we sailed from hence, directing our course to the westward. Being now past the SE. part of the island, we coasted down on the south side, and we saw abundance of canoes a-fishing, and now and then a small village. Neither were these inhabitants afraid of us as the former, but came aboard; yet we could not understand them, nor they us, but by signs; and when we mentioned the word Mindanao, they would point towards it. The 18th of July we arrived before the River of Mindanao. We anchored right against the river in fifteen fathoms water, clear hard sand, about two miles from the shore. We fired seven or nine guns, I remember not well which, and were answered again with three from the shore, for which we gave one again. Immediately after our coming to an anchor, Raja Laut and one of the Sultan's sons came off in a canoe, being rowed with ten oars, and demanded in Spanish what we were and from whence we came. Mr Smith (he who was taken prisoner at Leon in Mexico) answered in the same language that we were English, that we had been a great while out of England. They told us that we were welcome, and asked us a great many questions about England, especially concerning our East India merchants, and whether we were sent by them to settle a factory here. Mr Smith told them that we came hither only to buy provision. They seemed a little discontented when they understood that we were not come to settle among them; for they had heard of our arrival on the east side of the island a great while before, and entertained hopes that we were sent purposely out of England hither to settle a trade with them; which it should seem they are very desirous of, for Captain Goodlud had been here not long before to treat with them about it, and when he went away he told them, as they said, that in a short time they might expect an ambassador from England to make a full bargain with them. Indeed, upon mature thoughts I should think we could not have done better than to have complied with the desire they seemed to have of our settling here, and to have taken up our quarters among them. For as thereby we might better have consulted our own profit and satisfaction than by the other loose roving way of life; so it might probably have proved of public benefit to our nation, and been a means of introducing an English settlement and trade, not only here, but through several of the spice islands which lie in its neighbourhood. For the Islands Meangis, which I mentioned in the beginning of this Chapter, lie within twenty leagues of Mindanao. These are three small islands that abound with gold and cloves, if I may credit my author,[1] Prince Jeoly, who was born on one of them, and was at this time a slave in the city of Mindanao. He might have been purchased by us of his master for a small matter, as he was afterwards by Mr Moody, who came hither to trade, and laded a ship with clove bark; and by transporting him home to his own country we might have gotten a trade there. But of Prince Jeoly I shall speak more hereafter. These islands are as yet probably unknown to the Dutch, who, as I said before, endeavour to engross all the spice into their own hands. There was another opportunity offered us here of settling on another spice island that was very well inhabited; for the inhabitants fearing the Dutch, and understanding that the English were settling at Mindanao, their Sultan sent his nephew to Mindanao while we were there to invite us thither. Captain Swan conferred with him about it divers times, and I do believe he had some inclination to accept the offer; and I am sure most of the men were for it; but this never came to a head for want of a true understanding between Captain Swan and his men, as may be declared hereafter. Besides the benefit which might accrue from this trade with Meangis and other spice islands, the Philippine Islands themselves, by a little care and industry, might have afforded us a very beneficial trade; and all these trades might have been managed from Mindanao by settling there first. For that island lies very convenient for trading either to the Spice Islands or to the rest of the Philippine Islands; since, as its soil is much of the same nature with either of them, so it lies, as it were, in the centre of the gold and spice trade in these parts; the islands north of Mindanao abounding most in gold, and those south of Meangis in spice. . . . As to the capacity we were then in of settling ourselves at Mindanao, although we were not sent out of any such design of settling, yet we were as well provided, or better, considering all circumstances, than if we had. For there was scarce any useful trade but some or others of us understood it. We had sawyers, carpenters, joiners, brickmakers, bricklayers, shoemakers, tailors, &c.; we only wanted a good smith for great work, which we might have had at Mindanao. We were very well provided with iron, lead, and all sorts of tools, as saws, axes, hammers, &c. We had powder and shot enough, and very good small arms. If we had designed to build a fort, we could have spared eight or ten guns out of our ship, and men enough to have managed it, and any affair of trade beside. We had also a great advantage above raw men that are sent out of England into these places, who proceed usually too cautiously, coldly, and formally, to compass any considerable design, which experience better teaches than any rules whatsoever; besides the danger of their lives in so great and sudden a change of air, whereas we were all inured to hot climates, hardened by many fatigues, and in general daring men, and such as would not be easily baffled. To add one thing more, our men were almost tired, and began to desire a *quictus est;* and therefore they would gladly have seated themselves anywhere. We had a good ship, too, and enough of us (besides what might have been spared to manage our new settlement) to bring the news with the effects to the owners in England; for Captain Swan had already £5000 in gold, which he and his merchants received for goods sold mostly to Captain Harris and his men, which if he had laid but part of it out in spice, as probably he might have done, would have satisfied the merchants to their hearts' content. So much by way of digression.

To proceed therefore with our first reception at Mindanao. Raja Laut and his nephew sat still in their canoe and would not come aboard us, because, as they said, they had no

[1] Authority, informant.

orders for it from the Sultan. After about half-an-hour's discourse they took their leaves, first inviting Captain Swan ashore, and promising him to assist him in getting provision, which they said at present was scarce, but in three or four months' time the rice would be gathered in, and then he might have as much as he pleased, and that in the meantime he might secure his ship in some convenient place for fear of the westerly winds, which they said would be very violent at the latter end of this month and all the next, as we found them. We did not know the quality of these two persons till after they were gone, else we should have fired some guns at their departure. When they were gone, a certain officer under the Sultan came aboard and measured our ship, a custom derived from the Chinese, who always measure the length and breadth and the depth of the hold of all ships that come to load there, by which means they know how much each ship will carry. But for what reason this custom is used either by the Chinese or Mindanao men I could never learn, unless the Mindanayans design by this means to improve their skill in shipping, against they have a trade. Captain Swan, considering that the season of the year would oblige us to spend some time at this island, thought it convenient to make what interest he could with the Sultan, who might afterwards either obstruct or advance his designs. He therefore immediately provided a present to send ashore to the Sultan, viz., three yards of scarlet cloth, three yards of broad gold lace, a Turkish scimitar, and a pair of pistols; and to Raja Laut he sent three yards of scarlet cloth and three yards of silver lace. This present was carried by Mr Henry More in the evening. He was first conducted to Raja Laut's house, where he remained till report thereof was made to the Sultan, who immediately gave order for all things to be made ready to receive him. About 9 o'clock at night a messenger came from the Sultan to bring the present away. Then Mr More was conducted all the way, with torches and armed men, till he came to the house where the Sultan was. The Sultan, with eight or ten men of his Council, were seated on carpets waiting his coming. The present that Mr More brought was laid down before them, and was very kindly accepted by the Sultan, who caused Mr More to sit down by them, and asked a great many questions of him. The discourse was in Spanish by an interpreter. This conference lasted about an hour, and then he was dismissed, and returned again to Raja Laut's house. There was a supper provided for him and the boat's crew, after which he returned aboard.

The next day the Sultan sent for Captain Swan. He immediately went ashore, with a flag flying in the boat's head, and two trumpets sounding all the way. When he came ashore he was met at his landing by two principal officers, guarded along with soldiers, and abundance of people gazing to see him. The Sultan waited for him in his chamber of audience, where Captain Swan was treated with tobacco and betel, which was all his entertainment. The Sultan sent for two English letters for Captain Swan to read, purposely to let him know that our East India merchants did design to settle here, and that they had already sent a ship hither. One of these letters was sent to the Sultan from England by the East India merchants. The chief thing contained in it, as I remember—for I saw it afterwards in the Secretary's hand, who was very proud to show it to us—was to desire some privileges in order to the building of a fort there. This letter was written in a very fair hand, and between each line there was a gold line drawn. The other letter was left by Captain Goodlud, directed to any Englishmen who should happen to come thither. This related wholly to trade, giving an account at what rate he had agreed with them for goods of the island, and how European goods should be sold to them; with an account of their weights and measures, and their difference from ours. Captain Goodlud's letter concluded thus: "Trust none of them, for they are all thieves, but *tace* is Latin for a candle." We understood afterwards that Captain Goodlud was robbed of some goods by one of the General's men, and that he that robbed him was fled into the mountains, and could not be found while Captain Goodlud was here. But the fellow returning to the city some time after our arrival here, Raja Laut brought him bound to Captain Swan, and told him what he had done, desiring him to punish him for it as he pleased; but Captain Swan excused himself and said it did not belong to him, therefore he would have nothing to do with it. However the General Raja Laut would not pardon him, but punished him according to their own custom. He was stripped stark naked in the morning at sun-rising, and bound to a post, so that he could not stir hand nor foot but as he was moved, and was placed with his face eastward against the sun. In the afternoon they turned his face towards the west that the sun might still be in his face, and thus he stood all day parched in the sun, which shines here excessively hot, and tormented with the mosquitoes or gnats; after this the General would have killed him if Captain Swan had consented to it. Their common way of punishment is to strip them in this manner and place them in the sun, but sometimes they lay them flat on their backs on the sand, which is very hot, where they remain a whole day in the scorching sun, with the mosquitoes biting them all the time. This action of the General in offering Captain Swan the punishment of the thief, caused Captain Swan afterwards to make him the same offer of his men when any had offended the Mindanao men, but the General left such offenders to be punished by Captain Swan as he thought convenient. So that for the least offence Captain Swan punished his men, and that in the sight of the Mindanayans; and I think sometimes only for revenge, as he did once punish his chief mate Mr Tait, he that came captain of the bark to Mindanao. Indeed at that time Captain Swan had his men as much under command as if he had been in a king's ship; and had he known how to use his authority, he might have led them to any settlement and have brought them to assist him in any design he had pleased.

Captain Swan being dismissed from the Sultan with abundance of civility, after about two hours' discourse with him, went thence to Raja Laut's house. Raja Laut had then some difference with the Sultan, and therefore he was not present at the Sultan's reception of our Captain, but waited his return, and treated him and all his men with boiled rice and fowls. He then told Captain Swan again, and urged it to him, that it would be best to get his ship into the river as soon as he could, because of the usual tempestuous weather at this time of the year, and that he should want no assistance to further him in anything. He told him also that as we must of necessity stay here some time, so our men would often come ashore; and he therefore desired him to warn his men to be careful to give no affront to the natives, who, he said, were very revengeful. That their customs being different from ours, he feared that Captain Swan's men might some time or other offend them, though ignorantly; that therefore he gave him this friendly warning to prevent it; that his house should always be open to receive him or any of his men; and that he, knowing our customs, would never be offended at anything. After a great deal of such discourse he dismissed the Captain and his company, who took their leave and came aboard. Captain Swan having seen the two letters, did not doubt that the English did design to settle a factory here; therefore he did not much scruple[1] the honesty of these people, but immediately ordered us to get the ship into the river. The river upon which the city of Mindanao stands is but

[1] Doubt, suspect.

small, and has not above ten or eleven feet of water on the bar at a spring tide; therefore we lightened our ship, and the spring coming on, we with much ado got her into the river, being assisted by fifty or sixty Mindanayan fishermen who lived at the mouth of the river, Raja Laut himself being aboard our ship to direct them. We carried her about a quarter of a mile up within the mouth of the river, and there moored her head and stern in a hole, where we always rode afloat. After this the citizens of Mindanao came frequently aboard to invite our men to their houses and to offer us pagallies. It was a long time since any of us had received such friendship, and therefore we were the more easily drawn to accept of their kindnesses; and in a very short time most of our men got a comrade or two, and as many pagallies, especially such of us as had good clothes and store of gold, as many had who were of the number of those that accompanied Captain Harris over the Isthmus of Darien, the rest of us being poor enough. Nay, the very poorest and meanest of us could hardly pass the streets but we were even hauled by force into their houses to be treated by them, although their treats were but mean, viz., tobacco or betel-nut, or a little sweet-spiced water. Yet their seeming sincerity, simplicity, and the manner of bestowing these gifts, made them very acceptable. When we came to their houses they would always be praising the English, as declaring that the English and Mindanayans were all one. This they expressed by putting their two forefingers close together, and saying that the English and Mindanayans were *samo, samo*—that is, all one. Then they would draw their forefingers half a foot asunder and say the Dutch and they were *bugeto*, which signifies that they were at such distance in point of friendship. And for the Spaniards, they would make a greater representation of distance than for the Dutch, fearing these, but having felt and smarted from the Spaniards, who had once almost brought them under.

Captain Swan did seldom go into any house at first but into Raja Laut's; there he dined commonly every day; and as many of his men as were ashore, and had no money to entertain themselves, resorted thither about 12 o'clock, where they had rice enough boiled and well dressed, and some scraps of fowls or bits of buffalo dressed very nastily. Captain Swan was served a little better, and his two trumpeters sounded all the time that he was at dinner. After dinner Raja Laut would sit and discourse with him most part of the afternoon. It was now the Ramdam time, therefore the General excused himself that he could not entertain our Captain with dances and other pastimes as he intended to do when this solemn time was past, besides, it was the very height of the wet season, and therefore not so proper for pastimes. . . .

When the Ramdam time was over, and the dry time set in a little, the General, to oblige Captain Swan, entertained him every night with dances. The dancing-women that are purposely bred up to it, and make it their trade, I have already described. But besides them, all the women in general are much addicted to dancing. They dance forty or fifty at once, and that standing all round in a ring joined hand in hand, and singing and keeping time. But they never budge out of their places, nor make any motion till the chorus is sung; then all at once they throw out one leg and bawl out aloud, and sometimes they only clap their hands when the chorus is sung. Captain Swan, to retaliate the General's favours, sent for his violins, and some that could dance English dances, wherewith the General was very well pleased. They commonly spent the biggest part of the nights in these sort of pastimes. Among the rest of our men that did use to dance thus before the General, there was one John Thacker, who was a seaman bred, and could neither write nor read, but had formerly learnt to dance in the music-houses about Wapping. This man came into the South Seas with Captain Harris;

and getting with him a good quantity of gold, and being a pretty good husband of his share, had still some left, besides what he laid out in a very good suit of clothes. The General supposed by his garb and his dancing that he had been of noble extraction, and, to be satisfied of his quality, asked of one of our men if he did not guess aright of him. The man of whom the General asked this question told him he was much in the right, and that most of our ship's company were of the like extraction, especially all those that had fine clothes, and that they came abroad only to see the world, having money enough to bear their expenses wherever they came; but that for the rest, those that had but mean clothes, they were only common seamen. After this the General showed a great deal of respect to all that had good clothes, but especially to John Thacker, till Captain Swan came to know the business, and marred all, undeceiving the General, and drubbing the nobleman; for he was so much incensed against John Thacker that he could never endure him afterwards, though the poor fellow knew nothing of the matter.

About the middle of November we began to work on our ship's bottom, which we found very much eaten with the worm; for this is a horrid place for worms. . . . Having ripped off all our worm-eaten plank and clapped on new, by the beginning of December 1686, our ship's bottom was sheathed and tallowed; and the 10th we went over the bar, and took aboard the iron and lead that we could not sell, and began to fill our water and fetch aboard rice for our voyage. I was at that time a-hunting with the General for beef, which he had a long time promised us; but now I saw that there was no credit to be given to his word, for I was a week out with him and saw but four cows, which were so wild that we did not get one. There were five or six more of our company with me; these, who were young men, and had Delilahs there, which made them fond of the place, all agreed with the General to tell Captain Swan that there were beeves enough, only they were wild. But I told him the truth, and advised him not to be too credulous of the General's promises. He seemed to be very angry, and stormed behind the General's back, but in his presence was very mute, being a man of small courage. It was about the 20th of December when we returned from hunting, and the General designed to go again to another place to hunt for beef; but he stayed till after Christmas Day, because some of us designed to go with him, and Captain Swan had desired all his men to be aboard that day, that we might keep it solemnly together; and accordingly he sent aboard a buffalo the day before, that we might have a good dinner. So the 25th, about 10 o'clock, Captain Swan came aboard, and all his men who were ashore; for you must understand that near a third of our men lived constantly ashore with their comrades and pagallies, and some with women-servants whom they hired of their masters for concubines. Some of our men also had houses, which they hired or bought (for houses are very cheap) for five or six dollars; for many of them having more money than they knew what to do with, eased themselves here of the trouble of telling it, spending it very lavishly, their prodigality making the people impose upon them, to the making the rest of us pay the dearer for what we bought, and to the endangering the like impositions upon such Englishmen as may come here hereafter. For the Mindanayans knew how to get our squires' gold from them (for we had no silver), and when our men wanted silver they would change now and then an ounce of gold, and could get for it no more than ten or eleven dollars for a Mindanao ounce, which they would not part with again under eighteen dollars. Yet this, and the great prices the Mindanayans set on their goods, were not the only way to lessen their stocks; for their pagallies and comrades would often be begging somewhat of them, and our men were generous enough, and would bestow half-an-ounce of gold at a time in a ring for their pagallies, or in a silver wristband or hoop to come about their arms, in hopes to get a night's lodging with them. When we were all aboard on Christmas Day, Captain Swan and his two merchants, I did expect that Captain Swan would have made some proposals, or have told us his designs; but he only dined and went ashore again without speaking anything of his mind. Yet even then I think that he was driving on a design of going to one of the Spice Islands to load with spice; for the young man before mentioned, who I said was sent by his uncle, the Sultan of a spice island near Ternate, to invite the English to their island, came aboard at this time, and after some private discourse with Captain Swan they both went ashore together. This young man did not care that the Mindanayans should be privy to what he said. I have heard Captain Swan say that he offered to load his ship with spice, provided he would build a small fort and leave some men to secure the island from the Dutch; but I am since informed that the Dutch have now got possession of the island.

The next day after Christmas the General went away again, and five or six Englishmen with him, of whom I was one, under pretence of going a-hunting; and we all went together by water in his proa, together with his women and servants, to the hunting-place. The General always carried his wives and children, his servants, his money and goods with him; so we all embarked in the morning, and arrived there before night. I have already described the fashion of their proas, and the rooms made in them. We were entertained in the General's room or cabin. Our voyage was not so far but that we reached our port before night. At this time one of the General's servants had offended, and was punished in this manner: He was bound fast, flat on his belly, on a bamboo belonging to the proa, which was so near the water that by the vessel's motion it frequently delved under water, and the man along with it; and sometimes when hoisted up he had scarce time to blow before he would be carried under water again. When we had rowed about two leagues we entered a pretty large, deep river, and rowed up a league farther; the water salt all the way. There was a pretty large village, the houses built after the country fashion. We landed at this place, where there was a house made ready immediately for us. The General and his women lay at one end of the house, and we at the other end; and in the evening all the women in the village danced before the General. While he stayed here, the General with his men went out every morning betimes, and did not return till four or five o'clock in the afternoon; and he would often compliment us by telling us what good trust and confidence he had in us, saying that he left his women and goods under our protection, and that he thought them as secure with us six (for we had all our arms with us) as if he had left a hundred of his own men to guard them. Yet for all this great confidence he always left one of his principal men, for fear some of us should be too familiar with his women. They did never stir out of their own room when the General was at home; but as soon as he was gone out they would presently come into our room, and sit with us all day, and ask a thousand questions of us concerning our English women and our customs. You may imagine that before this time some of us had attained so much of their language as to understand them and give them answers to their demands. I remember that one day they asked how many wives the King of England had. We told them but one, and that our English laws did not allow of any more. They said it was a very strange custom that a man should be confined to one woman; some of them said it was a very bad law, but others again said it was a good law; so there was a great dispute among them about it. But one of the General's women said positively that our law was better than theirs, and made them all silent by the reason which she gave for it. This was the War Queen, as we called her, for she did always accompany the General whenever he was called out to engage his enemies, but the rest did not. By this familiarity among the women, and by often discoursing with them, we came to be acquainted with their customs and privileges. The General lies with his wives by turns, but she by whom he had the first son has a double portion of his company; for when it comes to her turn, she has him two nights, whereas the rest have him but one. She with whom he is to lie at night seems to have a particular respect shown her by the rest all the preceding day, and for a mark of distinction wears a striped silk handkerchief about her neck, by which we knew who was queen that day.

We lay here about five or six days, but did never in all that time see the least sign of any beef, which was the business we came about; neither were we suffered to go out with the General to see the wild kine, but we wanted for nothing else. However, this did not please us, and we often importuned him to let us go out among the cattle. At last he told us that he had provided a jar of rice-drink to be merry with us, and after that we should go with him. This rice-drink is made of rice boiled and put into a jar, where it remains a long time steeping in water. I know not the manner of making it, but it is very strong pleasant drink. The evening when the General designed to be merry, he caused a jar of this drink to be brought into our room, and he began to drink first himself, then afterwards his men; so they took turns till they were all as drunk as swine, before they suffered us to drink. After they had enough, then we drank, and they drank no more, for they will not drink after us. The General leaped about our room a little while; but, having his load, soon went to sleep. The next day we went out with the General into the savannah, where he had near 100 men making a large pen to drive the cattle into, for that is the manner of their hunting, having no dogs. But I saw not above eight or ten cows, and those as wild as deer, so that we got none this day; yet the next day some of his men brought in three heifers which they killed in the savannah. With these we returned aboard, they being all that we got there. Captain Swan was much vexed at the General's actions; for he promised to supply us with as much beef as we should want, but now either could not or would not make good his promise. Besides he failed to perform his promise in a bargain of rice that we were to have for the iron which he sold him, but he put us off still from time to time, and would not come to any account. Neither were these all his tricks; for a little before his son was circumcised, he pretended a great strait for money to defray the charges of that day; and therefore desired Captain Swan to lend him about twenty ounces of gold; for he knew that Captain Swan had a considerable quantity of gold in his possession, which the General thought was his own, but indeed had none but what belonged to the merchants. However, he lent it the General; but when he came to an account with Captain Swan he told him that it was usual at such solemn times to make presents, and that he received it as a gift. He also demanded payment for the victuals that our Captain and his men did eat at his house. These things startled Captain Swan, yet how to help himself he knew not. But all this, with other inward troubles, lay hard on our Captain's spirits, and put him very much out of humour; for his own company also were pressing him every day to be gone, because now was the height of the easterly monsoon, the only wind to carry us farther into the Indies.

About this time some of our men, who were weary and tired with wandering, ran away into the country and absconded, they being assisted, as was generally believed, by Raja Laut. There were others also, who, fearing we should not go to an English port, bought a canoe and designed to go in her to Borneo; for not long before a Mindanao vessel came from thence and brought a letter directed to the chief of the English factory at Mindanao. This letter the General would have Captain Swan to have opened; but he thought it might come from some of the East India merchants, whose affairs he would not intermeddle with, and therefore did not open it. I since met with Captain Bowry at Achin, and telling him this story he said that he sent that letter, supposing that the English were settled there at Mindanao; and by this letter we also thought that there was an English factory at Borneo; so here was a mistake on both sides. But this canoe wherewith some of them thought to go to Borneo, Captain Swan took from them, and threatened the undertakers very hardly. However, this did not so far discourage them, for they secretly bought another; but their designs taking air, they were again frustrated by Captain Swan. The whole crew were at this time under a general disaffection, and full of very different projects; and all for want of action. The main division was between those that had money and those that had none. There was a great difference in the humours of these; for they that had money lived ashore, and did not care for leaving Mindanao; whilst those that were poor lived aboard and urged Captain Swan to go to sea. These began to be unruly as well as dissatisfied, and sent ashore the merchants' iron to sell for rack and honey to make punch, wherewith they grew drunk and quarrelsome; which disorderly actions deterred me from going aboard, for I did ever abhor drunkenness, which now our men that were aboard addicted themselves wholly to. Yet these disorders might have been crushed if Captain Swan had used his authority to suppress them; but he with his merchants living always ashore, there was no command, and therefore every man did what he pleased and encouraged each other in his villanies. Now Mr Harton, who was one of Captain Swan's merchants, did very much importune him to settle his resolutions and declare his mind to his men; which at last he consented to do; therefore he gave warning to all his men to come aboard the 13th of January 1687.

We did all earnestly expect to hear what Captain Swan would propose, and therefore were very willing to go aboard; but unluckily for him, two days before this meeting was to be, Captain Swan sent aboard his gunner to fetch something ashore out of his cabin. The gunner rummaging to find what he was sent for, among other things took out the captain's journal from America to the Island of Guam, and laid it down by him. This journal was taken up by one John Reed, a Bristol man. He was a pretty ingenious young man, and of a very civil carriage and behaviour. He was also accounted a good artist, and kept a journal, and was now prompted by his curiosity to peep into Captain Swan's journal to see how it agreed with his own; a thing very usual among seamen that keep journals, when they have an opportunity, and especially young men who have no great experience. At the first opening of the book, he lighted on a place in which Captain Swan had inveighed bitterly against most of his men, especially against another John Reed, a Jamaica-man. This was such stuff as he did not seek after; but hitting so pat on the subject, his curiosity led him to pry further; and therefore while the gunner was busy, he conveyed the book away, to look over it at his leisure. The gunner having despatched his business, locked up the cabin-door, not missing the book, and went ashore; then John Reed shewed it to his name-sake, and to the rest that were aboard, who were by this time the biggest part of them ripe for mischief, only wanting some fair pretence to set themselves to work about it. Therefore looking on what was written in this journal to be matter sufficient for them to accomplish their ends, Captain Tait, who, as I said before, had been abused by Captain Swan, laid hold on this opportunity to be revenged for his injuries, and aggravated the matter to the height, persuading the men to turn out Captain Swan from being commander, in hopes to have commanded the ship himself. As for the seamen, they were easily persuaded to anything, for they were quite tired with this long and tedious voyage, and most of them despaired of ever getting home, and therefore did not care what they did or whither they went. It was only want of being busied in some action that made them so uneasy; therefore they consented to what Tait proposed, and immediately all that were aboard bound themselves by oath to turn Captain Swan out, and to conceal this design from those that were ashore, until the ship was under sail; which would have been presently, if the surgeon or his mate had been aboard: but they were both ashore, and they thought it no prudence to go to sea without a surgeon. Therefore the next morning they sent ashore one John Cookworthy, to hasten off either the surgeon or his mate, by pretending that one of the men in the night broke his leg by falling into the hold. The surgeon told him that he intended to come aboard the next day with the Captain, and would not come before, but sent his mate Herman Coppinger. This man, some time before this, was sleeping at his pagally's, and a snake twisted himself about his neck, but afterwards went away without hurting him. In this country it is usual to have the snakes come into the houses, and into the ships too; for we had several came aboard our ship when we lay in the river. But to proceed: Herman Coppinger provided to go aboard; and the next day, being the time appointed for Captain Swan and all his men to meet aboard, I went aboard with him, neither of us mistrusting what was designed by those aboard till we came thither. Then we found it was only a trick to get the surgeon off; for now, having obtained their desires, the canoe was sent ashore again immediately, to desire as many as they could meet to come aboard, but not to tell the reason, lest Captain Swan should come to hear of it.

The 13th, in the morning, they weighed, and fired a gun. Captain Swan immediately sent aboard Mr Nelly, who was now his chief mate, to see what the matter was; to him they told all their grievances, and showed him the journal. He persuaded them to stay till the next day for an answer from Captain Swan and the merchants; so they came to an anchor again, and the next morning Mr Hartop came aboard. He persuaded[1] them to be reconciled again, or at least to stay and get more rice; but they were deaf to it, and weighed again while he was aboard. Yet at Mr Hartop's persuasion they promised to stay till 2 o'clock in the afternoon for Captain Swan and the rest of the men, if they would come aboard; but they suffered no man to go ashore except one William Williams that had a wooden leg, and another that was a sawyer. If Captain Swan had yet come aboard, he might have dashed all their designs; but he neither came himself, as a captain of any prudence and courage would have done, nor sent till the time was expired. So we left Captain Swan and about thirty-six men ashore in the city, and six or eight that ran away; and about sixteen we had buried there, the most of which died by poison. The natives are very expert at poisoning, and do it upon small occasions: nor did our men want for given offence, through their general rogueries, and sometimes by dallying too familiarly with their women even before their faces. Some of their poisons are slow and lingering; for we had some now aboard who were poisoned there, but died not till some months after.

---

[1] Advised.

CHAPTER XIV.

The 14th of January 1687, at 3 o'clock in the afternoon, we sailed from the River of Mindanao, designing to cruise before Manilla. It was during our stay at Mindanao that we were first made sensible of the change of time in the course of our voyage. For having travelled so far westward, keeping the same course with the sun, we must consequently have gained something insensibly in the length of the particular days, but have lost in the tale, the bulk, or number, of the days or hours. According to the different longitudes of England and Mindanao, this isle being west from the Lizard, by common computation, about 210 degrees, the difference of time at our arrival at Mindanao ought to be about fourteen hours: and so much we should have anticipated our reckoning, having gained it by bearing the sun company. Now the natural day in every particular place must be consonant to itself: but this going about with or against the sun's course will of necessity make a difference in the calculation of the civil day between any two places. Accordingly, at Mindanao and all other places in the East Indies, we found them reckoning a day before us, both natives and Europeans; for, the Europeans coming eastward by the Cape of Good Hope, in a course contrary to the sun and us, wherever we met they were a full day before us in their accounts. So, among the Indian Mahometans here, their Friday, the day of their Sultan's going to their mosques, was Thursday with us; though it was Friday also with those who came eastward from Europe. Yet at the Ladrone Islands we found the Spaniards of Guam keeping the same computation with ourselves; the reason of which I take to be, that they settled that colony by a course westward from Spain; the Spaniards going first to America, and thence to the Ladrones and Philippines. . . .

We coasted to the westward on the south side of the Island Mindanao, keeping within four or five leagues off the shore. The land from hence trends away W. by S.; it is of a good height by the sea and very woody; and in the country we saw high hills. The next day we were abreast of Chambongo,[1] a town in this island, thirty leagues from the River of Mindanao. Here is said to be a good harbour and a great settlement, with plenty of beef and buffalo. It is reported that the Spaniards were formerly fortified here also. About six leagues before we came to the west end of the Island Mindanao, we fell in with a great many small low islands or keys; and about two or three leagues to the southward of these keys there is a long island, stretching NE. and SW. about twelve leagues.[2] This island is low by the sea on the north side, and has a ridge of hills in the middle running from one end to the other. Between this island and the small keys there is a good large channel. The 17th, we anchored on the east side of all these keys in eight fathoms water, clean sand. Here are plenty of green turtle, whose flesh is as sweet as any in the West Indies; but they are very shy. A little to the westward of these keys, on the Island Mindanao, we saw abundance of cocoa-nut trees. Therefore we sent our canoe ashore, thinking to find inhabitants, but found none, nor sign of any, but great tracks of hogs and great cattle; and close by the sea there were the ruins of an old fort; the walls thereof were of a good height, built with stone and lime, and, by the workmanship, seemed to be Spanish. We weighed again the 14th, and went through between the keys, but met such uncertain tides that we were forced to anchor again.

[1] Chambongo, or Zamboanga, stands at the south end of the great jut of land which forms the western portion of the Island of Mindanao; the bay enclosed in the curve of the coast between Mindanao and Zamboanga being called the Bay of Llana or Illana.

[2] Evidently the Basilian group of islands to the south of Zamboanga.

The 22d, we got about the westernmost point of all Mindanao, and stood to the northward, plying under the shore, and having the wind at NNE., a fresh gale. Here we met with two proas belonging to the Sologus, one of the Mindanayan nations before mentioned. They came from Manilla laden with silks and calicoes. We kept on this western part of the island, steering northerly, till we came abreast of some other of the Philippine islands that lay to the northward of us, then steered away towards them, but still keeping on the west side of them, and we had the winds at NNE. The 3d of February we anchored in a good bay on the west side of an island in Lat. 9° 55', where we had thirteen fathoms water, good soft ooze. This island has no name that we could find in any book,[1] but lies on the west side of Island Sebo. It is about eight or ten leagues long, mountainous and woody. At this place Captain Reed, who was the same Captain Swan had so much railed against in his journal, and was now made captain in his room (as Captain Tait was made master, and Mr Henry More quarter-master), ordered the carpenters to cut down our quarter-deck, to make the ship snug and the fitter for sailing. When that was done we heeled her, scrubbed her bottom, and tallowed it; then we filled all our water, for here is a delicate small run of water. The land was pretty low in this bay, the mould black and fat, and the trees of several kinds, very thick and tall. In some places we found plenty of canes, such as we use in England for walking-canes. These were short-jointed, not above two feet and a half or two feet ten inches the longest, and most of them not above two feet. They run along on the ground like a vine, or taking hold of the trees they climb up to their very tops. They are fifteen or twenty fathoms long, and much of a bigness from the root till within five or six fathoms of the end. They are of a pale green colour, clothed over with a coat of short thick hairy substance of a dun colour, but it comes off by only drawing the cane through your hand. We did cut many of them, and they proved very tough heavy canes. We saw no houses, nor sign of inhabitants. In the middle of this bay, about a mile from the shore, there is a small low woody island not above a mile in circumference; our ship rode about a mile from it. This island was the habitation of an incredible number of great bats, with bodies as big as ducks or larger fowl, and with vast wings; for I saw at Mindanao one of this sort, and I judge that the wings, stretched out in length, could not be less asunder than seven or eight feet from tip to tip, for it was much more than any of us could fathom with our arms extended to the utmost.

We stayed here till the 10th of February 1687, and then, having completed our business, we sailed hence with the wind at north; but going out we struck on a rock, where we lay two hours. It was very smooth water, and the tide of flood, or else we should there have lost our ship. We struck off a great piece of our rudder, which was all the damage that we received; but we more narrowly missed losing our ship this time than in any other in the whole voyage. This is a very dangerous shoal, because it does not break, unless probably it may appear in foul weather. After we were passed this shoal, we coasted along by the rest of the Philippine Islands, keeping on the west side of them. Some of them appeared to be very mountainous dry land. We saw many fires in the night as we passed by Panay,[2] a great island settled by Spaniards; and by the fires up and down it seems to be well settled by them; for this is a Spanish custom, whereby they give notice of any danger, or the like, from sea; and it is probable they had seen our ship the day before. The

[1] It seems to be the Island of Negros, which lies to the west of Zebu, or, as Dampier calls it, Sebo.

[2] Lying to the north-west of Negros.

18th of February we anchored at the NW. end of the Island Mindoro, in ten fathoms water, about three-quarters of a mile from the shore. Mindoro is a large island, the middle of it lying in Lat. 13°, about forty leagues long, stretching NW. and SE. It is high and mountainous, and not very woody. Here we saw great tracks of hogs and beef, and we saw some of each, and hunted them; but they were wild, and we could kill none. While we lay here, there was a canoe with four Indians came from Manilla. They were very shy of us a while; but at last, hearing us speak Spanish, they came to us, and told us that they were going to a friar that lived at an Indian village towards the SE. end of the island. They told us also that the harbour of Manilla is seldom or never without twenty or thirty sail of vessels, most Chinese, some Portuguese, and some few the Spaniards have of their own. They said that when they had done their business with the friar, they would return to Manilla, and hoped to be back again at this place in four days' time. We told them that we came for a trade with the Spaniards at Manilla, and should be glad if they would carry a letter to some merchant there, which they promised to do. But this was only a pretence of ours, to get out of them what intelligence we could as to their shipping, strength, and the like, under colour of seeking a trade; for our business was to pillage. Now if we had really designed to have traded here, this was as fair an opportunity as men could have desired, for these men could have brought us to the friar that they were going to, and a small present to him would have engaged him to do any kindness for us in the way of trade; for the Spanish Governors do not allow of it, and we must trade by stealth.

The 21st, we went from hence with the wind at ENE., a small gale. The 23d, in the morning, we were fair by the SE. end of the Island Luconia,[1] the place that had been so long desired by us. We presently saw a sail coming from the northward, and making after her, we took her in two hours' time. She was a Spanish bark that came from a place called Pangasanam, a small town on the N. end of Luconia, as they told us; probably the same with Pongassinay, which lies on a bay at the NW. side of the island. She was bound to Manilla, but had no goods aboard; and therefore we turned her away. The 23d we took another Spanish vessel that came from the same place as the other. She was laden with rice and cotton cloth, and bound for Manilla also. These goods were purposely for the Acapulco ship; the rice was for the men to live on while they lay there, and in their return; and the cotton cloth was to make sails. The master of this prize was boatswain of the Acapulco ship, which escaped us at Guam, and was now at Manilla. It was this man that gave us the relation of what strength it had, how they were afraid of us there, and of the accident that happened to them, as is before mentioned in the tenth Chapter. We took these two vessels within seven or eight leagues of Manilla.

Luconia I have spoken of already; but I shall now add this further account of it. It is a great island, taking up between six and seven degrees of Latitude in length, and its breadth near the middle is about sixty leagues, but the ends are narrow. The north end lies in about 19° N., and the south end in about 12° 30′. This great island has abundance of small keys or islands lying about it, especially at the north end. The south side fronts towards the rest of the Philippine Islands; of these that are its nearest neighbours, Mindoro, lately mentioned, is the chief, and gives name to the sea or

[1] Not of the whole island, which stretches away south-east of Manilla, into a long jagged peninsula; Dampier evidently means at the southern point of what we may call the mainland

strait that parts it and the other islands from Luconia, being called the Straits of Mindoro. The body of the Island Luconia is composed of many spacious plain savannahs, and large mountains. The north end seems to be more plain and even, I mean freer from hills, than the south end; but the land is all along of a good height. It does not appear so flourishing and green as some of the other islands in this range, especially that of St John, Mindanao, Bat Island, &c.; yet in some places it is very woody. Some of the mountains of this island afford gold, and the savannahs are well stocked with herds of cattle, especially buffaloes. These cattle are in great plenty all over the East Indies; and therefore it is very probable that there were many of these here even before the Spaniards came hither. But now there are also plenty of other cattle, as I have been told, as bullocks, horses, sheep, goats, hogs, &c., brought hither by the Spaniards. It is pretty well inhabited with Indians, most of them, if not all, under the Spaniards, who now are masters of it. The native Indians do live together in towns; and they have priests among them to instruct them in the Spanish religion. Manilla, the chief, or perhaps only city, lies at the foot of a ridge of high hills, facing upon a spacious harbour near the SW. point of the island, in about 14° N. It is environed with a high strong wall, and very well fortified with forts and breastworks. The houses are large, strongly built, and covered with pantile. The streets are large and pretty regular, with a parade[1] in the midst, after the Spanish fashion. There are a great many fair buildings, besides churches and other religious houses, of which there are not a few. The harbour is so large, that some hundreds of ships may ride here; and is never without many, both of their own, and strangers. I have already given you an account of the two ships going and coming between this place and Acapulco. Besides them, they have some small vessels of their own; and they do allow the Portuguese to trade here; but the Chinese are the chief merchants, and they drive the greatest trade; for they have commonly twenty or thirty, or forty junks in the harbour at a time, and a great many merchants constantly residing in the city, beside shop-keepers and handicraftsmen in abundance. Small vessels run up near the town, but the Acapulco ships, and others of greater burthen lie a league short of it, where there is a strong fort also, and storehouses to put goods in. I had the major part of this relation two or three years after this time, from Mr Coppinger our surgeon; for he made a voyage hither from Porto Novo, a town on the coast of Coromandel, in a Portuguese ship, as I think. We were not within sight of this town, but I was shown the hills that overlooked it, and drew a draught of them as we lay off at sea.[2]

The time of the year being now too far spent to do anything here, it was concluded to sail from hence to Pulo Condore, a little parcel of islands on the coast of Cambodia, and carry this prize with us, and there careen if we could find any convenient place for it; designing to return hither again by the latter end of May, and wait for the Acapulco ship that comes about that time. By our draughts (which we were guided by, being strangers to these parts) this seemed to us, then, to be a place out of the way, where we might lie snug for a while, and wait the time of returning for our prey. For we avoided as much as we could, going to lie by at any great place of commerce, lest we should become too much exposed, and perhaps be assaulted by a force

[1] Plaza.

[2] In the edition from which the present text is printed, there is a shaded skeleton drawing, about four inches long by three-quarters high, entitled "A Prospect of ye Coast of ye I. Luconia, near Manila, at 6 L. off shore, ye highest Pike bearing East."

greater than our own. So having set our prisoners ashore, we sailed from Luconia the 26th of February. In our way we went pretty near the shoals of Pracel,[1] and other shoals which are very dangerous. We were very much afraid of them, but escaped them without so much as seeing them, only at the very south end of the Pracel shoals we saw three little sandy islands or spots of sand, standing just above water, within a mile of us. It was the 13th of March before we came in sight of Pulo Condore, or the Island Condore, as "Pulo" signifies. The 14th about noon we anchored on the north side of the island, against a sandy bay two miles from the shore, in ten fathoms clean hard sand, with both ship and prize. Pulo Condore is the principal of a heap of islands, and the only inhabited one of them. They lie in Lat. 8° 40′ N. and about twenty leagues south and by east from the mouth of the River of Cambodia.[2] These islands lie so near together, that at a distance they appear to be but one island. Two of these islands are pretty large, and of a good height; they may be seen fourteen or fifteen leagues at sea; the rest are but little spots. The biggest of the two (which is the inhabited one) is about four or five leagues long, and lies east and west. It is not above three miles broad at the broadest place, in most places not above a mile wide. The other large island is about three miles long, and half-a-mile wide. This island stretches north and south. There are no more islands on the north side, but five or six on the south side of the great island. The mould of these islands for the biggest part is blackish, and pretty deep; only the hills are somewhat stony. The eastern part of the biggest island is sandy, yet all clothed with trees of divers sorts. The trees do not grow so thick as I have seen them in some places, but they are generally large and tall, and fit for any uses. There is one sort of tree much larger than any other on this island, and which I have not seen anywhere else. It is about three or four feet diameter in the body, from whence is drawn a sort of clammy juice, which being boiled a little becomes perfect tar; and if you boil it much it will become hard as pitch.[3] The fruit trees that Nature has bestowed on these isles are mangoes, and trees bearing a sort of grape, and other trees bearing a kind of wild or bastard nutmegs. These all grow wild in the woods, and in very great plenty. The mangoes here grow on trees as big as apple trees. Those at Fort St George are not so large. The fruit of these is as big as a small peach, but long and smaller towards the top. It is of a yellowish colour when ripe; it is very juicy, and of a pleasant smell and delicate taste. When the mango is young, they cut them in two pieces, and pickle them with salt and vinegar, in which they put some cloves and garlic. The grape tree grows with a straight body, of a diameter about a foot or more, and has but few limbs or boughs. The fruit grows in clusters, all about the body of the tree, like the jack, durian, and cacao fruits. There are of them both red and white. They are much like such grapes as grow on our vines, both in shape and colour. The wild nutmeg tree is as big as a walnut tree; but it does not spread so much. The boughs are gross,[4] and the fruit grows among the boughs as the walnut and other fruits. The animals of these islands are some hogs, lizards, guanas, and some of those creatures mentioned in Chapter XI., which are

[1] The Paracel Islands and reefs at the mouth of the Gulf of Tonquin.

[2] Or Mai-Kiang, which on its way to the coast traverses the whole extent of the empire of Annam; Pulo Condore is directly south of its main embouchure, at the mouth of which stands Saigon, chief town of the French colony of Cochin China.

[3] Well known in commerce and for nautical purposes as Cambodia pitch.

[4] Thick.

like, but much bigger, than the guana. Here are many sorts of birds, as parrots, paroquets, doves, and pigeons. Here are also a sort of wild cocks and hens, which crow like ours, but much more small and shrill; and by their crowing we do first find them out in the woods where we shoot them. Their flesh is very white and sweet. There are a great many limpets and mussels, and plenty of green turtle. These islands are pretty well watered with small brooks of fresh water, that run slush[1] into the sea for ten months in the years. The latter end of March they begin to dry away, and in April you shall have none in the brooks but what is lodged in deep holes; but you may dig wells in some places. In May, when the rain comes, the land is again replenished with water, and the brooks run out into the sea.

These islands lie very commodiously in the way to and from Japan, China, Manilla, Tonquin, Cochin China, and in general all this most easterly coast of the Indian continent, whether you go through the Straits of Malacca or the Straits of Sunda between Sumatra and Java; and one of them you must pass in the common way from Europe, or other parts of the East Indies, unless you mean to fetch a great compass round most of the East India islands, as we did. Any ship in distress may be refreshed and recruited here very conveniently, and, besides ordinary accommodations, be furnished with masts, yards, pitch, and tar. The inhabitants are by nation Cochin Chinese, as they told us, for one of them spoke good Malay, which language we learnt a smattering of, and some of us so as to speak it pretty well while we lay at Mindanao; and this is the common tongue of trade and commerce (though it be not in several of them the native language) in most of the East India islands, being the *lingua franca*, as it were, of these parts. I believe it is the vulgar tongue at Malacca, Sumatra, Java, and Borneo; but at Celebes, the Philippine Islands, and the Spice Islands, it seems borrowed for the carrying on of trade. The inhabitants of Pulo Condore are but a small people in stature, well enough shaped, and of a darker colour than the Mindanayans. They are pretty long-visaged, their hair is black and straight, their eyes are but small and black, their noses of a mean bigness and pretty high, their lips thin, their teeth white, and little mouths. They are very civil people, but extraordinary poor. Their chief employment is to draw the juice of those trees that I have described to make tar. They preserve it in wooden troughs, and when they have their cargo they transport it to Cochin China, their mother country. Some others of them employ themselves to catch turtle, and boil up their fat to oil, which they also transport home. These people have great large nets with wide meshes to catch the turtle. The Jamaica turtlers have such, and I did never see the like nets but at Jamaica and here. They are so free of their women that they would bring them aboard and offer them to us, and many of our men hired them for a small matter. This is a custom used by several nations in the East Indies, as at Pegu, Siam, Cochin China, and Cambodia, as I have been told. It is used at Tonquin also to my knowledge, for I did afterwards make a voyage thither, and most of our men had women aboard all the time of their abode there. In Africa also, on the coast of Guinea, our merchants, factors, and seamen that reside there have their black misses. It is accounted a piece of policy to do it, for the chief factors and captains of ships have the great men's daughters offered them, the Mandarin's or noblemen's at Tonquin, and even the King's wives in [New] Guinea; and by this sort of alliance the country people are engaged to a greater friendship. And if there should arise any difference about trade, or anything else, which might provoke the natives to seek some treacherous revenge (to which all these heathen nations are very prone), then these Delilahs would

[1] Full.

certainly declare it to their white friends, and so hinder their countrymen's designs.

These people are idolaters; but their manner of worship I know not. There are a few scattering houses and plantations on the great island, and a small village on the south side of it; where there is a little idol temple, and an image of an elephant, about five feet high, and in bigness proportionable, placed on one side of the temple, and a horse, not so big, placed on the other side of it: both standing with their heads towards the south. The temple itself was low and ordinary, built of wood, and thatched, like one of their houses, which are but very meanly. The images of the horse and the elephant were the most general idols that I observed in the temple of Tonquin when I travelled there. There were other images also, of beasts, birds, and fish; I do not remember I saw any human shape there, nor any such monstrous representations as I have seen among the Chinese. Wherever the Chinese seamen or merchants come (and they are very numerous all over the seas), they have always hideous idols on board their junks or ships, with altars, and images burning before them. These idols they bring ashore with them. And besides those they have in common; every man has one in his own house. Upon some particular solemn days I have seen their Bonzes, or priests, bring whole armfuls of painted papers, and burn them with a great deal of ceremony, being very careful to let no piece escape them. The same day they killed a goat, which had been purposely fatting a month before: this they offer or present before their idol, and then dress it and feast themselves with it. I have seen them do this in Tonquin, where I have at the same time been invited to their feasts: and at Bencoolen, in the Isle of Sumatra, they sent a shoulder of the sacrificed goat to the English, who ate of it and asked me to do so too; but I refused.

When I was at Madras, or Fort St George, I took notice of a great ceremony used for several nights successively by the idolaters inhabiting the suburbs. Both men and women (these very well clad) in a great multitude went in solemn procession with lighted torches, carrying their idols about with them. I know not the meaning of it. I observed some went purposely carrying oil to sprinkle into the lamps, to make them burn the brighter. They began their round about 11 o'clock at night; and having paced it gravely about the streets till 2 or 3 o'clock in the morning, their idols were carried with much ceremony into the temple by the chief of the procession, and some of the women I saw enter the temple particularly. Their idols were different from those of Tonquin, Cambodia, &c., being in human shape.

I have said already that we arrived at these islands the 14th of March 1687. The next day we searched about for a place to careen in; and the 16th we entered the harbour, and immediately provided to careen. Some men were set to fell great trees to saw into plank; others went to unrigging the ship: some made a house to put our goods in, and for the sailmaker to work in. The country people resorted to us, and brought us of the fruits of the island, with hogs, and sometimes turtle; for which they received rice in exchange, which we had a shipload of, taken at Manilla. We bought of them also a good quantity of their pitchy liquor, which we boiled, and used about our ship's bottom. We mixed it first with lime, which we made here; and it made an excellent coat, and stuck on very well. We stayed in this harbour from the 16th of March till the 16th of April; in which time we made a new suit of sails of the cloth that was taken in the prize. We cut a spare main-topmast, and sawed plank to sheathe the ship's bottom; for she was not sheathed all over at Mindanao, and that old plank that was left on then we now ripped off, and clapped on new. While we lay here, two of our men died, who were poisoned at

Mindanao: they told us of it when they found themselves poisoned, and had lingered ever since. They were opened by our doctor, according to their own request before they died, and their livers were black, light and dry, like pieces of cork. Our business being finished here, we left the Spanish prize taken at Manilla, and most of the rice, taking out enough for ourselves: and on the 17th we went from hence to the place where we first anchored, on the north side of the great island, purposely to water; for there was a great stream when we first came to the island, and we thought it was so now. But we found it dried up, only it stood in holes, two or three hogsheads or a tun in a hole; therefore we did immediately cut bamboos, and made spouts, through which we conveyed the water down to the sea-side by taking it up in bowls, and pouring it into these spouts or troughs. We conveyed some of it thus near half-a-mile. While we were filling our water, Captain Reed engaged an old man, one of the inhabitants of this island (the same who, I said, could speak the Malay language), to be his pilot to the Bay of Siam: for he had often been telling us, that he was well acquainted there, and that he knew some islands there where there were fishermen lived, who he thought could supply us with salt-fish to eat at sea; for we had nothing but rice to eat. The easterly monsoon was not yet done; therefore it was concluded to spend some time there, and then take the advantage of the beginning of the western monsoon to return to Manilla again.

The 21st of April 1687, we sailed from Pulo Condore, directing our course W. by S. for the Bay of Siam. The 23d, we arrived at Pulo Uby.[1] The island is about forty leagues to the westward of Pulo Condore; it lies just at the entrance of the Bay of Siam, and the SW. point of land that makes the bay, namely, the Point of Cambodia. This island is about seven or eight leagues round, and it is higher land than any of the Pulo Condore isles. Against the south-east part of it there is a small key, about a cable's length from the main island. This Pulo Uby is very woody. At Pulo Uby we found two small barks laden with rice. They belonged to Cambodia, from whence they came not above two or three days before; and they touched here to fill water. Rice is the general food of all these countries; therefore it is transported by sea from one country to another, as corn is in these parts of the world. For in some countries they produce more than enough for themselves, and send what they can spare to those places where there is but little. The 24th, we went into the Bay of Siam. This is a large deep bay, of which and of this kingdom I shall at present speak but little.[2] We run down into the Bay of Siam till we came to the islands that our Pulo Condore pilot told us of, which lie about the middle of the bay;[3] but as good a pilot as he was, he run us aground; yet we had no damage. Captain Reed went ashore at these islands, where he found a small town of fishermen; but they had no fish to sell, and so we returned empty. We had yet fair weather and very little wind; so that being often becalmed, we were till the 13th of May before we got to Pulo Uby again. There we found two small vessels at anchor on the east side: they were laden with rice and lacquer, which is used in japanning of cabinets. One of these came from Champa, bound to the town of Malacca, which belongs to the Dutch, who took it from the Portuguese; and this shows that they have a trade with Champa. This was a very pretty neat vessel, her bottom very clean and curiously coated; she had about forty men all armed with cortans or broadswords, lances, and some guns

[1] Pulo Obi, off the extreme southern point of the Cambodian peninsula.

[2] Reserving a more particular account to Appendix I. (see Introductory Note on page 115).

[3] Probably Pulo Way, in Lat. 10° N.

that went with a swivel upon their gunwales. They were of the idolaters, natives of Champa, and some of the briskest, most sociable, without fearfulness or shyness, and the most neat and dexterous about their shipping, of any such I have met with in all my travels.[1] The other vessel came from the River of Cambodia and was bound towards the Straits of Malacca. Both of them stopped here, for the westerly winds now began to blow, which were against them, being somewhat belated. We anchored also on the east side, intending to fill water.

The 21st of May we went back from hence towards Pulo Condore. In our way we overtook a great junk that came from Palembang, a town on the Island of Sumatra. She was full laden with pepper which they bought there, and was bound to Siam; but it blowing so hard, she was afraid to venture into that bay, and therefore came to Pulo Condore with us, where we both anchored May 24th. This vessel was of the Chinese make, full of little rooms or partitions like our well-boats. I shall describe them in the next Chapter. The men of this junk told us that the English were settled on the Island of Sumatra, at a place called Sillabar; and the first knowledge we had that the English had any settlement on Sumatra was from these. When we came to an anchor, we saw a small bark at anchor near the shore; therefore Captain Reed sent a canoe aboard her to know from whence they came; and supposing that it was a Malay vessel, he ordered the men not to go aboard, for they are accounted desperate fellows, and their vessels are commonly full of men, who all wear cressets or little daggers by their sides. The canoe's crew, not minding the Captain's orders, went aboard, all but one man that stayed in the canoe. The Malays, who were about twenty of them, seeing our men all armed, thought that they came to take their vessel; therefore at once, on a signal given, they drew out their cressets and stabbed five or six of our men before they knew what the matter was. The rest of our men leaped overboard, some into the canoe and some into the sea, and so got away. Among the rest, one Daniel Wallis leaped into the sea, who could never swim before nor since; yet now he swam very well a good while before he was taken up. When the canoe came aboard, Captain Reed manned two canoes and went to be revenged on the Malays; but they, seeing him coming, cut a hole in their vessel's bottom and went ashore in their boat. Captain Reed followed them, but they ran into the woods and hid themselves.

Here we stayed ten or eleven days, for it blew very hard all the time. While we stayed here, Herman Coppinger our surgeon went ashore, intending to live here; but Captain Reed sent some men and fetched him again. I had the same thoughts, and would have gone ashore too, but waited for a more convenient place. For neither he nor I when we went last on board at Mindanao had any knowledge of the plot that was laid to leave Captain Swan and run away with the ship; and being sufficiently weary of this mad crew, we were willing to give them the slip at any place from whence we might hope to get a passage to an English factory. There was nothing else of moment happened whilst we stayed here.

CHAPTER XV.

HAVING filled our water, cut our wood, and got our ship in a sailing posture while the blustering hard winds lasted, we took the first opportunity of a settled gale to sail towards Manilla. Accordingly, June the 4th 1687, we loosed from Pulo Condore with the wind at SW., fair weather, at a brisk gale. The pepper junk bound

[1] One is tempted to find in this graphic account traces of the Japanese, then little if at all known to even our most experienced navigators.

to Siam remained there waiting for an easterly wind; but one of his men, a kind of bastard Portuguese, came aboard our ship and was entertained for the sake of his knowledge in the several languages of these countries. The wind continued in the SW. but twenty-four hours, or a little more, and then came about to the N. and then to the NE., and the sky became exceeding clear. Then the wind came at E., and lasted betwixt E. and SE. for eight or ten days. Yet we continued plying to windward, expecting every day a shift of wind, because these winds were not according to the season of the year. We were now afraid lest the currents might deceive us and carry us on the shoals of Pracel, which were near us, a little to the NW.; but we passed on to the eastward without seeing any sign of them. Yet we were kept much to the northward of our intended course, and the easterly winds still continuing, we despaired of getting to Manilla, and therefore began to project some new design; and the result was, to visit the Island of Prata,[1] about the Lat. of 20° 40′ N., and not far from us at this time. It is a small low island environed with rocks clear round it, by report. It lieth so in the way between Manilla and Canton, the head of a province and a town of great trade in China, that the Chinese do dread the rocks about it more than the Spaniards did formerly dread Bermudas,[2] for many of their junks coming from Manilla have been lost there, and with abundance of treasure in them, as we were informed by all the Spaniards that ever we conversed with in these parts. They told us also that in these wrecks most of the men were drowned, and that the Chinese did never go thither to take up any of the treasure that was lost there for fear of being lost themselves. But the danger of the place did not daunt us, for we were resolved to try our fortunes there if the winds would permit; and we did beat for it five or six days, but at last were forced to leave that design also for want of winds, for the SE. winds continuing, forced us on the coast of China.

It was the 25th of June when we made the land, and running in towards the shore, we came to an anchor the same day on the NE. end of St John's Island.[3] This island is in Lat. about 22° 30′ N., lying on the S. coast of the province of Quan Tung, or Canton, in China. It is of an indifferent height and pretty plain, and the soil fertile enough. It is partly woody, partly savannahs or pasturage for cattle, and there is some moist arable land for rice. The skirts or outer part of the island, especially that part of it which borders on the main sea, is woody. The middle part of it is good thick grassy pasture, with some groves of trees; and that which is cultivated land is low wet land, yielding plentiful crops of rice, the only grain that I did see here. The tame cattle which this island affords are China hogs, goats, buffaloes, and some bullocks. The hogs of this island are all black; they have but small heads, very short thick necks, great bellies commonly touching the ground, and short legs. They eat but little food, yet they are most of them very fat, probably because they sleep much. The tame fowls are ducks and cocks and hens. I saw no wild fowl but a few small birds.

The natives of this island are Chinese. They are subject to the crown of China, and consequently at this time to the Tartars.[4] The Chinese in general are tall, straight-bodied,

---

[1] Pratos, lying in the north of the Chinese Sea, about equidistant from Canton, Formosa, and the northern extremity of Luzon.

[2] "The vext Bermoothes."

[3] Called in Chinese Chang-cheun, which is evidently an assimilation of the name given by the Portuguese; it lies nearly a degree south-west of Macao.

[4] The Manchoo Tartars, after a war lasting nearly thirty years, had established their dynasty more than forty years before the time of which Dampier writes.

raw-boned men. They are long-visaged, and their foreheads are high; but they have little eyes. Their noses are pretty large, with a rising in the middle. Their mouths are of a mean size, pretty thin lips. They are of an ashy complexion; their hair is black, and their beards thin and long, for they pluck the hair out by the roots, suffering only some few very long straggling hairs to grow about their chin, in which they take great pride, often combing them and sometimes tying them up in a knot; and they have such hairs too growing down from each side of their upper lip like whiskers. The ancient Chinese were very proud of the hair of their heads, letting it grow very long, and stroking it back with their hands curiously, and then winding the plats all together round a bodkin thrust through it at the hinder part of the head; and both men and women did thus. But when the Tartars conquered them, they broke them off this custom they were fond of by main force, insomuch that they resented this imposition worse than their subjection, and rebelled upon it; but being still worsted, were forced to acquiesce; and to this day they follow the fashion of their masters the Tartars, and shave all their heads, only reserving one lock, which some tie up, others let it hang down to a great or small length, as they please. The Chinese in other countries still keep their old custom, but if any of the Chinese is found wearing long hair in China, he forfeits his head; and many of them have abandoned their country to preserve their liberty of wearing their hair, as I have been told by themselves. The Chinese have no hats, caps, or turbans; but when they walk abroad they carry a small umbrella in their hands, wherewith they fence their heads from the sun or the rain by holding it over their heads. If they walk but a little way, they carry only a large fan made of paper or silk, of the same fashion as those our ladies have, and many of them are brought over hither; one of these every man carries in his hand if he do but cross the street, screening his head with it if he has not an umbrella with him. The common apparel of the men is a loose frock and breeches. They seldom wear stockings, but they have shoes, or a sort of slippers rather. The men's shoes are made diversely. The women have very small feet, and consequently but little shoes, for from their infancy their feet are kept swathed up with bands as hard as they can possibly endure them; and from the time they can go till they have done growing, they bind them up every night. This they do purposely to hinder them from growing, esteeming little feet to be a great beauty. But by this unreasonable custom they do in a manner lose the use of their feet, and instead of going, they only stumble about their houses, and presently squat down again, being, as it were, confined to sitting all the days of their lives. They seldom stir abroad; and one would be apt to think that, as some have conjectured, their keeping up their fondness for this fashion were a stratagem of the men's to keep them from gadding and gossiping about and confine them at home. They are kept constantly to their work, being fine needle-women, and making many curious embroideries, and they make their own shoes; but if any stranger be desirous to bring away any for novelty's sake, he must be a great favourite to get a pair of shoes of them, though he give twice their value. The poorer sort of women trudge about the streets, and to the market, without shoes or stockings; and these cannot afford to have little feet, being to get their living with them.

The Chinese, both men and women, are very ingenious, as may appear by the many curious things that are brought from thence, especially the porcelain or China earthenware. The Spaniards of Manilla, that we took on the coast of Luconia, told me that this commodity is made of conch shells, the inside of which looks like mother-of-pearl. But the Portuguese, lately mentioned, who had lived in China, and spoke that and the neighbouring languages very well, said that it was made of a fine sort of clay that was dug in the province of Canton. I have often made inquiry about it, but could never be well satisfied in it; but while I was on the coast of Canton I forgot to inquire about it. They make very fine lacquer ware also, and good silks; and they are curious at painting and carving. China affords drugs in great abundance, especially China root; but this is not peculiar to that country alone, for there is much of this root growing in Jamaica, particularly at Sixteen Mile Walk; and in the Bay of Honduras it is very plentiful. There is a great store of sugar made in this country; and tea in abundance is brought from thence, being much used there, and in Tonquin and Cochin China as common drinking, women sitting in the streets and selling dishes of tea hot and ready made; they call it Chau, and even the poorest people sip it. But the tea at Tonquin or Cochin China seems not so good, or of so pleasant a bitter, or of so fine a colour, or such virtue, as this in China; for I have drank of it in these countries, unless the fault be in their way of making it, for I made none there myself; and by the high red colour it looks as if they made a decoction of it, or kept it stale. Yet, at Japan, I was told there is a great deal of pure tea, very good.[1]

The Chinese are very great gamesters, and they will never be tired with it, playing night and day, till they have lost all their estates, then it is usual with them to hang themselves. This was frequently done by the Chinese factor at Manilla, as I was told by Spaniards that lived there. The Spaniards themselves are much addicted to gaming, and are very expert at it; but the Chinese are too subtle for them, being in general a very cunning people. But a particular account of them and their country would fill a volume; nor does my short experience of them qualify me to say much of them. Wherefore, to confine myself chiefly to what I observed at St John's Island, where we lay some time, and visited the shore every day to buy provision, as hogs, fowl, and buffalo. Here was a small town standing in a wet swampy ground, with many filthy ponds amongst the houses, which were built on the ground as ours are, not on posts as at Mindanao. In these ponds were plenty of ducks; the houses were small and low, and covered with thatch, and inside were but ill furnished, and kept nastily; and I have been told by one who was there, that most of the houses in the city of Canton itself are but poor and irregular. The inhabitants of this village seem to be most husbandmen; they were at this time very busy in sowing their rice, which is their chief commodity. The land in which they choose to sow the rice is low and wet, and when ploughed, the earth was like a mass of mud. They ploughed their land with a small plough drawn by one buffalo, and one man both holds the plough and drives the beast. When the rice is ripe and gathered in, they tread it out of the ear with buffaloes, in a large round place made with a hard floor fit for that purpose, where they chain three or four of these beasts, one at the tail of the other; and driving them round in a ring, as in a horse-mill, they so order it that the buffaloes may tread upon it all. I was once ashore at this island, with seven or eight Englishmen more, and having occasion to stay some time, we killed a small "shore" or young porker, and roasted it for our dinners. While we were busy dressing of our pork, one of the natives came and sat down by us; and when our dinner was ready, we cut a good piece and gave it him, which he willingly received. But by signs he begged more, and withal pointed into the woods; yet we did not understand his meaning, nor much mind him, till our hunger was pretty well assuaged, although he did still make signs, and walking a little way from us, he beckoned to us to come to him, which at last I did,

[1] Tea had been introduced in England, though only as a rare luxury, some thirty years before Dampier wrote.

and two or three more. He, going before, led the way in a small blind path through a thicket into a small grove of trees, in which there was an old idol-temple about ten feet square. The walls of it were about nine feet high, and two feet thick, made of bricks. The floor was paved with broad bricks, and in the middle of the floor stood an old rusty iron bell on its brims. This bell was about two feet high, standing flat on the ground; the brims on which it stood were about sixteen inches diameter. From the brims it did taper away a little towards the head, much like our bells, but that the brims did not turn out so much as ours do. On the head of the bell there were three iron bars as big as a man's arm, and about ten inches long from the top of the bell, where the ends joined as in a centre, and seemed of one mass with the bell, as if cast together. These bars stood all parallel to the ground; and their further ends, which stood triangularly and opening from each other at equal distances, like the flyers of our kitchen-jacks, were made exactly in the shape of the paw of some monstrous beast, having sharp claws on it. This, it seems, was their god; for as soon as our zealous guide came before the bell, he fell flat on his face, and beckoned to us, seeming very desirous to have us do the like. At the inner side of the temple, against the walls, there was an altar of white hewn stone. The table of the altar was about three feet long, sixteen inches broad, and three inches thick. It was raised about two feet from the ground, and supported by three small pillars of the same white stone. On this altar there were several small earthen vessels; one of them was full of small sticks that had been burned at one end. Our guide made a great many signs for us to fetch and to leave some of our meat there, and seemed very importunate; but we refused. We left him there, and went aboard. I did see no other temple nor idol here.

While we lay at this place, we saw several small China junks sailing in the lagoon between the islands and the main; one came and anchored by us. I and some more of our men went aboard to view her. She was built with a square flat head as well as stern, only the head or fore-part was not so broad as the stern. On her deck she had little thatched houses like hovels, covered with palmetto leaves, and raised about three feet high, for the seamen to creep into. She had a pretty large cabin, wherein there was an altar and a lamp burning; I did but just look in, and saw not the idol. The hold was divided into many small partitions, all of them made so tight, that if a leak should spring up in any one of them, it could go no farther, and so could do but little damage, but only to the goods in the bottom of that room where the leak springs up. Each of these rooms belongs to one or two merchants, or more; and every man freights his goods in his own room, and probably lodges there if he be on board himself. These junks have only two masts, a mainmast and a foremast. The foremast has a square yard and a square sail; but the mainmast has a sail narrow aloft, like a sloop's sail; and in fair weather they use a topsail, which is to haul down on the deck in foul weather, yard and all; for they do not go up to furl it. The mainmast in their biggest junks seemed to me as big as any third-rate man-of-war's mast in England, and yet not pieced as ours, but made of one grown tree; and in all my travels I never saw any single tree masts so big in the body, and so long, and yet so well tapered, as I have seen in the Chinese junks.

Some of our men went over to a pretty large town on the continent of China, where we might have furnished ourselves with provision, which was a thing we were always in want of, and was our chief business here; but we were afraid to lie in this place any longer, for we had some signs of an approaching storm, this being the time of the year in which storms are expected on this coast; and here was no safe riding.

It was now the time of the year for the SW. monsoon; but the wind had been whiffling about from one part of the compass to another for two or three days, and sometimes it would be quite calm. This caused us to put to sea, that we might have sea-room at least; for such fluttering weather is commonly the forerunner of a tempest. Accordingly we weighed anchor and set out; yet we had very little wind all the next night. But the day ensuing, which was the 4th of July, about 4 o'clock in the afternoon, the wind came to the NE. and freshened upon us, and the sky looked very black in that quarter, and the black clouds began to rise apace and move towards us, having hung all the morning in the horizon. This made us take in our topsails; and the wind still increasing, about 9 o'clock we reefed our mainsail and foresail. At ten we furled our foresail, keeping under a mainsail and mizzen. At 11 o'clock we furled our mainsail, and ballasted our mizzen, at which time it began to rain, and by 12 o'clock at night it blew exceeding hard, and the rain poured down as through a sieve. It thundered and lightened prodigiously, and the sea seemed all of a fire about us; for every sea that broke sparkled like lightning. The violent wind raised the sea presently to a great height, and it ran very short and began to break in on our deck. One sea struck away the rails of our head; and our sheet anchor, which was stowed with one fluke, or bending of the iron over the ship's gunwale, and lashed very well down to the side, was violently washed off, and had like to have struck a hole in our bow as it lay beating against it. Then we were forced to put right before the wind, to stow our anchor again, which we did with much ado; but afterwards we durst not adventure to bring our ship to the wind again, for fear of foundering, for the turning the ship either to or from the wind is dangerous in such violent storms. The fierceness of the weather continued till 4 o'clock that morning, in which time we cut away two canoes that were towing astern. After 4 o'clock the thunder and the rain abated, and then we saw a *Corpus Sant*[1] at our main-topmast head, on the very top of the truck of the spindle. This sight rejoiced our men exceedingly; for the height of the storm is commonly over when the *Corpus Sant* is seen aloft; but when they are seen lying on the deck it is generally accounted a bad sign. A *Corpus Sant* is a certain small glittering light. When it appears, as this did, on the very top of the mainmast or at a yard-arm, it is like a star; but when it appears on the deck it resembles a great glow-worm. The Spaniards have another name for it (though I take even this to be a Spanish or Portuguese name, and a corruption only of *Corpus Sanctum*); and I have been told that when they see them they presently go to prayers, and bless themselves for the happy sight. I have heard some ignorant seamen discoursing how they have seen them creep, or, as they say, travel about in the scuppers, telling many dismal stories that happened at such times; but I did never see any one stir out of the place where it first was fixed, except upon deck, where every sea washes it about. Neither did I ever see any but when we have had hard rain as well as wind, and therefore do believe it is some jelly: but enough of this. We continued scudding right before wind and sea from 2 till 7 o'clock in the morning; and then the wind being much abated, we set our mizzen again, and brought our ship to the wind, and lay under a mizzen till eleven. Then it fell flat calm,

[1] "Corposant. A name given to the luminous appearance often beheld in a dark tempestuous night about the decks and rigging of a ship, especially about the mast-heads, yard-arms, &c., caused by the electric fluid passing upwards and downwards, 'by means of the humidity on the masts and rigging,' and 'most frequent in heavy rain accompanied by lightning.'"—*Young's Nautical Dictionary.*

and it continued so for about two hours; but the sky looked very black and rueful, especially in the SW., and the sea tossed us about like an eggshell for want of wind. About 1 o'clock in the afternoon, the wind sprung up at SW., out of the quarter from whence we did expect it;[1] therefore we presently brailed up our mizzen and wore our ship; but we had no sooner put our ship before the wind but it blew a storm again, and it rained very hard, though not so violently as the night before; but the wind was altogether as boisterous, and so continued till 10 or 11 o'clock at night. All which time we scudded, or run, before the wind very swift, though only with our bare poles, that is, without any sail abroad. Afterwards the wind died away by degrees, and before day we had but little wind and fine clear weather.

I was never in such a violent storm in all my life; so said all the company. This was near the change of the moon; it was two or three days before the change. The 6th, in the morning, having fine handsome weather, we got up our yards again, and began to dry ourselves and our clothes, for we were all well sopped. This storm had deadened the hearts of our men so much that, instead of going to buy more provision at the same place from whence we came before the storm, or of seeking any more from the Island of Prata, they thought of going somewhere to shelter before the full moon, for fear of another such storm at that time; for commonly, if there is any very bad weather in the month it is about two or three days before or after the full or change of the moon. These thoughts, I say, put our men on thinking where to go; and the draughts or sea-plats[2] being first consulted, it was concluded to go to certain islands lying in Lat. 23° N., called Pescadores. For there was not a man aboard that was anything acquainted on these coasts; and therefore all our dependence was on the draughts, which only pointed out to us where such and such places or islands were, without giving us any account what harbour, roads, or bays there were, or the produce, strength, or trade of them. These we were forced to seek after ourselves. The Pescadores are a great many inhabited islands, lying near the Island of Formosa, between it and China, in or near Lat. 23° N., almost as high as the Tropic of Cancer.[3] These Pescadore Islands are moderately high, and appear much like our Dorsetshire and Wiltshire Downs in England. They produce thick short grass and a few trees. They are pretty well watered, and they feed abundance of goats and some great cattle. There are abundance of mounts[4] and old fortifications on them, but of no use now, whatever they have been. Between the two easternmost islands there is a very good harbour, which is never without junks riding in it; and on the west side of the easternmost island there is a large town and fort commanding the harbour. The houses are but low, yet well built, and the town makes a fine prospect. This is a garrison of the Tartars, wherein are also three or four hundred soldiers, who live here three years, and then they are removed to some other place. On the island on the west side of the harbour, close by the sea, there is a small town of Chinese, and most of the other islands have some Chinese living on them, more or less.

Having, as I said before, concluded to go to these islands, we steered away for them. The 20th of July we had first sight of them, and steered in among them, finding no place to anchor in till we came into the harbour before mentioned. We blundered in, knowing little of our way, and we admired[5] to see so many

[1] It had been in the NE. before; and thus, though Dampier knew nothing about modern theories of storms, it seems clear that in the two hours' lull he had passed through the vortex of a tornado.

[2] Plans or charts.

[3] They really lie about 20′ to the northward of the Tropic.

[4] Mounds.

[5] Wondered.

junks going and coming, and some at anchor, and so great a town as the neighbouring easternmost town, the Tartarian garrison; for we did not expect nor desire to have seen any people, being in care to lie concealed in these seas. However, seeing we were here, we boldly ran into the harbour, and presently sent ashore our canoe to the town. Our people were met by an officer at their landing, and our quartermaster, who was the chief man in the boat, was conducted before the Governor and examined, of what nation we were, and what was our business here. He answered that we were English, and were bound to Amoy or Anhay, which is a city standing on a navigable river in the province of Fo-kien in China, a place of vast trade, there being a huge multitude of ships there, and in general on all these coasts, as I have heard of several that have been there. He said also, that having received some damage by a storm, we therefore put in here to refit before we would adventure to go farther, and that we did intend to lie here till after the full moon, for fear of another storm. The Governor told him that we might better refit our ship at Amoy than here, and that he heard that two English vessels were arrived there already, and that he should be very ready to assist us in anything, but we must not expect to trade there, but must go to the places allowed to entertain merchant strangers, which were Amoy and Macao. (Macao is a town of great trade also, lying in an island at the very mouth of the River of Canton. It is fortified and garrisoned by a large Portuguese colony, but yet under the Chinese Governor, whose people inhabit one moiety of the town, and lay on the Portuguese what tax they please; for they dare not disoblige the Chinese for fear of losing their trade.) However, the Governor very kindly told our quartermaster that whatsoever we wanted, if that place could furnish us, we should have it; yet that we must not come ashore on that island, but he would send aboard some of his men to know what we wanted, and they should also bring it off to us; that nevertheless we might go on shore on the other islands, to buy refreshments of the Chinese. After the discourse was ended, the Governor dismissed him with a small jar of flour and three or four large cakes of very fine bread, and about a dozen pine-apples and water-melons (all very good in their kind) as a present to the Captain.

The next day an eminent officer came aboard with a great many attendants. He wore a black silk cap of a particular make, with a plume of black and white feathers standing up almost round his head behind, and all his outside clothes were black silk. He had a loose black coat which reached to his knees, and his breeches were of the same, and underneath his coat he had two garments more of other coloured silk. His legs were covered with small black limber boots. All his attendants were in a very handsome garb of black silk, all wearing those small black boots and caps. These caps were like the crown of a hat made of palmetto leaves, like our straw-hats, but without brims, and coming down but to their ears. These had no feathers, but had an oblong button on the top, and from between the button and the cap there fell down all round their head, as low as the cap reached, a sort of coarse hair like horse-hair, dyed (as I suppose) of a light red colour. The officer brought aboard, as a present from the Governor, a young heifer, the fattest and kindliest beef that I did ever taste in any foreign country; it was small yet full grown; two large hogs, four goats, two baskets of fine flour, twenty great flat cakes of fine well-tasted bread, two great jars of arrack (made of rice as I judged), called by the Chinese Sam-Shu, and fifty-five jars of Hog-Shu, as they call it, and our Europeans from them. This is a strong liquor, made of wheat, as I have been told. It looks like mum,[1]

[1] Described in Bailey as "a strong liquor brought from Brunswick, in Germany"—a drink so potent as to

and tastes much like it, and is very pleasant and hearty. Our seamen love it mightily, and will lick their lips with it; for scarce a ship goes to China but the men come home fat with soaking this liquor, and bring store of jars of it home with them. It is put into small, white, thick jars that hold near a quart; the double jars hold about two quarts. These jars are small below, and thence rise up with a pretty full belly, closing in pretty short at top, with a small thick mouth. Over the mouth of the jar they put a thin chip cut round just so as to cover the mouth, over that a piece of paper, and over that they put a great lump of clay, almost as big as the bottle or jar itself, with a hollow in it to admit the neck of the bottle, made round and about four inches long; this is to preserve the liquor. If the liquor take any vent, it will be sour presently; so that when we buy any of it of the ships from China returning to Madras or Fort St George, where it is then sold, or of the Chinese themselves, of whom I have bought it at Achin and Bencoolen in Sumatra, if the clay be cracked, or the liquor mothery,[1] we make them take it again. A quart jar there is worth sixpence. Besides this present from the Governor, there was a captain of a junk sent two jars of arrack, and abundance of pine-apples and water-melons. Captain Reed sent ashore, as a present to the Governor, a curious Spanish silver-hilted rapier, an English carbine, and a gold chain; and when the officer went ashore three guns were fired. In the afternoon the Governor sent off the same officer again, to compliment the Captain for his civility, and promised to retaliate his kindness before we departed; but we had such blustering weather afterwards, that no boat could come aboard.

We stayed here till the 29th, and then sailed from hence, with the wind at SW., and pretty fair weather. We now directed our course for some islands we had chosen to go to that lie between Formosa and Luconia. They are laid down in our plots[2] without any name, only with a figure of 5, denoting the number of them. It was supposed by us that these islands had no inhabitants, because they had not any name by our hydrographers; therefore we thought to lie there secure, and be pretty near the Island of Luconia, which we did still intend to visit. In going to them we sailed by the SW. end of Formosa, leaving it on our larboard side. The 6th of August we arrived at the five islands that we were bound to, and anchored on the east side of the northernmost island, in fifteen fathoms, a cable's length from the shore. Here, contrary to our expectation, we found abundance of inhabitants in sight; for there were three large towns all within a league of the sea, and another larger town than any of the three on the back side of a small hill close by also, as we found afterwards. These islands having no particular names in the draughts, some or other of us made use of the seamen's privilege to give them what names we pleased. Three of the islands were pretty large; the westernmost is the biggest. This the Dutchmen who were among us called the Prince of Orange's Island, in honour of his present majesty. The other two great islands are about four or five leagues to the eastward of this. The northernmost of them, where we first anchored, I called the Duke of Grafton's Isle as soon as we landed on it; having married my wife out of his Duchess's family, and leaving her at Arlington House at my going abroad. The other great isle our seamen called the Duke of Monmouth's Island; this is about a league to the southward of Grafton Isle. Between Monmouth and the south end of Orange Island there are two small islands of a roundish form, lying east and west. The easternmost island of the two our men unanimously called Bashee Island, from

make "mum" the word with the imbiber.

[1] Mouldy, muddy.

[2] Plats; maps, charts, or plates.

a liquor which we drank there plentifully every day after we came to an anchor at it. The other, which is the smallest of all, we called Goat Island, from the great number of goats there; and to the northward of them all are two high rocks. Orange Island, which is the biggest of them all, is not inhabited. It is high land, flat and even on the top, with steep cliffs against the sea; for which reason we could not go ashore there, as we did on all the rest. Monmouth and Grafton Isles are very hilly, with many of those steep inhabited precipices on them that I shall describe particularly. The two small islands are flat and even; only the Bashee Island has one steep, scraggy hill, but Goat Island is all flat and very even. The mould of these islands in the valleys is blackish in some places, but in most red. The hills are very rocky; the valleys are well watered with brooks of fresh water, which run into the sea in many different places. The soil is indifferent fruitful, especially in the valleys, producing pretty great plenty of trees (though not very big) and thick grass. The sides of the mountains have also short grass, and some of the mountains have mines within them; for the natives told us that the yellow metal they showed us (as I shall speak more particularly) came from these mountains; for when they held it up they would point towards them.

The fruit of the islands are a few plantains, bananas, pine-apples, pumpkins, sugar-canes, &c.; and there might be more if the natives would, for the ground seems fertile enough. Here are great plenty of potatoes and yams, which is the common food for the natives for bread kind; for those few plantains they have are only used as fruit. They have some cotton growing here, of the small plants. Here are plenty of goats and abundance of hogs, and few fowls, either wild or tame. For this I have always observed in my travels, both in the East and West Indies, that in those places where there is plenty of grain, that is, of rice in the one and maize in the other, there are also found great abundance of fowls; but on the contrary, few fowls in those countries where the inhabitants feed on fruits and roots only. The few wild fowls that are here are paroquets and some other small birds. Their tame fowl are only a few cocks and hens.

Monmouth and Grafton Islands are very thick inhabited; and Bashee Island has one town on it. The natives of these islands are short, squat people; they are generally round-visaged, with low foreheads and thick eyebrows; their eyes of a hazel colour and small, yet bigger than the Chinese; short low noses, and their lips and mouths middle proportioned. Their teeth are white; their hair is black, and thick, and lank, which they wear but short; it will just cover their ears, and so it is cut round very even. Their skins are of a very dark copper colour. They wear no hat, cap, or turbat,[1] nor anything to keep off the sun. The men for the biggest part have only a small clout to cover their nakedness; some of them have jackets made of plantain leaves, which were as rough as any bear's skin. I never saw such rugged things. The women have a short petticoat made of cotton, which comes a little below their knees. It is a thick sort of stubborn cloth, which they make themselves of their cotton. Both men and women wear large earrings, made of that yellow metal before mentioned. Whether it were gold or no I cannot positively say; I took it to be so, it was heavy, and of the colour of our paler gold. I would fain have brought away some to have satisfied my curiosity, but I had nothing wherewith to buy any. Captain Reed bought two of these rings with some iron, of which the people are very greedy; and he would have bought more, thinking he was come to a very fair market, but that the paleness of the metal made him and his crew distrust its being right gold. For my part, I should have ventured on the purchase of some;

[1] Turban.

but having no property in the iron, of which we had great store on board, sent from England by the merchants along with Captain Swan, I durst not barter it away. These rings when first polished look very gloriously; but time makes them fade, and turn to a pale yellow. Then they make a soft paste of red earth, and, smearing it over their rings, they cast them into a quick fire, where they remain till they be red-hot; then they take them out and cool them in water, and rub off the paste; and they look again of a glorious colour and lustre. These people make but small low houses. The sides, which are made of small posts, wattled with boughs, are not above four feet and a half high: the ridge pole is about seven or eight feet high. They have a fireplace at one end of their houses, and boards placed on the ground to lie on. They inhabit together in small villages, built on the sides and tops of rocky hills; three or four rows of houses one above another, and on such steep precipices, that they go up to the first row with a wooden ladder, and so with a ladder still from every story up to that above it: there being no [other] way to ascend. The plain on the first precipice may be so wide as to have room both for a row of houses that stand all along on the edge or brink of it, and a very narrow street running along before their doors; between the row of houses and the foot of the next precipice, the plain of which is in a manner level to the tops of the houses below; and so for the rest. The common ladder to each row or street comes up at a narrow passage left purposely about the middle of it; and the street being bounded with a precipice also at each end, it is but drawing up the ladder, if they be assaulted, and then there is no coming at them from below, but by climbing up as against a perpendicular wall; and that they may not be assaulted from above, they take care to build on the side of such a hill whose back side hangs over the sea, or is some high, steep, perpendicular precipice, altogether inaccessible. These precipices are natural; for the rocks seem too hard to work on; nor is there any sign that art has been employed about them. On Bashee Island there is one such, and built upon, with its back next the sea. Grafton and Monmouth Isles are very thick set with these hills and towns; and the natives, whether for fear of pirates, or foreign enemies, or factions among their own clans, care not for building but in these fastnesses, which I take to be the reason that Orange Isle, though the largest, and as fertile as any, yet, being level and exposed, has no inhabitants. I never saw the like precipices and towns.

These people are pretty ingenious also in building boats. Their small boats are much like our Deal yawls, but not so big; and they are built with very narrow plank, pinned with wooden pins and some nails. They have also some pretty large boats, which will carry forty or fifty men; these they row with twelve or fourteen oars of a side. They are built much like the small ones, and they row double-banked; that is, two men sitting on one bench, but one rowing on one side, the other on the other side, of the boat. They understand the use of iron, and work it themselves. Their bellows are like those at Mindanao. The common employment for the men is fishing; but I did never see them catch much: whether it is more plenty at other times of the year I know not. The women do manage their plantations.

I did never see them kill any of their goats or hogs for themselves; yet they would beg the paunches of the goats that they themselves did sell to us: and if any of our surly seamen did heave them into the sea, they would take them up again, and the skins of the goats also. They would not meddle with hogs' guts, if our men threw away any besides what they made chitterling and sausages off. The goats' skins these people would carry ashore, and making a fire they would singe off all the hair, and afterwards let the skin lie and parch on the coals, till they thought it eatable; and then they would gnaw it, and tear it to pieces with their teeth, and at last swallow it. The paunches of the goats would make them an excellent dish: they dressed it in this manner. They would turn out all the chopped grass and crudities found in the maw[1] into their pots, and set it over the fire, and stir it about often; this would smoke, and puff, and heave up as it was boiling; wind breaking out of the ferment, and making a very savoury stink. While this was doing, if they had any fish, as commonly they had two or three small fish, these they would make very clean (as hating nastiness belike) and cut the flesh from the bone, and then mince the flesh as small as possibly they could; and when that in the pot was well boiled, they would take it up, and strewing a little salt into it they would eat it, mixed with their raw minced fish. The dung in the maw would look like so much boiled herbs minced very small; and they took up their mess with their fingers, as the Moors do their pillau, using no spoons. They had another dish made of a sort of locusts, whose bodies were about an inch and a half long, and as thick as the top of one's little finger; with large thin wings, and long and small legs. At this time of the year these creatures came in great swarms to devour their potato-leaves and other herbs; and the natives would go out with small nets, and take a quart at one sweep. When they had enough, they would carry them home, and parch them over the fire in an earthen pan; and then their wings and legs would fall off, and their heads and backs would turn red like boiled shrimps, being before brownish. Their bodies being full, would eat very moist, their heads would crackle in one's teeth. I did once eat of this dish, and liked it well enough; but their other dish my stomach would not take.

Their common drink is water; as it is of all other Indians. Besides which, they make a sort of drink with the juice of the sugar-cane, which they boil and put some small black sort of berries among it. When it is well boiled, they put it into great jars, and let it stand three or four days, and work. Then it settles and becomes clear, and is presently fit to drink. This is an excellent liquor, and very much like English beer, both in colour and taste. It is very strong, and I do believe very wholesome: for our men, who drank briskly of it all day for several weeks, were frequently drunk with it, and never sick after it. The natives brought a vast deal of it every day to those aboard and ashore: for some of our men were ashore at work on Bashee Island; which island they gave that name to from their drinking this liquor there, that being the name which the natives called this liquor by: and as they sold it to our men very cheap, so they did not spare to drink it as freely. And indeed, from the plenty of this liquor, and their plentiful use of it, our men called all these islands the Bashee Islands.

What language those people speak I know not: for it had no affinity in sound to the Chinese, which is spoken much through the teeth; nor yet to the Malay language. They called the metal that their earrings were made of, *Bullawan*, which is the Mindanao word for gold; therefore probably they may be related to the Philippine Indies: for that is the general name for gold among all those Indians. I could not learn whence they have their iron; but it is most likely they go in their great boats to the north end of Luconia, and trade with the Indians of that island for it. Neither did I see anything besides iron, and pieces of buffaloes' hides, which I could judge that they bought of strangers. Their clothes were of their own growth and manufacture. These men had wooden lances, and a few lances headed with iron; which are all the weapons that they have. Their armour is a piece of buffalo hide, shaped like our carters' frocks, being without sleeves, and sewed both

[1] Stomach.

sides together, with holes for the head and the arms to come forth. This buff-coat reaches down to their knees; it is close about their shoulders, but below it is three feet wide, and as thick as a board.

I could never perceive them to worship anything, neither had they any idols; neither did they seem to observe any one day more than another. I could never perceive that one man was of greater power than another; but they seemed to be all equal: only every man ruling in his own house, and the children respecting and honouring their parents. Yet it is probable that they have some law, or custom, by which they are governed: for while we lay here we saw a young man buried alive in the earth; and it was for theft, as far as we could understand from them. There was a great deep hole dug, and abundance of people came to the place to take their last farewell of him. Among the rest, there was one woman who made great lamentation, and took off the condemned person's earrings. We supposed her to be his mother. After he had taken his leave of her and some others, he was put into the pit, and covered over with earth. He did not struggle, but yielded very quietly to his punishment; and they crammed the earth close upon him, and stifled him.

They have but one wife, with whom they live and agree very well; and their children live very obediently under them. The boys go out a-fishing with their fathers, and the girls live at home with their mothers: and when the girls are grown pretty strong, they send them to their plantations, to dig yams and potatoes, of which they bring home on their heads every day enough to serve the whole family: for they have no rice nor maize. Their plantations are in the valleys, at a good distance from their houses: where every man has a certain spot of land, which is properly his own. This he manages himself for his own use; and provides enough, that he may not be beholden to his neighbour. Notwithstanding the seeming nastiness of their dish of goat's maw, they are in their persons a very neat cleanly people, both men and women: and they are withal the quietest and civilest people that I did ever meet with. I could never perceive them to be angry with one another. I have admired to see twenty or thirty boats aboard our ship at a time, and yet no difference among them, but all civil and quiet, endeavouring to help each other on occasion: no noise, nor appearance of distaste: and although sometimes cross accidents would happen, which might have set other men together by the ears, yet they were not moved by them. They have no sort of coin: but they have small crumbs of the metal before described, which they bind up very safe in plantain-leaves, or the like. This metal they exchange for what they want, giving a small quantity of it, about two or three grains, for a jar of drink that would hold five or six gallons. They have no scales, but give it by guess. Thus much in general.

To proceed, therefore, with our affairs. I have said before that we anchored here the 6th of August. While we were furling our sails, there came near 100 boats of the natives aboard, with three or four men in each, so that our deck was full of men. We were at first afraid of them, and therefore got up twenty or thirty small arms on our poop, and kept three or four men as sentinels, with guns in their hands, ready to fire on them if they had offered to molest us. But they were pretty quiet, only they picked up such old iron as they found on our deck; and they also took out our pump-bolts, and linch-pins out of the carriages of our guns, before we perceived them. At last one of our men perceived one of them very busy getting out one of our linch-pins, and took hold of the fellow, who immediately bawled out; and all the rest presently leaped overboard—some into their boats, others into the sea—and they all made away for the shore. But when we perceived their fright we made much of him that was in hold, who stood trembling all the while; and at last we gave him a small piece of iron, with which he immediately leaped overboard and swam to his consorts, who hovered about our ship to see the issue. Then we beckoned to them to come aboard again, being very loth to lose a commerce with them. Some of the boats came aboard again, and they were always very honest and civil afterwards. We presently after this sent a canoe ashore to see their manner of living, and what provision they had. The canoe's crew were made very welcome with Bashee drink, and saw abundance of hogs, some of which they bought and returned aboard. After this the natives brought aboard both hogs and goats to us in their own boats; and every day we should have fifteen or twenty hogs and goats in boats aboard by our side. These we bought for a small matter. We could buy a good fat goat for an old iron hoop, and a hog of seventy or eighty pounds' weight for two or three pounds of iron. Their drink also they brought off in jars, which we bought for old nails, spikes, and leaden bullets. Besides the forementioned commodities, they brought aboard great quantities of yams and potatoes, which we purchased for nails, spikes, or bullets. It was one man's work to be all day cutting out bars of iron into small pieces with a cold chisel, and these were for the great purchases of hogs and goats, which they would not sell for nails, as their drink and roots. We never let them know what store we had, that they might value it the more. Every morning, as soon as it was light, they would thus come aboard with their commodities, which we bought as we had occasion. We did commonly furnish ourselves with as many goats and roots as served us all the day; and their hogs we bought in large quantities as we thought convenient, for we salted them. Their hogs were very sweet, but I never saw so many measled ones.

We filled all our water at a curious brook close by us in Grafton Isle, where we first anchored. We stayed there about three or four days before we went to other islands. We sailed to the southward, passing on the east side of Grafton Island; and then passed through between that and Monmouth Island, but we found no anchoring till we came to the north end of Monmouth Island, and there we stopped during one tide. When we went from hence, we coasted about two leagues to the southward on the west side of Monmouth Island; and finding no anchor ground, we stood over to Bashee Island, and came to an anchor on the north-east part of it against a small sandy bay in seven fathom clean hard sand, and about a quarter of a mile from the shore. We presently built a tent ashore to mend our sails in, and stayed all the rest of our time here, viz., from the 13th of August till the 26th of September. In which time we mended our sails and scrubbed our ship's bottom very well; and every day some of us went to their towns and were kindly entertained by them. Their boats also came aboard with their merchandise to sell, and lay aboard all day; and if we did not take it off their hands one day, they would bring the same again the next. We had yet the winds at SW. and SSW., mostly fair weather. In October we did expect the winds to shift to the NE., and therefore we provided to sail (as soon as the eastern monsoon was settled) to cruise off Manilla. Accordingly we provided a stock of provision. We salted seventy or eighty good fat hogs, and bought yams and potatoes good store to eat at sea.

About the 24th of September the winds shifted about to the E., and thence to the NE., fine fair weather. The 25th it came at N. and began to grow fresh, and the sky began to be clouded, and the wind freshened on us. At 12 of the clock at night it blew a very fierce storm. We were then riding with our best bower ahead, and though our yards and topmast were down, yet we drove. This obliged us to let go our sheet anchor, veering out a good scope of cable, which stopped us till 10 or 11 of the clock the next day. Then the wind came on so fierce that she drove again with both anchors ahead. The wind was now at N. by W., and we kept driving till 3 or 4 of the clock in the afternoon; and it was well for us that there were no islands, rocks, or sands in our way, for if there had been, we must have been driven upon them. We used our utmost endeavours to stop her, being loth to go to sea, because we had six of our men ashore who could not get off now. At last we were driven out into deep water, and then it was in vain to wait any longer; therefore we hove in our sheet cable, and got up our sheet anchor, and cut away our best bower (for to have heaved her up then would have gone near to have foundered us), and so put to sea. We had very violent weather the night ensuing, with very hard rain; and we were forced to scud with our bare poles till 3 o'clock in the morning. Then the wind slackened, and we brought our ship to under a mizzen, and lay with our head to the westward. The 27th the wind abated much, but it rained very hard all day and the night ensuing. The 28th the wind came about to the NE., and it cleared up and blew a hard gale; but it stood not there, for it shifted about to the eastward, thence to the SE., then to the S.; at last it settled at SW., and then we had a moderate gale and fair weather. It was the 29th when the wind came to the SW. Then we made all the sail we could for the island again. The 30th we had the wind at W., and saw the islands, but could not get in before night. Therefore we stood off to the southward till 2 of the clock in the morning, then we tacked and stood in all the morning; and about 12 of the clock, the 1st of October, we anchored again at the place whence we were driven.

Then our six men were brought aboard by the natives, to whom we gave three whole bars of iron for their kindness and civility, which was an extraordinary present to them. Mr Robert Hall was one of the men that were left ashore; I shall speak more of him hereafter. He and the rest of them told me that after the ship was out of sight the natives began to be more kind to them than they had been before, and persuaded them to cut their hair short, as theirs was; offering to each of them, if they would do it, a young woman to wife, and a small hatchet and other iron utensils fit for a planter, in dowry; and withal showed them a piece of land for them to manage. They were courted thus by several of the town where they then were, but they took up their headquarters at the house of him with whom they first went ashore. When the ship appeared in sight again, then they importuned them for some iron, which is the chief thing that they covet, even above their earrings. We might have bought all their earrings or other gold they had, with our iron bars, had we been assured of its goodness; and yet when it was touched and compared with other gold, we could not discern any difference, though it looked so pale in the lump; but the seeing them polish it so often was a new discouragement.

This last storm put our men quite out of heart; for although it was not altogether so fierce as that which we were in on the coast of China, which was still fresh in memory, yet it wrought more powerfully, and frighted them from their design of cruising before Manilla, fearing another storm there. Now every man wished himself at home, as they had done a hundred times before; but Captain Reed, and Captain Tait, the master, persuaded them to go towards Cape Comorin, and then they would tell them more of their minds, intending, doubtless, to cruise in the Red Sea; and they easily prevailed with the crew. The eastern monsoon was now at hand, and the best way had been to go through the Straits of Malacca; but Captain Tait said it was dangerous, by reason of many islands and shoals there, with which none of us were acquainted. Therefore he thought it best to go round on the east side of all the Philippine Islands, and so keeping south toward the Spice

Islands, to pass out into the East Indian Ocean about the Island Timor. This seemed to be a very tedious way about, and as dangerous altogether for shoals; but not for meeting with English or Dutch ships, which was their greatest fear. I was well enough satisfied, knowing that the farther we went, the more knowledge and experience I should get, which was the main thing that I regarded; and I should also have the more variety of places to attempt an escape from them, being fully resolved to take the first opportunity of giving them the slip.

CHAPTER XVI.

THE 3d of October 1687, we sailed from these islands standing to the southward, intending to sail through among the Spice Islands. We had fair weather, and the wind at W. We first steered SSW., and passed close by certain small islands that lie just by the north end of the Island Luconia. We left them all on the west of us, and passed on the east side of it, and the rest of the Philippine Islands, coasting to the southward. The NE. end of the Island Luconia appears to be good champaign land, of an indifferent height, plain and even for many leagues, only it has some pretty high hills standing upright by themselves in these plains; but no ridges of hills, or chains of mountains joining one to another. The land on this side seems to be most savannah, or pasture; the SE. part is more mountainous and woody. Leaving the Island Luconia, and with it our golden projects, we sailed on the southward, passing on the east side of the rest of the Philippine Islands. These appear to be more mountainous and less woody, till we came in sight of the Island St John, the first of that name I mentioned; the other I spoke of on the coast of China. This I have already described to be a very woody island. Here the wind coming southerly, forced us to keep farther from the islands. The 14th of October we came close by a small, low, woody island, that lies east from the SE. end of Mindanao, distant from it about twenty leagues. I do not find it set down in any sea-chart. The 15th we had the wind at NE., and steered west for the Island Mindanao, and arrived at the SE. end again on the 16th. There we went in and anchored between two small islands. Here we found a fine small cove on the NW. end of the easternmost island, fit to careen in or haul ashore; so we went in there, and presently unrigged our ship, and provided to haul our ship ashore, to clean her bottom.

These islands are about three or four leagues from the Island Mindanao; they are about four or five miles in circumference, and of a pretty good height. The mould is black and deep, and there are two small brooks of fresh water. They are both plentifully stored with great high trees; therefore our carpenters were sent ashore to cut down some of them for our use; for here they made a new boltsprit,[1] which we did set here also, our old one being very faulty. They made a new foreyard too, and a foretopmast; and our pumps being faulty and not serviceable, they did cut a tree to make a pump. They first squared it, then sawed it in the middle, and then hollowed each side exactly. The two hollow sides were made big enough to contain a pump-box in the midst of them both, when they were joined together, and it required their utmost skill to close them exactly to the making a tight cylinder for the pump-box; being unaccustomed to such work. We learnt this way of pump-making from the Spaniards, who make their pumps that they use in their ships in the South Seas after this manner; and I am confident that there are no better

---

[1] Bowsprit, so called, probably, from the meaning of the word "bolt," as something projected or thrust out from the bow of the ship.

hand-pumps in the world than they have.

While we lay here, the young Prince that I mentioned in Chapter XIII., came aboard.[1] He, understanding that we were bound farther to the southward, desired us to transport him and his men to his own island. He showed it to us in our draught, and told us the name of it, which we put down in our draught, for it was not named there; but I quite forgot to put it into my journal. This man told us, that not above six days before this he saw Captain Swan, and several of his men that we left there, and named the names of some of them, who, he said, were all well, and now they were at the city of Mindanao; but that they had been all of them out with Raja Laut, fighting under him in his wars against his enemies the Alfoores; and that most of them fought with undaunted courage, for which they were highly honoured and esteemed, as well by the Sultan, as by the General Raja Laut. That now Captain Swan intended to go with his men to Fort St George,[2] and that in order thereto, he had proffered forty ounces of gold for a ship, but the owner and he were not yet agreed; and that he feared the Sultan would not let him go away till the wars were ended. All this the Prince told us in the Malay tongue, which many of us had learnt; and when he went away he promised to return to us again in three days' time, and so long Captain Reed promised to stay for him (for we had now almost finished our business), and he seemed very glad of the opportunity of going with us.

After this I endeavoured to persuade our men to return with the ship to the River of Mindanao and offer their service again to Captain Swan. I took an opportunity when they were filling water, there being then half the ship's company ashore, and I found these all very willing to do it. I desired them to say nothing till I had tried the minds of the other half, which I intended to do the next day, it being their turn to fill water then; but one of these men, who seemed most forward to invite back Captain Swan, told Captain Reed and Captain Tait of the project, and they presently dissuaded the men from any such designs. Yet, fearing the worst, they made all possible haste to be gone. I have since been informed that Captain Swan and his men stayed there a great while afterward, and that many of the men got passage thence in Dutch sloops to Ternate, particularly Mr Rofy and Mr Nelly. There they remained a great while, and at last got to Batavia (where the Dutch took their journals from them), and so to Europe; and some of Captain Swan's men died at Mindanao, of which number Mr Harthope and Mr Smith, Captain Swan's merchants, were two. At last Captain Swan and his surgeon, going in a small canoe aboard of a Dutch ship then in the road, in order to get passage to Europe, were overset by the natives at the mouth of the river, who waited their coming purposely to do it, but unsuspected by them, where they both were killed in the water. This was done by the General's order, as some think, to get his gold, which he did immediately seize on. Others say it was because the General's house was burnt a little before, and Captain Swan was suspected to be the author of it; and others say that it was Captain Swan's threats occasioned his own ruin, for he would often say, passionately, that he had been abused[3] by the General, and that he would have satisfaction for it; saying also, that now he was well acquainted with their rivers, and knew how to come in at any time; that he also knew their manner of fighting and the weakness of their country; and therefore he would go away and get a band of men to assist him, and returning thither again he

[1] Who had been sent by his uncle, the Sultan of a spice island, to Mindanao, with an invitation to Captain Swan to come and trade.

[2] Madras.

[3] Dealt falsely with.

would spoil and take all that they had, and their country too. When the General has been informed of these discourses he would say, "What, is Captain Swan made of iron, and able to resist a whole kingdom? or does he think that we are afraid of him that he speaks thus?" Yet did he never touch him till now the Mindanayans killed him. It is very probable there might be somewhat of truth in all this, for the Captain was passionate, and the General greedy of gold. But whatever was the occasion, so he was killed, as several have assured me, and his gold seized on, and all his things; and his journal also from England, as far as Cape Corrientes on the coast of Mexico. This journal was afterwards sent away from thence by Mr Moody (who was there both a little before and a little after the murder), and he sent it into England by Mr Goddard, chief mate of the Defence.

But to our purpose. Seeing I could not persuade them to go to Captain Swan again, I had a great desire to have had the Prince's company; but Captain Reed was afraid to let his fickle crew lie long. That very day that the Prince had promised to return to us, which was November 2, 1687, we sailed hence, directing our course SW. and having the wind at NW. This wind continued till we came in sight of the Island Celebes, then it veered about to the W. and to the S. of W. We came up with the NE. end of the Island Celebes on the 9th, and there we found the current setting to the W. so strongly that we could hardly get on the E. side of that island.

The Island Celebes is a very large island, extended in length from north to south about seven degrees of Latitude, and in breadth about three degrees. It lies under the Equator, the north end being in Lat. 1° 30′ N., and the south end in Lat. 5° 30′ S.; and by common account the bulk of this island lies nearest north and south, but at the north-east end there runs out a long narrow point, stretching NE. about thirty leagues; and about thirty leagues to the eastward of this long slip is the Island Gilolo, on the west side of which are four small islands close by it, which are very well stored with cloves. The two chief are Ternate and Tidore. And as the Isle of Ceylon is reckoned the only place for cinnamon, and that of Banda for nutmegs; so these are thought by some to be the only clove islands in the world; but this is a great error, as I have already shown. At the south end of the Island Celebes there is a sea or gulf of about seven or eight leagues wide, and forty or fifty long, which runs up the country almost directly to the north; and this gulf has several small islands along the middle of it. On the west side of the island, almost at the south end of it, the town of Macassar is seated—a town of great strength and trade belonging to the Dutch. There are great inlets and lakes on the east side of the island, as also abundance of small islands and shoals lying scattered about it. We saw a high-peaked hill at the north end, but the land on the east side is low all along, for we cruised almost the length of it. The mould on this side is black and deep, and extraordinarily fat and rich, and full of trees; and many brooks of water run out into the sea. Indeed all this east side of the island seems to be but one large grove of extraordinary great high trees.

Having with much ado got on this east side, coasting along to the southward, and yet having but little wind, and even that little against us at SSW. and sometimes calm, we were a long time going about the island. The 22d we were in Lat. 1° 20′ S., and being about three leagues from the island, standing to the southward, with a very gentle land wind, about 2 or 3 of the clock in the morning, we heard clashing in the water, like boats rowing; and fearing some sudden attack, we got up all our arms and stood ready to defend ourselves. As soon as it was day we saw a great proa, built like the Mindanayan proas, with about sixty men in her, and six smaller proas. They lay still about a mile to windward of us to view us, and probably designed to make a prey of us when they first came out, but they were now afraid to venture on us. At last we showed them Dutch colours, thinking thereby to allure them to come to us, for we could not go to them; but they presently rowed in towards the island and went into a large opening, and we saw them no more; nor did we ever see any other boats or men but only one fishing canoe while we were about this island, neither did we see any house on all the coast.

About five or six leagues to the south of this place there is a great range of both large and small islands, and many shoals also that are not laid down in our draughts, which made it extremely troublesome for us to get through. But we passed between them all and the Island Celebes, and anchored against a sandy bay in eight fathoms sandy ground about half-a-mile from the main island, being then in Lat. 1° 50′ S. Here we stayed several days, and sent out our canoes a-striking of turtle every day, for here is great plenty of them; but they were very shy, as they were generally wherever we found them in the East India Seas. I know not the reason of it, unless the natives go very much a-striking here; for even in the West Indies they are shy in places that are much disturbed; and yet on New Holland we found them shy, as I shall relate, though the natives there do not molest them. On the shoals without us we went and gathered shell-fish at low water. There were a monstrous sort of cockles—the meat of one of them would suffice seven or eight men. It was very good wholesome meat. We did also beat about in the woods on the island, but found no game. One of our men, who was always troubled with sore legs, found a certain vine that supported itself by climbing about other trees. The leaves reached six or seven feet high, but the strings or branches eleven or twelve. It had a very green leaf, pretty broad and roundish, and of a thick substance. These leaves pounded small, and boiled with hog's lard, make an excellent salve. Our men, knowing the virtues of it, stocked themselves here; there was scarce a man in the ship but got a pound or two of it, especially such as were troubled with old ulcers, who found great benefit by it. The man that discovered these leaves here had his first knowledge of them in the Isthmus of Darien, he having had this receipt from one of the Indians there; and he had been ashore in divers places since purposely to seek these leaves, but did never find any but here.

Among the many vast trees hereabouts there was one exceeded all the rest. This Captain Reed caused to be cut down in order to make a canoe, having lost our boats, all but one small one, in the late storms; so six lusty men, who had been logwood cutters in the Bays of Campeachy and Honduras (as Captain Reed himself, and many more of us had), and so were very expert at this work, undertook to fell it, taking their turns—three always cutting together; and they were one whole day and half the next before they got it down. This tree, though it grew in a wood, was yet eighteen feet in circumference and forty-four feet of clean body, without knot or branch; and even there it had no more than one or two branches, and then ran clean again ten feet higher; there it spread itself into many great limbs and branches like an oak, very green and flourishing; yet it was perished at the heart, which marred it for the service intended. So, leaving it, and having no more business here, we weighed and went from hence the next day, it being the 29th of November. We had the wind at NE. when we weighed, and we steered off SSW. In the afternoon we saw a shoal ahead of us, and altered our course to the SSE. In the evening, at 4 of the clock, we were close by another great shoal; therefore we tacked and stood in for the Island Celebes again for fear of running on some of the shoals in the night. By day a man might avoid them well enough, for they had all beacons on them, like huts built on tall posts, above high-water mark, probably set up by the natives of the Island Celebes or those of some other neighbouring islands; and I never saw any such elsewhere.

The 30th we had a fresh land wind, and steered away south, passing between the two shoals which we saw the day before. Being past them, the wind died away, and we lay becalmed till the afternoon; then we had a hard tornado out of the SW., and towards the evening we saw two or three spouts, the first I had seen since I came into the East Indies: in the West Indies I had often met with them. A spout is a small ragged piece, or part of a cloud, hanging down about a yard, seemingly from the blackest part thereof. Commonly it hangs down sloping from thence, or sometimes appearing with a small bending or elbow in the middle. I never saw any hang perpendicularly down. It is small at the lower end, seeming no bigger than one's arm; but it is fuller towards the cloud, whence it proceeds. They seem terrible enough: the rather because they come upon you while you lie becalmed like a log in the sea, and cannot get out of their way; but though I have seen and been beset by them often, yet the fright was always the greatest of the harm.

December the 1st, we had a gentle gale at ESE. We steered south; and at noon I was by observation in Lat. 3° 34′ S. Then we saw the Island Bouton, bearing south-west, and about ten leagues distant. We had very uncertain and unconstant winds. The 5th, we got close by the NW. end of the Island Bouton, and in the evening, it being fair weather, we hoisted out our canoe, and sent the Mosquito men, of whom we had two or three, to strike turtle, for here are plenty of them; but they being shy, we chose to strike them in the night (which is customary in the West Indies also) for every time they come up to breathe, which is once in eight or ten minutes, they blow so hard, that one may hear them at thirty or forty yards' distance; by which means the striker knows where they are, and may more easily approach them than in the day, for the turtle sees better than he hears: but, on the contrary, the manatee's hearing is quickest. In the morning they returned with a very large turtle, which they took near the shore; and withal an Indian of the island came aboard with them. He spake the Malay language, by which we did understand him. He told us, that two leagues farther to the southward of us there was a good harbour, in which we might anchor: so having a fair wind, we got thither by noon.

This harbour is in Lat. 4° 54′ S., lying on the east side of the Island Bouton. Which island lies near the SE. end of the Island Celebes, distant from it about three or four leagues. It is of a long form, stretching SW. and NE. about twenty-five leagues, and ten broad. It is pretty high land, and appears pretty even, and flat, and very woody. There is a large town within a league of the anchoring-place, called Callasusung, being the chief, if there were more; which we knew not. It is about a mile from the sea, on the top of a small hill, in a very fair plain, encompassed with cocoa-nut trees. Without the trees there is a strong stone wall, clear round the town. The houses are built like the houses at Mindanao, but more neat; and the whole town was very clean and delightsome. The inhabitants are small and well shaped. They are much like the Mindanayans in shape, colour, and habit; but more neat and tight. They speak the Malay language, and are all Mahometans. They are very obedient to the Sultan, who is a little man, about forty or fifty years old, and has a great many wives and children. About an hour after we came to an anchor, the Sultan sent a messenger aboard, to know what we were, and what our business. We gave him an account, and he returned ashore, and in a short time after he came aboard again, and told us that the Sultan was very well pleased when he heard that we were English, and said, that we should have anything the island afforded; and that he himself would come aboard in the morning. Therefore the ship was made clean, and everything put in the best order to receive him.

The 6th, in the morning betimes, a great many boats and canoes came aboard, with fowls, eggs, plantains, potatoes, &c., but they would dispose of none till they had order for it from the Sultan, at his coming. About 10 of the clock the Sultan came aboard in a very neat proa, built after the Mindanao fashion. There was a large white silk flag at the head of the mast, edged round with a deep red for about two or three inches broad, and in the middle there was neatly drawn a green griffin, trampling on a winged serpent that seemed to struggle to get up, and threatened his adversary with open mouth, and with a long sting that was ready to be darted into his legs. Other East Indian princes have their devices also. The Sultan, with three or four of his nobles, and three of his sons, sat in the house of the proa. His guards were ten musketeers, five standing on one side of the proa, and five on the other side: and before the door of the proa-house stood one with a great broad sword and a target, and two more such at the after-part of the house; and in the head and stern of the proa stood four musketeers more, two at each end. The Sultan had a silk turban, laced with narrow gold lace by the sides, and broad lace at the end; which hung down on one side the head, after the Mindanayan fashion. He had a sky-coloured silk pair of breeches, and a piece of red silk thrown across his shoulders, and hanging loose about him; the greatest part of his back and waist appearing naked. He had neither stocking nor shoe. One of his sons was about fifteen or sixteen years old; the other two were young things, and they were always in the arms of one or other of his attendants.

Captain Reed met him at the side, and led him into his small cabin, and fired five guns for his welcome. As soon as he came aboard he gave leave to his subjects to traffic with us. and then our people bought what they had a mind to. The Sultan seemed very well pleased to be visited by the English; and said he had coveted to have a sight of Englishmen, having heard extraordinary characters of their just and honourable dealings: but he exclaimed against the Dutch (as all the Mindanayans, and all the Indians we met with, do) and wished them at a greater distance. For Macassar is not very far from hence, one of the chief towns that the Dutch have in those parts. Thence the Dutch come sometimes hither to purchase slaves. The slaves that these people get here and sell to the Dutch are some of the idolatrous natives of the island, who, not being under the Sultan, and having no head, live straggling in the country, flying from one place to another to preserve themselves from this prince and his subjects, who hunt after them to make them slaves. For the civilising Indians of the maritime places, who trade with foreigners, if they cannot reduce the inland people to the obedience of their prince, catch all they can of them and sell them for slaves; accounting them to be but as savages, just as the Spaniards do the poor Americans.

After two or three hours' discourse, the Sultan went ashore again, and five guns were fired at his departure also. The next day he sent for Captain Reed to come ashore; and he, with seven or eight men, went to wait on the Sultan. I could not slip an opportunity of seeing the place; and so accompanied them. We were met at the landing-place by two of the chief men, and guided to a pretty neat house, where the Sultan waited our coming. The house stood at the farther end of all the town before mentioned, which we passed through; and abundance of people were gazing on us as we passed by. When we came near the house, forty poor, naked soldiers with muskets made a line for us to pass through. This house was not built on posts, as the rest were, after the Mindanayan way; but the room in which we were entertained was on the ground, covered with mats to sit on. Our entertainment was tobacco and betel-nut, and young cocoa-nuts; and the house was beset with men, women, and children, who thronged to get near the windows to look on us. We did not tarry above an hour before we took our leave and departed. The next day the Sultan came aboard again, and presented Captain Reed with a little boy; but he was too small to be serviceable on board; and so Captain Reed returned thanks, and told him he was too little for him. Then the Sultan sent for a bigger boy, which the Captain accepted. This boy was a very pretty tractable boy; but what was wonderful in him, he had two rows of teeth, one within another, on each jaw. None of the other people were so, nor did I ever see the like. The Captain was presented also with two he-goats, and was promised some buffalo, but I do believe that they have but few of either on the island. We did not see any buffalo, nor many goats; neither have they much rice; but their chief food is roots. We bought here about a thousand pound weight of potatoes. Here our men bought also abundance of crockadores and fine large paroquets, curiously coloured, and some of the finest I saw. The crockador is as big as a parrot, and shaped much like it, with such a bill; but is as white as milk, and has a bunch of feathers on his head like a crown. At this place we bought a proa also of the Mindanayan make, for our own use, which our carpenters afterwards altered, and made a delicate boat fit for any service. She was sharp at both ends; but we sawed off one, and made that end flat, fastening a rudder to it; and she rowed and sailed incomparably.

We stayed here but till the 12th, because it was a bad harbour and foul ground, and a bad time of the year too, for the tornadoes began to come in thick and strong. When we went to weigh our anchor, it was hooked in a rock, and we broke our cable, and could not get our anchor, though we strove hard for it; so we went away and left it there. We had the wind at NNE., and we steered towards the SE., and fell in with four or five small islands, that lie in 5° 40′ S. Lat., and about five or six leagues from Callasusung harbour. These islands appeared very green with cocoa-nut trees, and we saw two or three towns on them, and heard a drum all night, for we were got in among shoals, and could not get out again till the next day. We know not whether the drum were for fear of us, or that they were making merry, as it is usual in these parts to do all the night, singing and dancing till morning. At last we passed between the islands, and tried for a passage on the east side. We met with divers shoals on this side also, but found channels to pass through; so we steered away for the Island Timor, intending to pass out by it. The 16th, we got clear of the shoals, and steered S. by E., with the wind at WSW., but veering every half hour, sometimes at SW., and then again at W., and sometimes at NNW., bringing much rain, with thunder and lightning. The 20th we passed by the Island Omba, which is a pretty high island, lying in Lat. 8° 20′, and not above five or six leagues from the NE. part of the Island Timor. It is about thirteen or fourteen leagues long, and five or six leagues wide. About seven or eight leagues to the west of Omba is another pretty large island, but it had no name in our plans; yet by the situation it should be that which in some maps is called Pentare.[1] We saw on it abundance of smokes by day, and fires by night, and a large town on the north side of it, not far from the sea; but it was such bad weather that we did not go ashore. Between Omba and Pentare, and in the mid-channel, there is a small, low, sandy island, with great shoals on either side; but there is a very good channel close by Pentare between them and the shoals about the small isle. We were three days beating off and on, not having a wind, for it was at SSW.

[1] Or Pantor; a small island about midway between Timor and Floris.

The 23d, in the evening, having a small gale at north, we got through, keeping close by Pentare. The tide of ebb here set out to the southward, by which we were helped through, for we had but little wind; but this tide, which did us a kindness in setting us through, had like to have ruined us afterwards. For there are two small islands lying at the south end of the channel we came through; and towards these islands the tide hurried so swiftly, that we very narrowly escaped being driven ashore; for the little wind we had before at north died away; we had not one breath of wind when we came there, neither was there an anchor-ground. But we got out our oars and rowed, yet all in vain; for the tide set wholly on one of the small islands, that we were forced by might and main strength to bear off the ship, by thrusting with our oars against the shore, which was a steep bank, and by this means we presently drove away, clear of danger; and having a little wind in the night at north, we steered away SSW. In the morning again we had the wind at WSW., and steered S.; and the wind coming to the WNW., we steered SW. to get clear of the SW. end of the Island Timor. The 26th, we saw the NW. point of Timor, SE. by E., distant about eight leagues. Timor is a long high mountainous island, stretching NE. and SW. It is about seventy leagues long, and fifteen or sixteen wide; the middle of the island is in Latitude about 9° S. I have been informed that the Portuguese trade to this island, but I know nothing of its produce, besides Coir, for making cables.[1]

Being now clear of all the islands, we stood off south, intending to touch at New Holland, a part of Terra Australis Incognita, to see what that country would afford us. Indeed, as the winds were, we could not now keep our intended course (which was first westerly, and then northerly) without going to New Holland, unless we had gone back again among the islands; but this was not a good time of the year to be among any islands to the south of the Equator, unless in a good harbour. The 31st, we were in Lat. 13° 20′, still standing to the southward, the wind bearing commonly very hard at W., and we keeping upon it under two courses, and our mizzen, and sometimes a maintopsail reefed. About 10 of the clock at night we tacked and stood to the northward, for fear of running on a shoal, which is laid down on our draughts in Lat. 13° 50′ or thereabouts. At 3 of the clock we tacked again, and stood S. by W. and SSW. In the morning, as soon as it was day, we saw the shoal right ahead. We stemmed right with the middle of it, and stood within half-a-mile of the rocks, and sounded, but found no ground. Then we went about and stood to the north two hours; and then tacked and stood to the southward again, thinking to weather it, but could not. So we bore away on the north side, till we came to the east point, giving the rocks a small berth; then we trimmed sharp, and stood to the southward, passing close to it, and sounded again, but found no ground. . . .

The 4th of January, 1688, we fell in with the land of New Holland in Lat. 16° 50′, having made our course due south from the shoal that we passed by the 31st of December. We ran in close by it, and finding no convenient anchoring, because it lies open to the NW., we ran along shore to the eastward, steering NE. by E., for so the land lies. We steered thus about twelve leagues, and then came to a point of land, whence the land trends east and southerly for ten or twelve leagues, but how afterwards I know not. About three leagues to eastward of this point, there is a pretty deep bay, with abundance of islands in it, and a very good place to anchor in, or to haul ashore. About a league to the eastward of that point we anchored, January the 5th, 1688, two miles from the shore, in twenty-nine

[1] Cordage made of cocoa-nut fibre.

fathoms, good hard sand, and clean ground.

New Holland is a very large tract of land. It is not yet determined whether it is an island or a main continent; but I am certain that it joins neither to Asia, Africa, nor America. This part of it that we saw is all low even land, with sandy banks against the sea; only the points are rocky, and so are some of the islands in this bay. The land is of a dry sandy soil, destitute of water, except you make wells; yet producing divers sorts of trees; but the woods are not thick, nor the trees very big. Most of the trees that we saw are dragon trees, as we supposed; and these two are the largest trees of any there. They are about the bigness of our large apple trees, and about the same height, and the rind is blackish, and somewhat rough. The leaves are of a dark colour; the gum distils out of the knots or cracks that are in the bodies of the trees. We compared it with some gum-dragon, or dragon's blood, that was aboard, and it was of the same colour and taste. The other sorts of trees were not known by any of us. There was pretty long grass growing under the trees; but it was very thin. We saw no trees that bore fruit or berries. We saw no sort of animal, nor any track of beast, but once, and that seemed to be the tread of a beast as big as a great mastiff dog. Here are a few small land-birds, but none bigger than a black-bird, and but few sea-fowls. Neither is the sea very plentifully stored with fish, unless you reckon the manatee and turtle as such. Of these creatures there is plenty; but they are extraordinary shy, though the inhabitants cannot trouble them much, having neither boats nor arrows.

The inhabitants of this country are the miserablest people in the world. The Hodmadods of Monomatapa,[1] though a nasty people, yet for wealth are gentlemen to these, who have no houses and skin garments, sheep, poultry, and fruits of the earth, ostrich eggs, &c., as the Hodmadods have; and setting aside their human shape, they differ but little from brutes. They are tall, straight-bodied, and thin, with small long limbs. They have great heads, round foreheads, and great brows. Their eyelids are always half-closed, to keep the flies out of their eyes, they being so troublesome here, that no fanning will keep them from coming to one's face; and without the assistance of both hands to keep them off, they will creep into one's nostrils, and mouth too, if the lips are not shut very close. So that from their infancy, being thus annoyed with these insects, they do never open their eyes as other people, and therefore they cannot see far, unless they hold up their heads, as if they were looking at somewhat over them. They have great bottle noses, pretty full lips, and wide mouths. The two fore teeth of their upper jaw are wanting in all of them, men and women, old and young; whether they draw them out, I know not, neither have they any beards. They are long-visaged, and of a very unpleasing aspect, having no one graceful feature in their faces. Their hair is black, short and curled, like that of the Negroes, and not long and lank like the common Indians. The colour of their skins, both of their faces and the rest of their body, is coal black, like that of the Negroes of Guinea. They have no sort of clothes, but a piece of the rind of a tree, tied like a girdle about their waists, and a handful of long grass, or three or four small green boughs, full of leaves, thrust under their girdle to cover their nakedness.

They have no houses, but lie in the open air, without any covering, the earth being their bed, and the heaven their canopy. Whether they cohabit one man to one woman, or promiscuously, I know not: but they do live in companies, twenty or thirty men, women, and children together. Their only food is a small sort of fish, which

[1] The Hottentots of the Cape. See Chapter XX.

they get by making wears[1] of stone across little coves, or branches of the sea; every tide bringing in the small fish, and there leaving them for a prey to these people, who constantly attend there, to search for them at low water. This small fry I take to be the top of their fishery: they have no instruments to catch great fish, should they come; and such seldom stay to be left behind at low water: nor could we catch any fish with our hooks and lines all the while we lay there. In other places at low water they seek for cockles, mussels, periwinkles. Of these shell-fish there are fewer still; so that their chief dependence is upon what the sea leaves in their wears, which, be it much or little, they gather up, and march to the places of their abode. There the old people, that are not able to stir abroad by reason of their age, and the tender infants, wait their return; and what Providence has bestowed on them, they presently broil on the coals, and eat it in common. Sometimes they get as many fish as make them a plentiful banquet; and at other times they scarce get every one a taste: but be it little or much that they get, every one has his part, as well the young and tender, as the old and feeble, who are not able to go abroad, and the strong and lusty. When they have eaten, they lie down till the next low water, and then all that are able to march out, be it night or day, rain or shine, it is all one: they must attend the wears, or else they must fast. For the earth affords them no food at all. There is neither herb, root, pulse, nor any sort of grain, for them to eat, that we saw: nor any sort of bird or beast that they can catch, having no instruments wherewithal to do so.

I did not perceive that they did worship anything. These poor creatures have a sort of weapon to defend their wear, or fight with their enemies if they have any that will interfere with their poor fishery. They did at first endeavour with their weapons to frighten us, who lying ashore deterred them from one of their fishing-places. Some of them had wooden swords, others had a sort of lances. The sword is a piece of wood, shaped somewhat like a cutlass. The lance is a long straight pole, sharp at one end, and hardened afterwards by heat. I saw no iron, nor any other sort of metal: therefore it is probable they use stone hatchets, as some Indians in America do.[2] How they get their fire I know not: but probably, as Indians do, out of wood. I have seen the Indians of Buen Ayre[3] do it, and have myself tried the experiment. They take a flat piece of wood, that is pretty soft, and make a small dent in one side of it: then they take another hard round stick, about the bigness of one's little finger, and sharping it at one end like a pencil, they put that sharp end in the hole or dent of the flat soft piece; then rubbing or twirling the hard piece between the palms of their hands, they drill the soft piece till it smokes and at last takes fire.

These people speak somewhat through the throat; but we could not understand one word that they said. We anchored, as I said before, January the 5th, and seeing men walking on the shore, we presently sent a canoe to get some acquaintance with them: for we were in hopes to get some provision among them. But the inhabitants, seeing our boat coming, ran away and hid themselves. We searched afterwards three days, in hopes to find their houses, but found none; yet we saw many places where they had made fires. At last, being out of hopes to find their habitations, we searched no farther; but left a great many toys ashore, in such places where we thought that they would come. In all our search we found no water, but old wells on the sandy bays. At last we went over to the islands, and there we

[1] Dams or embankments.

[2] Mentioned in Chapter IV., page 158.

[3] One of the Windward Islands, visited by the Author in 1681.

found a great many of the natives; I do believe there were forty on one island, men, women, and children. The men, at our first coming ashore, threatened us with their lances and swords; but they were frighted by firing one gun, which we fired purposely to scare them. The island was so small, that they could not hide themselves; but they were much disordered at our landing, especially the women and children: for we went directly to their camp. The lustiest of the women, snatching up their infants, ran away howling, and the little children ran after squeaking and bawling; but the men stood still. Some of the women, and such people as could not go from us, lay still by a fire, making a doleful noise, as if we had been coming to devour them. But when they saw we did not intend to harm them, they were pretty quiet; and the rest, that fled from us at our first coming, returned again. This their place of dwelling was only a fire, with a few boughs before it, set up on that side the wind was off. After we had been here a little while, the men began to be familiar, and we clothed some of them, designing to have had some service of them for it; for we found some wells of water here, and intended to carry two or three barrels of it aboard. But it being somewhat troublesome to carry to the canoes, we thought to have made these men to have carried it for us, and therefore we gave them some clothes; to one an old pair of breeches, to another a ragged shirt, to a third a jacket that was scarce worth owning: which yet would have been very acceptable at some places where we had been, and so we thought they might have been with these people. We put them on them, thinking that this finery would have brought them to work heartily for us; and our water being filled in small long barrels, about six gallons in each, which were made purposely to carry water in, we brought these our new servants to the wells, and put a barrel on each of their shoulders for them to carry to the canoe. But all the signs we could make were to no purpose, for they stood like statues, without motion, but grinned like so many monkeys, staring one upon another: for these poor creatures seem not accustomed to carry burthens: and I believe that one of our ship-boys of ten years old would carry as much as one of them. So we were forced to carry our water ourselves; and they very fairly put the clothes off again, and laid them down, as if clothes were only to work in. I did not perceive that they had any great liking to them at first; neither did they seem to admire[1] anything that we had.

At another time our canoe being among these islands seeking for game, espied a drove of these men swimming from one island to another; for they have no boats, canoes, or bark-logs. They took up four of them, and brought them aboard; two of them were middle-aged, the other two were young men about eighteen or twenty years old. To these we gave boiled rice, and with it turtle and manatee boiled. They did greedily devour what we gave them, but took no notice of the ship, or anything in it; and when they were set on land again, they ran away as fast as they could. At our first coming, before we were acquainted with them, or they with us, a company of them who lived on the main came just against our ship, and, standing on a pretty high bank, threatened us with their swords and lances, by shaking them at us: at last the Captain ordered the drum to be beaten, which was done of a sudden with much vigour, purposely to scare the poor creatures. They, hearing the noise, ran away as fast as they could drive, and when they ran away in haste, they would cry "Gurry, Gurry," speaking deep in the throat. Those inhabitants also that live on the main would always run away from us; yet we took several of them: for, as I have already observed, they had such bad eyes that they could not see us till we came close to them. We did always give them victuals, and let

[1] Wonder, be surprised at.

them go again; but the islanders, after our first time of being among them, did not stir for us.

When we had been here about a week, we hauled our ship into a small sandy cove, at a spring-tide, as far as she would float: and at low water she was left dry. All the neap-tides we lay wholly aground, for the sea did not come near us by about a hundred yards. We had therefore time enough to clean our ship's bottom, which we did very well. Most of our men lay ashore in a tent, where our sails were mending: and our strikers brought home turtle and manatee every day, which was our constant food. While we lay here, I did endeavour to persuade our men to go to some English factory; but was threatened to be turned ashore and left here for it. This made me desist, and patiently wait for some more convenient place and opportunity to leave them, than here: which I did hope I should accomplish in a short time; because they did intend, when they went hence, to bear down towards Cape Comorin. In their way thither they designed also to visit the Island Cocos, which lies in Lat. 12° 12′ N. by our draughts: hoping there to find of that fruit, the island having its name from thence.[1]

CHAPTER XVII.

March the 12th, 1688, we sailed from New Holland, with the wind at NNW., and fair weather. We directed our course to the northward, intending, as I said, to touch at the Island Cocos. It was the 26th of March before we were in the Latitude of the island, which is in 12° 12′; and then, by judgment, we were forty or fifty leagues to the east of it; and the wind was now at SW.: therefore we did rather choose to bear away towards some islands on the west side of Sumatra, than to beat against the wind for the Island Cocos. I was very glad of this, being in hopes to make my escape from them to Sumatra, or some other place. We met nothing of remark in this voyage, besides the catching two great sharks, till the 28th. Then we fell in with a small woody island, in Lat. 10° 30′. Its Longitude from New Holland, whence we came, was by my account 12° 6′ W. It was deep water about the island, and therefore no anchoring; but we sent two canoes ashore—one of them with the carpenters, to cut a tree to make another pump—the other canoe went to search for fresh water, and found a fine, small brook near the SW. point of the island; but there the sea fell in on the shore so high, that they could not get it off. At noon both our canoes returned aboard, and the carpenters brought aboard a good tree, which they afterwards made a pump with, such as they made at Mindanao. The other canoe brought aboard as many boobies and men-of-war birds as sufficed all the ship's company, when they were boiled. They got also a sort of land animal somewhat resembling a large craw-fish without its great claws. These creatures lived in holes in the dry, sandy ground like rabbits. Sir Francis Drake, in his Voyage round the World, makes mention of such that he found at Ternate, or some other of the Spice Islands, or near them.[1] They were very good sweet meat, and so large that two of them were more than a man could eat, being almost as thick as one's leg. Their shells were of a dark brown, but red when boiled.

About 1 o'clock in the afternoon we made sail from this island, with the wind at SW., and we steered NW. We met nothing of remark till the 7th of April, and then, being in Lat. 7° S., we saw the land of

[1] Dampier lays it down, despite his text, in his "Map of the East Indies," in something between 12° and 13° south of the Line, and to the SSW. of the Strait of Sunda. The island is subsequently several times mentioned.

[1] See page 65.

Sumatra at a great distance, bearing north. The 8th we saw the east end of the Island Sumatra very plainly, we being then in Lat. 6° S. The 10th, being in Lat. 5° 11′, and about seven or eight leagues from the Island Sumatra, on the west side of it, we saw abundance of cocoa-nuts swimming in the sea, and we hoisted out our boat and took up some of them, as also a small hatch, or scuttle rather, belonging to some bark. The nuts were very sound, and the kernel sweet; and in some the milk or water was yet sweet and good. The 13th we came to a small island called Triste, in Lat. (by observation) 4° S. It is about fourteen or fifteen leagues to the west of the Island Sumatra. From hence to the northward there are a great many small uninhabited islands lying much at the same distance from Sumatra. This Island Triste is not a mile round, and so low, that the tide flows clear over it. It is of a sandy soil, and full of cocoa-nut trees. The nuts are but small, yet sweet enough, full, and more ponderous than I ever felt any of that bigness, notwithstanding that every spring-tide the salt water goes clear over the island. We sent ashore our canoes for cocoa-nuts, and they returned aboard laden with them three times. Our strikers also went out and struck some fish, which was boiled for supper. They also killed two young alligators, which we salted for the next day.

I had no opportunity at this place to make my escape, as I would have done, and gone over hence to Sumatra, could I have kept a boat with me. But there was no compassing this; and so on the 15th we went from hence, steering to the northward on the west side of Sumatra. Our food now was rice and the meat of the cocoa-nuts rasped and steeped in water, which made a sort of milk, into which we put our rice, making a pleasant mess enough. After we parted from Triste, we saw other small islands that were also full of cocoa-nut trees. The 19th, being in Lat. 3° 25′ S., the SW. point of the Island Nassau bore N. about five miles distant. This is a pretty large uninhabited island, in Lat. 3° 20′ S., and is full of high trees. About a mile from the Island Nassau, there is a small island full of cocoa-nut trees. There we anchored the 20th to replenish our stock of cocoa-nuts. A reef of rocks lies almost round this island, so that our boats could not go ashore, nor come aboard at low water, yet we got aboard four boat loads of nuts. The 21st we went from hence, and kept to the northward, coasting still on the west side of the Island Sumatra. The 25th we crossed the Equator, still coasting to the northward between the Island Sumatra and a range of small islands lying fourteen or fifteen leagues off it. Among all these islands, Hog Island is the most considerable. It lies in Lat. 3° 40′ N. It is pretty high even land, clothed with tall, flourishing trees; we passed by it on the 28th.

The 29th we saw a sail to the north of us, which we chased; but it being little wind, we did not come up with her till the 30th. Then, being within a league of her, Captain Reed went in a canoe and took her, and brought her aboard. She was a proa with four men in her, belonging to Achin,[1] whither she was bound. She came from one of these cocoa-nut islands that we passed by, and was laden with cocoa-nuts and cocoa-nut oil. Captain Reed ordered his men to take aboard all the nuts, and as much of the oil as he thought convenient, and then cut a hole in the bottom of the proa, and turned her loose, keeping the men prisoners. It was not for the lucre of the cargo that Captain Reed took this boat, but to hinder me and some others from going ashore; for he knew that we were ready to make our escape if an opportunity presented itself, and he thought that by his abusing and robbing the natives, we should be afraid to trust ourselves among them. But yet this

[1] Or Acheen, a native town at the extreme north point of Sumatra, which now carries on an extensive trade with Hindostan.

proceeding of his turned to our great advantage, as shall be declared hereafter.

May the 1st we ran down by the north-west end of the Island Sumatra, within seven or eight leagues of the shore. All this west side of Sumatra which we thus coasted along, our Englishmen at Fort St George call the West Coast, simply without adding the name of Sumatra. The prisoners who were taken the day before showed us the islands that lie off Achin harbour, and the channels through which ships go in, and told us also that there was an English factory at Achin. I wished myself there, but was forced to wait with patience till my time was come. We were now directing our course towards the Nicobar Islands, intending there to clean the ship's bottom, in order to make her sail well. The 4th, in the evening, we had sight of one of the Nicobar Islands. The southernmost of them lies about forty leagues NNW. from the NW. end of the Island Sumatra. This most southerly of them is Nicobar itself,[1] but all the cluster of islands lying south of the Andaman Islands are called by our seamen the Nicobar Islands.

The inhabitants of these islands have no certain converse with any nation; but as ships pass by them they will come aboard in their proas, and offer their commodities for sale, never inquiring of what nation they are: for all white people are alike to them. Their chief commodities are ambergris and fruits. Ambergris is often found by the native Indians of these islands, who know it very well, as also know how to cheat ignorant strangers with a certain mixture like it. Several of our men bought such of them for a small purchase. Captain Weldon also about this time touched at some of these islands to the north of the island where we lay, and I saw a great deal of such ambergris that one of his men bought there, but it was not good, having no smell at all. Yet I saw some there very good and fragrant. At that island where Captain Weldon was, there were two friars sent thither to convert the Indians. One of them came away with Captain Weldon, the other remained there still. He that came away with Captain Weldon gave a very good character of the inhabitants of that island, that they were very honest, civil, harmless people; that they were not addicted to quarrelling, theft, or murder; that they did marry, or at least live as man and wife, one man with one woman, never changing till death made the separation; that they were punctual and honest in performing their bargains; and that they were inclined to receive the Christian religion. This relation I had afterwards from the mouth of a priest at Tonquin, who told me that he received this information by a letter from the friar that Captain Weldon brought away from thence. But to proceed.

The 5th of May we ran down on the west side of the Island Nicobar properly so called, and anchored at the NW. end of it, in a small bay, in eight fathoms water, not half-a-mile from the shore. The body of this island is in 7° 30′ N. Lat.; it is about twelve leagues long, and three or four broad. The south end of it is pretty high, with steep cliffs against the sea; the rest of the island is low, flat, and even. The mould of it is black and deep, and it is very well watered with small running streams. It produces abundance of tall trees fit for any uses: for the whole bulk of it seems to be but one entire grove. But that which adds most to its beauty off at sea are the many spots of cocoa-nut trees which grow round it in every small bay. The bays are half-a-mile or a mile long, more or less, and these bays are intercepted or divided from each other with as many little rocky points of woodland. As the cocoanut trees do thus grow in groves fronting to the sea in the bays, so there is another sort of fruit tree in the bays bordering on the back side of the cocoa trees farther from the sea. It is called by the natives a melory

[1] Great Nicobar.

tree. This tree is as big as our large apple trees, and as high. It has a blackish rind, and a pretty broad leaf. The fruit is as big as the bread-fruit at Guam, or a large penny loaf. It is shaped like a pear, and has a pretty tough smooth rind of a light green colour. The inside of the fruit is in substance much like an apple, but full of small strings as big as a brown thread. I did never see these trees anywhere but here.

The natives of this island are tall, well-limbed men; pretty long-visaged, with black eyes; their noses middle proportioned, and the whole symmetry of their faces agreeing very well. Their hair is black and lank, and their skin of a dark copper colour. The women have no hair on their eyebrows. I do believe it is plucked up by the roots; for the men had hair growing on their eyebrows, as other people. [The men all go naked, save a long, narrow strip of cloth round their waist. The women wear a short petticoat reaching from their waist to the knee. Their houses are described as small, square, and low, and curiously thatched with palmetto leaves. Their canoes are commonly manned by twenty or thirty natives, and seldom fewer than nine or ten.] . . .

But to proceed with our affairs. It was, as I said, before the 5th of May, about ten in the morning, when we anchored at this island. Captain Reed immediately ordered his men to heel the ship, in order to clean her, which was done this day and the next. All the water vessels were filled, they intending to go to sea at night; for the winds being yet at NNE., the Captain was in hopes to get over to Cape Comorin before the wind shifted, otherwise it would have been somewhat difficult for him to get thither, because the westerly monsoon was now at hand. I thought now was my time to make my escape, by getting leave, if possible, to stay here. For it seemed not very feasible to do it by stealth; and I had no reason to despair of getting leave, this being a place where my stay could probably do our crew no harm, should I design it. Indeed, one reason that put me on the thoughts of staying at this particular place, besides the present opportunity of leaving Captain Reed, which I did always intend to do as soon as I could, was, that I had here also a prospect of advancing a profitable trade for ambergris with these people, and of gaining a considerable fortune for myself; for in a short time I might have learned their language, and by accustoming myself to row with them in their proas or canoes, especially by conforming myself to their customs and manners of living, I should have seen how they got their ambergris, and have known what quantities they got, and the time of the year when most is found. And then afterwards, I thought it would be easy for me to have transported myself from thence, either in some ship that passed this way, whether English, Dutch, or Portuguese, or else to have got some of the young men of the island to have gone with me in one of their canoes to Achin, and there to have furnished myself with such commodities as I found most coveted by them; and therewith, at my return, to have bought their ambergris.

I had, till this time, made no open show of going ashore here. But now, the water being filled, and the ship in readiness to sail, I desired Captain Reed to set me ashore on this island. He, supposing that I could not go ashore in a place less frequented by ships than this, gave me leave, which probably he would have refused, if he thought I should have got from hence in any short time; for fear of my giving an account of him to the English or Dutch. I soon got up my chest and bedding, and immediately got some to row me ashore, for fear lest his mind should change again. The canoe that brought me ashore landed me on a small sandy bay, where there were two houses, but no person in them. For the inhabitants were removed to some other house, probably for fear of us, because the ship was close by; and yet both men and women came aboard the ship without any sign of fear. When our ship's canoe was going aboard again, they met the owner of the houses coming ashore in his boat. He made a great many signs to them to fetch me off again; but they would not understand him. Then he came to me and offered his boat to carry me off; but I refused. Then he made signs for me to go up into the house, and according as I did understand him by his signs, and a few Malay words that he used, he intimated that somewhat would come out of the woods in the night, when I was asleep, and kill me, meaning probably some wild beast. Then I carried my chest and clothes up into the house.

I had not been ashore an hour, before Captain Tait, and one John Damarell, with three or four armed men more, came to fetch me aboard again. They need not have sent an armed *posse* for me, for had they but sent the cabin-boy ashore for me, I would not have denied going aboard. For though I could have hid myself in the woods, yet then they would have abused or have killed some of the natives, purposely to incense them against me. I told them, therefore, that I was ready to go with them, and went aboard with all my things. When I came aboard I found the ship in an uproar, for there were three men more, who, taking courage by my example, desired leave also to accompany me. One of them was the surgeon, Mr Coppinger, the others were Mr Robert Hall, and one named Ambrose; I have forgot his surname. These men had always harboured the same designs as I had. The two last were not much opposed; but Captain Reed and his crew would not part with the surgeon. At last the surgeon leaped into the canoe, and taking up my gun, swore he would go ashore, and if any man did oppose it, he would shoot him. But John Oliver, who was then quarter-master, leaped into the canoe, taking hold of him, took away the gun, and with the help of two or three more, they dragged him again into the ship. Then Mr Hall, and Ambrose, and I were again set ashore; and one of the men that rowed us ashore stole an axe and gave it to us, knowing it was a good commodity with the Indians. It was now dark, therefore we lighted a candle, and I being the oldest stander in our new country, conducted them into one of the houses, where we did presently hang up our hammocks. We had scarce done this, before the canoe came ashore again, and brought the four Malay men belonging to Achin, which we took in the proa we took off Sumatra, and the Portuguese that came to our ship out of the Siam junk at Pulo Condore, the crew having no occasion for these, being leaving the Malay parts, where the Portuguese served as an interpreter; and not fearing now that the Achinese could be serviceable to us in bringing us over to their country, forty leagues off. Nor imagining that we durst make such an attempt, as, indeed, it was a bold one. Now we were men enough to defend ourselves against the natives of this island, if they should prove our enemies; though if none of these men had come ashore to me, I should not have feared any danger. Nay, perhaps less, because I should have been cautious of giving any offence to the natives; and I am of the opinion, that there are no people in the world so barbarous as to kill a single person that falls accidentally into their hands, or comes to live among them, except they have before been injured by some outrage or violence committed against them. Yet even then, or afterwards, if a man could but preserve his life from their first rage, and come to treat with them (which is the hardest thing, because their way is usually to abscond,[1] and rushing suddenly upon their enemy, to kill him at unawares), one might, by some sleight, insinuate one's self into their favour again; especially by showing some toy or knack that they did never see before, which any European that has seen the world might soon contrive to amuse

[1] Conceal themselves.

them withal, as might be done, generally, even with a little fire struck with a flint and steel.

As for the common opinion of Anthropophagi, or man-eaters, I did never meet with any such people. All nations or families in the world that I have seen or heard of, having some sort of food to live on, either fruit, grain, pulse, or roots, which grow naturally, or else planted by them, if not fish, and land animals besides (yea, even the people of New Holland had fish amidst all their penury), would scarce kill a man purposely to eat him. I know not what barbarous customs may formerly have been in the world: and to sacrifice their enemies to their gods is a thing that has been much talked of with relation to the savages of America. I am a stranger to that also, if it be or have been customary in any nation there; and yet, if they sacrifice their enemies, it is not necessary they should eat them too. After all, I will not be peremptory in the negative, but I speak as to the compass of my own knowledge, and know some of these cannibal stories to be false; and many of them have been disproved since I first went to the West Indies. At that time how barbarous were the poor Florida Indians accounted, which now we find to be civil enough? What strange stories have we heard of the Indians whose islands were called the Isles of Cannibals? Yet we find that they trade very civilly with the French and Spaniards, and have done so with us. I own that they have formerly endeavoured to destroy our plantations at Barbadoes, and have since hindered us from settling the Island Santa Lucia, by destroying two or three colonies successively of those that were settled there; and even the Island Tobago has been often annoyed and ravaged by them, when settled by the Dutch, and still lies waste (though a delicate fruitful island) as being too near the Caribbees on the continent, who visit it every year. But this was to preserve their own right, by endeavouring to keep out any that would settle themselves on those islands where they had planted themselves: yet even these people would not hurt a single person, as I have been told by some that have been prisoners among them. I could instance also the Indians of Bocca Toro and Bocca Drago, and many other places where they do live, as the Spaniards call it, wild and savage; yet there they have been familiar with privateers, but by abuses have withdrawn their friendship again. As for these Nicobar people, I found them affable enough, and therefore did not fear them; but I did not much care whether I had gotten any more company or no. But, however, I was very well satisfied, and the rather because we were now men enough to row ourselves over to the Island Sumatra; and accordingly we presently consulted how to purchase a canoe of the natives.

It was a fine clear moonlight night in which we were left ashore. Therefore we walked on the sandy bay, to watch when the ship would weigh and be gone, not thinking ourselves secure in our new-gotten liberty till then. About 11 or 12 o'clock we saw her under sail, and then we returned to our chamber, and so to sleep. This was the 6th of May. The next morning betimes, our landlord, with four or five of his friends, came to see his new guests, and was somewhat surprised to see so many of us, for he knew of no more but myself. Yet he seemed to be very well pleased, and entertained us with a large calabash of toddy, which he brought with him. Before he went away again (for wheresoever we came, they left their houses to us, but whether out of fear or superstition I know not), we bought a canoe of him for an axe, and did presently put our chests and clothes in it, designing to go to the south end of the island, and lie there till the monsoon shifted, which we expected every day. When our things were stowed away, we with the Achinese entered with joy into our new frigate, and launched off from the shore. We were no sooner off, but our canoe overset, bottom upwards. We preserved our lives well enough by swimming, and dragged also our chests and clothes ashore; but all our things were wet. I had nothing of value but my journal, and some draughts of land of my own taking, which I much prized, and which I had hitherto carefully preserved. Mr Hall had also such another cargo of books and draughts, which were now like to perish. But we presently opened our chests, and took out our books, which, with much ado, we did afterwards dry; but some of our draughts that lay loose in our chests were spoiled. We lay here afterwards three days, making great fires to dry our books. The Achinese in the meantime fixed our canoe with outlagers on each side; and they also cut a good mast for her, and made a substantial sail with mats.

The canoe being now very well fixed, and our books and clothes dry, we launched out the second time, and rowed towards the east side of the island, leaving many islands to the north of us. The Indians of the island accompanied us with eight or ten canoes, against our desire; for we thought that these men would make provision dearer at that side of the island we were going to, by giving an account what rates we gave for it at the place whence we came, which was owing to the ship's being there; for the ship's crew were not so thirsty in bargaining (as they seldom are) as single persons or a few men might be apt to be, who would keep to one bargain. Therefore to hinder them from going with us, Mr Hall scared one canoe's crew by firing a shot over them. They all leaped overboard, and cried out; but seeing us row away, they got into their canoes again, and came after us. The firing of that gun made all the inhabitants of the island our enemies. For presently after this we put ashore, at a bay where were four houses and a great many canoes: but they all went away, and came near us no more, for several days. We had then a great loaf of melory, which was our constant food; and if we had a mind to cocoa-nuts, or toddy, our Malays of Achin would climb the trees, and fetch as many nuts as we would have, and a good pot of toddy every morning. Thus we lived till our melory was almost spent; being still in hopes that the natives would come to us, and sell it as they had formerly done. But they came not to us: nay, they opposed us wherever we came, and often shaking their lances at us, made all the show of hatred that they could invent. At last, when we saw that they stood in opposition to us, we resolved to use force to get some of their food, if we could not get it other ways. With this resolution, we went in our canoe to a small bay on the north part of the island, because it was smooth water there, and good landing; but on the other side, the wind being yet on that quarter, we could not land without jeopardy of oversetting our canoe and wetting our arms, and then we must have lain at the mercy of our enemies, who stood 200 or 300 men in every bay where they saw us coming, to keep us off.

When we set out, we rowed directly to the north end, and presently were followed by seven or eight of their canoes. They keeping at a distance, rowed away faster than we did, and got to the bay before us: and there, with about twenty more canoes full of men, they all landed and stood to hinder us from landing. But we rowed in within a hundred yards of them; then we lay still, and I took my gun, and presented at them: at which they all fell down flat on the ground. But I turned myself about, and, to show that we did not intend to harm them, I fired my gun off to sea, so that they might see the shot graze on the water. As soon as my gun was loaded again, we rowed gently in; at which some of them withdrew. The rest, standing up, did still cut and hew the air, making signs of their hatred; till I once more frighted them with my gun, and discharged it as before. Then more of them sneaked away, leaving only five or six men on the bay. Then we rowed in again, and Mr Hall, taking his sword in his hand, leaped ashore; and I stood ready with my gun to fire at the Indians, if they had injured him. But they did not stir, till he came to them, and saluted them. He shook them by the hand, and by such signs of friendship as he made, the peace was concluded, ratified and confirmed by all that were present; and others that were gone were again called back, and they all very joyfully accepted of a peace. This became universal over all the island, to the great joy of the inhabitants. There was no ringing of bells, nor bonfires made, for that is not the custom here; but gladness appeared in their countenances, for now they could go out and fish again without fear of being taken. This peace was not more welcome to them than to us; for now the inhabitants brought their melory again to us; which we bought for old rags, and small stripes of cloth, about as broad as the palm of one's hand. I did not see above five or six hens, for they have but few on the island. At some places we saw some small hogs, which we could have bought of them reasonably; but we would not offend our Achinese friends, who were Mahometans.

We stayed here two or three days, and then rowed toward the south end of the island, keeping on the east side, and we were kindly received by the natives wherever we came. When we arrived at the south end of the island, we fitted ourselves with melory and water. We bought three or four loaves of melory, and about twelve large cocoa-nut shells, that had all the kernel taken out, yet were preserved whole, except only a small hole at one end; and all these held for us about three gallons and a half of water. We bought also two or three bamboos, that held about four or five gallons more: this was our sea-store. We now designed to go to Achin, a town on the NW. end of the Island Sumatra, distant from hence about forty leagues, bearing SSE. We only waited for the western monsoon, which we had expected a great while, and now it seemed to be at hand; for the clouds began to hang their heads to the eastward, and at last moved gently that way; and though the wind was still at east, yet this was an infallible sign that the western monsoon was nigh.

CHAPTER XVIII.

It was the 15th of May 1688, about 4 o'clock in the afternoon, when we left Nicobar Island, directing our course toward Achin, being eight men of us in company—viz., three English, four Malays who were born at Achin, and the mongrel Portuguese. Our vessel, the Nicobar canoe, was not one of the biggest nor of the least size. She was much about the burthen of one of our London wherries below bridge, and built sharp at both ends, like the forepart of a wherry. She was deeper than a wherry, but not so broad, and was so thin and light that when empty, four men could launch her, or haul her ashore on a sandy bay. We had a good substantial mast and a mat sail, and good outlagers lashed very fast and firm on each side the vessel, being made of strong poles. So that while these continued firm the vessel could not overset, which she would easily have done without them, and with them too, had they not been made very strong; and we were therefore much beholden to our Achinese companions for this contrivance. These men were none of them so sensible of the danger as Mr Hall and myself, for they all confided so much in us that they did not so much as scruple anything that we did approve of. Neither was Mr Hall so well provided as I was, for before we left the ship I had purposely consulted our draught of the East Indies (for we had but one in the ship), and out of that I had written in my pocket-book an account of the bearing and distance of all the Malacca coast, and that of Sumatra, Pegu, and Siam; and also brought away with me a pocket compass for my direction in any enterprize that I should undertake.

The weather at our setting out was very fair, clear, and hot. The wind was still at SE., a very small breeze just fanning the air; and the clouds were moving gently from west to east, which gave us hopes that the winds were either at west already, abroad at sea, or would be so in a very short time. We took this opportunity of fair weather, being in hopes to accomplish our voyage to Achin before the western monsoon was set in strong, knowing that we should have very blustering weather after this fair weather, especially at the first coming of the western monsoon. We rowed, therefore, away to the southward, supposing that when we were clear from the island we should have a true wind, as we call it, for the land hauls the wind; and we often find the wind at sea different from what it is near the shore. We rowed with four oars, taking our turns; Mr Hall and I steered also by turns, for none of the rest were capable of it. We rowed the first afternoon, and the night ensuing, about twelve leagues, by my judgment. Our course was SSE., but the 16th, in the morning, when the sun was an hour high, we saw the island whence we came, bearing NW. by N. Therefore I found we had gone a point more to the east than I intended, for which reason we steered S. by E. In the afternoon, at 4 o'clock, we had a gentle breeze at WSW., which continued so till 9, all which time we laid down our oars and steered away SSE. I was then at the helm, and I found by the rippling of the sea that there was a strong current against us. It made a great noise that might be heard near half-a-mile. At 9 o'clock it fell calm, and so continued till 10. Then the wind sprung up again, and blew a fresh breeze all night.

The 17th, in the morning, we looked out for the Island Sumatra, supposing that we were now within twenty leagues of it, for we had rowed and sailed, by our reckoning, twenty-four leagues from Nicobar Island; and the distance from Nicobar to Achin is about forty leagues. But we looked in vain for the Island Sumatra, for, turning ourselves about, we saw, to our grief, Nicobar Island lying WNW., and not above eight leagues distant. By this it was visible that we had met a very strong current against us in the night. But the wind freshened on us, and we made the best of it while the weather continued fair. The 18th, the wind freshened on us again, and the sky began to be clouded. It was indifferent clear till noon, and we thought to have had an observation; but we were hindered by the clouds that covered the face of the sun when it came on the meridian. We had then also a very ill presage by a great circle about the sun (five or six times the diameter of it), which seldom appears but storms of wind or much rain ensue. Such circles about the moon are more frequent, but of less import. We commonly take great notice of those that are about the sun, observing if there be any breach in the circle, and in what quarter the breach is, for thence we commonly find the greatest stress of the wind will come. I must confess that I was a little anxious at the sight of this circle, and wished heartily that we were near some land. Yet I showed no signs of it to discourage my consorts, but made a virtue of necessity and put a good countenance on the matter. I told Mr Hall that if the wind became too strong and violent, as I feared it would, it being even then very strong, we must of necessity steer away before the wind and sea till better weather presented; and that, as the winds were now, we should, instead of about twenty leagues to Achin, be driven sixty or seventy leagues to the coast of Cudda or Queda,[1] a kingdom and town and harbour of trade on the coast of Malacca.

The winds therefore bearing very

[1] Quedah, on the western coast of the Malayan Peninsula, a little to the north of the British settlement of Pulo Penang.

hard, we rolled up the foot of our sail on a pole fastened to it, and settled our yard within three feet of the canoe sides, so that we had now but a small sail; yet it was still too big, considering the wind, for the wind being on our broadside, pressed her down very much, though supported by her outlagers, insomuch that the poles of the outlagers going from the sides of their vessel bent as if they would break; and should they have broken, our overturning and perishing had been inevitable. Besides, the sea increasing would soon have filled the vessel this way. Yet thus we made a shift to bear up with the side of the vessel against the wind for a while; but the wind still increasing, about 1 o'clock in the afternoon we put away right before wind and sea, continuing to run thus all the afternoon and part of the night ensuing. The wind continued increasing all the afternoon, and the sea still swelled higher and often broke, but did us no damage; for the ends of the vessel being very narrow, he that steered received and broke the sea on his back, and so kept it from coming in so much as to endanger the vessel; though much water would come in, which we were forced to keep heaving out continually. And by this time we saw it was well that we had altered our course, every wave would else have filled and sunk us, taking the side of the vessel; and though our outlagers were well lashed down to the canoe's bottom with rattans, yet they must probably have yielded to such a sea as this, when even before they were plunged under water and bent like twigs.

The evening of this 18th was very dismal. The sky looked very black, being covered with dark clouds; the wind blew hard and the seas ran high. The sea was already roaring in a white foam about us, a dark night coming on, no land in sight to shelter us, and our little ark in danger to be swallowed by every wave; and what was worst of all, none of us thought ourselves prepared for another world. The reader may better guess than I can express the confusion that we were all in. I had been in many imminent dangers before now, some of which I have already related; but the worst of them all was but a play-game in comparison with this. I must confess that I was in great conflicts of mind at this time. Other dangers came not upon me with such a leisurely and dreadful solemnity: a sudden skirmish or engagement or so was nothing when one's blood was up and pushed forward with eager expectations. But here I had a lingering view of approaching death, and little or no hopes of escaping it; and I must confess that my courage, which I had hitherto kept up, failed me here; and I made very sad reflections on my former life, and looked back with horror and detestation on actions which before I disliked, but now I trembled at the remembrance of. I had long before this repented me of that roving course of life, but never with such concern as now. I did also call to mind the many miraculous acts of God's providence towards me in the whole course of my life, of which kind I believe few men have met with the like. For all these I returned thanks in a peculiar manner, and this once more desired God's assistance, and composed my mind as well as I could in the hopes of it; and, as the event showed, I was not disappointed of my hopes.

Submitting ourselves therefore to God's good providence, and taking all the care we could to preserve our lives, Mr Hall and I took turns to steer, and the rest took turns to heave out the water, and thus we provided to spend the most doleful night I ever was in. About 10 o'clock it began to thunder, lighten, and rain; but the rain was very welcome to us, having drunk up all the water we brought from the island. The wind at first blew harder than before; but within half-an-hour it abated, and became more moderate, and the sea also assuaged of its fury; and then by a lighted match, of which we kept a piece burning on purpose, we looked on our compass to see how we steered, and found our course to be still east. We had no occasion to look on the compass before, for we steered right before the wind, which if it had shifted, we had been obliged to have altered our course accordingly. But now it being abated, we found our vessel lively enough, with that small sail which was then aboard, to haul to our former course, SSE., which accordingly we did, being now in hopes again to get to the Island Sumatra. But about 2 o'clock in the morning of the 19th, we had another gust of wind, with much thunder, lightning, and rain, which lasted till day, and obliged us to put before the wind again, steering thus for several hours. It was very dark, and the hard rain soaked us so thoroughly, that we had not one dry thread about us. The rain chilled us extremely; for any fresh water is much colder than that of the sea. For even in the coldest climates the sea is warm, and in the hottest climates the rain is cold and unwholesome for man's body. In this wet starveling plight we spent the tedious night. Never did poor mariners on a lee-shore more earnestly long for the dawning light, than we did now. At length the day appeared; but with such dark black clouds near the horizon, that the first glimpse of the dawn appeared thirty or forty degrees high, which was dreadful enough. For it is a common saying among seamen, and true, as I have experienced, that a high dawn will have high winds, and a low dawn, small winds.

We continued our course still east, before wind and sea, till about 8 o'clock in the morning of this 19th, and then one of our Malay friends cried out, "Pulo Way." Mr Hall, and Ambrose, and I, thought the fellow had said "Pull away," an expression usual among English seamen when they are rowing; and we wondered what he meant by it, till we saw him point to his consorts, and then we looking that way, saw land appearing like an island, and all our Malays said it was an island at the NW. end of Sumatra, called Way, for Pulo Way is the Island Way. We, who were dripping with wet, cold and hungry, were all overjoyed at the sight of the land, and presently marked its bearing. It bore south, and the wind was still at west, a strong gale; but the sea did not run so high as in the night. Therefore we trimmed our small sail no bigger than an apron, and steered with it. Now our outlagers did us a great kindness again; for although we had but a small sail, yet the wind was strong, and pressed down our vessel's side very much; but being supported by the outlagers, we could brook it well enough, which otherwise we could not have done. About noon we saw more land, beneath the supposed Pulo Way; and steering towards it, before night we saw all the coast of Sumatra, and found the errors of our Achinese; for the high land that we first saw, which then appeared like an island, was not Pulo Way, but a great high mountain on the Island Sumatra, called by the English the Golden Mountain. Our wind continued till about 7 o'clock at night, then it abated, and at 10 o'clock it died away. And then we stuck to our oars again, though all of us quite tired with our former fatigues and hardships.

The next morning, being the 20th, we saw all the low land plain, and judged ourselves not above eight leagues off. About 8 o'clock in the morning we had the wind again at west, a fresh gale; and steering in still for the shore, at 5 o'clock in the afternoon we ran to the mouth of a river on the Island Sumatra, called Passange Jonca. It is thirty-four leagues to the eastward of Achin, and six leagues to the west of Diamond Point. Our Malays were very well acquainted here, and carried us to a small fishing village, within a mile of the river's mouth, called also by the name of the River Passange Jonca. The hardships of this voyage, with the scorching heat of the sun at our first setting out, and the cold rain, and our continuing wet for the last two days, cast us all into fevers, so that now we were not able to help each other, nor so much as to get our canoe up to the village; but our Malays got some of the townsmen to bring her up.

The news of our arrival being noised abroad, one of the Oramkais, or noblemen of the island, came in the night to see us. We were then lying in a small hut at the end of the town, and it being late, this lord only viewed us, and having spoken with our Malays, went away again; but he returned to us the next day, and provided a large house for us to live in, till we should be recovered of our sickness; ordering the town's-people to let us want for nothing. The Achinese Malays that came with us, told them all the circumstances of our voyage; how they were taken by our ship, and where, and how we that came with them were prisoners aboard the ship, and had been set ashore together at Nicobar, as they were. It was for this reason, probably, that the gentlemen of Sumatra were thus extraordinary kind to us, to provide everything that we had need of; nay, they would force us to accept of presents from them, that we knew not what to do with, as young buffaloes, goats, &c., for these we would turn loose at night, after the gentlemen that gave them to us were gone, for we were prompted by our Achinese consorts to accept of them for fear of disobliging by our refusal. But the cocoa-nuts, plantains, fowls, eggs, fish, and rice, we kept for our use. The Malays that accompanied us from Nicobar separated themselves from us now, living at one end of the house by themselves, for they were Mahometans, as all those of the kingdom of Achin are; and though during our passage by sea together we made them be contented to drink their water out of the same cocoa-shell with us, yet, being now no longer under that necessity, they again took up their accustomed nicety and reservedness. They all lay sick, and as their sickness increased, one of them threatened us, that if any of them died, the rest should kill us, for having brought them this voyage; yet I question whether they would attempted, or the country people have suffered it. We made a shift to dress our own food; for none of these people, though they were very kind in giving us anything that we wanted, would yet come near us to assist us in dressing our victuals; nay, they would not touch anything that we used. We had all fevers, and therefore took turns to dress victuals, according as we had strength to do it, or stomachs to eat it. I found my fever to increase, and my head so distempered, that I could scarce stand, therefore I whetted and sharpened my penknife, in order to let myself blood; but I could not, for my knife was too blunt. We stayed here ten or twelve days, in hopes to recover our health; but finding no amendment, we desired to go to Achin. But we were delayed by the natives, who had a desire to have kept Mr Hall and myself, to sail in their vessels to Malacca, Cudda, or other places whither they trade. But finding us more desirous to be with our countrymen in our factory at Achin, they provided a large proa to carry us thither, we not being able to manage our own canoe. Besides, before this, three of our Malay comrades were gone very sick into the country, and only one of them and the Portuguese remained with us, accompanying us to Achin, and they both as sick as we.

It was the beginning of June 1688, when we left Passange Jonca. We had four men to row, one to steer, and a gentleman of the country that went purposely to give information to the Government of our arrival. We were but three days and nights in our passage, having sea-breezes by day and land winds by night, and very fair weather. When we arrived at Achin, I was carried before the Shabander, the chief magistrate in the city. One Mr Dennis Driscall, an Irishman, and a resident in the factory which our East India Company had there then, was interpreter. I, being weak, was suffered to stand in the Shabander's presence; for it is their custom to make men sit on the floor, as they do, cross-legged like tailors; but I had not strength then to pluck up my heels in that manner. The Shabander asked of me several questions, especially how we durst adventure to come in a canoe from Nicobar Island to Sumatra. I told him that I had been accustomed to hardships and hazards, therefore I did with much freedom undertake it. He inquired also concerning our ship, whence she came, &c. I told him, from the South Seas; that she had ranged about the Philippine Islands, &c., and was now gone towards Arabia and the Red Sea. The Malays also and Portuguese were afterwards examined, and confirmed what I declared; and in less than half-an-hour I was dismissed with Mr Driscall, who then lived in the English East India Company's factory. He provided a room for us to lie in, and some victuals.

Three days after our arrival here, our Portuguese died of a fever. What became of our Malays I know not. Ambrose lived not long after. Mr Hall also was so weak, that I did not think he would recover. I was the best, yet still very sick of a fever, and little likely to live. Therefore Mr Driscall and some other Englishmen persuaded me to take some purging physic of a Malay doctor. I took their advice, being willing to get ease; but after three dozes, each a large calabash of nasty stuff, finding no amendment, I thought to desist from more physic, but was persuaded to take one doze more; which I did, and it wrought so violently, that I thought it would have ended my days. I thought my Malay doctor, whom they so much commended, would have killed me outright. I continued extraordinary weak for some days after his drenching me thus; but my fever left me for above a week, after which it returned upon me again for a twelvemonth, and a flux with it. However, when I was a little recovered from the effects of my drench, I made a shift to go abroad; and having been kindly invited to Captain Bowry's house there, my first visit was to him, who had a ship in the road, but lived ashore. This gentleman was extraordinary kind to us all, particularly to me, and importuned me to go as his boatswain to Persia, whither he was bound, with a design to sell his ship there, as I was told, though not by himself. Thence he intended to pass with the caravan to Aleppo, and so home for England. His business required him to stay some time longer at Achin, I judge, to sell some commodities that he had not yet disposed of. Yet he chose rather to leave the disposal of them to some merchant there, and make a short trip to the Nicobar Islands in the meantime, and on his return to take in his effects, and so proceed towards Persia. This was a sudden resolution of Captain Bowry's presently after the arrival of a small frigate from Siam, with an Ambassador from the King of Siam to the Queen of Achin. The Ambassador was a Frenchman by nation. The vessel that he came in was but small, yet very well manned, and fitted for a fight. Therefore it was generally supposed here that Captain Bowry was afraid to lie in Achin Road, because the Siamese were now at war with the English, and he was not able to defend his ship if he should be attacked by them. But whatever made him think of going to the Nicobar Islands, he provided to sail, and took me, Mr Hall, and Ambrose with him, though all of us so sick and weak that we could do him no service. It was some time about the beginning of June when we sailed out of Achin Road; but we met with the winds at NW., with turbulent weather, which forced us back again in two days' time. Yet he gave us each twelve "mess" apiece—a gold coin, each of which is about the value of fifteenpence English. So he gave over that design, and some English ships coming into Achin Road, he was not afraid of the Siamese who lay there. After this he again invited me to his house at

Achin, and treated me always with wine and good cheer, and still importuned me to go with him to Persia. But I being very weak, and fearing the westerly winds would create a great deal of trouble, did not give him a positive answer, especially because I thought I might get a better voyage in the English ships newly arrived, or some others now expected here.[1]

A short time after this, Captain Welden arrived here from Fort St George, in a ship called the Cartana, bound to Tonquin. This being a more agreeable voyage than to Persia, at this time of the year, besides that the ship was better accommodated, especially with a surgeon, and I being still sick, I therefore chose rather to serve Captain Welden than Captain Bowry. But to go on with a particular account of that expedition were to carry my reader back again; whom having brought thus far towards England in my circumnavigation of the Globe, I shall not now weary him with new rambles, nor so much swell this volume, as I must, to describe the tour I made in those remote parts of the East Indies from and to Sumatra. So that my voyage to Tonquin at this time, as also another to Malacca afterwards, with my observations in them, and the descriptions of those and the neighbouring countries; as well as the description of the Island Sumatra itself, and therein the kingdom and city of Achin, Bencouli,[2] &c., I shall refer to another place, where I may give a particular relation of them.[3] In short, it may suffice that I set out to Tonquin with Captain Welden about July 1688, and returned to Achin in the April following. I stayed there till the latter end of September 1689, and making a short voyage to Malacca, came thither again about Christmas. Soon after that I went to Fort St George, and staying there about five months, I returned once more to Sumatra; not to Achin, but to Bencouli, an English factory on the west coast, of which I was gunner about five months more. So that, having brought my reader to Sumatra, without carrying him back, I shall bring him on next way from thence to England. And of all that occurred between my first setting out from this island in 1688, and my final departure from it at the beginning of the year 1691, I shall only take notice at present of two passages which I think I ought not to omit.

The first is, that at my return from Malacca, a little before Christmas 1689, I found at Achin one Mr Morgan, who was one of our ship's crew that left me ashore at Nicobar, now mate of a Danish ship of Trangambar,[4] which is a town on the Coast of Coromandel, near Cape Comorin, belonging to the Danes: and receiving an account of our crew from him and others, I thought it might not be amiss to gratify the reader's curiosity therewith, who would probably be desirous to know the success of those ramblers in their new intended expedition towards the Red Sea. And withal I thought it might not be unlikely that these papers might fall into the hands of some of our London merchants, who were concerned in fitting out that ship; which, I said formerly, was called the Cygnet of London, sent on a trading voyage into the South Seas, under the command of Captain Swan. To proceed therefore with Morgan's relation. He told me, that when they in the Cygnet went away from Nicobar, in pursuit of their intended voyage to Persia,

[1] Captain Bowry was the writer of the letter from Borneo to the "English factory at Mindanao," referred to in Chapter XIII.

[2] Bencoolen, where the English had settled in 1685, and where a year or two later the East India Company built a fort, which was called Fort York.

[3] This Dampier does in Appendix No. I to his greater work. See Introductory Note to "The Author's Account of Himself.

[4] Tranquebar, then capital of the Danish possessions in India.

they directed their course towards Ceylon. But not being able to weather it, the westerly monsoon bearing hard against them, they were obliged to seek refreshment on the Coast of Coromandel. Here this mad, fickle crew were upon new projects again; their designs meeting with such delays and obstructions, that many of them grew weary of it, and about half of them went ashore. Of this number, Mr Morgan, who told me this, and Mr Herman Coppinger the surgeon, went to the Danes at Trangambar, who kindly received them. There they lived very well; and Mr Morgan was employed as a mate in a ship of theirs at this time to Achin; and Captain Knox tells me, that he since commanded the Curtana, the ship that I went in to Tonquin, which Captain Welden having sold to the Mogul's subjects, they employed Mr Morgan as captain to trade in her for them; and it is an usual thing for the trading Indians to hire Europeans to go officers on board their ships, especially captains and gunners. About two or three more of these that were set ashore went to Fort St George; but the main body of them were for going into the Mogul's service. Our seamen are apt to have great notions of I know not what profit and advantages to be had in serving the Mogul; nor do they want for fine stories to encourage one another to it. It was what these men had long been thinking and talking of as a fine thing; but now they went upon it in good earnest. The place where they went ashore was at a town of the Moors; which name our seamen give to all the subjects of the Great Mogul, but especially his Mahometan subjects; calling the idolaters Gentoos or Rashbouts. At this Moors' town they got a peon to be their guide to the Mogul's nearest camp: for he has always several armies in his vast empire.

These peons are some of the Gentoos or Rashbouts, who in all places along the coast, especially in seaport towns, make it their business to hire themselves to wait upon strangers, be they merchants, seamen, or what they will. To qualify them for such attendance, they learn the European languages, English, Dutch, French, Portuguese, &c., according as they have any of the factories of these nations in their neighbourhood, or are visited by their ships. No sooner does any such ship come to an anchor and the men come ashore, but a great many of these peons are ready to proffer their service. It is usual for the strangers to hire their attendance during their stay there, giving them about a crown a month of our money, more or less. The richest sort of men will ordinarily hire two or three peons to wait upon them; and even the common seamen, if able, will hire one a-piece to attend them, either for convenience or ostentation; or sometimes one peon between two of them. These peons serve them in many capacities, as interpreters, brokers, servants to attend at meals, and go to market and on errands, &c. Nor do they give any trouble, eating at their own homes, and lodging there, when they have done their masters' business for them; expecting nothing but their wages, except that they have a certain allowance of about a "fanam," or threepence in a dollar, which is an eighteenth part profit, by way of brokerage for every bargain they drive: they being generally employed in buying and selling. When the strangers go away, their peons desire them to give them their names in writing, with a certificate of their honest and diligent serving them: and these they show to the next comers, to get into business; some being able to produce a large scroll of such certificates.

But to proceed. The Moors' town where these men landed was not far from Cunnimere, a small English factory on the Coromandel Coast. The Governor whereof having intelligence by the Moors of the landing of these men, and their intended march to the Mogul's camp, sent out a captain with his company to oppose it. He came up with them and gave them hard words, but they being thirty or forty resolute fellows not easily daunted, he durst not attack them, but returned to the Governor; and the news of it was soon carried to Fort St George. During their march, John Oliver, who was one of them, privately told the peon who guided them, that himself was their captain. So when they came to the camp the peon told this to the General; and when their stations and pay were assigned them, John Oliver had a greater respect paid him than the rest; and whereas their pay was ten pagodas a month each man (a pagoda is two dollars, or 9s. English), his pay was twenty pagodas. Which stratagem and usurpation of his, occasioned him no small envy and indignation from his comrades. Soon after this, two or three of them went to Agra to be of the Mogul's guard. A while after, the Governor of Fort St George sent a message to the main body of them, and a pardon, to withdraw them from thence, which most of them accepted, and came away. John Oliver and the small remainder continued in the country, but leaving the camp, went up and down plundering the villages, and fleeing when they were pursued; and this was the last news I heard of them. This account I had partly by Mr Morgan from some of those deserters he met with at Trangambar, and partly from others of them whom I met with myself afterwards at Fort St George. And these were the adventures of those who went up into the country.

Captain Reed having thus lost the best half of his men sailed away with the rest of them, after having filled his water and got rice, still intending for the Red Sea. When they were near Ceylon they met with a Portuguese ship richly laden, out of which they took what they pleased, and then turned her away again. From thence they pursued their voyage, but the westerly winds bearing hard against them, and making it hardly feasible for them to reach the Red Sea, they stood away for Madagascar. There they entered into the service of one of the petty princes of that island, to assist him against his neighbours, with whom he was at war. During this interval, a small vessel from New York came hither to purchase slaves, which trade is driven here, as it is upon the Coast of Guinea, one nation or clan selling others that are their enemies. Captain Reed, with about five or six more, stole away from their crew and went aboard this New York ship, and Captain Tait was made commander of the residue. Soon after which, a brigantine from the West Indies, Captain Knight commander, coming thither with design to go to the Red Sea also, these of the Cygnet consorted with them, and they went together to the Island Johanna.[1] Thence going together towards the Red Sea, the Cygnet proving leaky, and sailing heavily, as being much out of repair, Captain Knight grew weary of her company; and giving her the slip in the night, went away for Achin; for, having heard that there was plenty of gold there, he went thither with a design to cruise; and it was from one Mr Humes belonging to the Ann of London, Captain Freke commander, who had gone aboard Captain Knight, and whom I saw afterwards at Achin, that I had this relation. Some of Captain Freke's men, their own ship being lost, had gone aboard the Cygnet at Johanna; and after Captain Knight had left her, she still pursued her voyage towards the Red Sea. But the winds being against them, and the ship in so ill a condition, they were forced to bear away for Coromandel, where Captain Tait and his own men went ashore to serve the Mogul. But the strangers of Captain Freke's ship, who kept still aboard the Cygnet, undertook to carry her for England; and the last news I heard of the Cygnet was from Captain Knox, who tells me that she now lies sunk in St Augustine's Bay in Madagascar.[2] This digression I have made to give an account of our ship.

The other passage I shall speak of

[1] One of the Comoro group, between Madagascar and Mozambique.

[2] On the south-west of the island.

that occurred during this interval of the tour I made from Achin is with relation to the Painted Prince whom I brought with me into England, and who died at Oxford. For while I was at Fort St George, about April 1690, there arrived a ship called the Mindanao Merchant, laden with clove-bark from Mindanao. Three of Captain Swan's men that remained there when we went from thence came in her, from whom I had the account of Captain Swan's death, as is before related. There was also one Mr Moody, who was supercargo of the ship. This gentleman bought at Mindanao the Painted Prince Jeoly,[1] and his mother, and brought them to Fort St George, where they were much admired by all that saw them. Some time after this, Mr Moody, who spoke the Malay language very well, and was a person very capable to manage the Company's affairs, was ordered by the Governor of Fort St George to prepare to go to Indrapore, an English factory on the west coast of Sumatra, in order to succeed Mr Gibbons, who was chief of that place. By this time I was very intimately acquainted with Mr Moody, and was importuned by him to go with him, and to be gunner of the fort there. I always told him I had a great desire to go to the Bay of Bengal, and that I had now an offer to go thither with Captain Metcalf, who wanted a mate, and had already spoken to me. Mr Moody, to encourage me to go with him, told me that if I would go with him to Indrapore he would buy a small vessel there, and send me to the Island Meangis as commander of her; and that I should carry Prince Jeoly and his mother with me (that being their country), by which means I might gain a commerce with his people for cloves. This was a design that I liked very well, therefore I consented to go thither. It was some time in July 1690 when we went from Fort St George in a small ship called the Diamond, Captain Howel commander. We were about fifty or sixty passengers in all; some ordered to be left at Indrapore, and some at Bencouli; five or six of us were officers, the rest soldiers to the Company. We met nothing in our voyage that deserves notice till we came abreast of Indrapore; then the wind came at NW., and blew so hard that we could not get in, but were forced to bear away to Bencouli, another English factory on the same coast, lying fifty or sixty leagues to the southward of Indrapore.

Upon our arrival at Bencouli we saluted the fort, and were welcomed by them. The same day we came to an anchor, and Captain Howel and Mr Moody, with the other merchants, went ashore, and were all kindly received by the Governor of the fort. It was two days after before I went ashore, and then I was importuned by the Governor to stay there to be gunner of this fort, because the gunner was lately dead; and this being a place of greater import than Indrapore, I should do the Company more service here than there. I told the Governor, if he would augment my salary, which by agreement with the Governor of Fort St George I was to have had at Indrapore, I was willing to serve him, provided Mr Moody would consent to it. As to my salary, he told me I should have twenty-four dollars per month, which was as much as he gave to the old gunner. Mr Moody gave no answer till a week after, and then, being ready to be gone to Indrapore, he told me I might use my own liberty, either to stay here or go with him to Indrapore. He added, that if I went with him, he was not certain as yet to perform his promise in getting a vessel for me to go to Meangis with Jeoly and his mother; but he would be so fair to me, that because I left Madras on his account, he would give me the half share of the two painted people, and leave them in my possession and at my disposal. I accepted of the offer, and writings were immediately drawn between us.

Thus it was that I came to have

[1] Who was a slave at Mindanao during Dampier's stay there. See Chapter XIII.

this Painted Prince, whose name was Jeoly, and his mother. They were born on a small island called Meangis. I saw the island twice, and two more close by it. Each of the three seemed to be about four or five leagues round, and of a good height. Jeoly himself told me that they all three abounded with gold, cloves, and nutmegs; for I showed him some of each sort several times, and he told me in the Malay language, which he spake indifferent well, "*Meangis hadda madochala se bullawan;*" that is, "There is abundance of gold at Meangis." "*Bullawan*" I have observed to be the common word for gold at Mindanao; but whether the proper Malay word I know not; for I found much difference between the Malay language as it was spoken at Mindanao, and the language on the coast of Malacca and Achin. When I showed him spice, he would not only tell me that there was *madochala*, that is, abundance; but, to make it appear more plain, he would also show me the hair of his head, a thing frequent among all the Indians that I have met with, to show their hair when they would express more than they can number. He told me also that his father was Raja of the island where they lived; that there were not above thirty men on the island, and about one hundred women; that he himself had five wives and eight children, and that one of his wives painted him. He was painted all down his breast; between his shoulders behind; on his thighs mostly before; and in the form of several broad rings, or bracelets, round his arms and legs. I cannot liken the drawings to any figures of animals, or the like; but they were very curious, full of great variety of lines, flourishes, chequered work, &c., keeping a very graceful proportion, and appearing very artificial,[1] even to wonder, especially that upon and between his shoulder blades. By the account he gave me of the manner of doing it, I understood that the painting was done in the same manner as the Jerusalem Cross is made in men's arms, by pricking the skin and rubbing in a pigment.[2] But whereas powder is used in making the Jerusalem Cross, they at Meangis use the gum of a tree beaten to powder, called by the English drammer, which is used instead of pitch in many parts of India. He told me that most of the men and women on the island were thus painted; and also that they had all earrings made of gold, and gold shackles about their legs and arms; that their common food, of the produce of the land, was potatoes and yams; that they had plenty of cocks and hens, but no other tame fowl. He said that fish (of which he was a great lover, as wild Indians generally are) was very plentiful about the island; and that they had canoes, and went a-fishing frequently in them; and that they often visited the other two small islands, whose inhabitants speak the same language as they did; which was so unlike the Malay, which he had learnt while he was a slave at Mindanao, that when his mother and he were talking together in their Meangian tongue I could not understand one word they said. And indeed all the Indians who speak Malay, who are the trading and politer sort, looked on these Meangians as a kind of barbarians; and, upon any occasion of dislike, would call them "bobby," that is, "hogs," the greatest expression of contempt that can be, especially from the mouth of Malays, who are generally Mahometans. And yet the Malays everywhere call a woman babby, by a name not much different; and mamma signifies a man: though these two last words properly denote male and female; and as "ejam" signifies a fowl, so "ejam mamma" is a cock, and "ejam babby" is a hen. But this by the way.

He said also, that the customs of those other isles, and their manner of living, was like theirs, and that they

[1] Skilful, ingenious.

[2] That is, by tattooing.

were the only people with whom they had any converse; and that one time, as he, with his father, mother, and brother, with two or three men more, were going to one of these other islands, they were driven by a strong wind on the coast of Mindanao, where they were taken by the fishermen of that island, and carried ashore and sold as slaves, they being first stripped of their gold ornaments. I did not see any of the gold that they wore; but there were great holes in their ears, by which it was manifest that they had worn some ornaments in them. Jeoly was sold to one Michael, a Mindanayan, that spoke good Spanish, and commonly waited on Raja Laut, serving him as our interpreter where the Raja was at a loss in any word, for Michael understood it better. He did often beat and abuse his painted servant, to make him work, but all in vain; for neither fair means, threats, nor blows would make him work as he would have him. Yet he was very timorous, and could not endure to see any sort of weapons; and he often told me that they had no arms at Meangis, they having no enemies to fight with. I knew this Michael very well while we were at Mindanao. I suppose that name was given him by the Spaniards, who baptized many of them at the time when they had footing at that island; but, at the departure of the Spaniards, they were Mahometans again as before. Some of our people lay at this Michael's house, whose wife and daughter were pagallies to some of them. I often saw Jeoly at his master Michael's house; and when I came to have him so long after, he remembered me again. I did never see his father nor brother, nor any of the others that were taken with them; but Jeoly came several times aboard our ship when we lay at Mindanao, and gladly accepted of such victuals as we gave him; for his master kept him at very short commons.

Prince Jeoly lived thus a slave at Mindanao four or five years, till at last Mr Moody bought him and his mother for sixty dollars, and, as is before related, carried him to Fort St George, and thence along with me to Bencouli. Mr Moody stayed at Bencouli about three weeks, and then went back with Captain Howel to Indrapore, leaving Jeoly and his mother with me. They lived in a house by themselves without the fort. I had no employment for them, but they both employed themselves. She used to make and mend their own clothes, at which she was not very expert, for they wear no clothes at Meangis, but only a cloth about their waists; and he busied himself in making a chest with four boards and a few nails that he begged of me. It was but an ill-shaped, odd thing, yet he was as proud of it as if it had been the rarest piece in the world. After some time they were both taken sick, and though I took as much care of them as if they had been my brother and sister, yet she died. I did what I could to comfort Jeoly; but he took on extremely, insomuch that I feared him also.[1] Therefore I caused a grave to be made presently, to hide her out of his sight. I had her shrouded decently in a piece of new calico; but Jeoly was not so satisfied, for he wrapped all her clothes about her, and two new pieces of chintz that Mr Moody gave her, saying that they were his mother's, and she must have them. I would not disoblige him, for fear of endangering his life; and I used all possible means to recover his health; but I found little amendment while we stayed here. In the little printed relation that was made of him when he was shown for a sight in England, there was a romantic story of a beautiful sister of his, a slave with them at Mindanao, and of the Sultan's falling in love with her; but these were stories indeed. They reported also that his paint was of such virtue, that serpents and venomous creatures would flee from him; for which reason, I suppose, they represented so many serpents scampering about in the

[1] That is, I feared for his life also, so profound was his grief.

printed picture that was made of him. But I never knew any paint of such virtue; and as for Jeoly, I have seen him as much afraid of snakes, scorpions, or centipedes as myself.

Having given this account of the ship that left me at Nicobar, and of my Painted Prince whom I brought with me to Bencouli, I shall now proceed with the relation of my voyage thence to England, after I have given this short account of the occasion of it, and the manner of my getting away. To say nothing, therefore, now of that place, and my employment there as gunner of the fort, the year 1690 drew towards an end; and not finding the Governor keep to his agreement with me, nor seeing by his carriage towards others any great reason I had to suspect he would, I began to wish myself away again. I saw so much ignorance in him with respect to his charge, being much fitter to be a book-keeper than governor of a fort; and yet so much insolence and cruelty with respect to those under him, and rashness in his management of the Malay neighbourhood, that I soon grew weary of him, not thinking myself very safe, indeed, under a man whose humours were so brutish and barbarous. I had other motives also for my going away. I began to long after my native country, after so tedious a ramble from it; and I proposed no small advantage to myself from my Painted Prince, whom Mr Moody had left entirely to my disposal, only reserving to himself his right to one half share in him. For besides what might be gained by showing him in England, I was in hopes that when I had got some money, I might there obtain what I had in vain sought for in the Indies—a ship from the merchants, wherewith to carry him back to Meangis, and reinstate him there in his own country, and by his favour and negotiation to establish a traffic for the spice and other products of those islands.

Upon these projects, I went to the Governor and Council, and desired that I might have my discharge to go for England with the next ship that came. The Council thought it reasonable, and they consented to it; he also gave me his word that I should go. Upon the 2d of January 1691, there came to an anchor in Bencouli Road the Defence, Captain Heath commander, bound for England, in the service of the Company. They had been at Indrapore, where Mr Moody then was; and he had made over his share in Prince Jeoly to Mr Goddard, chief mate of the ship. Upon his coming on shore, he showed me Mr Moody's writings, and looked upon Jeoly, who had been sick for three months; in all which time I tended him as carefully as if he had been my brother. I agreed matters with Mr Goddard, and sent Jeoly on board, intending to follow him as I could, and desiring Mr Goddard's assistance to fetch me off and conceal me aboard the ship if there should be occasion; which he promised to do, and the captain promised to entertain me. For it proved, as I had foreseen, that upon Captain Heath's arrival, the Governor repented him of his promise, and would not suffer me to depart. I importuned him all I could, but in vain; so did Captain Heath also, but to no purpose. In short, after several essays, I shipped away at midnight (understanding the ship was to sail the next morning, and that they had taken leave of the fort); and, creeping through one of the portholes of the fort, I got to the shore, where the ship's boat waited for me, and carried me on board. I brought with me my journal, and most of my written papers; but some papers and books of value I left in haste and all my furniture, being glad I was myself at liberty, and had hopes of seeing England again.

CHAPTER XIX.

BEING thus got on board the Defence, I was concealed there till a boat which came from the fort laden with pepper was gone off again. And then we set sail for the Cape of Good Hope, January 25th, 1691, and made the best of our way, as wind and weather would permit, expecting there to meet three English ships more, bound home from the Indies: for the war with the French having been proclaimed at Fort St George a little before Captain Heath came from thence, he was willing to have company home if he could.

A little before this war was proclaimed, there was an engagement in the road of Fort St George between some French men-of-war and some Dutch and English ships at anchor in the road; which, because there is such a plausible story made of it in Monsieur Duquesne's late Voyage to the East Indies, I shall give a short account of, as I had it particularly related to me by the gunner's mate of Captain Heath's ship, a very sensible man, and several others of his men who were in the action. The Dutch have a fort on the Coast of Coromandel, called Pullicat, about twenty leagues to the northward of Fort St George. Upon some occasion or other the Dutch sent some ships thither to fetch away their effects, and transport them to Batavia. Acts of hostility were already begun between the French and Dutch; and the French had at this time a squadron newly arrived in India, and lying at Pondicherry, a French fort on the same coast southward of Fort St George. The Dutch, in returning to Batavia, were obliged to coast it along by Fort St George and Pondicherry for the sake of the wind; but when they came near this last, they saw the French men-of-war lying at anchor there, and should they have proceeded along the shore, or stood out to sea, expected to be pursued by them. They therefore turned back again; for though their ships were of a pretty good force, yet were they unfit for fight, as having great loads of goods, and many passengers, women, and children on board: so they put in at Fort St George, and, desiring the Governor's protection, had leave to anchor in the road, and to send their goods and useless people ashore. There were then in the road a few small English ships, and Captain Heath, whose ship was a very stout merchantman, and which the French relater calls the English Admiral, was just come from China, but very deep laden with goods, and the deck full of canisters of sugar, which he was preparing to send ashore; but before he could do it, the French appeared, coming into the road with their lower sails and topsails, and had with them a fireship. With this they thought to have burnt the Dutch Commodore, and might probably enough have done it as she lay at anchor, if they had had the courage to come boldly on; but they fired their ship at a distance, and the Dutch sent and towed her away, where she spent herself without any execution. Had the French men-of-war also come boldly up and grappled with their enemies, they might have done something considerable; for the fort could not have played on them without damaging our ships as well as theirs. But instead of this, the French dropped anchor out of reach of the shot of the fort, and there lay exchanging shot with their enemies' ships, with so little advantage to themselves, that after about four hours' fighting they cut their cables and went away in haste and disorder, with all their sails loose, even their topgallant sails, which is not usual but when ships are just next to running away. Captain Heath, notwithstanding his ship was so heavy and encumbered, behaved himself very bravely in the fight; and upon the going off of the French, went on board the Dutch Commodore, and told him that if he would pursue them he would stand out with them to sea, though he had very little water aboard. But the Dutch commander excused himself, saying he had orders to defend himself from the French, but none to chase them, or go out of his way to seek them. And this was the exploit which the French have thought fit to brag of. I hear that the Dutch have taken from them since their fort of Pondicherry.

But to proceed with our voyage. We had not been at sea long before our men began to droop in a sort of a distemper that stole insensibly on them, and proved fatal to above thirty, who died before we arrived at the Cape. We had sometimes two and once three men thrown overboard in a morning. This distemper might probably arise from the badness of the water which we took in at Bencouli, for I did observe while I was there that the river water, wherewith our ships were watered, was very unwholesome, it being mixed with the water of many small creeks that proceeded from low land, and whose streams were always very black, they being nourished by the water that drained out of the low swampy unwholesome ground. I have observed, not only there but in other hot countries also, both in the East and West Indies, that the land-floods which pour into the channels of the rivers about the season of the rains are very unwholesome. This happens chiefly, as I take it, where the water drains through thick woods and savannahs of long grass and swampy grounds, with which some hot countries abound; and I believe it receives a strong tincture from the roots of several kind of trees, herbs, &c.; and especially where there is any stagnancy of the water, it soon corrupts; and possibly the serpents and other poisonous vermin and insects may not a little contribute to its bad qualities; at such times it will look very deep coloured, yellow, red, or black, &c. The season of the rains was over, and the land-floods were abating upon the taking up this water in the River of Bencouli; but would the seamen have given themselves the trouble, they might have filled their vessels with excellent good water at a spring on the back side of the fort, not above 200 or 300 paces from the landing-place, and with which the fort is served. Beside the badness of our water, it was stowed among the pepper in the hold, which made it very hot. Every morning when we came to take our allowance, it was so hot that a man could hardly suffer his hands in it, or hold a bottleful of it in his hand. I never anywhere felt the like, nor could I have thought it possible that water should heat to that degree in a ship's hold. It was exceeding black, too, and looked more like ink than water. Whether it grew so black with standing, or was tinged with the pepper, I know not; for this water was not so black when it was first taken up. Our food also was very bad, for the ship had been out of England upon this voyage above three years; and the salt provision brought from thence, which we fed on, having been so long in salt, was but ordinary food for sickly men to feed on. Captain Heath, when he saw the misery of his company, ordered his own tamarinds —of which he had some jars aboard— to be given some to each mess to eat with their rice. This was a great refreshment to the men, and I do believe it contributed much to keep us on our legs. This distemper was so universal that I do believe there was scarce a man in the ship but languished under it; yet it stole so insensibly on us, that we could not say we were sick, feeling little or no pain, only a weakness, and but little stomach. Nay, most of those that died in this voyage would hardly be persuaded to keep their cabins or hammocks till they could not stir about; and when they were forced to lie down, they made their wills, and piked off[1] in two or three days.

The loss of these men, and the weak languishing condition that the rest of us were in, rendered us incapable to govern our ship when the wind blew more than ordinary. This often happened when we drew near the Cape, and as oft put us to our trumps[2] to

---

[1] "Peaked off;" gradually dwindled and died. The word is used in the witches' curse in Macbeth, Act 1, s. 3:

> "Weary seven nights, nine times nine,
> Shall he dwindle, peak, and pine."

[2] Forced us to our utmost efforts; drove us to our wits' end.

manage the ship. Captain Heath, to encourage his men to their labour, kept his watch as constantly as any man, though sickly himself, and lent a helping hand on all occasions. But at last, almost despairing of gaining his passage to the Cape by reason of the winds coming southerly, and we having now been sailing eight or nine weeks, he called all our men to consult about our safety, and desired every man, from the highest to the lowest, freely to give his real opinion and advice what to do in this dangerous juncture; for we were not in a condition to keep out long, and could we not get to land quickly, must have perished at sea. He consulted, therefore, whether it were best to beat still for the Cape or bear away for Johanna, where we might expect relief, that being a place where our outward-bound East India ships usually touch, and whose natives are very familiar; but other places, especially St Lawrence or Madagascar,[1] which was nearer, were unknown to us. We were now so nigh the Cape that, with a fair wind, we might expect to be there in four or five days; but as the wind was now, we could not hope to get thither. On the other side, this wind was fair to carry us to Johanna; but then Johanna was a great way off; and if the wind should continue as it was, to bring us into a true trade-wind, yet we could not get thither under a fortnight; and if we should meet calms, as we might probably expect, it might be much longer. Besides, we should lose our passage about the Cape till October or November, this being about the latter end of March; for after the 10th of May it is not usual to beat about the Cape to come home. All circumstances therefore being weighed and considered, we at last unanimously agreed to prosecute our voyage towards the Cape, and with patience wait for a shift of wind. But Captain Heath, having thus far sounded the inclination of his weak men, told them that it was not enough that they all consented to beat for the Cape, for our desires were not sufficient to bring us thither, but that there would need a more than ordinary labour and management from those that were able; and withal, for their encouragement, he promised a month's pay *gratis* to every man that would engage to assist on all occasions, and be ready upon call, whether it were his turn to watch or not; and this money he promised to pay at the Cape. This offer was first embraced by some of the officers, and then as many of the men as found themselves in a capacity listed themselves in a roll to serve their commander. This was wisely contrived of the captain, for he could not have compelled them in their weak condition, neither would fair words alone, without some hopes of a reward, have engaged them to so much extraordinary work; for the ship, sail, and rigging were much out of repair. For my part, I was too weak to enter myself in that list; for else our common safety, which I plainly saw lay at stake, would have prompted me to do more than any such reward would do. In a short time after this it pleased God to favour us with a fine wind, which, being improved to the best advantage by the incessant labour of these new-listed men, brought us in a short time to the Cape.

The night before we entered the harbour, which was about the beginning of April, being near the land, we fired a gun every hour to give notice that we were in distress. The next day, a Dutch captain came aboard in his boat; who, seeing us so weak as not to be able to trim our sails to turn into the harbour, though we did tolerably well at sea before the wind, and being requested by our captain to assist him, sent ashore for a hundred lusty men, who immediately came aboard, and brought our ship in to an anchor. They also unbent our sails, and did everything for us that they were required to do, for which Captain Heath gratified them

[1] Which received the name of St Lawrence from its Portuguese discoverer, Emanuel de Meneses, in 1506.

to the full. These men had better stomachs than we, and ate freely of such food as the ship afforded; and they having the freedom of our ship, to go to and fro between decks, made prize of what they could lay their hands on, especially salt beef, which our men, for want of stomachs in the voyage, had hung up, six, eight, or ten pieces in a place. This was conveyed away before we knew it or thought of it; besides, in the night, there was a bale of muslins broken open, and a great deal conveyed away; but whether the muslins were stolen by our own men or the Dutch I cannot say, for we had some very dexterous thieves in our ship. Being thus got safe to an anchor, the sick were presently sent ashore, to quarters provided for them, and those that were able remained aboard and had good fat mutton or fresh beef sent aboard every day. I went ashore, also, with my Painted Prince, where I remained with him till the time of sailing again, which was about six weeks, in which time I took the opportunity to inform myself of what I could concerning this country, which I shall in this next place give a brief account of, and so make what haste I can home.

The Cape of Good Hope is the utmost bounds of the continent of Africa towards the south, lying in Lat. 34° 30′ S., in a very temperate climate. I look upon this Latitude to be one of the mildest and sweetest, for its temperature, of any whatsoever.[1] . . .

This large promontory consists of high and very remarkable land; and off at sea it affords a very pleasant and agreeable prospect. And without doubt the prospect of it was very agreeable to those Portuguese who first found out this way by sea to the East Indies, when after coasting along the vast continent of Africa, towards the South Pole, they had the comfort of seeing the land and their course end in this promontory, which therefore they called the Cape de Bon Esperance, or of Good Hope, finding that they might now proceed eastward.[2] The most remarkable land at sea is a high mountain, steep to the sea, with a flat even top, which is called the Table Land. On the west side of the Cape, a little to the northward of it, there is a spacious harbour,[3] with a low flat island lying off it, which you may leave on either hand, and pass in or out securely at either end. Ships that anchor here ride near the mainland, leaving the island at a farther distance without them. The land by the sea against the harbour is low, but backed with high mountains a little way in, to the southward of it.

The soil of this country is of a brown colour; not deep, yet indifferently productive of grass, herbs, and trees. The grass is short, like that which grows on our Wiltshire or Dorsetshire Downs. The trees hereabouts are but small and few; the country also farther from the sea does not much abound in trees, as I have been informed. The mould or soil also is much like this near the harbour, which though it cannot be said to be very fat or rich land, yet it is very fit for cultivation, and yields good crops to the industrious husbandman; and the country is pretty well settled with farms, Dutch families and French refugees, for twenty or thirty leagues up in the country; but there are but few farms near the harbour. Here grows plenty of wheat, barley, pease, &c. Here are also fruits of many kinds, as apples, pears, quinces, and the largest pomegranates that I did ever see. The chief fruits are grapes. These thrive very well, and the

[1] A digression is here omitted, in which Dampier combats and explains a "common prejudice" among European seamen, who look upon the Cape as much colder than other places in the same Latitude to the north of the Line.

[2] A passage relating to soundings and signs of nearing the Cape is omitted.

[3] Table Bay.

country is of late years so well stocked with vineyards, that they make abundance of wine, of which they have enough and to spare, and do sell great quantities to ships that touch here. This wine is like a French high country white-wine, but of a pale yellowish colour; it is sweet, very pleasant, and strong.

The tame animals of this country are sheep, goats, hogs, cows, horses, &c. The sheep are very large and fat, for they thrive very well here. There is a very beautiful sort of wild ass in this country, whose body is curiously striped with equal lists[1] of white and black; the stripes coming from the ridge of his back, and ending under the belly, which is white. Here are a great many ducks, dunghill fowls, &c.; and ostriches are plentifully found in the dry mountains and plains. The sea hereabouts affords plenty of fish of divers sorts; especially a small sort of fish, not so big as a herring, whereof they have such great plenty, that they pickle great quantities yearly, and send them to Europe. Seals are also in great numbers about the Cape, which, as I have still observed, is a good sign of the plentifulness of fish, which is their food.

The Dutch have a strong fort by the seaside, against the harbour, where the Governor lives. At about 200 or 300 paces distance from thence, on the west side of the fort, there is a small Dutch town, in which I told about fifty or sixty houses, low, but well built, with stone walls, there being plenty of stone drawn out of a quarry close by. On the back side of the town, as you go towards the mountains, the Dutch East India Company have a large house, and a stately garden walled in with a high stone wall. This garden is full of divers sorts of herbs, flowers, roots, and fruits, with curious spacious gravel walks and arbours; and is watered with a brook that descends out of the mountains, which being cut into many channels is conveyed into all parts of the garden. The hedges which make the walks are very thick, and nine or ten feet high. They are kept exceeding neat and even by continual pruning. There are lower hedges within these again, which serve to separate the fruit trees from each other, but without shading them; and they keep each sort of fruit by themselves, as apples, pears, abundance of quinces, pomegranates, &c. These all prosper very well, and bear good fruit, especially the pomegranate. The roots and garden herbs have also their distinct places, hedged in apart by themselves; and all in such order, that it is exceeding pleasant and beautiful. There are a great number of Negro slaves brought from other parts of the world; some of which are continually weeding, pruning, trimming, and looking after it. All strangers are allowed the liberty to walk there; and, by the servant's leave, you may be admitted to taste of the fruit; but if you think to do it clandestinely you may be mistaken, as I knew one was when I was in the garden, who took five or six pomegranates, and was espied by one of the slaves, and threatened to be carried before the Governor. I believe it cost him some money to make his peace, for I heard no more of it. Farther up from the sea, beyond the garden, towards the mountains, there are several other small gardens and vineyards, belonging to private men; but the mountains are so nigh, that the number of them is but small.

The Dutch that live in the town get considerably by the ships that frequently touch here, chiefly by entertaining strangers that come ashore to refresh themselves: for you must give three shillings or a dollar a day for your entertainment; the bread and flesh is as cheap here as in England. Besides, they buy good pennyworths of the seamen, both outward and homeward bound, which the farmers up the country buy of them again at a dear rate; for they have not an opportunity of buying things at the best hand, but must buy of

[1] Rings, streaks.

those that live at the harbour; the nearest settlements, as I was informed, being twenty miles off. Notwithstanding the great plenty of corn and wine, yet the extraordinary high taxes which the Company lays on liquor make it very dear, and you can buy none but at the tavern, except it be by stealth. There are but three houses in the town that sell strong liquor, one of which is this wine-house or tavern; there they sell only wine; another sells beer and mum[1]; and the third sells brandy and tobacco, all extraordinary dear. A flask of wine which holds three quarts will cost eighteen stivers,[2] for so much I paid for it; yet I bought as much for eight stivers in another place, but it was privately, at an unlicensed house; and the person that sold it would have been ruined had it been known. And thus much for the country and the European inhabitants.

CHAPTER XX.

THE natural[3] inhabitants of the Cape are the Hodmadods, as they are commonly called, which is a corruption of the word Hottentot; for this is the name by which they call to one another, either in their dances, or on any occasion, as if every one of them had this for his name. The word probably has some signification or other in their language, whatever it is. The Hottentots are people of a middle stature, with small limbs and thin bodies, full of activity. Their faces are of a flat, oval figure, of the Negro make, with great eyebrows, black eyes; but neither are their noses so flat, nor their lips so thick, as the Negroes of Guinea. Their complexion is darker than the common Indians, though not so black as the Negroes or New Hollanders; neither is their hair so much frizzled. They besmear themselves all over with grease, as well to keep their joints supple, as to fence their half-naked bodies from the air by stopping up their pores. To do this the more effectually, they rub soot over the greased parts, especially their faces, which adds to their natural beauty as painting does in Europe; but withal sends from them a strong smell, which, though sufficiently pleasing to themselves, is very unpleasant to others. They are glad of the worst of kitchen stuff for this purpose, and use it as often as they can get it. This custom of anointing the body is very common in other parts of Africa, especially on the coast of Guinea, where they generally use palm oil, anointing themselves from head to foot; but when they want oil they make use of kitchen stuff, which they buy of the Europeans that trade with them. In the East Indies also, especially on the coast of Cudda and Malacca, and in general on almost all the easterly islands, as well on Sumatra, Java, &c., as on the Philippine and Spice Islands, the Indian inhabitants anoint themselves with cocoa-nut oil two or three times a day, especially mornings and evenings. They spend sometimes half-an-hour in chafing the oil, and rubbing it into their hair and skin, leaving no place unsmeared with oil but their face, which they daub not like these Hottentots. The Americans also in some places do use this custom, but not so frequently, perhaps for want of oil and grease to do it. Yet some American Indians in the North Seas frequently daub themselves with a pigment made with leaves, roots, or herbs, or with a sort of red earth, giving their skins a yellow, red, or green colour, according as the pigment is. And these smell unsavourly enough to people not accustomed to them; though not so rank as those who use oil or grease.

The Hottentots wear no covering

---

[1] A kind of strong beer, introduced into England from Brunswick in Germany.

[2] According to Bailey's Dictionary, a stiver was, about the beginning of the eighteenth century, equivalent to a penny and one-fifth English.

[3] Native, aboriginal.

on their heads, but deck their hair with small shells. Their garments are sheepskins wrapped about their shoulders like a mantle, with the woolly sides next their bodies. The men have, besides this mantle, a piece of skin like a small apron hanging before them. The women have another skin tucked about their waists which comes down to their knees like a petticoat: and their legs are wrapped round with sheepguts, two or three inches thick, some up as high as to their calves, others even from their feet to their knees; which at a small distance seems to be a sort of boots. These are put on when they are green; and so they grow hard and stiff on their legs, for they never pull them off again, till they have occasion to eat them; which is when they journey from home, and have no other food: then these guts, which have been worn, it may be, six, eight, ten, or twelve months, make them a good banquet. This I was informed of by the Dutch. They never pull off their sheepskin garments but to louse themselves; for by continual wearing them they are full of vermin, which obliges them often to strip and sit in the sun two or three hours together in the heat of the day, to destroy them. Indeed, most Indians that live remote from the Equator are molested with lice, though their garments afford less shelter for lice than these Hottentots' sheepskins do. For all those Indians who live in cold countries, as in the north and south parts of America, have some sort of skin or other to cover their bodies, as deer, otter, beaver, or seal skins, all which they as constantly wear, without shifting themselves, as these Hottentots do their sheepskins. And, hence they are lousy too, and strong scented, though they do not daub themselves at all, or but very little; for even by reason of their skins they smell strong.

The Hottentots' houses are the meanest that I did ever see. They are about nine or ten feet high, and ten or twelve from side to side. They are in a manner round, made with small poles stuck into the ground, and brought together at the top, where they are fastened. The sides and top of the house are filled up with boughs coarsely wattled between the poles, and all is covered over with long grass, rushes, and pieces of hides; and the house at a distance appears just like a haycock. They leave only a small hole on one side, about three or four feet high, for a door to creep in and out at; but when the wind comes in at this door they stop it up, and make another hole in the opposite side. They make the fire in the middle of the house, and the smoke ascends out of the crannies, from all parts of the house. They have no beds to lie on, but tumble down at night round the fire. Their household furniture is commonly an earthen pot or two to boil victuals, and they live very miserably and hard; it is reported that they will fast two or three days together when they travel about the country. Their common food is either herbs, flesh, or shellfish, which they get among the rocks, or other places at low water: for they have no boats, bark-logs, nor canoes to go a-fishing in; so that their chief subsistence is on land animals, or on such herbs as the land naturally produces. I was told by my Dutch landlord that they kept sheep and bullocks here before the Dutch settled among them: and that the inland Hottentots have still great stocks of cattle, and sell them to the Dutch for rolls of tobacco; and that the price for which they sell a cow or sheep, was as much twisted tobacco as will reach from the horns or head to the tail; for they are great lovers of tobacco, and will do anything for it. This their way of trucking[1] was confirmed to me by many others, who yet said that they could not buy their beef this cheap way, for they had not the liberty to deal with the Hottentots, that being a privilege which the Dutch East India Company reserve to themselves. My landlord, having a

[1] Bartering, exchanging.

great many lodgers, fed us most with mutton, some of which he bought of the butcher, and there is but one in the town; but most of it he killed in the night, the sheep being brought privately by the Hottentots, who assisted in skinning and dressing, and had the skin and guts for their pains. I judge these sheep were fetched out of the country a good way off; for he himself would be absent a day or two to procure them, and two or three Hottentots with him. These of the Hottentots that live by the Dutch town have their greatest subsistence from the Dutch: for there is one or more of them belonging to every house. These do all sorts of servile work, and there take their food and grease. Three or four more of their nearest relations sit at the doors or near the Dutch house, waiting for the scraps and fragments that come from the table; and if between meals the Dutch people have any occasion for them to go on errands or the like, they are ready at command, expecting little for their pains; but for a stranger they will not budge under a stiver.

Their religion, if they have any, is wholly unknown to me; for they have no temple nor idol, nor any place of worship that I did see or hear of. Yet their mirth and nocturnal pastimes at the new and full of the moon looked as if they had some superstition about it. For at the full especially they sing and dance all night, making a great noise. I walked out to their huts twice at these times, in the evening, when the moon arose above the horizon, and viewed them for an hour or more. They seem all very busy, both men, women, and children, dancing very oddly on the green grass by their houses. They traced to and fro promiscuously, often clapping their hands and singing aloud. Their faces were sometimes to the east, sometimes to the west; neither did I see any motion or gesture that they used when their faces were towards the moon, more than when their backs were towards it. After I had thus observed them for a while, I returned to my lodging, which was not above 200 or 300 paces from their huts; and I heard them singing in the same manner all night. In the grey of the morning I walked out again, and found many of the men and women still singing and dancing, who continued their mirth till the moon went down, and then they left off; some of them going into their huts to sleep, and others to their attendance in their Dutch houses. Other Negroes are less circumspect, in their night-dances, as to the precise time of the full moon, they being more general in these nocturnal pastimes, and use them oftener; as do many people also in the East and West Indies. Yet there is a difference between colder and warmer countries as to their divertisements.[1] The warmer climates being generally very productive of delicate fruits, &c., and these uncivilised people caring for little else than what is barely necessary, they spend the greatest part of their time in diverting themselves after their several fashions; but the Indians of colder climates are not so much at leisure, the fruits of the earth being scarce with them, and they necessitated to be continually fishing, hunting, or fowling for their subsistence; not as with us, for recreation. As for these Hottentots, they are a very lazy sort of people; and though they live in a delicate country, very fit to be manured, and where there is land enough for them, yet they choose rather to live as their forefathers, poor and miserable, than be at pains for plenty. And so much for the Hottentots: I shall now return to our own affairs.

Upon our arrival at the Cape, Captain Heath took a house to live in, in order to recover his health. Such of his men as were able did so too: for the rest he provided lodgings and paid their expenses. Three or four of our men, who came ashore very sick, died; but the rest, by the assistance of the doctors of the fort, a fine air, and good kitchen and cellar physic, soon recovered their health. Those

---

[1] Sports, diversions.

that subscribed to be at all calls, and assisted to bring in the ship, received Captain Heath's bounty, by which they furnished themselves with liquor for their homeward voyage. But we were now so few, that we could not sail the ship; therefore Captain Heath desired the Governor to spare him some men; and, as I was informed, had a promise to be supplied out of the homeward-bound Dutch East India ships, that were now expected every day; and we waited for them. In the meantime, in came the James and Mary, and the Josiah of London, bound home. Out of these we thought to have been furnished with men, but they had only enough for themselves; therefore we waited yet longer for the Dutch Fleet, which at last arrived: but we could get no men from them. Captain Heath was therefore forced to get men by stealth, such as he could pick up, whether soldiers or seamen. The Dutch knew our want of men; therefore near forty of them, those that had a design to return to Europe, came privately and offered themselves, and waited in the night at places appointed, where our boats went and fetched three or four aboard at a time, and hid them, especially when any Dutch boat came aboard our ship. Here at the Cape I met my friend Daniel Wallis, the same who leaped into the sea and swam at Pulo Condore.[1] After several traverses to Madagascar, Don Mascarin,[2] Pondicherry, Pegu, Cunnimere, Madras, and the River Hooghly, he was now got hither in a homeward-bound Dutch ship. I soon persuaded him to come over to us, and found means to get him aboard our ship.

About the 23d of May we sailed from the Cape in the company of the James and Mary and the Josiah, directing our course towards the Island Santa Helena. We met nothing of remark in this voyage except a great swelling sea out of the SW., which, taking us on the broadside, made us roll sufficiently. Such of our water-casks as were between decks, running from side to side, were in a short time all staved, and the deck well washed with the fresh water. The shot tumbled out of the lockers and garlands, and rung a loud peal, rumbling from side to side every roll that the ship made; neither was it an easy matter to reduce them again within bounds. The guns being carefully looked after and lashed fast, never budged, but the tackles or pulleys and lashings made great music too. The sudden and violent motion of the ship made us fearful lest some of the guns should have broken loose, which must have been very detrimental to the ship's sides. The masts were also in great danger to be rolled by the board; but no harm happened to any of us besides the loss of three or four butts of water, and a barrel or two of good Cape wine, which was staved in the great cabin. This great tumbling sea took us shortly after we came from the Cape. The violence of it lasted but one night; yet we had a continual swelling out of the SW. almost during all the passage to Santa Helena, which was an eminent token that the SW. winds were now violent in the higher latitudes towards the South Pole; for this was the time of year for those winds. Notwithstanding this boisterous sea coming thus obliquely upon us, we had fine clear weather, and a moderate gale at SE., or between that and the east, till we came to the Island Santa Helena, where we arrived the 20th of June. There we found the Princess Ann at anchor waiting for us.

The Island Santa Helena lies in about 16° S. Lat. The air is commonly serene and clear, except in the months that yield rain; yet we had one or two very rainy days even while we were here. Here are moist seasons to plant and sow; and the weather is temperate enough as to heat, though so near the Equator, and very healthy. The island is but small, not above nine or ten leagues in length, and stands 300 or 400 leagues from the

[1] Escaping from a murderous Malayan crew. See end of Chapter XIV., page 231.

[2] The Isle of France.

main land. It is bounded against the sea with steep rocks, so that there is no landing but at two or three places. The land is high and mountainous, and seems to be very dry and poor, yet there are fine valleys proper for cultivation. The mountains appear bare, only in some places you may see a few low shrubs; but the valleys afford some trees fit for building, as I was informed.

This island is said to have been first discovered and settled by the Portuguese,[1] who stocked it with goats and hogs; but it being afterwards deserted by them, it lay waste till the Dutch, finding it convenient to relieve their East India ships, settled it again; but they afterwards relinquished it for a more convenient place, I mean the Cape of Good Hope. Then the English East India Company settled their servants there, and began to fortify it; but they being yet weak, the Dutch about the year 1672 came thither and retook it, and kept it in their possession. This news being reported in England, Captain Monday was sent to retake it, who, by the advice and conduct of one that had formerly lived there, landed a party of armed men in the night in a small cove, unknown to the Dutch then in garrison, and climbing the rocks, got up into the island, and so came in the morning to the hills hanging over the fort which stands by the sea in a small valley. Thence firing into the fort, they soon made them surrender. There were at this time two or three Dutch East India ships either at anchor, or coming thither, when our ships were there. These, when they saw that the English were masters of the island again, made sail to be gone; but being chased by the English frigates, two of them became rich prizes to Captain Monday and his men. The island has continued ever since in the hands of the English East India Company, and has been greatly strengthened both with men and guns; so that at this day it is secure enough from the invasion of any enemy. For the common landing-place is a small bay, like a half-moon, scarce 500 paces wide, between the two points. Close by the seaside are good guns planted at equal distances, lying along from one end of the bay to the other; besides a small fort, a little farther in from the sea, near the midst of the bay: all which makes this bay so strong, that it is impossible to force it. The small cove where Captain Monday landed his men when he took the island from the Dutch, is scarce fit for a boat to land at, and yet that is now also fortified.

There is a small English town within the great bay, standing in a little valley between two high steep mountains. There may be about twenty or thirty small houses, whose walls are built with rough stones; the inside furniture is very mean. The Governor has a pretty tolerably handsome low house by the fort, where he commonly lives, having a few soldiers to attend him, and to guard the fort. But the houses in the town before mentioned stand empty, save only when ships arrive here; for their owners have all plantations farther in the island, where they constantly employ themselves. But when ships arrive, they all flock to the town, where they live all the time that the ships lie here; for then is their fair or market, to buy such necessaries as they want, and to sell off the produce of their plantations. Their plantations afford potatoes, yams, and some plantains and bananas. Their stock consists chiefly of hogs, bullocks, cocks and hens, ducks, geese, and turkeys, of which they have great plenty, and sell them at a low rate to the sailors; taking in exchange shirts, drawers, or any light clothes, pieces of calico, silks, or muslins. Arrack, sugar, and lime-juice are also much esteemed and coveted by them. But now they are

[1] By Juan de Nova, in 1501, who gave such a favourable account of the island, that the Portuguese Admirals were instructed in future to touch there for refreshments.

in hopes to produce wine and brandy in a short time; for they already begin to plant vines for that end, there being a few Frenchmen there to manage that affair. This I was told, but I saw nothing of it, for it rained so hard when I was ashore, that I had not the opportunity of seeing their plantations. I was also informed that they get manatee or sea-cows here, which seemed very strange to me. Therefore inquiring more strictly into the matter, I found the Santa Helena manatee to be, by their shapes and manner of lying ashore on the rocks, those creatures called sea-lions; for the manatee never come ashore, neither are they found near any rocky shores as this island is, there being no feeding for them in such places. Besides, in this island there is no river for them to drink at, though there is a small brook runs into the sea out of the valley by the fort.

We stayed here five or six days, all which time the islanders lived at the town, to entertain the seamen, who constantly flocked ashore to enjoy themselves among their country people. Our touching at the Cape had greatly drained the seamen of their loose coins, at which these islanders as greatly repined; and some of the poorer sort openly complained against such doings, saying it was fit that the East India Company should be acquainted with it, that they might hinder their ships from touching at the Cape. Yet they were extremely kind, in hopes to get what was remaining. They are most of them very poor; but such as could get a little liquor to sell to the seamen at this time got what the seamen could spare, for the punch-houses were never empty. But had we all come directly hither, and not touched at the Cape, even the poorest people among them would have gotten something by entertaining sick men. For commonly the seamen coming home are troubled more or less with scorbutic distempers, and their only hopes are to get refreshment and health at this island; and these hopes seldom or never fail them if once they get footing here: for the island affords abundance of delicate herbs, wherewith the sick are first bathed to supple their joints, and then the fruits and herbs and fresh food soon after cure them of their scorbutic humours; so that in a week's time men that have been carried ashore in hammocks, and they who were wholly unable to go, have been able to leap and dance. Doubtless the serenity and wholesomeness of the air contributes much to the carrying off of these distempers, for there is constantly a fresh breeze. While we stayed here, many of the seamen got sweethearts. One young man belonging to the James and Mary was married, and brought his wife to England with him. Another brought his sweetheart to England, they being each engaged by bonds to marry at their arrival in England; and several others of our men were over head and ears in love with the Santa Helena maids, who, though they were born there, yet very earnestly desired to be released from that prison, which they have no other way to compass but by marrying seamen or passengers that touch here. The young women born here are but one remove from English, being the daughters of such. They are well shaped, proper, and comely, were they in a dress to set them off.

My stay ashore here was but two days, to get refreshments for myself and Jeoly, whom I carried ashore with me; and he was very diligent to pick up such things as the island afforded, carrying ashore with him a bag, which the people of the isle filled with roots for him. They flocked about him, and seemed to admire him much. This was the last place where I had him at my own disposal; for the mate of the ship, who had Mr Moody's share in him, left him entirely to my management, I being to bring him to England. But I was no sooner arrived in the Thames, but he was sent ashore to be seen by some eminent persons; and I, being in want of money, was prevailed upon to sell first part of my share in him, and by degrees all of it. After this I heard that he was carried about to be shown as a sight, and that he died of the small-pox at Oxford.

But to proceed. Our water being filled, and the ships all stocked with fresh provision, we sailed hence in company of the Princess Ann, the James and Mary, and the Josiah, July the 2d, 1691, directing our course towards England, and designing to touch nowhere by the way. . . . In our passage before we got to the Line, we saw three ships, and making towards them, we found two of them to be Portuguese, bound to Brazil. The third kept on a wind, so that we could not speak with her; but we found by the Portuguese it was an English ship, called the Dorothy, Captain Thwayt commander, bound to the East Indies. After this we kept company still with our three consorts till we came near England, and then were separated by bad weather; but before we came within sight of land, we got together again, all but the James and Mary. She got into the Channel before us, and went to Plymouth, and there gave an account of the rest of us; whereupon our men-of-war who lay there came out to join us, and meeting us, brought us off Plymouth. There our consort the James and Mary came to us again; and thence we all sailed in company of several men-of-war towards Portsmouth. There our first convoy left us, and went in thither. But we did not want convoys, for our fleets were then repairing to their winter harbours to be laid up; so that we had the company of several English ships to the Downs, and a squadron also of Dutch sailed up the Channel, but kept off farther from our English coast, they being bound home to Holland. When we came as high as the South Foreland, we left them standing on their course, keeping on the back of the Goodwin Sands; and we luffed in for the Downs, where we anchored September the 16th, 1691.

END OF DAMPIER'S VOYAGE.

AN OLD SALT'S YARN.—ANSON'S VOYAGE ROUND THE WORLD.

(*Frontispiece.*)

A

VOYAGE ROUND THE WORLD

IN THE YEARS 1740-44

BY

GEORGE ANSON

EDITED, FROM THE ORIGINAL NARRATIVE, WITH NOTES, BY

D. LAING PURVES

Special Edition.

LONDON:
PUBLISHED BY THE LI-QUOR TEA COMPANY,
5 GEORGE STREET, TOWER HILL.
1879.

CONTENTS.

BIOGRAPHICAL SKETCH.

George Anson, Lord Anson, Baron Soberton, was the second son of William Anson, Esq., of Shugborough, in Staffordshire. His great-grandfather, who was an eminent barrister in the reign of James I., had purchased and founded the family mansion where he was born, 23d April 1697. Little is positively known about his early history and nautical training, save that his name was first found entered as a volunteer in the books of the Ruby, under date January 1712. His services being transferred from the Ruby to the Hampshire ship of war, he then received his acting orders as second lieutenant, on the 9th May 1716. From this date, up till 1724, his progress was as follows: Promoted to the command of the Weasel sloop in 1718, raised to the rank of post-captain in 1724, with the command of the Scarborough man-of-war. The Scarborough was at this time ordered to defend the coast of South Carolina against pirates, and to prevent illicit commerce with the Bahamas. His popularity among the settlers of South Carolina must have been considerable, as we find that his name was attached to several towns and districts, such as Anson's County, Anson's Ville, Anson's Mines, etc. He returned to England in 1730, was cruising again on the American coast in 1733, but returned again in 1735.

On the 9th December 1737, Captain Anson was appointed to the command of the Centurion, a ship of 60 guns, and despatched to the African coast, ostensibly with a view to the protection of our merchants engaged in the gum trade, from the annoyance of French ships of war. A resolution having been come to by the ministry to strike a blow against the Spanish power in the West Indies, South Seas, and at Manilla, two officers were selected for this purpose—Captain Anson and Captain James Cornwall. On Anson's arrival at Spithead, 10th November 1739, he found a letter awaiting him from Admiral Sir Charles Wager, ordering him to proceed at once to the Admiralty. The first programme submitted to him, to say the least of it, was both difficult and dangerous, and may be taken as a proof of the confidence entertained in his ability as a seaman. He was to attack and carry

Manilla with part of his squadron, while another part, under Cornwall, was to go round Cape Horn into the Southern Ocean, attacking and destroying the Spanish settlements on the South American coasts, then crossing the Pacific to join the previous squadron at Manilla, and there await further orders. This scheme was never fully carried out, the proposed expedition to Manilla being dropped; but the part of the plan which was to have been entrusted to Cornwall was eventually carried out by Anson.

On the 10th January 1740, Anson was appointed commodore of the squadron which was designed to share in the riches which they imagined Spain derived from her possessions in the South Seas. Before sailing, he made himself acquainted with the best printed and manuscript accounts of the Spanish settlements on the coasts of Chili, Peru, and Mexico. The victualling and manning of this squadron was a notorious example of avaricious and heartless jobbery. In addition to the fact that several of the vessels were scarcely seaworthy and badly manned, the troops sent on board were worn-out pensioners from Chelsea, not one of whom returned alive. It is a record of this voyage round the world which is here presented.

On the 3d May 1747, Anson achieved a brilliant victory over a French fleet bound for the Indies, off Cape Finisterre. In recognition of this service, he was created a peer under the title of Lord Anson, Baron of Soberton, in the county of Southampton, and shortly afterwards made Vice-Admiral of England. In 1751 he was appointed First Lord of the Admiralty, a position he held, except for a short interval, until his death. On 30th July 1761, he sailed from Harwich in the Charlotte yacht, to convey the future queen of George III. to England. In the month of February 1762, in assisting at the ceremony of accompanying the queen's brother, Prince George of Mecklenberg, to Portsmouth, he caught a cold which proved fatal on 6th June 1762. In April 1748, Lord Anson had married Elizabeth, eldest daughter of Lord Hardwicke, who died without issue, 1st June 1760.

In business Anson was slow to decide, but quick to execute. In matters of ceremony and correspondence he was awkward, and in writing showed marks of a defective education. This was more than compensated by other sterling qualities of mind and character. In society he was modest and reserved, it being said of him, "he had been round the world, but never in it." The Duke of Newcastle observed of him: "There never was a more able, a more upright, or a more useful servant of his king and country, or a more sincere and valuable friend."

ANSON'S VOYAGE ROUND THE WORLD.

DEDICATION.

To His Grace, John, Duke of Bedford, Marquis of Tavistock, Earl of Bedford, Baron Russel, Baron Russel of Thornhaugh, and Baron Howland of Streatham: one of His Majesty's Principal Secretaries of State, and Lord-Lieutenant and Custos Rotulorum of the County of Bedford.

My Lord,—The following narrative of a very singular naval achievement is addressed to Your Grace, both on account of the infinite obligations which the Commander-in-Chief at all times professes to have received from your friendship; and also, as the subject itself naturally claims the patronage of one under whose direction the British Navy has resumed its ancient spirit and lustre, and has in one summer ennobled itself by two victories, the most decisive and (if the strength and number of the captures be considered) the most important that are to be met with in our annals.[1] Indeed, an uninterrupted series of success, and a manifest superiority gained universally over the enemy, both in commerce and glory, seem to be the necessary effects of a revival of strict discipline, and of an unbiassed regard to merit and service. These are marks that must distinguish the happy period of time in which Your Grace presided, and afford a fitter subject for history than for an address of this nature. Very signal advantages of rank and distinction, obtained and secured to the naval profession by Your Grace's auspicious influence, will remain a lasting monument of your unwearied zeal and attachment to it, and be for ever remembered with the highest gratitude by all who shall be employed in it. As these were the generous rewards of past exploits, they will be likewise the noblest incentives and surest pledges of the future. That Your Grace's eminent talents, magna-

[1] In 1747, when Anson, then Rear-Admiral of the White, defeated the French Admiral, Ionquierre, near Cape Finisterre, capturing six ships of the line and a valuable convoy, and gaining, as his reward, a peerage, with the title of Lord Anson, Baron Soberton; and Hawke totally defeated the French fleet off Belleisle, also taking six ships, and winning promotion to the rank of Vice-Admiral of the Blue. Of the first-named victory, it is narrated that when M. St George, captain of one of the French vessels, gave up his sword to Anson, he addressed him thus, with allusion to the names of two of the ships that had surrendered: "Vous avez vaincu L'Invincible, et La Gloire vous suit"—"You have vanquished the Invincible, and Glory follows you." The Dedication was written in 1748.

nimity, and disinterested zeal, whence the public has already reaped such signal benefits, may in all times prove equally successful in advancing the prosperity of Great Britain, is the ardent wish of, My Lord, Your Grace's most obedient, most devoted, and most humble servant,

RICHARD WALTER.

INTRODUCTION.

NOTWITHSTANDING the great improvement of navigation within the last two centuries, a Voyage Round the World is still considered as an enterprise of a very singular nature; and the public have never failed to be extremely inquisitive about the various accidents and turns of fortune with which this uncommon attempt is generally attended. And though the amusement expected in a narration of this kind is doubtless one great source of this curiosity, and a strong incitement with the bulk of readers, yet the more intelligent part of mankind have always agreed that from these relations, if faithfully executed, the more important purposes of navigation, commerce, and national interest may be greatly promoted. For every authentic account of foreign coasts and countries will contribute to one or more of these great ends in proportion to the wealth, wants, or commodities of those countries, and our ignorance of those coasts; and therefore a Voyage Round the World promises a species of information of all others the most desirable and interesting, since great part of it is performed in seas and on coasts with which we are as yet but very imperfectly acquainted, and in the neighbourhood of a country renowned for the abundance of its wealth, though it is at the same time stigmatised for its poverty in the necessaries and conveniencies of a civilised life.

These considerations have occasioned the publication of the ensuing work, which, in gratifying the inquisitive turn of mankind, and contributing to the safety and success of future navigators and to the extension of our commerce and power, may doubtless vie with any narration of this kind hitherto made public. Since the circumstances of this undertaking already known to the world may be supposed to have strongly excited the general curiosity. For whether we consider the force of the squadron sent on this service, or the diversified distresses that each single ship was separately involved in, or the uncommon instances of various fortune which attended the whole enterprise, each part, I conceive, must, from its rude well-known outlines, appear worthy of a completer and more finished delineation.[1]

As there are hereafter occasionally interspersed some accounts of Spanish transactions, and many observations on the disposition of the American Spaniards, and on the condition of the countries bordering on the South Seas, and as herein I may appear to differ greatly from the opinions generally established, I think it incumbent on me particularly to recite the authorities I have been guided by on this occasion, that I may not be censured as having given way either to a thoughtless credulity on one hand, or, what would be a much more criminal imputation, to a wilful and deliberate misrepresentation on the other. Mr Anson, before he set sail upon this

---

[1] In the Introduction, and throughout the whole Narrative, all the descriptions and references which relate to the elaborate charts, plans, and drawings of the original edition, and which are cumbrous and unintelligible without them, have been omitted; as also many digressions of the Narrator on nautical, topographical, or historical points, which now serve little purpose but to delay the progress and enfeeble the interest of the main story. Those omissions, however, save where trivial in matter or in amount, have been mentioned in the notes.

expedition, besides the printed journals to those parts, took care to furnish himself with the best manuscript accounts he could procure of all the Spanish settlements upon the coasts of Chili, Peru, and Mexico. These he carefully compared with the examinations of his prisoners, and the informations of several intelligent persons who fell into his hands in the South Seas. He had likewise the good fortune, in some of his captures, to possess himself of a great number of letters and papers of a public nature, many of them written by the Viceroy of Peru to the Viceroy of Santa Fé, to the Presidents of Panama and Chili, to Don Blas de Lezo, Admiral of the Galleons, and to divers other persons in public employments; and in these letters there was usually inserted a recital of those they were intended to answer; so that they contained a considerable part of the correspondence between these officers for some time previous to our arrival on that coast. We took besides many letters, sent from persons employed by the Government to their friends and correspondents, which were frequently filled with narrations of public business, and sometimes contained undisguised animadversions on the views and conduct of their superiors. From these materials those accounts of the Spanish affairs are taken which may at first sight appear the most exceptionable. In particular, the history of the various casualties which befell Pizarro's squadron is for the most part composed from intercepted letters: though indeed the relation of the insurrection of Orellana and his followers is founded on rather a less disputable authority, for it was taken from the mouth of an English gentleman then on board Pizarro, who often conversed with Orellana; and it was, on inquiry, confirmed in its principal circumstances by others who were in the ship at the same time: so that the fact, however extraordinary, is, I conceive, not to be contested.

And on this occasion I cannot but mention, that though I have endeavoured, with my utmost care, to adhere strictly to truth in every article of the ensuing narration, yet I am apprehensive that in so complicated a work some oversights must have been committed, by the inattention to which at times all mankind are liable. However, I know of none but literal mistakes: and if there are other errors which have escaped me, I flatter myself they are not of moment enough to affect any material transaction, and therefore I hope they may justly claim the reader's indulgence.[1]

If what has been said merits the attention of travellers of all sorts, it is, I think, more particularly applicable to the gentlemen of the Navy; since without drawing and planning neither chart nor views of lands can be taken; and without these it is sufficiently evident that navigation is at a full stand. It is doubtless from a persuasion of the utility of these qualifications, that his Majesty has established a drawing-master at Portsmouth, for the instruction of those who are presumed to be hereafter intrusted with the command of his royal navy. And though some have been so far misled as to suppose that the perfection of sea-officers consisted in a turn of mind and temper resembling the boisterous element they had to deal with, and have condemned all literature and science as effeminate, and derogatory to that ferocity which, they would falsely persuade us, was the most unerring characteristic of courage: yet it is to be hoped that such absurdities as these have at no time been authorised by the public opinion, and that the belief of them daily diminishes. . . . Indeed,

---

[1] A long passage is here omitted, in which the Author animates his countrymen to "the encouragement and pursuit of all kinds of nautical and geographical observations, and every species of mechanical and commercial information," and especially insists on the advantage and necessity of a traveller's being able to draw, and possessing an acquaintance with the general principles of surveying.

when the many branches of science are considered of which even the common practice of navigation is composed, and the many improvements which men of skill have added to this practice within these few years, it would induce one to believe that the advantages of reflection and speculative knowledge were in no profession more eminent than in that of a sea-officer. For, not to mention some expertness in geography, geometry, and astronomy, which it would be dishonourable for him to be without (as his journal and his estimate of the daily position of the ship are no more than the practice of particular branches of these arts), it may be well supposed that the management and working of a ship, the discovery of her most eligible position in the water (usually styled her trim), and the disposition of her sails in the most advantageous manner, are articles wherein the knowledge of mechanics cannot but be greatly assistant: and perhaps the application of this kind of knowledge to naval subjects may produce as great improvements in sailing and working a ship, as it has already done in many other matters conducive to the ease and convenience of human life. For when the fabric of a ship and the variety of her sails are considered, together with the artificial contrivances of adapting them to her different motions, as it cannot be doubted but these things have been brought about by more than ordinary sagacity and invention, so neither can it be doubted but that a speculative and scientific turn of mind may find out the means of directing and disposing this complicated mechanism much more advantageously than can be done by mere habit, or by a servile copying of what others may perhaps have erroneously practised in the like emergency. But it is time to finish this digression, and to leave the reader to the perusal of the ensuing work; which, with how little art soever it may be executed, will yet, from the importance of the subject, and the utility and excellence of the materials, merit some share of the public attention.

BOOK I.

CHAPTER I.

THE squadron under the command of Mr Anson, of which I here propose to recite the most material proceedings, having undergone many changes in its destination, its force, and its equipment, in the ten months between its first appointment and its final sailing from St Helens, I conceive the history of these alterations is a detail necessary to be made public, both for the honour of those who first planned and promoted this enterprise, and for the justification of those who have been entrusted with its execution. Since it will from hence appear, that the accidents the expedition was afterwards exposed to, and which prevented it from producing all the national advantages, the strength of the squadron and the expectation of the public seemed to presage, were principally owing to a series of interruptions which delayed the commander in the course of his preparations, and which it exceeded his utmost industry either to avoid or to get removed.

When, in the latter end of the summer of the year 1739, it was foreseen that a war with Spain was inevitable,[1] it was the opinion of several

[1] A convention regulating the sum to be paid by Spain to England on account of damage sustained to English commerce through the arbitrary means taken by the Spaniards to protect their American trade, had been signed

considerable persons, then trusted with the administration of affairs, that the most prudent step the nation could take, on the breaking out of the war, was attacking that Crown in her distant settlements: for by this means (as at that time there was the greatest probability of success) it was supposed that we should cut off the principal resources of the enemy, and reduce them to the necessity of sincerely desiring a peace, as they would hereby be deprived of the returns of that treasure by which alone they could be enabled to carry on a war.[1] In pursuance of these sentiments, several projects were examined, and several resolutions taken in Council. And in these deliberations it was from the first determined that George Anson, Esq., then captain of the Centurion,[2] should be employed as commander-in-chief of an expedition of this kind; and he then being absent on a cruise, a vessel was dispatched to his station so early as the beginning of September, to order him to return with his ship to Portsmouth. And soon after he came there—that is, on the 10th of November following—he received a letter from Sir Charles Wager, ordering him to repair to London, and to attend the Board of Admiralty; where, when he arrived, he was informed by Sir Charles that two squadrons would be immediately fitted out for two secret expeditions, which, however, would have some connection with each other; that he, Mr Anson, was intended to command one of them, and Mr Cornwall (who has since lost his life gloriously in the defence of his country's honour) the other; that the squadron under Mr Anson was to take on board three independent companies of a hundred men each, and Bland's regiment of foot; that Colonel Bland was likewise to embark with his regiment, and to command the land forces; and that, as soon as this squadron could be fitted for the sea, they were to set sail, with express orders to touch at no place till they came to Java Head, in the East Indies; that there they were only to stop to take in water, and thence to proceed directly to the city of Manilla, situated on Luconia,[3] one of the Philippine Islands; that the other squadron was to be of equal force with this com-

---

at Madrid in January 1739. But the question of the Right of Search exercised by the Spanish Crown over English vessels trading to its western colonies, and other delicate subjects of dispute, were reserved for future negotiation; a fierce clamour of dissatisfaction with the Convention, and eagerness for war, arose among the British people and in Parliament; and Walpole, unable to stem the tide of popular desire, resolved on entering upon a conflict which he condemned and deplored. The War is sometimes known as that of "the Merchants," arising, as it did, purely out of trade disputes; it was declared in London, amid wild public rejoicing, on the 19th of October 1739.

[1] Compare the reasons assigned for Drake's fatal Puerto Rico expedition; *ante*, page 99.

[2] Earl Stanhope, in his "History of England," Chapter XXII., says of Anson: "George Anson deserves to be held forth as a model to British seamen of what may be accomplished by industry, by courage, by love of their profession. He was born of a family at that period new and obscure, nor had he the advantage of distinguished talents. After his expedition, it used to be said of him that he had been round the world but never in it: he was dull and unready on land, slow in business, and sparing of speech. But he had undaunted bravery, steady application, and cool judgment; he punctually followed his instructions, and zealously discharged his duty; and by these qualities —qualities within the attainment of all—did he rise to well-earned honours, and bequeath an unsullied renown."

[3] Or Luzon, the northernmost and largest of the group.

manded by Mr Anson, and was intended to pass round Cape Horn into the South Seas, and there to range along that coast; and after cruising upon the enemy in those parts, and attempting their settlements, this squadron in its return was to rendezvous at Manilla, and there to join the squadron under Mr Anson, where they were to refresh their men, and refit their ships, and perhaps receive further orders.[1]

This scheme was doubtless extremely well projected, and could not but greatly advance the public service, and at the same time the reputation and fortune of those concerned in its execution; for had Mr Anson proceeded for Manilla at the time and in the manner proposed by Sir Charles Wager, he would in all probability have arrived there before they had received any advice of the war between us and Spain, and consequently before they had been in the least prepared for the reception of an enemy, or had any apprehensions of their danger. The city of Manilla might be well supposed to have been at that time in the same defenceless condition with all the other Spanish settlements just at the breaking out of the war; that is to say, their fortifications neglected, and in many places decayed; their cannon dismounted, or useless by the mouldering of their carriages; their magazines, whether of military stores or provision, all empty; their garrisons unpaid, and consequently thin, ill-affected, and dispirited; and the royal chests in Peru, whence alone all these disorders could receive their redress, drained to the very bottom. This, from the intercepted letters of their Viceroys and Governors, is well known to have been the defenceless state of Panama and the other Spanish places on the coast of the South Seas, for near a twelvemonth after our declaration of war. And it cannot be supposed that the city of Manilla, removed still farther by almost half the circumference of the globe, should have experienced from the Spanish Government a greater share of attention and concern for its security than Panama, and the other important ports in Peru and Chili, on which their possession of that immense empire depends. Indeed, it is well known that Manilla was at that time incapable of making any considerable defence, and, in all probability, would have surrendered only on the appearance of our squadron before it. The consequence of this city, and the island it stands on, may be in some measure estimated from the healthiness of its air, the excellency of its port and bay, the number and wealth of its inhabitants, and the very extensive and beneficial[2] commerce which it carries on to the principal ports in the East Indies and China, and its exclusive trade to Acapulco, the returns for which,[3] being made in silver, are upon the lowest valuation not less than three millions of dollars per annum.

And on this scheme Sir Charles Wager was so intent, that in a few days after this first conference, that is, on November 18, Mr Anson received an order to take under his command the Argyle, Severn, Pearl, Wager, and Trial sloop; and other orders were issued to him in the same month, and in the December following, relating to the victualling of this squadron. But Mr Anson attending the Admiralty the beginning of January, he was informed by Sir Charles Wager that for reasons with which he, Sir Charles, was not acquainted, the expedition to Manilla was laid aside. It may be conceived that Mr Anson was extremely chagrined at losing the command of so infallible, so honourable, and in every respect so desirable an enterprise, especially, too, as he had already, at a very great expense, made the necessary provision for his own accommodation

[1] Ed. 1776: "And perhaps receive orders for other considerable enterprises."

[2] Profitable.

[3] That is, for the Acapulco trade alone.

in this voyage, which he had reason to expect would prove a very long one. However, Sir Charles, to render this disappointment in some degree more tolerable, informed him that the expedition to the South Seas was still intended; and that he, Mr Anson, and his squadron, as their first destination was now countermanded, should be employed in that service. And on the 10th of January [1740] he received his commission, appointing him commander-in-chief of the forementioned squadron, which (the Argyle being in the course of their preparation changed for the Gloucester) was the same he sailed with above eight months after from St Helens. On this change of destination, the equipment of the squadron was still prosecuted with as much vigour as ever; and the victualling, and whatever depended on the Commodore, was [soon] so far advanced, that he conceived the ships might be capable of putting to sea the instant he should receive his final orders, of which he was in daily expectation. And at last, on the 28th of June 1740, the Duke of Newcastle, Principal Secretary of State, delivered to him his Majesty's instructions, dated January 31, 1739, with an additional instruction from the Lords Justices, dated June 19, 1740. On the receipt of these, Mr Anson immediately repaired to Spithead, with a resolution to sail with the first fair wind, flattering himself that all his delays[1] were now at an end. For though he knew by the musters that his squadron wanted 300 seamen of their complement (a deficiency which, with all his assiduity, he had not been able to get supplied), yet as Sir Charles Wager informed him that an order from the Board of Admiralty was despatched to Sir John Norris to spare him the numbers which he wanted, he doubted not of his complying therewith. But on his arrival at Portsmouth he found himself greatly mistaken and disappointed in this persuasion; for, on his application, Sir John Norris told him he could spare him none, for he wanted men for his own fleet. This occasioned an inevitable and a very considerable delay; for it was the end of July before this deficiency was by any means supplied, and all that was then done was extremely short of his necessities and expectation. For Admiral Balchen, who succeeded to the command at Spithead after Sir John Norris had sailed to the westward, instead of 300 able sailors, which Mr Anson wanted of his complement, ordered on board the squadron 170 men only, of which thirty-two were from the hospital and sick quarters, thirty-seven from the Salisbury, with three officers of Colonel Lowther's regiment, and ninety-eight marines; and these were all that were ever granted to make up the forementioned deficiency.

But the Commodore's mortification did not end here. It has been already observed, that it was at first intended that Colonel Bland's regiment, and three independent companies of 100 men each, should embark as land forces on board the squadron. But this disposition was now changed, and all the land forces that were to be allowed were 500 invalids, to be collected from the out-pensioners of Chelsea College. As these out-pensioners consist of soldiers, who, from their age, wounds, or other infirmities, are incapable of service in marching regiments, Mr Anson was greatly chagrined at having such a decrepit detachment allotted to him; for he was fully persuaded that the greatest part of them would perish long before they arrived at the scene of action, since the delays he had already encountered necessarily confined his passage round Cape Horn to the most rigorous season of the year. Sir Charles Wager, too, joined in opinion with the Commodore that invalids were no ways proper for this service, and solicited strenuously to have them exchanged; but he was told, that persons who were supposed to be better judges of soldiers than he or Mr Anson thought them the properest men that could be employed on this

[1] Ed. 1776: "His difficulties."

occasion.[1] And upon this determination they were ordered on board the squadron on the 5th of August; but instead of 500 there came on board no more than 259; for all those who had limbs and strength to walk out of Portsmouth deserted, leaving behind them only such as were literally invalids, most of them being sixty years of age, and some of them upwards of seventy. Indeed, it is difficult to conceive a more moving scene than the embarkation of these unhappy veterans; they were themselves extremely averse to the service they were engaged in, and fully apprised of all the disasters they were afterwards exposed to; the apprehensions of which were strongly marked by the concern that appeared in their countenances, which was mixed with no small degree of indignation to be thus hurried from their repose into a fatiguing employ to which neither the strength of their bodies, nor the vigour of their minds, were any ways proportioned, and where, without seeing the face of an enemy, or in the least promoting the success of the enterprise they were engaged in, they would in all probability uselessly perish by lingering and painful diseases; and this, too, after they had spent the activity and strength of their youth in their country's service.

And I cannot but observe, on this melancholy incident, how extremely unfortunate it was, both to this aged and diseased detachment, and to the expedition they were employed in, that amongst all the out-pensioners of Chelsea Hospital, which were supposed to amount to 2000 men, the most crazy and infirm only should be culled out for so fatiguing and perilous an undertaking. For it was well known that, however unfit invalids in general might be for this service, yet by a prudent choice there might have been found amongst them 500 men who had some remains of vigour left. And Mr Anson fully expected that the best of them would have been allotted him; whereas the whole detachment that was sent to him seemed to be made up of the most decrepit and miserable objects that could be collected out of the whole body; and by the desertion abovementioned, [even] these were a second time cleared of that little health and strength which were to be found amongst them, and he was to take up with such as were much fitter for an infirmary than for any military duty.

And here it is necessary to mention another material particular in the equipment of this squadron. It was proposed to Mr Anson, after it was resolved that he should be sent to the South Seas, to take with him two persons under the denomination of agent-victuallers. Those who were mentioned for this employment had formerly been in the Spanish West Indies,[2] in the South Sea Company's service; and it was supposed that by their knowledge and intelligence on that coast, they might often procure provision for him by compact with the inhabitants, when it was not to be got by force of arms. These agent-victuallers were, for this purpose, to be allowed to carry to the value of £15,000 in merchandise on board the squadron; for they had represented that it would be much easier for them to procure provisions with goods, than with the value of the same goods in money. Whatever colours were given to this scheme, it was difficult to persuade the generality of mankind that it was not principally intended for the enrichment of the agents, by the beneficial commerce they proposed to carry on upon that coast. Mr Anson, from the beginning, objected both to the appoint-

[1] Sir John Barrow, in his Life of Anson, says—"The feelings of these excellent judges are not to be envied, when they were afterwards made acquainted with the fact, that not one of these unfortunate individuals, who went on the voyage, survived to reach their native land—every man had perished."

[2] Ed. 1776: "In the Spanish American colonies."

ment of agent-victuallers, and the allowing them to carry a cargo on board the squadron. For he conceived that in those few amicable ports where the squadron might touch he needed not their assistance to contract for any provisions the place afforded; and on the enemy's coast he did not imagine that they could ever procure him the necessaries he should want, unless (which he was resolved not to comply with) the military operations of his squadron were to be regulated by the ridiculous views of their trading projects. All that he thought the Government ought to have done on this occasion was to put on board to the value of £2000 or £3000 only of such goods as the Indians, or the Spanish planters in the less cultivated part of the coast, might be tempted with; since it was in such places only that he imagined it would be worth while to truck with the enemy for provisions. And in these places, it was sufficiently evident, a very small cargo would suffice.

But though the Commodore objected both to the appointment of these officers, and to their project;[1] yet, as they had insinuated that their scheme, besides victualling the squadron might contribute to settling a trade upon that coast, which might be afterwards carried on without difficulty, and might thereby prove a very considerable national advantage, they were much listened to by some considerable persons. And of the £15,000, which was to be the amount of their cargo, the Government agreed to advance them £10,000 upon imprest,[2] and the remaining £5000 they raised on bottomry bonds; and the goods purchased with this [latter] sum were all that were taken to sea by the squadron, how much soever the amount of them might be afterwards magnified by common report. This cargo was at first shipped on board the Wager store-ship, and one of the victuallers; no part of it being admitted on board the men-of-war. But when the Commodore was at St Catherine's, he considered, that in case the squadron should be separated, it might be pretended that some of the ships were disappointed of provisions for want of a cargo to truck with; and therefore he distributed some of the least bulky commodities on board the men-of-war, leaving the remainder principally on board the Wager, where it was lost. And more of the goods perishing, by various accidents to be recited hereafter, and no part of them being disposed of upon the coast, the few that came home to England did not produce, when sold, above a fourth part of the original price. So true was the Commodore's prediction about the event of this project, which had been by many considered as infallibly productive of immense gains. But to return to the transactions at Portsmouth.

To supply the place of the 240 invalids which had deserted, as is mentioned above, there were ordered on board 210 marines detached from different regiments. These were raw and undisciplined men, for they were just raised, and had scarcely anything more of the soldier than their regimentals, none of them having been so far trained as to be permitted to fire. The last detachment of these marines came on board the 8th of August, and on the 10th the squadron sailed from Spithead to St Helens, there to wait for a wind to proceed on the expedition. But the delays we had already suffered had not yet spent all their influence, for we were now advanced into a season of the year when the westerly winds are usually very constant and very violent; and it was thought proper that we should put to sea in company with the fleet commanded by Admiral Balchen, and the expedition under Lord Cathcart.[3]

[1] Ed. 1776 adds: "Of the ill-success of which he had no question."

[2] "Prest money" is money advanced on condition that it shall be "ready" when the lender demands it back. French, "prêt;" that is, it is lent "on call."

[3] This expedition was designed to

And as we made up in all twenty-one men-of-war, and a 124 sail of merchantmen and transports, we had no hopes of getting out of the Channel with so large a number of ships, without the continuance of a fair wind for some considerable time. This was what we had every day less and less reason to expect, as the time of the equinox drew near; so that our golden dreams, and our ideal possession of the Peruvian treasures, grew each day more faint, and the difficulties and dangers of the passage round Cape Horn in the winter season filled our imaginations in their room. For it was forty days from our arrival at St Helens to our final departure from thence. And even then (having orders to proceed without Lord Cathcart) we tided it down the Channel with a contrary wind. But this interval of forty days was not free from the displeasing fatigue of often setting sail, and being as often obliged to return; nor exempt from dangers greater than have been sometimes experienced in surrounding the globe. For the wind coming fair for the first time on the 23d of August, we got under sail, and [Admiral] Balchen showed himself truly solicitous to have proceeded to sea; but the wind, soon returning to its old quarter, obliged us to put back to St Helens, not without considerable hazard, and some damage received by two of the transports, which, in tacking, ran foul of each other. Besides this, we made two or three more attempts to sail, but without any better success; and on the 6th of September, being returned to an anchor at St Helens, after one of these fruitless efforts, the wind blew so fresh that the whole fleet struck their yards and topmasts to prevent their driving. And, notwithstanding this precaution, the Centurion drove the next evening, and brought both cables ahead, and we were in no small danger of driving foul of the Prince Frederick, a seventy-gun ship, moored at a small distance under our stern, which we happily escaped, by her driving at the same time, and so preserving our distance; nor did we think ourselves secure till we at last let go the sheet anchor, which fortunately brought us up.

However, on the 9th of September we were in some degree relieved from this lingering vexatious situation by an order which Mr Anson received from the Lords Justices, to put to sea the first opportunity with his own squadron only if Lord Cathcart should not be ready. Being thus freed from the troublesome company of so large a fleet, our Commodore resolved to weigh and tide it down Channel as soon as the weather should become sufficiently moderate; and this might easily have been done with our own squadron alone full two months sooner had the orders of the Admiralty for supplying us with seamen been punctually complied with, and had we met with none of those other delays mentioned in this narration. It is true, our hopes of a speedy departure were even now somewhat damped by a subsequent order which Mr Anson received on the 12th of September, for by that he was required to take under his convoy the St Albans, with the Turkey fleet, and to join the Dragon and the Winchester, with the Straits and the American trade,[1] at Torbay or Plymouth, and to proceed with them to sea as far as their

reinforce Admiral Vernon, who had captured Porto Bello in the preceding November; but through various delays it did not sail till the end of October, and after joining Vernon, the united force attacked Carthagena, only to be repulsed. Lord Cathcart, before that miscarriage, had died from the effects of the climate, General Wentworth succeeding him in command of the troops.

[1] That is, the merchant vessels proceeding to the Mediterranean through the Straits of Gibraltar and to the American colonies; the collective word "trade" being aptly enough used to denote the gathering of all the ships bound for the one or the other destination, under the care of their armed convoys.

way and ours lay together. This encumbrance of a convoy gave us some uneasiness, as we feared it might prove the means of lengthening our passage to Madeira. However, Mr Anson, now having the command himself, resolved to adhere to his former determination, and to tide it down the Channel with the first moderate weather; and that the junction of his convoy might occasion as little loss of time as possible, he immediately sent directions to Torbay that the fleets he was there to take under his care might be in readiness to join him instantly on his approach. And at last, on the 18th of September, he weighed from St Helens; and though the wind was at first contrary, had the good fortune to get clear of the Channel in four days, as will be more particularly related in the ensuing Chapter.

Having thus gone through the respective steps taken in the equipment of this squadron, it is sufficiently obvious how different an aspect this expedition bore at its first appointment in the beginning of January from what it had in the latter end of September when it left the Channel; and how much its numbers, its strength, and the probability of its success, were diminished by the various incidents which took place in that interval. For instead of having all our old and ordinary seamen exchanged for such as were young and able (which the Commodore was at first promised), and having our numbers completed to their full complement, we were obliged to retain our first crews, which were very indifferent; and a deficiency of 300 men in our numbers was no otherwise made up to us than by sending us on board 170 men, the greatest part composed of such as were discharged from hospitals, or new-raised marines who had never been at sea before. And in the land forces allotted us the change was still more disadvantageous, for there, instead of three independent companies of 100 men each, and Bland's regiment of foot, which was an old one, we had only 470 invalids and marines—one part of them incapable for action by age and infirmities, and the other part useless by their ignorance of their duty. But the diminishing the strength of the squadron was not the greatest inconvenience which attended these alterations, for the contests, representations, and difficulties which they continually produced (as we have above seen, that in these cases the authority of the Admiralty was not always submitted to), occasioned a delay and waste of time which in its consequences was the source of all the disasters to which this enterprise was afterwards exposed. For by this means we were obliged to make our passage round Cape Horn in the most tempestuous season of the year, whence proceeded the separation of our squadron, the loss of numbers of our men, and the imminent hazard of our total destruction. And by this delay, too, the enemy had been so well informed of our designs that a person who had been employed in the South Sea Company's service, and arrived from Panama three or four days before we left Portsmouth, was able to relate to Mr Anson most of the particulars of the destination and strength of our squadron from what he had learned amongst the Spaniards before he left them. And this was afterwards confirmed by a more extraordinary circumstance; for we shall find that when the Spaniards (fully satisfied that our expedition was intended for the South Seas) had fitted out a squadron to oppose us, which had so far got the start of us as to arrive before us off the Island of Madeira, the commander of this squadron was so well instructed in the form and make of Mr Anson's broad pennant, and had imitated it so exactly that he thereby decoyed the Pearl, one of our squadron, within gun-shot of him before the captain of the Pearl was able to discover his mistake.

CHAPTER II.

On the 18th of September 1740, the squadron, as we have observed in the preceding Chapter, weighed from St Helens with a contrary wind, the Commodore proposing to tide it down the Channel, as he dreaded less the inconveniences he should thereby have to struggle with than the risk he should run of ruining the enterprise by an uncertain and in all probability a tedious attendance for a fair wind.

The squadron allotted to this service consisted of five men-of-war, a sloop-of-war, and two victualling ships. They were the Centurion, of 60 guns, 400 men, George Anson, Esq., commander; the Gloucester, of 50 guns, 300 men, Richard Norris, commander; the Severn, of 50 guns, 300 men, the Honourable Edward Legg, commander; the Pearl, of 40 guns, 250 men, Matthew Mitchel, commander; the Wager, of 28 guns, 160 men, Dandy Kidd, commander; and the Trial sloop, of 8 guns, 100 men, the Honourable John Murray, commander. The two victuallers were pinks,[1] the largest about 400 and the other about 200 tons burthen; these were to attend us till the provisions we had taken on board were so far consumed as to make room for the additional quantity they carried with them, which when we had taken into our ships they were to be discharged. Besides the complement of men borne by the above-mentioned ships as their crews, there were embarked on board the squadron about 470 invalids and marines, under the denomination of land forces, as has been particularly mentioned in the preceding Chapter, which were commanded by Lieutenant-Colonel Crackerode. With this squadron, together with the St Albans and the Lark, and the [Turkey] trade under their convoy, Mr Anson, after weighing from St Helens, tided it down the Channel for the first forty-eight hours; and on the 20th, in the morning, we discovered off the Ram Head the Dragon, Winchester, South Sea Castle, and Rye, with a number of merchantmen[2] under their convoy. These we joined about noon the same day, our Commodore having orders to see them (together with the [convoy of the] St Albans and Lark) as far into the sea as their course and ours lay together. When we came in sight of this last-mentioned fleet, Mr Anson first hoisted his broad pennant, and was saluted by all the men-of-war in company.

When we had joined this last convoy, we made up eleven men-of-war, about 150 sail of merchantmen, consisting of the Turkey, the Straits, and the American trade. Mr Anson, the same day, made a signal for all the captains of the men-of-war to come on board him, where he delivered them their fighting and sailing instructions; and then, with a fair wind, we all stood towards the south-west; and the next day at noon, being the 21st, we had run forty leagues from the Ram Head; and being now clear of the land, our Commodore, to render our view more extensive, ordered Captain Mitchel, in the Pearl, to make sail two leagues ahead of the fleet every morning, and to repair to his station every evening. Thus we proceeded till the 25th, when the Winchester and the American convoy made the concerted signal for leave to separate, which being answered by the Commodore, they left us, as the St Albans and the Dragon, with the Turkey and Straits convoy, did on the 29th. After which separation, there remained in company only our own squadron and our two victuallers, with which we kept on our course for the Island of Madeira. But the winds were so contrary, that we had the mortification to be forty days in our passage thither from St Helens, though it is known to be often done in ten or twelve. This delay was a most unpleasing circumstance, productive

[1] French, "*Pinque;*" originally applied to sailing ships of small size available for reconnoitring, spying, or sounding purposes.

[2] Nearly 200, according to Mr Parcoe Thomas, the mathematical master on board the Centurion, who wrote an account of the voyage, from which many notes in this edition are derived.

of much discontent and ill-humour amongst our people, of which those only can have a tolerable idea who have had the experience of a like situation. And besides the peevishness and despondency which foul and contrary winds, and a lingering voyage, never fail to create on all occasions, we in particular had very substantial reasons to be greatly alarmed at this unexpected impediment. For as we had departed from England much later than we ought to have done, we had placed almost all our hopes of success in the chance of retrieving in some measure at sea the time we had so unhappily wasted at Spithead and St Helens.[1] However, at last, on Monday, October the 25th, at five in the morning, we, to our great joy, made the land, and in the afternoon came to an anchor in Madeira Road in forty fathoms water—the Brazen-Head bearing from us E. by S., the Loo NNW., and the Great Church NNE. We had hardly let go our anchor when an English privateer sloop ran under our stern and saluted the Commodore with nine guns, which we returned with five; and the next day, the [English] Consul of the island coming to visit the Commodore, we saluted him with nine guns on his coming on board.

This Island of Madeira, where we are now arrived, is famous through all our American settlements for its excellent wines, which seem to be designed by Providence for the refreshment of the inhabitants of the torrid zone. It is situated in a fine climate, in the Latitude of 32° 27′ N.; and in the Longitude from London of, by our different reckonings, from 18°30′ to 19° 30′ W., though laid down in the charts in 17°.[2] It is composed of one continued hill, of a considerable height, extending itself from east to west, the declivity of which, on the south side, is cultivated and interspersed with vineyards; and in the midst of this slope the merchants have fixed their country seats, which help to form an agreeable prospect. There is but one considerable town in the whole island, it is named Fonchiale [Funchal], and is seated on the south part of the island, at the bottom of a large bay. This is the only place of trade, and indeed the only one where it is possible for a boat to land. Fonchiale, towards the sea, is defended by a high wall, with a battery of cannon, besides a castle on the Loo, which is a rock standing in the water at a small distance from the shore. Even here the beach is covered with large stones, and a violent surf continually beats upon it: so that the Commodore did not care to venture the ships' long-boats to fetch the water off, as there was so much danger of their being lost; and therefore ordered the captains of the squadron to employ Portuguese boats on that service.

We continued about a week at this island, watering our ships, and providing the squadron with wine and other refreshments. And, on the 3d of November, Captain Richard Norris having signified by a letter to the Commodore his desire to quit his command on board the Gloucester, in order to return to England for the recovery of his health, the Commodore complied with his request; and thereupon was pleased to appoint Captain Matthew Mitchel to command the Gloucester in his room, and to remove Captain Kidd from the Wager to the Pearl, and Captain Murray from the Trial sloop to the Wager, giving the command of the Trial to Lieutenant Cheap. These promotions being

---

1 Thomas mentions, that on the 13th of October the first man lost on the voyage died—a common sailor, named Philip Meritt; and that next day, by an order from the Commodore, the ship's company went on short allowance—that is, one-third of the allowance granted by Government was kept back, to make the provisions hold out the longer. Anson was evidently disquieted and stimulated to foresight by the unpromising commencement of his voyage.

2 The charts, however, are right; the best most modern maps placing Madeira in 17°.

settled, with other changes in the lieutenancies, the Commodore, on the following day, gave to the captains their orders, appointing St Jago, one of the Cape Verd Islands, to be the first place of rendezvous in case of separation; and directing them, if they did not meet the Centurion there, to make the best of their way to the Island of St Catherine's on the coast of Brazil. The water for the squadron being the same day completed, and each ship supplied with as much wine and other refreshments as they could take in, we weighed anchor in the afternoon, and took our leave of the Island of Madeira. But, before I go on with the narration of our own transactions, I think it necessary to give some account of the proceedings of the enemy, and of the measures they had taken to render all our designs abortive.

When Mr Anson visited the Governor of Madeira, he received information from him, that for three or four days in the latter end of October there had appeared, to the westward of that island, seven or eight ships of the line, and a patache, which last was sent every day close in to make the land. The Governor assured the Commodore, upon his honour, that none upon the island had either given them intelligence, or had in any sort communicated with them; but that he believed them to be either French or Spanish, but was rather inclined to think them Spanish. On this intelligence, Mr Anson sent an officer in a clean sloop [2] eight leagues to the westward, to reconnoitre them, and, if possible, to discover what they were. But the officer returned without being able to get a sight of them, so that we still remained in uncertainty. However, we could not but conjecture that this fleet was intended to put a stop to our expedition; which, had they cruised to the eastward of the island instead of the westward, they could not but have executed with great facility. For as, in that case, they must have certainly fallen in with us, we should have been obliged to throw overboard vast quantities of provision to clear our ships for an engagement; and this alone, without any regard to the event of the action, would have effectually prevented our progress. This was so obvious a measure, that we could not help imagining reasons which might have prevented them from pursuing it. And we therefore supposed, that this French or Spanish squadron was sent out upon advice of our sailing in company with Admiral Balchen and Lord Cathcart's expedition: and thence, from an apprehension of being overmatched, they might not think it advisable to meet with us till we had parted company, which they might judge would not happen before our arrival at this island. These were our speculations at that time; and from hence we had reason to suppose, that we might still fall in with them in our way to the Cape Verd Islands. And afterwards, in the course of our expedition, we were many of us persuaded that this was the Spanish squadron commanded by Don Joseph Pizarro, which was sent out purposely to traverse the views and enterprises of our squadron, to which in strength they were greatly superior. As this Spanish armament then, was so nearly connected with our expedition, and as the catastrophe it underwent, though not effected by our force, was yet a considerable advantage to this nation produced in consequence of our equipment, I have, in the following Chapter, given a summary account of their proceedings, from their first setting out from Spain in the year 1740, till the Asia, the only ship which returned to Europe of the whole squadron, arrived at the Groyne [Corunna] in the beginning of the year 1746.

---

[2] Thomas, who put the suspicious squadron at sixteen or eighteen sail, and supposes that they were a junction of French and Spanish ships of war, says that Anson sent out "an English privateer which lay in the road."

CHAPTER III.

THE squadron fitted out by the Court of Spain to attend our motions, and traverse our projects, we supposed to have been the ships seen off Madeira, as mentioned in the preceding Chapter. And as this force was sent out particularly against our expedition, I cannot but imagine that the following history of the casualties it met with, as far as by intercepted letters and other information the same has come to my knowledge, is a very essential part of the present work. For by this it will appear we were the occasion that a considerable part of the naval power of Spain was diverted from the prosecution of the ambitious views of that Court in Europe; and the men and ships lost by the enemy in this undertaking were lost in consequence of the precautions they took to secure themselves against our enterprises. This squadron (besides two ships intended for the West Indies, which did not part company till after they had left Madeira) was composed of the following men-of-war, commanded by Don Joseph Pizarro:

The Asia, of 66 guns, and 700 men: this was the Admiral's ship.

The Guipuscoa, of 74 guns, and 700 men.

The Hermiona, of 54 guns, and 500 men.

The Esperanza, of 50 guns, and 450 men.

The St Estevan, of 40 guns, and 350 men.

And a patache of 20 guns.

These ships, over and above their complement of sailors and marines, had on board an old Spanish regiment of foot, intended to reinforce the garrisons on the coast of the South Seas. When this fleet had cruised for some days to the leeward of Madeira, as is mentioned in the preceding Chapter, they left that station in the beginning of November, and steered for the River of Plate, where they arrived the 5th of January, O.S.; and coming to an anchor in the Bay of Maldonado, at the mouth of that river, their Admiral, Pizarro, sent immediately to Buenos Ayres for a supply of provisions; for they had departed from Spain with only four months' provisions on board. While they lay here expecting this supply, they received intelligence, by the treachery of the Portuguese Governor of St Catherine's, of Mr Anson's having arrived at that island on the 21st of December preceding, and of his preparing to put to sea again with the utmost expedition. Pizarro, notwithstanding his superior force, had his reasons (and as some say his orders likewise) for avoiding our squadron anywhere short of the South Seas. He was, besides, extremely desirous of getting round Cape Horn before us, as he imagined that step alone would effectually baffle all our designs; and therefore, on hearing that we were in his neighbourhood, and that we should soon be ready to proceed for Cape Horn, he weighed anchor with the five large ships (the patache being disabled and condemned, and the men taken out of her), after a stay of seventeen days only, and got under sail without his provisions, which arrived at Maldonado within a day or two after his departure. But notwithstanding the precipitation with which he departed, we put to sea from St Catherine's four days before him; and in some part of our passage to Cape Horn the two squadrons were so near together, that the Pearl, one of our ships, being separated from the rest, fell in with the Spanish Fleet, and mistaking the Asia for the Centurion, had got within gun-shot of Pizarro before she discovered her error, and narrowly escaped being taken.

It being the 22d of January when the Spaniards weighed from Maldonado, they could not expect to get into the latitude of Cape Horn before the equinox; and as they had reason to apprehend very tempestuous weather in doubling it at that season, and as the Spanish sailors, being for the most part accustomed to a fair-weather country, might be expected to be very averse to so dangerous and fatiguing a navigation, the better to encourage them, some part of their pay was advanced to them in European goods, which they were to be permitted to dispose of in the South Seas; that so the hopes of the great profit each man was to make on his small venture might animate him in his duty, and render him less disposed to repine at the labour, the hardships, and the perils he would in all probability meet with before his arrival on the coast of Peru.

Pizarro with his squadron having, towards the latter end of February, run the length of Cape Horn, he then stood to the westward in order to double it; but in the night of the last day of February, O.S., while with this view they were turned to windward, the Guipuscoa, the Hermiona, and the Esperanza were separated from the Admiral. On the 6th of March following, the Guipuscoa was separated from the other two; and on the 7th (being the day after we had passed Straits le Maire) there came on a most furious storm at NW., which, in despite of all their efforts, drove the whole squadron to the eastward, and obliged them, after several fruitless attempts, to bear away for the River of Plate, where Pizarro in the Asia arrived about the middle of May, and a few days after him the Esperanza and the St Estevan. The Hermiona was supposed to founder at sea, for she was never heard of more; and the Guipuscoa was run ashore and sunk on the coast of Brazil. The calamities of all kinds which this squadron underwent in this unsuccessful navigation can only be paralleled by what we ourselves experienced in the same climate when buffeted by the same storms. There was indeed some diversity in our distresses, which rendered it difficult to decide whose situation was most worthy of commiseration. For to all the misfortunes we had in common with each other, as shattered rigging, leaky ships, and the fatigues and despondency which necessarily attend these disasters, there was superadded on board our squadron the ravage of a most destructive and incurable disease, and on board the Spanish squadron the devastation of famine.

For this squadron, either from the hurry of their outset,[1] their presumption of a supply at Buenos Ayres, or from other less obvious motives, departed from Spain, as has been already observed, with no more than four months' provision, and even that, as it is said, at short allowance only; so that, when by the storms they met with off Cape Horn their continuance at sea was prolonged a month or more beyond their expectation, they were thereby reduced to such infinite distress, that rats, when they could be caught, were sold for four dollars a-piece; and a sailor, who died on board, had his death concealed for some days by his brother, who during that time lay in the same hammock with the corpse, only to receive the dead man's allowance of provisions. In this dreadful situation they were alarmed (if their horrors were capable of augmentation) by the discovery of a conspiracy among the marines on board the Asia, the Admiral's ship. This had taken its rise chiefly from the miseries they endured. For though no less was proposed by the conspirators than the massacring the officers and the whole crew, yet their motive for this bloody resolution seemed to be no more than their desire of relieving their hunger, by appropriating the whole ship's provisions to themselves. But their designs were prevented, when just upon the point of execution, by means of one of their confessors, and three of their ringleaders were immediately put to death. However, though the conspiracy was suppressed, their other calamities admitted of no alleviation, but grew each day more and more destructive; so that by the complicated distress of fatigue, sickness, and hunger, the three ships which escaped lost the greatest part of their men. The Asia, their Admiral's ship, arrived

[1] Ed. 1776: "Outfit."

at Monte Video in the River of Plate, with half her crew only; the St Estevan had lost in like manner half her hands when she anchored in the Bay of Barragan. The Esperanza, a fifty-gun ship, was still more unfortunate, for of 450 hands which she brought from Spain, only fifty-eight remained alive; and the whole regiment of foot perished except sixty men.[1] . . .

The Asia having considerably suffered in this second unfortunate expedition (*see Note* 1), the Esperanza, which had been left behind at Monte Video, was ordered to be refitted, the command of her being given to Mindinuetta, who was captain of the Guipuscoa when she was lost. He, in the November of the succeeding year, that is, in November 1742, sailed from the River of Plate for the South Seas, and arrived safe on the coast of Chili, where his Commodore, Pizarro, passing overland from Buenos Ayres, met him. There were great animosities and contests between these two gentlemen at their meeting, occasioned principally by the claim of Pizarro to command the Esperanza, which Mindinuetta had brought round; for Mindinuetta refused to deliver her up to him, insisting that as he came into the South Seas alone, and under no superior, it was not now in the power of Pizarro to resume that authority which he had once parted with. However, the President of Chili interposing, and declaring for Pizarro, Mindinuetta, after a long and obstinate struggle, was obliged to submit.

But Pizarro had not yet completed the series of his adventures; for when he and Mindinuetta came back by land from Chili to Buenos Ayres, in the year 1745, they found at Monte Video the Asia, which near three years before they had left there. This ship they resolved, if possible, to carry to Europe, and with this view they refitted her in the best manner they could; but their great difficulty was to procure a sufficient number of hands to navigate her, for all the remaining sailors of the squadron to be met with in the neighbourhood of Buenos Ayres did not amount to 100 men. They endeavoured to supply this defect by pressing many of the inhabitants of Buenos Ayres, and putting on board besides all the English prisoners then in their custody, together with a number of Portuguese smugglers whom they had taken at different times, and some of the Indians of the country. Among these last there was a chief and ten of his followers, who had been surprised by a party of Spanish soldiers about three months before. The name of this chief was Orellana: he belonged to a very powerful tribe, which had committed great ravages in the neighbourhood of Buenos Ayres. With this motley crew (all of them, except the European Spaniards, extremely averse to the voyage) Pizarro set sail from Monte Video in the River of Plate, about the beginning of November 1745; and the native Spaniards, being no strangers to the dissatisfaction of their forced men, treated both those, the English prisoners and the Indians, with great insolence and barbarity, but more particularly the Indians; for it was common for the meanest officers in the ship to beat

[1] The fate of the Guipuscoa was little better. On being separated from the Hermiona and Esperanza in a fog on March 6th, they met with a severe storm while SE. from Staten Island. They were driven out of their course, and did not reach the shore on the coast of Brazil till 24th April, when those on board were reduced to one ounce and a half of biscuit a man per day. Many died through the hardships of the voyage; the remainder of the crew, to the number of 400, got safely to land, when the vessel sank shortly afterwards. The three remaining ships of the squadron which got into the River Plate sent an advice boat to Rio Janeiro for provisions and help, and an express across the Continent to Santiago. An attempt was made to round Cape Horn, in the Asia, in October following, but they were driven back to the River Plate in great distress.

them most cruelly on the slightest pretences, and oftentimes only to exert their superiority. Orellana and his followers, though in appearance sufficiently patient and submissive, meditated a severe revenge for all these inhumanities. As he conversed very well in Spanish (these Indians having in time of peace a great intercourse with Buenos Ayres), he affected[1] to talk with such of the English as understood that language, and seemed very desirous of being informed how many Englishmen there were on board, and which they were. As he knew that the English were as much enemies to the Spaniards as himself, he had doubtless an intention of disclosing his purposes to them, and making them partners in the scheme he had projected for revenging his wrongs and recovering his liberty; but having sounded them at a distance, and not finding them so precipitate and vindictive as he expected, he proceeded no further with them, but resolved to trust alone to the resolution of his ten faithful followers. These, it should seem, readily engaged to observe his directions, and to execute whatever commands he gave them; and having agreed on the measures necessary to be taken, they first furnished themselves with Dutch knives sharp at the point, which, being the common knives used in the ship, they found no difficulty in procuring. Besides this, they employed their leisure in secretly cutting out thongs from raw hides, of which there were great numbers on board, and in fixing to each end of these thongs the double-headed shot of the small quarter-deck guns: this, when swung round their heads according to the practice of their country, was a most mischievous weapon, in the use of which the Indians about Buenos Ayres are trained from their infancy, and consequently are extremely expert. These particulars being in good forwardness, the execution of their scheme was perhaps precipitated by a particular outrage committed on Orellana himself. For one of the officers, who was a very brutal fellow, ordered Orellana aloft; which being what he was incapable of performing, the officer, under pretence of his disobedience, beat him with such violence that he left him bleeding on the deck, and stupefied for some time with his bruises and wounds. This usage undoubtedly heightened his thirst for revenge, and made him eager and impatient till the means of executing it were in his power; so that within a day or two after this incident he and his followers opened[2] their desperate resolves in the ensuing manner.

It was about nine in the evening, when many of the principal officers were on the quarter-deck indulging in the freshness of the night air; the waist of the ship was filled with live cattle, and the forecastle was manned with its customary watch. Orellana and his companions, under cover of the night, having prepared their weapons, and thrown off their trousers and the more cumbrous part of their dress, came all together on the quarter-deck, and drew towards the door of the great cabin. The boatswain immediately reprimanded them, and ordered them to be gone. On this Orellana spoke to his followers in his native language, when four of them drew off, two towards each gangway, and the chief and the six remaining Indians seemed to be slowly quitting the quarter-deck. When the detached Indians had taken possession of the gangways, Orellana placed his hands hollow to his mouth, and bellowed out the war-cry used by those savages, which is said to be the harshest and most terrifying sound known in nature. This hideous yell was the signal for beginning the massacre: for on this the [Indians] all drew their knives, and brandished their prepared double-headed shot, and the six, with their chief, who remained on the quarter-deck, immediately fell on the Spaniards who were inter-

[1] "Affect" is here used, not in the sense of making an ostentatious pretence or show, but in that of preferring or making a practice of something.

[2] Ed. 1776: "Began to execute."

mingled with them, and laid near forty of them at their feet, of whom above twenty were killed on the spot, and the rest disabled. Many of the officers, in the beginning of the tumult, pushed into the great cabin, where they put out the lights, and barricaded the door. And of the others, who had avoided the first fury of the Indians, some endeavoured to escape along the gangways into the forecastle; but the Indians placed there on purpose stabbed the greatest part of them as they attempted to pass by, or forced them off the gangways into the waist. Others threw themselves voluntarily over the barricades into the waist, and thought themselves happy to lie concealed amongst the cattle; but the greatest part escaped up the main-shrouds, and sheltered themselves either in the tops or rigging. And though the Indians attacked only the quarter-deck, yet the watch in the forecastle finding their communication cut off, and being terrified by the wounds of the few who, not being killed on the spot, had strength sufficient to force their passage along the gangways, and not knowing either who their enemies were or what were their numbers, they likewise gave all over for lost, and in great confusion ran up into the rigging of the foremast and bowsprit.

Thus these eleven Indians, with a resolution perhaps without example, possessed themselves almost in an instant of the quarter-deck of a ship mounting sixty-six guns, with a crew of nearly 500 men, and continued in peaceable possession of this post a considerable time: for the officers in the great cabin (amongst whom were Pizarro and Mindinuetta), the crew between decks, and those who had escaped into the tops and rigging, were only anxious for their own safety, and were for a long time incapable of forming any project for suppressing the insurrection and recovering the possession of the ship. It is true, the yells of the Indians, the groans of the wounded, and the confused clamours of the crew, all heightened by the obscurity of the night, had at first greatly magnified their danger, and had filled them with the imaginary terrors which darkness, disorder, and an ignorance of the real strength of an enemy never fail to produce. For as the Spaniards were sensible of the disaffection of their pressed hands, and were also conscious of their barbarity to their prisoners, they imagined the conspiracy was general, and considered their own destruction as infallible; so that, it is said, some of them had once taken the resolution of leaping into the sea, but were prevented by their companions.

However, when the Indians had entirely cleared the quarter-deck, the tumult in a great measure subsided; for those who had escaped were kept silent by their fears, and the Indians were incapable of pursuing them to renew the disorder. Orellana, when he saw himself master of the quarter-deck, broke open the arm chest, which, on a slight suspicion of mutiny, had been ordered there a few days before, as to a place of the greatest security. Here, he took it for granted, he should find cutlasses sufficient for himself and his companions, in the use of which weapon they were all extremely skilful, and with these, it was imagined, they proposed to have forced the great cabin; but on opening the chest there appeared nothing but fire-arms, which to them were of no use. There were indeed cutlasses in the chest, but they were hid by the fire-arms being laid over them. This was a sensible disappointment to them, and by this time Pizarro and his companions in the great cabin were capable of conversing aloud, through the cabin windows and port-holes, with those in the gun-room and between decks; and from hence they learned that the English (whom they principally suspected) were all safe below, and had not intermeddled in this mutiny; and by other particulars they at last discovered that none were concerned in it but Orellana and his people. On this Pizarro and the officers resolved to attack them on the quarter-deck, before any of the discontented on board should so far recover their first surprise as to reflect on the facility and certainty of seizing the ship by a junction with the Indians in the present emergency. With this view Pizarro got together what arms were in the cabin, and distributed them to those who were with him; but there were no other fire-arms to be met with but pistols, and for these they had neither powder nor ball. However, having now settled a correspondence with the gun-room, they lowered down a bucket out of the cabin window, into which the gunner, out of one of the gun-room ports, put a quantity of pistol cartridges. When they had thus procured ammunition, and had loaded their pistols, they set the cabin-door partly open, and fired some shot amongst the Indians on the quarter-deck, at first without effect. But at last Mindinuetta, whom we have often mentioned, had the good fortune to shoot Orellana dead on the spot; on which his faithful companions, abandoning all thoughts of further resistance, instantly leaped into the sea, where they every man perished. Thus was this insurrection quelled, and the possession of the quarter-deck regained, after it had been full two hours in the power of this great and daring chief and his gallant and unhappy countrymen.

Pizarro, having escaped this imminent peril, steered for Europe, and arrived safe on the coast of Gallicia in the beginning of the year 1746, after having been absent between four and five years, and having, by his attendance on our expedition, diminished the naval power of Spain by above 3000 hands (the flower of their sailors) and by four considerable ships of war and a patache. For we have seen that the Hermiona foundered at sea; the Guipuscoa was stranded and sunk on the coast of Brazil; the St Estevan was condemned and broken up in the River of Plate; and the Esperanza, being left in the South Seas, is doubtless by this time incapable of returning to Spain. So that the Asia only, with less than 100 hands, may be considered as all the remains of that squadron with which Pizarro first put to sea. And whoever attends to the very large proportion which this squadron bore to the whole navy of Spain, will, I believe, confess that had our undertaking been attended with no other advantages than that of ruining so great a part of the sea force of so dangerous an enemy, this alone would be a sufficient equivalent for our equipment, and an incontestable proof of the service which the nation has thence received. Having thus concluded this summary of Pizarro's adventures, I shall now return again to the narration of our own transactions.

CHAPTER IV.

I HAVE already mentioned, that on the 3d of November we weighed from Madeira, after orders had been given to the captains to rendezvous at Santiago, one of the Cape Verd Islands, in case the squadron was separated. But the next day, when we were got to sea, the Commodore, considering that the season was far advanced, and that touching at Santiago would create a new delay, he for this reason thought proper to alter his rendezvous, and to appoint the Island of St Catherine's, on the coast of Brazil, to be the first place to which the ships of the squadron were to repair in case of separation. In our passage to the Island of St Catherine's, we found the direction of the trade-winds to differ considerably from what we had reason to expect, both from the general histories given of these winds, and the experience of former navigators.[1]

On the 16th of November, one of our victuallers made a signal to speak with the Commodore, and we shortened sail for her to come up with us. The master came on board, and ac-

[1] Omission is here made of some technical and obsolete observations on the trade-winds.

quainted Mr Anson that he had complied with the terms of his charter-party, and desired to be unloaded and dismissed. Mr Anson, on consulting the captains of the squadron, found all the ships had still such quantities of provision between their decks, and were withal so deep, that they could not without great difficulty take in their several proportions of brandy from the Industry pink, one of the victuallers only; and consequently he was obliged to continue the other of them, the Anna pink, in the service of attending the squadron. And the next day the Commodore made a signal for the ships to bring to, and to take on board their shares of the brandy from the Industry pink; and in this the long-boats of the squadron were employed the three following days, that is, till the 19th in the evening, when the pink being unloaded, she parted company with us, being bound for Barbadoes, there to take in a freight for England. Most of the officers of the squadron took the opportunity of writing to their friends at home by this ship; but she was afterwards, as I have been since informed, unhappily taken by the Spaniards.

On the 20th of November, the captains of the squadron represented to the Commodore that their ships' companies were very sickly, and that it was their own opinion as well as their surgeons' that it would tend to the preservation of the men to let in more air between decks; but that their ships were so deep they could not possibly open their lower ports. On this representation the Commodore ordered six air-scuttles to be cut in each ship, in such places where they would least weaken it. . . .

We crossed the Equinoctial, with a fine fresh gale at SE., on Friday the 28th of November, at four in the morning, being then in the Longitude of 27° 59′ W. from London. And on the 2d of December, in the morning, we saw a sail in the NW. quarter, and made the Gloucester's and Trial's signals to chase; and half-an-hour after we let [out] our reefs and chased with the squadron; and about noon a signal was made for the Wager to take our remaining victuallar, the Anna pink, in tow. But at seven in the evening, finding we did not near the chase, and that the Wager was very far astern, we shortened sail, and made a signal for the cruisers to join the squadron. The next day but one we again discovered a sail, which, on the nearer approach, we judged to be the same vessel. We chased her the whole day, and though we rather gained upon her, yet night came on before we could overtake her, and obliged us to give over the chase, to collect our scattered squadron. We were much chagrined at the escape of this vessel, as we then apprehended her to be an advice boat sent from Old Spain to Buenos Ayres with notice of our expedition. But we have since learned that we were deceived in this conjecture, and that it was our East India Company's packet bound to St Helena.

On the 10th of December, being by our accounts in the Latitude of 20° S., and 36° 30′ Longitude W. from London, the Trial fired a gun to denote soundings. We immediately sounded, and found sixty fathoms water, the bottom coarse ground with broken shells. The Trial, being ahead of us, had at one time thirty-seven fathoms, which afterwards increased to ninety: and then she found no bottom, which happened to us too at our second trial, though we sounded with 150 fathoms of line. This is the shoal which is laid down in most charts by the name of the Abrollos;[1] and it appeared we were upon the very edge of it; perhaps farther in it may be extremely dangerous. We were then, by our different accounts, from ninety to sixty leagues east of the coast of Brazil. The next day but one we spoke with a Portuguese brigantine from Rio Janeiro, bound to Bahia de todos los Santos, who informed us that we were sixty-four leagues from Cape St Thomas, and forty leagues from Cape Frio, which last bore from us

[1] The Abrolhos; a small group of islets or reefs off the coast of Brazil, in about Lat. 18° S., Long. 39° W.

WSW. By our accounts we were near eighty leagues from Cape Frio; and though, on the information of this brigantine, we altered our course and stood more to the southward, yet by our coming in with the land afterwards we were fully convinced that our reckoning was much correcter than our Portuguese intelligence. We found a considerable current setting to the southward after we had passed the Latitude of 16° S. And the same took place all along the coast of Brazil, and even to the southward of the River of Plate, it amounting sometimes to thirty miles in twenty-four hours, and once to above forty miles. . . .

We now began to grow impatient for a sight of land, both for the recovery of our sick, and for the refreshment and security of those who as yet continued healthier. When we departed from St Helens, we were in so good a condition, that we lost but two men on board the Centurion in our long passage to Madeira. But in this present run between Madeira and St Catherine's we have been very sickly, so that many died, and great numbers were confined to their hammocks, both in our own ship and in the rest of the squadron; and several of these past all hopes of recovery. The disorders they in general labour under are such as are common to the hot climates, and what most ships bound to the southward experience in a greater or less degree. These are those kind of fevers which they usually call calentures: a disease which was not only terrible in its first instance, but even the remains of it often proved fatal to those who considered themselves as recovered from it. For it always left them in a very weak and helpless condition, and usually afflicted with fluxes and tenesmuses. And by our continuance at sea all our complaints were every day increasing, so that it was with great joy that we discovered the coast of Brazil on the 18th of December, at seven in the morning.

The coast of Brazil appeared high and mountainous land, extending from W. to WSW., and when we first saw it, it was about seventeen leagues distant. At noon we perceived a low double land bearing WSW., about ten leagues distant, which we took to be the Island of St Catherine's. That afternoon and the next morning, the wind being NNW., we gained very little to windward, and were apprehensive of being driven to the leeward of the island; but a little before noon the next day the wind came about to the southward, and enabled us to steer in between the north point of St Catherine's and the neighbouring Island of Alvoredo. As we stood in for the land, we had regular soundings, gradually decreasing from thirty-six to twelve fathoms, all muddy ground. In this last depth of water we let go our anchor at 5 o'clock in the evening of the 19th, the north-west point of the Island of St Catherine's bearing SSW. distant three miles; and the Island Alvoredo NNE. distant two leagues. Here we found the tide to set SSE. and NNW., at the rate of two knots, the tide of flood coming from the southward. We could from our ships observe two fortifications at a considerable distance within us, which seemed designed to prevent the passage of an enemy between the Island of St Catherine's and the main. And we could soon perceive that our squadron had alarmed the coast, for we saw the two forts hoist their colours, and fire several guns, which we supposed to be intended for assembling the inhabitants. To prevent any confusion, the Commodore immediately sent a boat with an officer on shore, to compliment the Governor, and to desire a pilot to carry us into the road. The Governor returned a very civil answer, and ordered us a pilot. On the morning of the 20th we weighed and stood in, and towards noon the pilot came on board us, who the same afternoon brought us to an anchor in five fathoms and a half, in a large commodious bay on the continent side, called by the French Bon Port. In standing from our last anchorage to this place, we everywhere found an oozy bottom, with a depth of water first regularly decreasing to five fathoms, and then increasing to seven, after which we had six and five fathoms alternately. The next morning we weighed again with the squadron, in order to run above the two fortifications we have mentioned, which are called the castles of Santa Cruz and St Juan. And now the soundings between the island and the main were four, five, and six fathoms, with muddy ground. As we passed by the castle of Santa Cruz, we saluted it with eleven guns, and were answered by an equal number; and at one in the afternoon the squadron came to an anchor in five fathoms and a half, the Governor's Island bearing NNW., St Juan's castle NE. half E., and the Island of St Antonio S. In this position we moored at the Island of St Catherine's on Sunday the 21st of December, the whole squadron being, as I have already mentioned, sickly and in great want of refreshments: both which inconveniencies we hoped to have soon removed at this settlement, celebrated by former navigators for its healthiness and its [abundance of] provisions, and for the freedom, indulgence, and friendly assistance there given to the ships of all European nations in amity with the Crown of Portugal.

CHAPTER V.[1]

Our first care, after having moored our ships, was to send our sick men on shore, each ship being ordered by the Commodore to erect two tents for that purpose; one of them for the reception of the diseased, and the other for the accommodation of the surgeon and his assistants. We sent about eighty sick from the Centurion, and the other ships I believe sent nearly as many in proportion to the number of their hands. As soon as we had performed this necessary duty, we scraped our decks, and gave our ship a thorough cleansing; then smoked it between decks, and after all washed every part well with vinegar. These operations were extremely necessary for correcting the noisome stench on board, and destroying the vermin: for from the number of our men, and the heat of the climate, both these nuisances had increased upon us to a very loathsome degree, and, besides being most intolerably offensive, they were doubtless in some sort productive of the sickness we had laboured under for a considerable time before our arrival at this island. Our next employment was wooding and watering our squadron, calking our ships' sides and decks, overhauling our rigging, and securing our mast against the tempestuous weather we were, in all probability, to meet with in our passage round Cape Horn in so advanced and inconvenient a season. . . .

When we first arrived at St Catherine's we were employed in refreshing our sick on shore, in wooding and watering the squadron, cleansing our ships, and examining and securing our masts and rigging, as I have already observed in the foregoing Chapter. At the same time, Mr Anson gave directions that the ships' companies should be supplied with fresh meat, and that they should be victualled with whole allowance of all kinds of provision. In consequence of these orders, we had fresh beef sent on board us continually for our daily expense,[2] and what was wanting to make up our allowance we received from our victualler, the Anna pink, in order to preserve the provisions on board our squadron entire for our future service.[3] The season of the

[1] The description of the island, except one passage of political interest, and the account of Brazil, is here omitted, as needlessly hindering the course of the narrative.

[2] Consumption.

[3] Thomas says that "the agents for victualling, of which we had two with us, were ordered to procure what fresh provisions we could expend during our stay here, which they accordingly did; but though their meat, which is altogether beef, was both cheap and plenty, it was for the

year growing each day less favourable for our passage round Cape Horn, Mr Anson was very desirous of leaving this place as soon as possible; and we were at first in hopes that our whole business would be done, and we should be in readiness to sail in about a fortnight from our arrival; but, on examining the Trial's masts, we, to our no small vexation, found inevitable employment for twice that time. For, on a survey, it was found that the mainmast was sprung at the upper woulding,[1] though it was thought capable of being secured by a couple of fishes; but the foremast was reported to be unfit for service, and thereupon the carpenters were sent into the woods to endeavour to find a stick proper for a foremast. But after a search of four days they returned without having been able to meet with any tree fit for the purpose. This obliged them to come to a second consultation about the old foremast, when it was agreed to endeavour to secure it by casing it with three fishes; and in this work the carpenters were employed till within a day or two of our sailing. In the meantime, the Commodore, thinking it necessary to have a clean vessel on our arrival in the South Seas, ordered the Trial to be hove down, as this would not occasion any loss of time, but might be completed while the carpenters were refitting her masts, which was done on shore.

---

greatest part miserably bad, and scarce fit to be eaten. The men throughout the whole squadron began now to drop off apace with fevers and fluxes, occasioned, I believe, by the violent heat of the climate, and the bad air; the country being so very woody that the air must thereby be stagnated, and rendered unhealthful."

[1] Or "woolding;" explained in Bailey—"The winding of ropes hard about a yard or mast of a ship, after it hath been strengthened by some piece of timber nailed thereto." Young's "Nautical Dictionary," *sub voce*, also suggests the idea of previous "fishing" or repair.

On the 27th of December we discovered a sail in the offing; and not knowing but she might be a Spaniard, the eighteen-oared boat was manned and armed, and sent under the command of our second lieutenant to examine her before she arrived within the protection of the forts. She proved to be a Portuguese brigantine from Rio Grande. And though our officer, as it appeared on inquiry, had behaved with the utmost civility to the master, and had refused to accept a calf which the master would have forced on him as a present, yet the Governor took great offence at our sending our boat, and talked of it in a high strain, as a violation of the peace subsisting between the Crowns of Great Britain and Portugal. We at first imputed this ridiculous blustering to no deeper a cause than Don Jose's insolence; but as we found he proceeded so far as to charge our officer with behaving rudely and opening letters, and particularly with an attempt to take out of the vessel by violence the very calf which we knew he had refused to receive as a present (a circumstance which we were satisfied the Governor was well acquainted with), we had hence reason to suspect that he purposely sought this quarrel, and had more important motives for engaging in it than the mere captious bias of his temper. What these motives were, it was not so easy for us to determine at that time; but as we afterwards found, by letters which fell into our hands in the South Seas, that he had despatched an express to Buenos Ayres, where Pizarro then lay, with an account of our squadron's arrival at St Catherine's, together with the most ample and circumstantial intelligence of our force and condition, we thence conjectured that Don Jose had raised this groundless clamour only to prevent our visiting the brigantine when she should put to sea again, lest we might there find proofs of his perfidious behaviour, and perhaps at the same time discover the secret of his smuggling correspondence with his neighbouring Governors, and the Spaniards at Buenos Ayres.

It was near a month before the Trial was refitted; for not only her lower masts were defective, as has been already mentioned, but her main-topmast and foreyard were likewise decayed and rotten. While this work was carrying on, the other ships of the squadron fixed new standing rigging, and set up a sufficient number of preventer shrouds to each mast to secure them in the most effectual manner. And in order to render the ships stiffer, and to enable them to carry more sail abroad, and to prevent their labouring in hard gales of wind, each captain had orders given him to strike down some of their great guns into the hold. These precautions being complied with, and each ship having taken in as much wood and water as there was room for, the Trial was at last completed, and the whole squadron was ready for the sea; on which the tents on shore were struck, and all the sick were received on board. And here we had a melancholy proof how much the healthiness of this place had been overrated by former writers, for we found that though the Centurion alone had buried no less than twenty-eight men since our arrival, yet the number of our sick was in the same interval increased from eighty to ninety-six.

And now our crews being embarked, and everything prepared for our departure, the Commodore made a signal for all captains, and delivered them their orders, containing the successive places of rendezvous from hence to the coast of China.[1] And then on the next day, being the 18th of January 1741, the signal was made for weighing, and the squadron put to sea, leaving without regret this Island of St Catherine's, where we had been so extremely disappointed in our refreshments, in our accommodations, and in the humane and friendly offices which we had been taught to expect in a place which has been so much celebrated for its hospitality, freedom, and conveniency.

[1] Ed. 1776: "Chili;" an obvious blunder, as the opening paragraph of Chapter VI. shows.

CHAPTER VI.

In leaving St Catherine's, we left the last amicable port we proposed to touch at, and were now proceeding to an hostile, or at best a desert and inhospitable coast. And as we were to expect a more boisterous climate to the southward than any we had yet experienced, not only our danger of separation would by this means be much greater than it had been hitherto, but other accidents of a more pernicious nature were likewise to be apprehended, and as much as possible to be provided against. And therefore Mr Anson, in appointing the various stations at which the ships of the squadron were to rendezvous, had considered that it was possible his own ship might be disabled from getting round Cape Horn, or might be lost; and had given proper directions that even in that case the expedition should not be abandoned. For the orders delivered to the captains the day before we sailed from St Catherine's, were, that in case of separation—which they were with the utmost care to endeavour to avoid—the first place of rendezvous should be the Bay of Port St Julian, describing the place from Sir John Narborough's account of it. There they were to supply themselves with as much salt as they could take in, both for their own use and the use of the squadron; and if, after a stay there of ten days, they were not joined by the Commodore, they were then to proceed through Straits le Maire round Cape Horn into the South Seas, where the next place of rendezvous was to be the Island of Nuestra Señora del Socoro,[2] in the Latitude of 45° S., and Longitude from the Lizard, 71° 12′ W. They were to bring this island to bear ENE., and to cruise from five to twelve leagues distance from it, as long as their store of wood and water would permit, both which they were to ex-

[2] One of the smaller outer islands of the Chonos Archipelago, on the western coast of Patagonia.

pend with the utmost frugality. And when they were under an absolute necessity of a fresh supply, they were to stand in, and endeavour to find out an anchoring-place; and in case they could not, and the weather made it dangerous to supply their ships by standing off and on, they were then to make the best of their way to the Island of Juan Fernandez, in the Latitude of 33° 37′ S. And as soon as they had recruited their wood and water, they were to continue cruising off the anchoring-place of that island for fifty-six days, in which time, if they were not joined by the Commodore, they might conclude that some accident had befallen him; and they were forthwith to put themselves under the command of the senior officer, who was to use his utmost endeavours to annoy the enemy both by sea and land. That with these views their new Commodore was to continue in those seas as long as his provisions lasted, or as long as they were recruited by what he should take from the enemy, reserving only a sufficient quantity to carry him and the ships under his command to Macao at the entrance of the River Tigris, near Canton on the coast of China, where, having supplied himself with a new stock of provisions, he was thence without delay to make the best of his way to England. And as it was found impossible as yet to unload our victualler, the Anna pink, the Commodore gave the master of her the same rendezvous, and the same orders to put himself under the command of the remaining senior officer.

Under these orders the squadron sailed from St Catherine's on Sunday the 18th of January, as has been already mentioned in the preceding Chapter. The next day we had very squally weather, attended with rain, lightning, and thunder; but it soon became fair again, with light breezes, and continued thus till Wednesday evening, when it blew fresh again; and increasing all night, by eight the next morning it became a most violent storm, and we had with it so thick a fog that it was impossible to see at the distance of two ships' lengths, so that the whole squadron disappeared.[1] On this a signal was made by firing guns, to bring to with the larboard tacks, the wind being then due east. We ourselves immediately handed the topsails, bunted the mainsail, and lay to under a reefed mizzen till noon, when the fog dispersed; and we soon discovered all the ships of the squadron, except the Pearl, which did not join us till near a month afterwards. The Trial sloop was a great way to leeward, having lost her mainmast in this squall, and having been obliged, for fear of bilging, to cut away the raft.[2] We bore down with the squadron to her relief, and the Gloucester was ordered to take her in tow, for the weather did not entirely abate till the day after, and even then a great swell continued from the eastward in consequence of the preceding storm. After this accident we stood to the southward with little interruption, and here we experienced the same setting of the current which we had observed before our arrival at St Catherine's, that is, we generally found ourselves to the southward of our reckoning by about twenty miles each day. This error continued, with a little variation, till we had passed the Latitude of the River of Plate; and even then we found that the same current, however difficult to be accounted for, did yet undoubtedly take place, for we were not satisfied in deducing it from the error in our reckoning, but we actually tried it more than once when a calm made it practicable.

When we had passed the Latitude of the River of Plate we had soundings all along the coast of Patagonia. These soundings, when well ascertained, being of great use in determining the position of the ship, and we having tried them more frequently in greater depths, and with more attention than I believe had been done before us, I

[1] That is, was lost sight of by the Centurion, on board of which the Narrator sailed.

[2] Ed. 1776: "The wreck."

shall recite our observations as succinctly as I can. In the Latitude of 36° 52′ we had sixty fathoms of water, with a bottom of fine black and grey sand; from thence to 39° 55′ we varied our depths from fifty to eighty fathoms, though we had constantly the same bottom as before; between the last-mentioned Latitude and 43° 16′ we had only fine grey sand, with the same variation of depths, except that we once or twice lessened our water to forty fathoms. After this we continued in forty fathoms for about half a degree, having a bottom of coarse sand and broken shells, at which time we were in sight of land, and not above seven leagues from it. As we edged from the land we met with variety of soundings; first black sand, then muddy, and soon after rough ground with stones; but then increasing our water to forty-eight fathoms we had a muddy bottom to the Latitude of 46° 10′. We then returned again into thirty-six fathoms, and kept shoaling our water, till at length we came into twelve fathoms, having constantly small stones and pebbles at the bottom. Part of this time we had a view of Cape Blanco, which lies in about the Latitude of 46° 52′, and Longitude W. from London 66° 43′. This is the most remarkable land upon the coast. Steering from hence S. by E. nearly, we, in a run of about thirty leagues, deepened our water to fifty fathoms without once altering the bottom; and then drawing towards the shore with a SW. course, varying rather to the westward, we had everywhere a sandy bottom, till our coming into thirty fathoms, where we had again a sight of land distant from us about eight leagues, lying in the Latitude of 48° 31′. We made this land on the 17th of February, and at five in the afternoon we came to an anchor upon the same bottom in the Latitude of 48° 58′, the southernmost land then in view bearing SSW., the northernmost N. half E., a small island NW., and the westernmost hummock WSW. In this station we found the tide to set S. by W.

Weighing again at five the next morning, we an hour afterwards discovered a sail, upon which the Severn and Gloucester were both directed to give chase; but we soon perceived it to be the Pearl, which separated from us a few days after we left St Catherine's; and on this we made a signal for the Severn to rejoin the squadron, leaving the Gloucester alone in the pursuit. And now we were surprised to see that, on the Gloucester's approach, the people on board the Pearl increased their sail and stood from her. However, the Gloucester came up with them, but found them with their hammocks in their nettings, and everything ready for an engagement. At two in the afternoon the Pearl joined us, and running up under our stern, Lieutenant Salt hailed the Commodore, and acquainted him that Captain Kidd died on the 31st of January. He likewise informed him that he had seen five large ships on the 10th instant, which he for some time imagined to be our squadron: that he suffered the commanding ship, which wore a red broad pennant exactly resembling that of the Commodore, at the main-topmast head, to come within gun-shot of him before he discovered his mistake; but then, finding it not to be the Centurion, he hauled close upon the wind, and crowded from them with all his sail, and standing across a rippling,[1] where they hesitated to follow him, he happily escaped. He made them [out] to be five Spanish men-of-war, one of them exceedingly like the Gloucester, which was the occasion of his apprehensions when the Gloucester chased him. By their appearance he thought they consisted of two ships of 70 guns, two of 50, and one of 40 guns. The whole squadron continued in chase of him all that day, but at night, finding they could not get near him, they gave over the chase, and directed their course to the southward.

And now, had it not been for the

[1] A broken piece of water, due to a current, a violent tide, or some other perturbing cause.

necessity we were under of refitting the Trial, this piece of intelligence would have prevented our making any stay at St Julian; but as it was impossible for that sloop to proceed round the Cape in her present condition, some stay there was inevitable; and therefore the same evening we came to an anchor again in twenty-five fathoms water, the bottom a mixture of mud and sand, and the high hummock bearing SW. by W. And weighing at nine in the morning, we soon after sent the two cutters belonging to the Centurion and Severn in shore to discover the harbour of St Julian, while the ships kept standing along the coast at about the distance of a league from the land. At 6 o'clock we anchored in the Bay of St Julian,[1] in nineteen fathoms, the bottom muddy ground with sand, the northernmost land in sight bearing N. and by E., the southernmost S. half E., and the high hummock—to which Sir John Narborough formerly gave the name of Wood's Mount—WSW. Soon after the cutters returned on board, having discovered the harbour, which did not appear to us in our situation, the northernmost point shutting in upon the southernmost, and in appearance closing the entrance.

Being come to an anchor in this bay of St Julian, principally with a view of refitting the Trial, the carpenters were immediately employed in that business, and continued so during our whole stay at the place. The Trial's mainmast having been carried away about twelve feet below the cap, they contrived to make the remaining part of the mast serve again; and the Wager was ordered to supply her with a spare main-topmast, which the carpenters converted into a new foremast. And I cannot help observing, that this accident to the Trial's mast, which gave us so much uneasiness at that time on account of the delay it occasioned, was in all probability the means of preserving the sloop and all her crew: for before this her masts, how well soever proportioned to a better climate, were much too lofty for these high southern latitudes; so that had they weathered the preceding storm, it would have been impossible for them to have stood against those seas and tempests we afterwards encountered in passing round Cape Horn; and the loss of masts in that boisterous climate would scarcely have been attended with less than the loss of the vessel and of every man on board her, since it would have been impracticable for the other ships to have given them any relief during the continuance of those impetuous storms.

While we stayed at this place, the Commodore appointed the Honourable Captain Murray to succeed to the Pearl, and Captain Cheap to the Wager; and he promoted Mr Charles Saunders, his first lieutenant, to the command of the Trial sloop. But Captain Saunders lying dangerously ill of a fever on board the Centurion, and it being the opinion of the surgeons that the removing him on board his own ship in his present condition might tend to the hazard of his life, Mr Anson gave an order to Mr Saumarez, first lieutenant of the Centurion, to act as master and commander of the Trial during the illness of Captain Saunders. Here the Commodore, too, in order to ease the expedition of all unnecessary expense, held a further consultation with his captains about unloading and discharging the Anna pink; but they represented to him that they were so far from being in a condition of taking any part of her loading on board, that they had still great quantities of provisions in the way of their guns between decks, and that their ships were withal so very deep[2] that they were not fit for action without being cleared. This put the Commodore under a necessity of retaining the

[1] So called by Drake. It was the scene of Doughty's trial and execution; "whence," Thomas says, "a small island within the harbour is to this day called the Island of True Justice."

[2] Ed. 1776: "And so lumbered."

pink in the service; and as it was apprehended we should certainly meet with the Spanish squadron in passing the Cape, Mr Anson thought it advisable to give orders to the captains to put all their provisions which were in the way of their guns on board the Anna pink, and to remount such of their guns as had formerly, for the ease of their ships, been ordered into the hold.[1] . . .

We, on our first arrival [at St Julian[2]] sent an officer on shore to a salt pond, in order to procure a quantity of salt for the use of the squadron; Sir John Narborough having observed, when he was here, that the salt produced in that place was very white and good, and that in February there was enough of it to fill a thousand ships. But our officer returned with a sample which was very bad, and he told us that even of this there was but little to be got; I suppose the weather had been more rainy than ordinary, and had destroyed it.[3]

CHAPTER VII.

The Trial being nearly refitted, which was our principal occupation at this Bay of St Julian, and the sole occasion of our stay, the Commodore thought it necessary, as we were now directly bound for the South Seas and the enemy's coasts, to regulate the plan of his future operations. And therefore, on the 24th of February, a signal was made for all captains, and a council of war was held on board the Centurion, at which were present the Honourable Edward Legg, Captain Matthew Mitchel, the Honourable George Murray, Captain David Cheap, together with Colonel Mordaunt Cracherode, commander of the land forces. At this council Mr Anson proposed that their first attempt, after their arrival in the South Seas, should be the attack of the town and harbour of Baldivia, the principal frontier [place] of the district of Chili; Mr Anson informing them, at the same time,[4]

---

[1] Thomas naïvely remarks with reference to their stay in the Bay of St Julian: "Sir John Narborough and some others write that they have often seen and conversed with the inhabitants in this and other parts of Patagonia, and have given wonderful descriptions of them; but as we saw none of them, I have nothing to say of that sort, nor indeed do I think there is anything in this wild part of the world worthy of the least notice."

[2] The district round Port St Julian is described as destitute of wood, Sir John Narborough, in the time of Charles the Second, making the sweeping assertion that he never saw a stick of wood in the country large enough to make the handle of a hatchet. It is, however, good pasture land, feeding immense herds of cattle, of which many thousands are annually slain by the hunters there for the hides and tallow alone. The method of taking them alive is by the *lasso*, in the use of which the native Indians and Spaniards are very dexterous. The plains also abound with wild horses and Peruvian sheep. The lengthy account of the above is here omitted.

[3] Ed. 1776: "Or prevented its fermentation." Thomas adds some particulars of interest with regard to the doings at St Julian: "Having lost the hopes of a supply of water here, we were put to the allowance of one quart a man for one day, and three pints for another, alternately; but, considering our passage had hitherto proved extremely stormy and cold, and a dead time of the year coming on very fast, it was thought proper, in order to keep the people in as good heart as possible, to give them whole allowance of all other provisions, which was ordered accordingly. Here we further secured our lower deck guns, by nailing quoins under the trucks, in case the tackles, breechings, or iron-work, might give way, or fail in the stormy weather which we had much reason to expect."

[4] Ed. 1776: "As an inducement for this enterprise."

that it was an article contained in his Majesty's instructions to him, to endeavour to secure some port in the South Seas where the ships of the squadron might be careened and refitted. To this proposition made by the Commodore, the council unanimously and readily agreed; and in consequence of this resolution new instructions were given to the captains of the squadron, by which, though they were still directed, in case of separation, to make the best of their way to the Island of Nuestra Señora del Socoro, yet (notwithstanding the orders they had formerly given them at St Catherine's) they were to cruise off that island only ten days; from whence, if not joined by the Commodore, they were to proceed and cruise off the harbour of Baldivia, making the land between the Latitudes of 40° and 40° 30′, and taking care to keep to the southward of the port; and if in fourteen days they were not joined by the rest of the squadron, they were then to quit this station, and to direct their course to the Island of Juan Fernandez, after which they were to regulate their further proceedings by their former orders. The same directions were also given to the master of the Anna pink, and he was particularly instructed to be very careful in answering the signals made by any ship of the squadron, and likewise to destroy his papers and orders if he should be so unfortunate as to fall into the hands of the enemy. And as the separation of the squadron might prove of the utmost prejudice to his Majesty's service, each captain was ordered to give it in charge to the respective officers of the watch not to keep their ship at a greater distance from the Centurion than two miles, as they would answer it at their peril; and if any captain should find his ship beyond the distance specified, he was to acquaint the Commodore with the name of the officer who had thus neglected his duty.

These necessary regulations being established, and the Trial sloop completed, the squadron weighed on Friday the 27th of February, at seven in the morning, and stood to sea; the Gloucester indeed found a difficulty in purchasing her anchor, and was left a considerable way astern, so that in the night we fired several guns as a signal to her captain to make sail, but he did not come up to us till the next morning, when we found that they had been obliged to cut their cable and leave their best bower behind them. At ten in the morning, the day after our departure, Wood's Mount, the high land over St Julian, bore from us N. by W., distant ten leagues, and we had fifty-two fathoms of water. And now, standing to the southward, we had great expectation of falling in with Pizarro's squadron; for during our stay at Port St Julian there had generally been hard gales between the WNW. and SW., so that we had reason to conclude the Spaniards had gained no ground upon us in that interval. And it was the prospect of meeting with them that had occasioned our Commodore to be so very solicitous to prevent the separation of our ships; for had we been solely intent on getting round Cape Horn in the shortest time, the properest method for this purpose would have been to have ordered each ship to have made the best of her way to the rendezvous, without waiting for the rest.

From our departure from St Julian to the 4th of March we had little wind, with thick, hazy weather and some rain; and our soundings were generally from forty to fifty fathoms, with a bottom of black and grey sand, sometimes intermixed with pebble stones. On the 4th of March we were in sight of Cape Virgin Mary, and not more than six or seven leagues distant from it. This is the northern cape of the Straits of Magellan; it lies in the Latitude of 52° 21′ S., and Longitude from London 71° 44′ W., and seems to be a low, flat land, ending in a point. Off this cape our depth of water was from thirty-five to forty-eight fathoms. The afternoon of this day was very bright and clear, with small breezes of wind, inclinable to a calm; and most of the captains took the opportunity of this favourable weather to pay a visit to the Commodore; but while they were in company together, they were all greatly alarmed by a sudden flame, which burst out on board the Gloucester, and which was succeeded by a cloud of smoke. . However, they were soon relieved from their apprehensions by receiving information that the blast was occasioned by a spark of fire from the forge, lighting on some gunpowder and other combustibles which an officer on board was preparing for use in case we should fall in with the Spanish fleet; and that it had been extinguished without any damage to the ship.

We here found, what was constantly verified by all our observations in these high [southern] latitudes, that fair weather was always of an exceeding short duration, and that when it was remarkably fine it was a certain presage of a succeeding storm; for the calm and sunshine of our afternoon ended in a most turbulent night, the wind freshening from the SW. as the night came on, and increasing its violence continually till nine in the morning the next day, when it blew so hard that we were obliged to bring to with the squadron, and to continue under a reefed mizzen till eleven at night, having in that time from forty-three to fifty-seven fathoms water, with black sand and gravel; and by an observation we had at noon, we concluded a current had set us twelve miles to the southward of our reckoning. Towards midnight, the wind abating, we made sail again; and steering south, we discovered in the morning for the first time the land called Tierra del Fuego, stretching from the S. by W. to the SE. half E. This indeed afforded us but a very uncomfortable prospect, it appearing of a stupendous height, covered everywhere with snow.[1] We steered along this shore[2] all day, having soundings from forty to fifty fathoms, with stones and gravel. And as we intended to pass through Straits Le Maire next day, we lay to at night that we might not overshoot them, and took this opportunity to prepare ourselves for the tempestuous climate we were soon to be engaged in; with which view we employed ourselves good part of the night in bending an entire new suit of sails to the yards. At four the next morning, being the 7th of March, we made sail, and at eight saw the land; and soon after we began to open the straits, at which time Cape St James bore from us ESE., Cape St Vincent SE. half E., the middlemost of the Three Brothers S. by W., Montegorda S., and Cape St Bartholomew, which is the southernmost point of Staten Land, ESE. Though Tierra del Fuego had an aspect extremely barren and desolate, yet this Island of Staten Land far surpasses it in the wildness and horror of its appearance; it seeming to be entirely composed of inaccessible rocks, without the least mixture of earth or mould between them. These rocks terminate in a vast number of ragged points, which spire up to a prodigious height, and are all of them covered with everlasting snow; the points themselves are on every side surrounded with frightful precipices, and often overhang in a most astonishing manner; and the hills which bear them are generally separated from each other by narrow clefts, which appear as if the country had been rent by earthquakes; for these chasms are nearly perpendicular, and extend through the substance of the main rocks, almost to their very bottoms; so that nothing can be imagined more savage and gloomy than the whole aspect of this coast.

I have above mentioned, that on the 7th of March, in the morning, we opened Straits Le Maire; and soon

[1] "So that the whole," says Thomas, "may not improperly be termed the Land of Desolation; and I much question whether a more dreary aspect is to be seen in any other part of the habitable earth."

[2] Ed. 1776: "This uncouth and rugged coast."

after, or about 10 o'clock, the Pearl and the Trial being ordered to keep ahead of the squadron, we entered them with fair weather and a brisk gale, and were hurried through by the rapidity of the tide in about two hours though they are between seven and eight leagues in length. As these Straits are often considered as the boundary between the Atlantic and Pacific Oceans, and as we presumed we had nothing now before us but an open sea, till we arrived on those opulent coasts where all our hopes and wishes centred, we could not help flattering ourselves that the greatest difficulty of our passage was now at an end, and that our most sanguine dreams were upon the point of being realised; and hence we indulged our imaginations in those romantic schemes which the fancied possession of the Chilian gold and Peruvian silver might be conceived to inspire. These joyous ideas were heightened by the brightness of the sky, and the serenity of the weather, which was indeed most remarkably pleasing; for though the winter was now advancing apace, yet the morning of this day, in its brilliancy and mildness, gave place to none we had seen since our departure from England. Thus animated by these delusions, we traversed these memorable Straits, ignorant of the dreadful calamities that were then impending, and just ready to break upon us; ignorant that the time drew near when the squadron would be separated never to unite again; and that this day of our passage was the last cheerful day that the greatest part of us would ever live to enjoy.

CHAPTER VIII.

WE had scarcely reached the southern extremity of the Straits of Le Maire, when our flattering hopes were instantly lost in the apprehensions of immediate destruction. For before the sternmost ships of the squadron were clear of the Straits, the serenity of the sky was suddenly changed, and gave us all the presages of an impending storm; and immediately the wind shifted to the southward, and blew in such violent squalls, that we were obliged to hand our topsails and reef our mainsail. The tide, too, which had hitherto favoured us, now turned against us,[1] and drove us to the eastward with prodigious rapidity, so that we were in great anxiety for the Wager and the Anna pink, the two sternmost vessels, fearing they would be dashed to pieces against the shore of Staten Land. Nor were our apprehensions without foundation, for it was with the utmost difficulty they escaped. And now the whole squadron, instead of pursuing their intended course to the SW., were driven to the eastward by the united force of the storm and of the currents; so that next day in the morning we found ourselves near seven leagues to the eastward of Staten Land, which then bore from us NW. The violence of the current, which had set us with so much precipitation to the eastward, together with the force and constancy of the westerly winds, soon taught us to consider the doubling of Cape Horn as an enterprise that might prove too mighty for our efforts; though some amongst us had lately treated the difficulties which former voyagers were said to have met with in this undertaking as little better than chimerical, and had supposed them to arise rather from timidity and unskilfulness than from the real embarrassments of the winds and seas. But we were severely convinced that these censures were rash and ill-grounded: for the distresses with which we struggled during the three succeeding months will not easily be paralleled in the relation of any former naval expedition. This will, I doubt not, be readily allowed by those who shall carefully peruse the ensuing narration.

From the storm which came on before we had well got clear of Straits Le Maire, we had a continual succes-

---

[1] Ed. 1776: "Turned furiously adverse."

sion of such tempestuous weather as surprised the oldest and most experienced mariners on board, and obliged them to confess, that what they had hitherto called storms were inconsiderable gales compared with the violence of these winds, which raised such short and at the same time such mountainous waves as greatly surpassed in danger all seas known in any other part of the globe. And it was not without great reason that this unusual appearance filled us with continual terror; for had any one of these waves broke fairly over us, it must in all probability have sent us to the bottom. Nor did we escape with terror only; for the ship, rolling incessantly gunwale-to, gave us such quick and violent motions, that the men were in perpetual danger of being dashed to pieces against the decks or sides of the ship. And though we were extremely careful to secure ourselves from these shocks by grasping some fixed body, yet many of our people were forced from their hold, some of whom were killed, and others greatly injured; in particular, one of our best seamen was canted overboard and drowned, another dislocated his neck, a third was thrown into the mainhold and broke his thigh, and one of our boatswain's mates broke his collar-bone twice; not to mention many other accidents of the same kind.

It was on the 7th of March, as has been already observed, that we passed Straits Le Maire, and were immediately afterwards driven to the eastward by a violent storm and the force of the current which set that way. For the four or five succeeding days we had hard gales of wind from the same quarter, with a most prodigious swell; so that though we stood, during all that time, towards the SW., yet we had no reason to imagine we had made any way to the westward. In this interval we had frequent squalls of rain and snow, and shipped great quantities of water; after which for three or four days, though the seas ran mountains high, yet the weather was rather more moderate. But, on the 18th, we had again strong gales of wind with extreme cold, and at midnight the main-topsail split, and one of the straps of the main dead-eyes broke. From hence to the 23d the weather was more favourable, though often intermixed with rain and sleet, and some hard gales: but as the waves did not subside, the ship, by labouring in this lofty sea, was now grown so loose in her upper works that she let in the water at every seam; so that every part within board was constantly exposed to the sea-water, and scarcely any of the officers ever lay in dry beds. Indeed it was very rare that two nights ever passed without many of them being driven from their beds by the deluge of water that came upon them.

On the 23d we had a most violent storm of wind, hail, and rain, with a very great sea; and though we handed the main-topsail before the height of the squall, yet we found the yard sprung; and soon after, the foot-rope of the mainsail breaking, the mainsail itself split instantly to rags, and in spite of our endeavours to save it, much the greater part of it was blown overboard. On this the Commodore made the signal for the squadron to bring to; and, the storm at length flattening to a calm, we had an opportunity of getting down our main-topsail yard to put the carpenters at work upon it, and of repairing our rigging; after which, having bent a new mainsail, we got under sail again with a moderate breeze. But in less than twenty-four hours we were attacked by another storm still more furious than the former; for it proved a perfect hurricane, and reduced us to the necessity of lying to under our bare poles. As our ship kept the wind better than any of the rest, we were obliged in the afternoon to wear ship, in order to join the squadron to the leeward, which otherwise we should have been in danger of losing in the night; and as we dared not venture any sail abroad, we were obliged to make use of an expedient which answered our purpose; this was putting the helm a-weather, and manning the fore-shrouds.

But though this method proved successful for the end intended, yet in the execution of it one of our ablest seamen was canted overboard; and notwithstanding the prodigious agitation of the waves, we perceived that he swam very strong, and it was with the utmost concern that we found ourselves incapable of assisting him; and we were the more grieved at his unhappy fate, since we lost sight of him struggling with the waves, and conceived from the manner in which he swam that he might continue sensible for a considerable time longer of the horror attending his irretrievable situation.[1]

Before this last-mentioned storm was quite abated, we found two of our main shrouds and one mizzen-shroud broken, all which we knotted and set up immediately; and from hence we had an interval of three or four days less tempestuous than usual, but accompanied with a thick fog, in which we were obliged to fire guns almost every half hour, to keep our squadron together. On the 31st we were alarmed by a gun fired from the Gloucester, and a signal made by her to speak with the Commodore. We immediately bore down to her, and were prepared to hear of some terrible disaster; but we were apprised of it before we joined her, for we saw that her mainyard was broke in the slings. This was a grievous misfortune to us all at this juncture; as it was obvious it would prove an hindrance to our sailing, and would detain us the longer in these inhospitable latitudes. But our future success and safety were not to be promoted by repining, but by resolution and activity; and therefore, that this unlucky incident might delay us as little as possible, the Commodore ordered several carpenters to be put on board the Gloucester from the other ships of the squadron, in order to repair her damage with the utmost expedition. And the captain of the Trial complaining at the same time that his pumps were so bad, and the sloop made so great a quantity of water, that he was scarcely able to keep her free, the Commodore ordered him a pump ready fitted from his own ship. It was very fortunate for the Gloucester and the Trial that the weather proved more favourable this day than for many days both before and after; since by this means they were enabled to receive the assistance which seemed essential to their preservation, and which they could scarcely have had at any other time, as it would have been extremely hazardous to have ventured a boat on board.[2]

The next day, that is, on the 1st of April, the weather returned again to its customary bias, the sky looked dark and gloomy, and the wind began to

[1] With reference to this affecting circumstance, Cowper composed his verses on "The Castaway." One is as follows:

"He long survives, who lives an hour
In ocean, self-upheld:
And so long he, with unspent power,
His destiny repell'd:
And ever, as the minutes flew,
Entreated help, or cried—'Adieu.'"

[2] Under this date, March 31st, Anson's Official Report of his voyage makes the first mention of the scurvy: "Men falling down every day with scorbutic complaints." Thomas also now notes, with some graphic details, the outbreak of the scurvy, which Mr Walter, with a sad want of dramatic instinct, defers to a period of comparatively trivial elemental peril: "And now, as it were to add the finishing stroke to our misfortunes, our people began to be universally afflicted with that most terrible, obstinate, and, at sea, incurable disease, the scurvy, which quickly made a most dreadful havoc among us, beginning at first to carry off two or three a day, but soon increasing, and at last carrying off eight or ten; and as most of the living were very ill of the same distemper, and the little remainder who preserved their healths better, in a manner quite worn out with incessant labour, I have sometimes seen four or five dead bodies, some sewn up in their hammocks, others not, washing about the decks, for want of help to bury them in the sea."

freshen and to blow in squalls; however, it was not yet so boisterous as to prevent our carrying our topsails close reefed; but its appearance was such as plainly prognosticated that a still severer tempest was at hand. And accordingly, on the 3d of April, there came on a storm which both in its violence and continuation (for it lasted three days) exceeded all that we had hitherto encountered. In its first onset, we received a furious shock from the sea which broke upon our larboard quarter, where it stove in the quarter gallery, and rushed into the ship like a deluge; our rigging, too, suffered extremely, for one of the straps of the main dead-eyes was broke, as was also a mainshroud and futtock-shroud, so that to ease the stress upon the masts and shrouds we lowered both our main and fore yards, and furled all our sails, and in this posture we lay to for three days, when, the storm somewhat abating, we ventured to make sail under our courses only. But even this we could not do long, for the next day, which was the 7th, we had another hard gale of wind, with lightning and rain, which obliged us to lie to again all night. It was wonderful that, notwithstanding the hard weather we had endured, no extraordinary accident had happened to any of the squadron since the breaking of the Gloucester's mainyard: but this wonder soon ceased;[1] for at three the next morning several guns were fired to leeward as signals of distress. And the Commodore making a signal for the squadron to bring to, we at daybreak saw the Wager a considerable way to leeward of any of the other ships; and we soon perceived that she had lost her mizzenmast and maintopsail yard. We immediately bore down to her, and found this disaster had arisen from the badness of her ironwork; for all the chain-plates to windward had given way upon the ship's fetching a deep roll. This proved the more unfortunate to the Wager, as her carpenter had been on board the Gloucester ever since the 31st of March, and the weather was now too severe to permit him to return. Nor was the Wager the only ship of the squadron that had suffered in the late tempest; for the next day a signal of distress was made by the Anna pink, and, upon speaking with the master, we learned that they had broken their forestay and the gammon of the bowsprit, and were in no small danger of having all the masts come by the board; so that we were obliged to bear away until they had made all fast, after which we hauled upon a wind again. . . .

By the latitude of the land we [next] fell in with, it was agreed to be a part of Tierra del Fuego, near the southern outlet described in Frazier's chart of the Straits of Magellan, and was supposed to be that point called by him Cape Noir. It was indeed most wonderful that the currents should have driven us to the eastward with such strength; for the whole squadron esteemed themselves upwards of ten degrees more westerly than this land, so that in running down, by our account, about nineteen degrees of longitude, we had not really advanced above half that distance. And now, instead of having our labours and anxieties relieved by approaching a warmer climate and more tranquil seas, we were [forced] to steer again to the southward, and again to combat those western blasts which had so often terrified us; and this, too, when we were weakened by our men falling sick and dying apace, and when our spirits, dejected by a long continuance at sea, and by our late disappointment, were much less capable of supporting us in the various difficulties which we could not but expect in this new [and arduous] undertaking. Add to all this, too, the discouragement we received by the diminution of the strength of the squadron; for three days before this we lost sight of the Severn and the Pearl in the morning; and though we spread our ships, and beat about for some time, yet we never saw them more; whence we had apprehensions that they too might have fallen in with this land in the night, and, being

[1] Ed. 1776: "This good fortune now no longer attended us."

less favoured by the wind and the moon than we were, might have run on shore and have perished. Full of these dejected thoughts and gloomy presages, we stood away to the SW., prepared by our late disaster to suspect, that how large soever an allowance we made in our westing for the drift of the eastern current, we might still upon a second trial perhaps find it insufficient.

CHAPTER IX.[1]

The improper season of the year in which we attempted to double Cape Horn, and to which is to be imputed the disappointment recited in the foregoing Chapter in falling in with Tierra del Fuego, when we reckoned ourselves at least a hundred leagues to the westward of that whole coast, and consequently well advanced into the Pacific Ocean; this unseasonable navigation, I say, to which we were necessitated by our too late departure from England, was the fatal source of all the misfortunes we afterwards encountered. For from hence proceeded the separation of our ships, the destruction of our people, the ruin of our project on Baldivia and of all our other views on the Spanish places, and the reduction of our squadron from the formidable condition in which it passed Straits Le Maire to a couple of shattered, half-manned cruisers, and a sloop, so far disabled that in many climates they scarcely durst have put to sea.

[1] This Chapter, of twenty pages in the original, is almost entirely devoted to "Observations and Directions for facilitating the Passage of our Future Cruisers round Cape Horn." But as its matter is purely technical, and, however curious as casting light on the state of nautical science a century and a quarter ago, possesses not the smallest popular interest now-a-days, the Chapter is omitted, with the exception of one or two introductory sentences which bear on the actual narrative.

CHAPTER X.

After the mortifying disappointment of falling in with the coast of Tierra del Fuego, when we esteemed ourselves ten degrees to the westward of it; after this disappointment, I say, recited in the eighth Chapter, we stood away to the SW. till the 22d of April, when we were in upwards of 60° S., and by our account near six degrees to the westward of Cape Noir. And in this run we had a series of as favourable weather as could well be expected in that part of the world, even in a better season; so that this interval, setting the inquietude of our thoughts aside, was by far the most eligible of any we enjoyed from Straits Le Maire to the west coast of America. This moderate weather continued with little variation till the 24th; but on the 24th in the evening the wind began to blow fresh, and soon increased to a prodigious storm; and the weather being extremely thick, about midnight we lost sight of the other ships of the squadron, which, notwithstanding the violence of the preceding storms, had hitherto kept in company with us. Nor was this our sole misfortune; for the next morning, endeavouring to hand the topsails, the clewlines and buntlines broke, and, the sheets being half-flown, every seam in the topsails was soon split from top to bottom, and the main-topsail shook so strongly in the wind, that it carried away the top lantern, and endangered the head of the mast. However, at length some of the most daring of our men ventured upon the yard, and cut the sail away close to the reefs, though with the utmost hazard of their lives. At the same time, the foretopsail beat about the yard with so much fury, that it was soon blown to pieces; and that we might have full employment, the mainsail blew loose, which obliged us to lower down the yard to secure the sail; and the fore-yard being likewise lowered, we lay to under a mizzen. And besides the loss of our topsails, we had much of our other rigging broke, and lost a main studding-sail boom out of the chains.

On the 25th, about noon, the weather became more moderate, which enabled us to sway up our yards, and to repair, in the best manner we could, our shattered rigging; but still we had no sight of the rest of our squadron, nor indeed were we joined by any of them again till after our arrival at Juan Fernandez, nor did any two of them, as we have since learned, continue in company together. And this total separation was the more wonderful, as we had hitherto kept together for seven weeks, through all the reiterated tempests of this turbulent climate. It must indeed be owned that this separation gave us room to expect that we might make our passage in a shorter time than if we had continued together, because we could now make the best of our way without being retarded by the misfortunes of the other ships; but then we had the melancholy reflection that we ourselves were hereby deprived of the assistance of others, and our safety would depend upon our single ship. So that, if a plank started, or any other accident of the same nature should take place, we must all irrecoverably perish; or, should we be driven on shore, we had the uncomfortable prospect of ending our days on some desolate coast, without any reasonable hope of ever getting away; whereas, with another ship in company, all these calamities are much less formidable, since in every kind of danger there would be some probability that one ship at least might escape, and might be capable of preserving or relieving the crew of the other.

The remaining part of this month of April we had generally hard gales, although we had been every day since the 22d edging to the northward; however, on the last day of the month we flattered ourselves with the hopes of soon terminating all our sufferings, for we that day found ourselves in the Latitude of 52° 13′, which, being to the northward of the Straits of Magellan, we were assured that we had completed our passage, and had arrived in the confines of the Southern Ocean; and this Ocean being nominated Pacific, from the equability of the seasons which are said to prevail there, and the facility and security with which navigation is there carried on, we doubted not but we should be speedily cheered with the moderate gales, the smooth water, and the temperate air, for which that tract of the globe has been so renowned. And under the influence of these pleasing circumstances we hoped to experience some kind of compensation for the complicated miseries which had so constantly attended us for the last eight weeks. But here we were again disappointed; for in the succeeding month of May our sufferings rose to a much higher pitch than they had ever yet done, whether we consider the violence of the storms, the shattering of our sails and rigging, or the diminishing and weakening of our crew by deaths and sickness, and the probable prospect of our total destruction. All this will be sufficiently evident from the following circumstantial account of our diversified misfortunes.

Soon after our passing Straits Le Maire, the scurvy began to make its appearance amongst us; and our long continuance at sea, the fatigue we underwent, and the various disappointments we met with, had occasioned its spreading to such a degree, that at the latter end of April there were but few on board who were not in some degree afflicted with it; and in that month no less than forty-three died of it on board the Centurion. But though we thought that the distemper had then risen to an extraordinary height, and were willing to hope that as we advanced to the northward its malignity would abate; yet we found, on the contrary, that in the month of May we lost nearly double that number. And as we did not get to land till the middle of June, the mortality went on increasing, and the disease extended itself so prodigiously, that after the loss of above 200 men we could not at last muster more than six foremast men in a watch capable of duty.

This disease, so frequently attending all long voyages, and so particularly destructive to us, is surely the most singular and unaccountable of any that affects the human body. For its symptoms are inconstant and innumerable, and its progress and effects extremely irregular; for scarcely any two persons have the same complaints, and where there has been found some conformity in the symptoms the order of their appearance has been totally different. However, though it frequently puts on the form of many other diseases, and is therefore not to be described by any exclusive and infallible criterions; yet there are some symptoms which are more general than the rest, and, therefore, occurring the oftenest, deserve a more particular enumeration. These common appearances are large discoloured spots dispersed over the whole surface of the body, swelled legs, putrid gums, and above all, an extraordinary lassitude of the whole body, especially after any exercise however inconsiderable; and this lassitude at last degenerates into a proneness to swoon [1] on the least exertion of strength, or even on the least motion. This disease is likewise usually attended with a strange dejection of the spirits, and with shiverings, tremblings, and a disposition to be seized with the most dreadful terrors on the slightest accident. Indeed it was most remarkable, in all our reiterated experience of this malady, that whatever discouraged our people, or at any time damped their hopes, never failed to add new vigour to the distemper; for it usually killed those who were in the last stage of it, and confined those to their hammocks who were before capable of some kind of duty; so that it seemed as if alacrity of mind and sanguine thoughts were no contemptible preservatives from its fatal malignity.

But it is not easy to complete the long roll of the various concomitants of this disease; for it often produced putrid fevers, pleurisies, the jaundice, and violent rheumatic pains, and sometimes it occasioned an obstinate costiveness, which was generally attended with a difficulty of breathing; and this was esteemed the most deadly of all the scorbutic symptoms. At other times the whole body, but more especially the legs, were subject to ulcers of the worst kind, attended with rotten bones, and such a luxuriance of fungus flesh, as yielded to no remedy. But a most extraordinary circumstance, and what would be scarcely credible upon any single evidence, is, that the scars of wounds which had been for many years healed were forced open again by this virulent distemper. Of this there was a remarkable instance in one of the invalids on board the Centurion, who had been wounded above fifty years before at the battle of the Boyne; for though he was cured soon after, and had continued well for a great number of years past, yet, on his being attacked by the scurvy, his wounds, in the progress of his disease, broke out afresh, and appeared as if they had never been healed. Nay, what is still more astonishing, the callus of a broken bone, which had been completely formed for a long time, was found to be hereby dissolved, and the fracture seemed as if it had never been consolidated. Indeed, the effects of this disease were in almost every instance wonderful; for many of our people, though confined to their hammocks, appeared to have no inconsiderable share of health, for they ate and drank heartily, were cheerful, and talked with much seeming vigour and with a loud, strong tone of voice; and yet on their being the least moved, though it was only from one part of the ship to the other, and that in their hammocks, they have immediately expired; and others who have confided in their seeming strength, and have resolved to get out of their hammocks, have died before they could well reach the deck; and it was no uncommon thing for those who were able to walk the deck, and

[1] Ed. 1776: "And even to die."

to do some kind of duty, to drop down dead in an instant, on any endeavours to act with their utmost vigour, many of our people having perished in this manner during the course of this voyage.

With this terrible disease we struggled the greatest part of the time of our beating round Cape Horn; and though it did not then rage with its utmost violence, yet we buried no less than forty-three men on board the Centurion in the month of April, as has been already observed. We still entertained hopes, that when we should have once secured our passage round the Cape, we should put a period to this and all the other evils which had so constantly pursued us. But it was our misfortune to find, that the Pacific Ocean was to us less hospitable than the turbulent neighbourhood of Tierra del Fuego and Cape Horn;[1] for being arrived, on the 8th of May, off the Island of Socoro, which was the first rendezvous appointed for the squadron, and where we hoped to have met with some of our companions, we cruised for them in that station several days. And here we were not only disappointed in our hopes of being joined by our friends, and thereby induced to favour the gloomy suggestions of their having all perished; but we were likewise perpetually alarmed with the fears of being driven on shore upon this coast, which appeared too craggy and irregular to give us the least hopes that in such a case any of us could possibly escape immediate destruction. For the land had indeed a most tremendous aspect; the most distant part of it, and which appeared far within the country, being the mountains usually called the Andes or Cordilleras, was extremely high, and covered with snow; and the coast itself seemed quite rocky and barren, and the water's edge skirted with precipices. In some places, indeed, there appeared several deep bays running into the land, but the entrance into them was generally blocked up by numbers of little islands; and though it was not improbable but there might be convenient shelter in some of those bays, and proper channels leading thereto; yet, as we were utterly ignorant of the coast, had we been driven ashore by the western winds which blew almost constantly there, we did not expect to have avoided the loss of our ship and of our lives.

And this continued peril, which

---

[1] Thomas dwells far more impressively on this disappointment: "Friday, May 8, at seven in the morning, saw the main land of Patagonia appearing in high mountains covered mostly with snow. We likewise saw several islands, one of which we took to be the Island del Soccoro, so called by Sir John Narborough, in his account of his voyage into those parts; and from the fine description this gentleman had given of this island (having been there in the very height of summer), this place was appointed for our first general rendezvous in the South Seas. An unhappy appointment it was in its consequences; for when the people, already reduced to the last extremity, found this to be the place of rendezvous, where they had hoped to meet the rest of their companions with joy, and what a miserable part of the world it appeared to be, their grief gave way to despair; they saw no end of their sufferings, nor any door open to their safety. Those who had hitherto been well and in heart, now full of despondency, fell down, sickened, and died; and, to sum up this melancholy part, I verily believe, that our touching on this coast, the long stay we made here, and our hindrance by cross winds, which we should have avoided in a direct course to Juan Fernandez, lost us at least sixty or seventy of as stout and able men as any in the navy. This unspeakable distress was still aggravated by the difficulties we found in working the ship, as the scurvy had by this time destroyed no less than 200 of our men, and had in some degree affected almost the whole crew."

lasted for above a fortnight, was greatly aggravated by the difficulties we found in working the ship; as the scurvy had by this time destroyed so great a part of our hands, and had in some degree affected almost the whole crew.[1] Nor did we, as we hoped, find the winds less violent as we advanced to the northward; for we had often prodigious squalls, which split our sails, greatly damaged our rigging, and endangered our masts. Indeed, during the greatest part of the time we were upon this coast, the wind blew so hard, that in another situation where we had sufficient sea-room we should certainly have lain to; but in the present exigency we were necessitated to carry both our courses and topsails, in order to keep clear of this lee-shore. In one of these squalls, which was attended by several violent claps of thunder, a sudden flash of fire darted along our decks, which, dividing, exploded with a report like that of several pistols, and wounded many of our men and officers as it passed, marking them in different parts of the body. This flame was attended with a strong sulphurous stench, and was doubtless of the same nature with the larger and more violent blasts of lightning which then filled the air.

It were endless to recite minutely the various disasters, fatigues, and terrors which we encountered on this coast; all these went on increasing till the 22d of May, at which time the fury of all the storms which we had hitherto encountered seemed to be combined, and to have conspired our destruction. In this hurricane almost all our sails were split, and great part of our standing rigging broken; and, about eight in the evening, a mountainous overgrown sea took us upon our starboard quarter, and gave us so prodigious a shock, that several of our shrouds broke with the jerk, by which our masts were greatly endangered; our ballast and stores, too, were so strangely shifted, that the ship heeled afterwards two streaks to port. Indeed, it was a most tremendous blow, and we were thrown into the utmost consternation from the apprehension of instantly foundering; and though the wind abated in a few hours, yet, as we had no more sails left in a condition to bend to our yards, the ship laboured very much in a hollow sea, rolling gunwale-to, for want of sail to steady her: so that we expected our masts, which were now very slenderly supported, to come by the board every moment. However, we exerted ourselves the best we could to stirrup our shrouds, to reeve new halyards, and to mend our sails; but while these necessary operations were carrying on, we ran great risk of being driven on shore on the Island of Chiloe, which was not far distant from us; but in the midst of our peril the wind happily shifted to the southward, and we steered off the land with the mainsail only, the master and myself undertaking the management of the helm, while every one else on board was busied in securing the masts, and bending the sails as fast as they could be repaired. This was the last effort of that stormy climate; for in a day or two after we got clear of the land, and found the weather more moderate than we had yet experienced since our passing Straits Le Maire. And now having cruised in vain for more than a fortnight in quest of the other ships of the squadron, it was resolved to take advantage of the present favourable season, and the offing we had made from this terrible coast, and to make the best of our way for the Island of Juan Fernandez. For though our next rendezvous was appointed off the harbour of Baldivia, yet as we had hitherto seen none of our companions at this first rendezvous, it was not to be supposed that any of them would be found at the second; indeed, we had the greatest reason to suspect that all but ourselves had perished. Besides, we

[1] Anson himself writes in his Official Report, under date May 8th, that he "had not men able to keep the deck sufficient to take in a topsail, all being violently afflicted with the scurvy, and every day lessening our number by six, eight, or ten."

were by this time reduced to so low a condition, that, instead of attempting to attack the places of the enemy, our utmost hopes could only suggest to us the possibility of saving the ship, and some part of the remaining enfeebled crew, by our speedy arrival at Juan Fernandez; for this was the only road in that part of the world where there was any probability of our recovering our sick, or refitting our vessel, and consequently our getting thither was the only chance we had left to avoid perishing at sea.

Our deplorable situation, then, allowing no room for deliberation, we stood for the Island of Juan Fernandez: and to save time, which was now extremely precious (our men dying four, five, and six in a day), and likewise to avoid being engaged again with a lee-shore, we resolved if possible to hit the island upon a meridian. And on the 28th of May, being nearly in the parallel upon which it is laid down, we had great expectations of seeing it; but not finding it in the position in which the charts had taught us to expect it, we began to fear that we had got too far to the westward; and therefore, though the Commodore himself was strongly persuaded that he saw it on the morning of the 28th, yet his officers believing it to be only a cloud, to which opinion the haziness of the weather gave some kind of countenance, it was on a consultation resolved to stand to the eastward in the parallel of the island; as it was certain that by this course we should either fall in with the island, if we were already to the westward of it, or should at least make the mainland of Chili, whence we might take a new departure, and assure ourselves, by running to the westward afterwards, of not missing the island a second time.

On the 30th of May we had a view of the continent of Chili, distant about twelve or thirteen leagues; the land made exceeding high and uneven, and appeared quite white; what we saw being doubtless a part of the Cordilleras, which are always covered with snow. Though by this view of the land we ascertained our position, yet it gave us great uneasiness to find that we had so needlessly altered our course when we were, in all probability, just upon the point of making the island; for the mortality amongst us was now increased to a most dreadful degree, and those who remained alive were utterly dispirited by this new disappointment and the prospect of their longer continuance at sea. Our water, too, began to grow scarce, so that a general dejection prevailed amongst us, which added much to the virulence of the disease, and destroyed numbers of our best men; and to all these calamities there was added this vexatious circumstance, that when, after having got a sight of the main, we tacked and stood to the westward in quest of the island, we were so much delayed by calms and contrary winds, that it cost us nine days to regain the westing which, when we stood to the eastward, we ran down in two. In this desponding condition, with a crazy ship, a great scarcity of fresh water, and a crew so universally diseased that there were not above ten foremast men in a watch capable of doing duty, and even some of these lame, and unable to go aloft: under these disheartening circumstances, I say, we stood to the westward; and, on the 9th of June, at daybreak, we at last discovered the long-wished-for Island of Juan Fernandez. And with this discovery I shall close this Chapter, and the First Book, after observing (which will furnish a very strong image of our unparalleled distresses) that by our suspecting ourselves to be to the westward of the island on the 28th of May, and, in consequence of this, standing in for the main, we lost between seventy and eighty of our men, whom we should doubtless have saved had we made the island that day, which, had we kept on our course for a few hours longer, we could not have failed to have done.

BOOK II.

CHAPTER I.

On the 9th of June, at daybreak, as is mentioned in the preceding Chapter, we first descried the Island of Juan Fernandez, bearing N. by E. half E., at eleven or twelve leagues' distance. And though, on this view, it appeared to be a mountainous place, extremely ragged and irregular; yet, as it was land, and the land we sought for, it was to us a most agreeable sight. For at this place only we could hope to put a period to those terrible calamities we had so long struggled with, which had already swept away above half our crew, and which, had we continued a few days longer at sea, would inevitably have completed our destruction. For we were by this time reduced to so helpless a condition, that out of 200 and odd men who remained alive, we could not, taking all our watches together, muster hands enough to work the ship on an emergency, though we included the officers, their servants, and the boys.

The wind being northerly when we first made the island, we kept plying all that day, and the next night, in order to get in with the land; and wearing the ship in the middle watch, we had a melancholy instance of the almost incredible debility of our people; for the lieutenant could muster no more than two quarter-masters and six foremast men capable of working; so that without the assistance of the officers, servants, and boys, it might have proved impossible for us to have reached the island after we had got sight of it; and even with this assistance they were two hours in trimming the sails. To so wretched a condition was a sixty-gun ship reduced, which had passed Straits Le Maire but three months before, with between 400 and 500 men, almost all of them in health and vigour.

However, on the 10th, in the afternoon, we got under the lee of the island, and kept ranging along it at about two miles' distance, in order to look out for the proper anchorage, which was described to be in a bay on the north side. And now, being nearer in with the shore, we could discover that the broken craggy precipices, which had appeared so unpromising at a distance, were far from barren, being in most places covered with woods; and that between them there were everywhere interspersed the finest valleys, clothed with a most beautiful verdure, and watered with numerous streams and cascades; no valley, of any extent, being unprovided of its proper rill. The water, too, as we afterwards found, was not inferior to any we had ever tasted, and was constantly clear; so that the aspect of this country would at all times have been extremely delightful, but in our distressed situation, languishing as we were for the land and its vegetable productions (an inclination constantly attending every stage of the sea-scurvy), it is scarcely credible with what eagerness and transport we viewed the shore, and with how much impatience we longed for the greens and other refreshments which were then in sight; and particularly for the water, for of this we had been confined to a very sparing allowance for a considerable time, and had then but five tons remaining on board. Those only who have endured a long series of thirst, and who can readily recall the desire and agitation which the ideas alone of springs and brooks have at that time raised in them, can judge of the emotion with which we eyed a large cascade of the most transparent water, which poured itself from a rock near 100 feet high into the sea, at a small distance from the ship. Even those amongst the diseased, who were not in the very last stages of the distemper, though they had long been confined to their hammocks, exerted the small remains of strength that were left them, and crawled up to the deck to feast themselves with this reviving prospect. Thus we coasted the shore, fully employed in the contemplation of this diversified landscape, which still improved upon us the farther we advanced. But at last the night closed upon us before we had satisfied ourselves which was the proper bay to anchor in; and therefore we resolved to keep in soundings all night (we having then from sixty-four to seventy fathoms), and to send our boat next morning to discover the road. However, the current shifted in the night, and set us so near the land, that we were obliged to let go the best bower in fifty-six fathoms, not half-a-mile from the shore. At four in the morning the cutter was despatched with our third lieutenant to find out the bay we were in search of, who returned again at noon with the boat laden with seals and grass; for though the island abounded with better vegetables, yet the boat's crew, in their short stay, had not met with them; and they well knew that even grass would prove a dainty, and, indeed, it was all soon and eagerly devoured. The seals, too, were considered as fresh provision; but as yet were not much admired, though they grew afterwards into more repute; for what rendered them less valuable at this juncture was the prodigious quantity of excellent fish which the people on board had taken during the absence of the boat.

The cutter, in this expedition, had discovered the bay where we intended to anchor, which we found was to the westward of our present station; and the next morning, the weather proving favourable, we endeavoured to weigh, in order to proceed thither. But though, on this occasion, we mustered all the strength we could, obliging even the sick, who were scarce able to keep on their legs, to assist us, yet the capstan was so weakly manned, that it was near four hours before we hove the cable right up and down; after which, with our utmost efforts, and with many surges and some purchases we made use of to increase our power, we found ourselves incapable of starting the anchor from the ground. However, at noon, as a fresh gale blew towards the bay, we were induced to set the sails, which fortunately tripped the anchor; on which we steered along shore till we came abreast of the point that forms the eastern part of the bay. On opening the bay, the wind, that had befriended us thus far, shifted, and blew from thence in squalls; but by means of the headway we had got, we luffed close in, till the anchor brought us up in fifty-six fathoms.[1] Soon after we had thus got to our new berth, we discovered a sail, which we made no doubt was one of our squadron; and on its nearer approach, we found it to be the Trial sloop. We immediately sent some of our hands on board her, by whose assistance she was brought to an anchor between us and the land. We soon found that the sloop had not been exempted from those calamities which we had so severely felt; for her commander, Captain Saunders, waiting on the Commodore, informed him, that out of his small complement he had buried thirty-four of his men; and those that remained were so universally afflicted with the scurvy, that only himself, his lieutenant, and three of his men, were able to stand by the sails.

The Trial came to an anchor within us on the 12th about noon, and we carried our hawsers on board her, in order to moor ourselves nearer in shore; but the wind, coming off the land in violent gusts, prevented our mooring in the berth we intended, especially as our principal attention was now employed on business rather of more importance. For we were

[1] "To our inexpressible joy," says Thomas, "having been from St Catherine's, in the Brazils, to this place 148 days, on such a dreadful and fatal a passage, as, I believe, very few persons ever experienced."

now extremely occupied in sending on shore materials to raise tents for the reception of the sick, who died apace on board; and doubtless the distemper was considerably augmented by the stench and filthiness in which they lay; for the number of the diseased was so great, and so few could be spared from the necessary duty of the sails to look after them, that it was impossible to avoid a great relaxation in the article of cleanliness, which had rendered the ship extremely loathsome between decks. But notwithstanding our desire of freeing the sick from their hateful situation, and their own extreme impatience to get on shore, we had not hands enough to prepare the tents for their reception before the 16th; but on that and the two following days we sent them all on shore, amounting to 167 persons, besides at least a dozen who died in the boats on their being exposed to the fresh air. The greatest part of our sick were so infirm, that we were obliged to carry them out of the ship in their hammocks, and to convey them afterwards in the same manner from the water-side to their tents, over a stony beach. This was a work of considerable fatigue to the few who were healthy; and therefore the Commodore, with his accustomed humanity, not only assisted herein with his own labour, but obliged his officers, without distinction, to give their helping hand. The extreme weakness of our sick may in some measure be collected from the numbers who died after they had got on shore; for it had generally been found that the land, and the refreshments it produces, very soon recover most stages of the sea-scurvy; and we flattered ourselves that those who had not perished on this first exposure to the open air, but had lived to be placed in their tents, would have been speedily restored to their health and vigour. But, to our great mortification, it was near twenty days after their landing before the mortality was tolerably ceased; and for the first ten or twelve days we buried rarely less than six each day, and many of those who survived recovered by very slow and insensible degrees. Indeed, those who were well enough, at their first getting on shore, to creep out of their tents and crawl about, were soon relieved, and recovered their health and strength in a very short time; but in the rest the disease seemed to have acquired a degree of inveteracy which was altogether without example.[1]

The excellence of the climate and the looseness of the soil render this place extremely proper for all kinds of vegetation; for if the ground be anywhere accidentally turned up it is immediately overgrown with turnips and Sicilian radishes; and therefore Mr Anson having with him garden seeds of all kinds, and stones of different sorts of fruits, he, for the better accommodation of his countrymen who should hereafter touch here, sowed both lettuces, carrots, and other garden plants, and set in the woods a great variety of plum, apricot, and peach stones. And these last, he has been informed, have since thriven to a very remarkable degree; for some gentlemen, who in their passage from Lima to Old Spain were taken and brought to England, having procured leave to wait upon Mr Anson to thank him for his generosity and humanity to his prisoners, some of whom were their relations, they in casual discourse with him about his transactions in the South Seas, particularly asked him if he had not planted a great number of fruit-stones on the Island of Juan Fernandez; for they told him their late navigators had discovered there numbers of peach trees and apricot trees, which being fruits before unobserved in that place,

[1] The Narrator here goes into a long and minute description of Juan Fernandez, for the advantage of future British cruisers in those seas; but the island has been described in Dampier's Voyage (page 158), and the Editor has omitted those parts of Mr Walter's account which do not bear on the actual proceedings of the squadron.

they concluded them to be produced from kernels set by him.

The spot where the Commodore pitched his tent, and which he made choice of for his own residence, was a small lawn that lay on a little ascent, at the distance of about half-a-mile from the sea.[1] In the front of his tent there was a large avenue cut through the woods to the seaside, which sloping to the water, with a gentle descent, opened a prospect of the bay and the ships at anchor. This lawn was screened behind by a tall wood of myrtle sweeping round it in the form of a theatre, the ground on which the wood stood rising with a much sharper ascent than the lawn itself, though not so much but that the hills and precipices within land towered up considerably above the tops of the trees, and added to the grandeur of the view. There were, besides, two streams of crystal water which ran on the right and left of the tent, within 100 yards' distance, and were shaded by the trees which skirted the lawn on either side, and completed the symmetry of the whole.

It remains now only that we speak of the animals and provisions which we met with at this place. Former writers have related that this island abounded with vast numbers of goats; and their accounts are not to be questioned, this place being the usual haunt of the buccaneers and privateers who formerly frequented those seas. And there are two instances—one of a Mosquito Indian, and the other of Alexander Selkirk, a Scotchman, who were left by their respective ships, and lived alone upon this island for some years, and consequently were no strangers to its produce. Selkirk, who was the last, after a stay of between four and five years, was taken off the place by the Duke and Duchess privateers, of Bristol, as may be seen at large in the journal of their voyage. His manner of life during his solitude was in most particulars very remarkable; but there is one circumstance he relates which was so strangely verified by our own observation, that I cannot help reciting it. He tells us, among other things, as he often caught more goats than he wanted, he sometimes marked their ears and let them go. This was about thirty-two years before our arrival at the island. Now it happened that the first goat that was killed by our people at their landing had his ears slit; whence we concluded that he had doubtless been formerly under the power of Selkirk. This was indeed an animal of a most venerable aspect, dignified with an exceeding majestic beard, and with many other symptoms of antiquity. During our stay on the island we met with others marked in the same manner, all the males being distinguished by an exuberance of beard, and every other characteristic of extreme age.

I remember we had once an opportunity of observing a remarkable dispute betwixt a herd of these animals and a number of dogs, for going in our boat into the eastern bay, we saw some dogs running very eagerly upon the foot, and being willing to discover what game they were after, we lay upon our oars some time to view them; and at last we saw them take to a hill, and looking a little farther we observed upon the ridge of it a herd of goats which seemed drawn up for their reception; there was a very narrow path, skirted on each side by precipices, on which the master of the herd posted himself fronting the enemy, the rest of the goats being all behind him, where the ground was more open. As this spot was inaccessible

[1] And was probably, as Thomas suggests, the very spot on which Shelvocke pitched his tent after his shipwreck on the island in May 1720. Shelvocke, as quoted in Kerr's Collection of Voyages, Part I., Book IV., chap. 12, sec. 22, says: "I now took some pains to find out a convenient place in which to set up my tent, and at length found a commodious spot of ground not half-a-mile from the sea, having a fine stream of water on each side, with trees close at hand for firing, and building our huts."

by any other path excepting where this champion had placed himself, the dogs, though they ran up-hill with great alacrity, yet when they came within about twenty yards of him durst not encounter him (for he would infallibly have driven them down the precipice), but gave over the chase, and quietly laid themselves down, panting at a great rate. The dogs, which, as I have mentioned, are masters of all the accessible parts of the island, are of various kinds, but some of them very large, and are multiplied to a prodigious degree. They sometimes came down to our habitations at night and stole our provision, and once or twice they set upon single persons, but assistance being at hand, they were driven off without doing any mischief. As at present it is rare for goats to fall in their way, we conceived that they lived principally upon young seals; and indeed some of our people had the curiosity to kill dogs sometimes and dress them, and they seemed to agree that they had a fishy taste.

Goats' flesh, as I have mentioned, being scarce, we rarely being able to kill above one a day, and our people growing tired of fish (which abounds at this place), they at last condescended to eat seals, which by degrees they came to relish, and called it lamb. The seal, numbers of which haunt this island, has been so often described by former writers that it is unnecessary to say anything particular about them in this place. But there is another amphibious creature to be met with here, called a sea-lion, that bears some resemblance to a seal, though it is much larger. This, too, we ate, under the denomination of beef. They are in size, when arrived at their full growth, from twelve to twenty feet in length, and from eight to fifteen in circumference; they are extremely fat, so that after having cut through the skin, which is about an inch in thickness, there is at least a foot of fat before you can come at either lean or bones; and we experienced more than once that the fat of some of the largest afforded us a butt of oil. They are likewise very full of blood, for if they are deeply wounded in a dozen places, there will instantly gush out as many fountains of blood, spouting to a considerable distance; and to try what quantity of blood they contained, we shot one first, and then cut its throat; and measuring the blood that came from him, we found that, besides what remained in the vessels—which, to be sure, was considerable—we got at least two hogsheads. . . . We killed many of them for food, particularly for their hearts and tongues, which we esteemed exceeding good eating, and preferable even to those of bullocks. And in general there was no difficulty in killing them, for they were incapable either of escaping or resisting, their motion being the most unwieldy that can be conceived, their blubber, all the time they are moving, being agitated in large waves under their skins. However, a sailor one day being carelessly employed in skinning a young sea-lion, the female from which he had taken it came upon him unperceived, and getting his head in her mouth, she with her teeth scored his skull in notches in many places, and thereby wounded him so desperately that, though all possible care was taken of him, he died in a few days.

But that which furnished us with the most delicious repasts at this island remains still to be described. This was the fish with which the whole bay was most plentifully stored, and with the greatest variety. For we found here cod of a prodigious size; and by the report of some of our crew, who had been formerly employed in the Newfoundland fishery, not in less plenty than is to be met with on the banks of that island. We caught also cavillies, gropers, large breams, maids, silver-fish, congers of a peculiar kind, and above all, a black fish which we most esteemed, called by some a chimney-sweeper, in shape resembling a carp. Indeed the beach is everywhere so full of rocks and loose stones that there is no possibility of hauling the seyne; but with hooks and lines we caught what numbers we pleased, so that a boat with two or three lines would return loaded with fish in about two or three hours' time. The only interruption we ever met with arose from great quantities of dog-fish and large sharks, which sometimes attended our boats and prevented our sport. Besides the fish we have already mentioned, we found here one delicacy in greater perfection, both as to size, flavour, and quantity, than is perhaps to be met with in any other part of the world. This was sea craw-fish; they generally weighed eight or nine pounds a-piece, were of an excellent taste, and lay in such abundance near the water's edge that the boat-hooks often struck into them in putting the boat to and from the shore.

These are the most material articles relating to the accommodations, soil, vegetables, animals, and other productions of the Island of Juan Fernandez.[1] By which it must appear how properly that place was adapted for recovering us from the deplorable situation to which our tedious and unfortunate navigation round Cape Horn had reduced us. And having thus given the reader some idea of the site and circumstances of this place, which was to be our residence for three months, I shall now proceed in the next Chapter to relate all that occurred to us in that interval, resuming my narration from the 18th day of June, being the day on which the Trial sloop, having by a squall been driven out to sea three days before, came again to her moorings, the day in which we finished the sending our sick on shore, and about eight days after our first anchoring at this island.

CHAPTER II.

The arrival of the Trial sloop at this island, so soon after we came there ourselves [in the Centurion], gave us great hopes of being speedily joined by the rest of the squadron; and we were for some days continually looking out, in expectation of their coming in sight. But near a fortnight being elapsed without any of them having appeared, we began to despair of ever meeting them again; as we knew that, had our ship continued so much longer at sea, we should every man of us have perished, and the vessel, occupied by dead bodies only, would have been left to the caprice of the winds and waves: and this we had great reason to fear was the fate of our consorts, as each hour added to the probability of these desponding suggestions.

But, on the 21st of June, some of our people, from an eminence on shore, discerned a ship to leeward,

---

[1] Thomas adds, in somewhat enthusiastic terms, another and a pleasant feature of the island: "It is astonishing, that among all the voyagers who have visited this fortunate island before us, and who have obliged the world with descriptions of it, none of them have mentioned a charming little bird that, with its wild, various, and irregular notes, enchants the ear, and makes the woods resound with its melody. This untutored chorister is somewhat less in size than the goldfinch, its plumage beautifully intermixed with red and other vivid colours, and the golden crown upon its head so bright and glowing when seen in the full light of the sun that it surpasses all description. These little birds are far from being uncommon or unfamiliar, for they perched upon the branches of the myrtle-trees so near us, and sung so cheerfully, as if they had been conscious we were strangers, and came to give us welcome. There is, besides the above, another little bird unnoticed by any former writer, and which seems likewise peculiar to the island, and consequently without a name; it is still less than the former in size, but not inferior in beauty, though not so musical; the back, wings, and head, are of a lively green, intermixed with fine shining golden spots, and the belly a snow-white ground, with ebony-coloured spots, so elegantly varied as no art can imitate."

with her courses even with the horizon; and they at the same time particularly observed, that she had no sail abroad except her courses and her main-topsail. This circumstance made them conclude that it was one of our squadron, which had probably suffered in her sails and rigging as severely as we had done: but they were prevented from forming more definite conjectures about her; for, after viewing her for a short time, the weather grew thick and hazy, and they lost sight of her. On this report, and no ship appearing for some days, we were all under the greatest concern, suspecting that her people were in the utmost distress for want of water, and so diminished and weakened by sickness as not to be able to ply up to windward; so that we feared that, after having been in sight of the island, her whole crew would notwithstanding perish at sea. However, on the 26th, towards noon, we discerned a sail in the NE. quarter, which we conceived to be the very same ship that had been seen before, and our conjectures proved true; and about 1 o'clock she approached so near that we could distinguish her to be the Gloucester. As we had no doubt of her being in great distress, the Commodore immediately ordered his boat to her assistance, laden with fresh water, fish, and vegetables, which was a very seasonable relief to them: for our apprehensions of their calamities appeared to be but too well grounded, as perhaps there never was a crew in a more distressed situation. They had already thrown overboard two-thirds of their complement, and of those that remained alive scarcely any were capable of doing duty, except the officers and their servants. They had been a considerable time at the small allowance of a pint of fresh water to each man for twenty-four hours; and yet they had so little left, that, had it not been for the supply we sent them, they must soon have died of thirst.

The ship plied in within three miles of the bay; but, the winds and currents being contrary, she could not reach the road. However, she continued in the offing the next day, but had no chance of coming to an anchor unless the wind and current shifted; and therefore the Commodore repeated his assistance, sending to her the Trial's boat manned with the Centurion's people, and a further supply of water and other refreshments. Captain Mitchel, the captain of the Gloucester, was under a necessity of detaining both this boat and that sent the preceding day; for without the help of their crews he had no longer strength enough to navigate the ship. In this tantalising situation the Gloucester continued for near a fortnight, without being able to fetch the road, though frequently attempting it, and at sometimes bidding very fair for it. On the 9th of July we observed her stretching away to the eastward at a considerable distance, which we supposed was with a design to get to the southward of the island; but as we soon lost sight of her, and she did not appear for near a week, we were prodigiously concerned, knowing that she must be again in extreme distress for want of water. After great impatience about her, we discovered her again on the 16th, endeavouring to come round the eastern point of the island; but the wind, still blowing directly from the bay, prevented her getting nearer than within four leagues of the land. On this Captain Mitchel made signals of distress; and our long-boat was sent to him with a store of water, and plenty of fish and other refreshments. And the long-boat being not to be spared, the cockswain had positive orders from the Commodore to return again immediately: but the weather proving stormy the next day, and the boat not appearing, we much feared she was lost, which would have proved an irretrievable misfortune to us all. But, the third day after, we were relieved from this anxiety by the joyful sight of the long-boat's sails upon the water; and we sent the cutter immediately to her assistance, which towed her alongside in a few hours. The crew of our long-boat had taken in six of the

Gloucester's sick men to bring them on shore, two of whom had died in the boat. And now we learned that the Gloucester was in a most dreadful condition, having scarcely a man in health on board, except those they received from us; and numbers of their sick dying daily, we found that, had it not been for the last supply sent by our long-boat, both the healthy and diseased must have all perished together for want of water. And these calamities were the more terrifying, as they appeared to be without remedy: for the Gloucester had already spent a month in her endeavours to fetch the bay, and she was now no farther advanced than at the first moment she made the island; on the contrary, the people on board her had worn out all their hopes of ever succeeding in it, by the many experiments they had made of its difficulty. Indeed, the same day her situation grew more desperate than ever; for after she had received our last supply of refreshments, we again lost sight of her; so that we in general despaired of her ever coming to an anchor.

Thus was this unhappy vessel bandied about within a few leagues of her intended harbour, whilst the neighbourhood of that place, and of those circumstances which could alone put an end to the calamities they laboured under, served only to aggravate their distress, by torturing them with a view of the relief it was not in their power to reach. But she was at last delivered from this dreadful situation, at a time when we least expected it; for, after having lost sight of her for several days, we were pleasingly surprised, on the morning of the 23d of July, to see her open the NW. point of the bay with a flowing sail; when we immediately despatched what boats we had to her assistance, and in an hour's time from our first perceiving her she anchored safe within us in the bay. And now we were more particularly convinced of the importance of the assistance and refreshments we so often sent them, and how impossible it would have been for a man of them to have survived had we given less attention to their wants; for notwithstanding the water, the greens, and fresh provisions which we supplied them with, and the hands we sent them to navigate the ship, by which the fatigue of their own people was diminished, their sick relieved, and the mortality abated: notwithstanding this indulgent care of the Commodore, they yet buried three-fourths of their crew, and a very small proportion of the remainder were capable of assisting in the duty of the ship. On their coming to an anchor, our first care was to assist them in mooring, and our next to send the sick on shore. These were now reduced by deaths to less than fourscore, of which we expected to lose the greatest part; but whether it was that those farthest advanced in the distemper were all dead, or that the greens and fresh provisions we had sent on board had prepared those who remained for a more speedy recovery, it happened, contrary to our expectations, that their sick were in general relieved and restored to their strength in a much shorter time than our own had been when we first came to the island, and very few of them died on shore.

I have thus given an account of the principal events relating to the arrival of the Gloucester, in one continued narration; I shall only add, that we never were joined by any other of our ships, except our victualler, the Anna pink, which came in about the middle of August, and whose history I shall more particularly relate hereafter.[1] . . .

---

[1] The sick were put ashore here and the Centurion was cleansed from the effects of the recent distress on board, and the water was filled. In addition to supplies of vegetables and fresh fish, new bread was also baked in order to revive the health of the crew. As soon as the health of the men was tolerably recovered, the strongest of them were employed in cutting down trees, and splitting them into billets; while the smiths had their forge sent ashore to mend the chain plates and other broken and decayed iron-work. A large tent was also set up on the

The occupations of cleaning and watering the ship (which was by this time pretty well completed), the attendance on our sick, and the frequent relief sent to the Gloucester, were the principal transactions of our infirm crew till the arrival of the Gloucester at an anchor in the bay. And then Captain Mitchel, waiting on the Commodore, informed him, that he had been forced by the winds, in his last absence, as far as the small island called Mas-a-fuera, lying about twenty-two leagues to the westward of Juan Fernandez; and that he endeavoured to send his boat on shore at this place for water, of which he could observe several streams, but the wind blew so strong upon the shore, and occasioned such a surf, that it was impossible for the boat to land; though the attempt was not altogether useless, as they returned with a boat-load of fish. This island had been represented by former navigators as a barren rock; but Captain Mitchel assured the Commodore that it was almost everywhere covered with trees and verdure, and was near four miles in length; and added that it appeared to him far from impossible but some small bay might be found on it, which might afford sufficient shelter for any ship desirous of refreshing there.

As four ships of our squadron were missing, this description of the Island of Mas-a-fuera gave rise to a conjecture that some of them might possibly have fallen in with that island, and have mistaken it for the true place of our rendezvous; and this suspicion was the more plausible, as we had no draught of either island that could be relied on. In consequence of this reasoning, Mr Anson determined to send the Trial sloop thither, as soon as she could be fitted for the sea, in order to examine all its bays and creeks, that we might be satisfied whether any of our missing ships were there or not. For this purpose, some of our best hands were sent on board the Trial the next morning, to overhaul and fix her rigging; and our long-boat was employed in completing her water; and whatever stores and necessaries she wanted were immediately supplied either from the Centurion or the Gloucester. But it was the 4th of August before the Trial was in readiness to sail, when having weighed, it soon after fell calm, and the tide set her very near the eastern shore. Captain Saunders hung out lights, and fired several guns to acquaint us with his danger; upon which all the boats were sent to his relief, who towed the sloop into the bay; where she anchored until the next morning, and then weighing again proceeded on her cruise with a fair breeze.

And now after the Gloucester's arrival we were employed in earnest in examining and repairing our rigging; but in stripping our foremast we were alarmed by discovering it was sprung just above the partners of the upper deck. The spring was two inches in depth and twelve in circumference; but the carpenters, inspecting it, gave it as their opinion that fishing it with two leaves of an anchor-stock would render it as secure as ever. But our greatest difficulty in refitting was the want of cordage and canvas; for though we had taken to sea much greater quantities of both than had ever been done before, yet the continued bad weather we met with had occasioned such a consumption of these stores, that we were driven to great straits. For after working up all our junk and old shrouds, to make twice laid cordage, we were at last obliged to unlay a cable to work into running rigging; and with all the canvas, and remnants of old sails, that could be mustered, we could only make up one complete suit.

Towards the middle of August, our men being indifferently recovered, they were permitted to quit their sick tents, and to build separate huts for themselves; as it was imagined that by living apart they would be much cleanlier, and consequently likely to recover their strength the sooner; but at the same time particular orders were given, that on the firing of a gun from the ship they should instantly repair to

beach for the use of the sail-makers in their repairs of the sails and rigging.

the waterside. Their employment on shore was now either the procuring of refreshments, the cutting of wood, or the making of oil from the blubber of the sea-lions. This oil served us for several uses, as burning in lamps, or mixing with pitch to pay the ship's sides, or, when mixed with wood-ashes, to supply the use of tallow, of which we had none left, to give the ship boot-hose tops.[1] Some of the men, too, were occupied in salting cod; for there being two Newfoundland fishermen in the Centurion, the Commodore made use of them in laying in a considerable quantity of salted cod for a sea-store; but very little of it was made use of, as it was afterwards thought to be as productive of the scurvy as any other kind of salt provisions.

I have before mentioned that we had a copper oven on shore to bake bread for the sick; but it happened that the greatest part of the flour for the use of the squadron was embarked on board our victualler, the Anna pink. And I should have mentioned that the Trial sloop, at her arrival, had informed us that on the 9th of May she had fallen in with our victualler not far distant from the continent of Chili, and had kept company with her for four days, when they were parted in a hard gale of wind. This gave us some room to hope that she was safe, and that she might soon join us; but all June and July being past without any news of her, we suspected she was lost; and at the end of July the Commodore ordered all the ships to a short allowance of bread. And it was not in our bread only that we feared a deficiency; for since our arrival at this island we discovered that our former purser had neglected to take on board large quantities of several kinds of provisions which the Commodore had expressly ordered him to receive; so that the supposed loss of our victualler was on all accounts a mortifying consideration. However, on Sunday the 16th of August, about noon, we espied a sail in the northern quarter, and a gun was immediately fired from the Centurion to call off the people from shore, who readily obeyed the summons, and repaired to the beach, where the boats waited to carry them on board. And now being prepared for the reception of this ship in view, whether friend or enemy, we had various speculations about her. At first many imagined it to be the Trial sloop returned from her cruise; but as she drew nearer this opinion was confuted by observing she was a vessel with three masts. And then other conjectures were eagerly canvassed, some judging it to be the Severn, others the Pearl, and several affirming that it did not belong to our squadron. But about three in the afternoon our disputes were ended by an unanimous persuasion that it was our victualler, the Anna pink. This ship, though, like the Gloucester, she had fallen in to the northward of the island, had yet the good fortune to come to an anchor in the bay at five in the afternoon. Her arrival gave us all the sincerest joy; for each ship's company was now restored to its full allowance of bread, and we were now freed from the apprehensions of our provisions falling short before we could reach some amicable port—a calamity which in these seas is of all others the most irretrievable. This was the last ship that joined us; and the dangers she encountered, and the good fortune which she afterwards met with, being matters worthy of a separate narration, I shall refer them, together with a short account of the other ships of the squadron, to the ensuing Chapter.

[1] Boot-topping in those days denoted the scraping of a ship's bottom, or that part of its side near the surface of the water, and paying it over with a mixture of tallow, sulphur, resin, &c., as a temporary protection to the plank from worms.

CHAPTER III.

On the first appearance of the Anna pink it seemed wonderful to us how the crew of a vessel which came to this rendezvous two months after us should be capable of working their ship in the manner they did, with so little appearance of debility and distress. But this difficulty was soon solved when she came to an anchor; for we then found that they had been in harbour since the middle of May, which was near a month before we arrived at Juan Fernandez: so that their sufferings (the risk they had run of shipwreck only excepted) were greatly short of what had been undergone by the rest of the squadron. It seems, on the 16th of May, they fell in with the land, which was then but four leagues distant, in the Latitude of 45° 15′ S. On the first sight of it they wore ship and stood to the southward; but their foretopsail splitting, and the wind being WSW., they drove towards the shore; and the captain at last, either unable to clear the land, or, as others say, resolved to keep the sea no longer, steered for the coast with a view of discovering some shelter amongst the many islands which then appeared in sight. And about four hours after the first view of the land, the pink had the good fortune to come to an anchor to the eastward of the Island of Inchin; but as they did not run sufficiently near to the east shore of that island, and had not hands to veer away the cable briskly, they were soon driven to the eastward, deepening their water from twenty-five fathoms to thirty-five, and still continuing to drive, they, the next day, the 17th of May, let go their sheet anchor; which though it brought them up for a short time, yet on the 18th they drove again, till they came into sixty-five fathoms water, and were now within a mile of the land, and expected to be forced on shore every moment, in a place where the coast was so very high and steep, that there was not the least prospect of saving the ship or cargo. And their boats being very leaky, and there being no appearance of a landing-place, the whole crew, consisting of sixteen men and boys, gave themselves over for lost; for they apprehended that if any of them by some extraordinary chance should get on shore, they would in all probability be massacred by the savages on the coast: for these knowing no other Europeans but Spaniards, it might be expected they would treat all strangers with the same cruelty which they had so often and so signally exerted against their Spanish neighbours. Under these terrifying circumstances the pink drove nearer and nearer to the rocks which formed the shore; but at last, when the crew expected each instant to strike, they perceived a small opening in the land, which raised their hopes; and immediately cutting away their two anchors, they steered for it, and found it to be a small channel betwixt an island and the main, which led them into a most excellent harbour, which, for its security against all winds and swells, and the smoothness of its waters, may perhaps compare with any in the known world. And this place being scarcely two miles distant from the spot where they deemed their destruction inevitable, the horrors of shipwreck and of immediate death which had so long and so strongly possessed them vanished almost instantaneously, and gave place to the more joyous ideas of security, repose, and refreshment. In this harbour, discovered in this almost miraculous manner, the pink came to an anchor in twenty-five fathoms water, with only a hawser and a small anchor of about three hundred-weight; and here she continued for near two months, refreshing her people, who were many of them ill of the scurvy, but were soon restored to perfect health by the fresh provisions of which they procured good store, and the excellent water with which the adjacent shore abounded.[1] . . .

[1] Anna Pink Bay is laid down in modern maps to the extreme north of the peninsula of Tres Montes, between that land and the southernmost island of the Chonos Archipelago, off the western coast of Patagonia.

It may be expected that I should relate the discoveries made by the [Anna's] crew on the adjacent coast, and the principal incidents during their stay there. But here I must observe, that being only a few in number, they did not dare to detach any of their people on distant discoveries; for they were perpetually terrified with the apprehension that they should be attacked either by the Spaniards or the Indians; so that their excursions were generally confined to that tract of land which surrounded the port, and where they were never out of view of the ship. But even had they at first known how little foundation there was for these fears, yet the country in the neighbourhood was so grown up with wood, and traversed with mountains, that it appeared impracticable to penetrate it; so that no account of the inland parts could be expected from them. Indeed, they were able to disprove the relations given by Spanish writers, who had represented this coast as inhabited by a fierce and powerful people; for they were certain that no such inhabitants were there to be found, at least during the winter season; since all the time they continued there they saw no more than one Indian family, which came into the harbour in a periagua about a month after the arrival of the pink, and consisted of an Indian near forty years old, his wife, and two children, one three years of age and the other still at the breast. They seemed to have with them all their property, which was a dog, a cat, a fishing-net, a hatchet, a knife, a cradle, some bark of trees intended for covering a hut, a reel, some worsted, a flint and steel, and a few roots of a yellow hue and a very disagreeable taste, which served them for bread. The master of the pink, as soon as he perceived them, sent his yawl, which brought them on board; and fearing lest they might discover him if they were permitted to go away, he took, as he conceived, proper precautions for securing them, but without any mixture of ill-usage or violence. For in the day-time they were permitted to go where they pleased about the ship, but at night were locked up in the forecastle. As they were fed in the same manner with the rest of the crew, and were often indulged with brandy, which they seemed greatly to relish, it did not at first appear that they were much dissatisfied with their situation; especially as the master took the Indian on shore when he went a-shooting (who always seemed extremely delighted when the master killed his game), and as all the crew treated them with great humanity. But it was soon perceived, that though the woman continued easy and cheerful, yet the man grew pensive and restless at his confinement. He seemed to be a person of good natural parts; and, though not capable of conversing with the pink's people otherwise than by signs, was yet very curious and inquisitive, and showed great dexterity in the manner of making himself understood. In particular, seeing so few people on board such a large ship, he let them know that he supposed they were once more numerous; and to represent to them what he imagined was become of their companions, he laid himself down on the deck, closing his eyes, and stretching himself out motionless, to imitate the appearance of a dead body. But the strongest proof of his sagacity was the manner of his getting away; for after being in custody on board the pink eight days, the scuttle of the forecastle, where he and his family were locked up every night, happened to be [left] unnailed, and the following night being extremely dark and stormy, he contrived to convey his wife and children through the unnailed scuttle, and then over the ship's side into the yawl; and, to prevent being pursued, he cut away the long-boat and his own periagua, which were towing astern, and immediately rowed ashore. All this he conducted with so much diligence and secrecy, that though there was a watch on the quarter-deck with loaded arms, yet he was not discovered by them till the noise of his oars in the water, after he had put off from the ship, gave them notice of his escape; and then it was too late either to prevent him or pursue him, for their boats being all adrift, it was a considerable time before they could contrive the means of getting on shore themselves to search for their boats. The Indian, too, by this effort, besides the recovery of his liberty, was in some sort revenged on those who had confined him, both by the perplexity they were involved in from the loss of their boats, and by the terror he threw them into at his departure; for on the first alarm of the watch, who cried out "The Indians!" the whole ship was in the utmost confusion, believing themselves to be boarded by a fleet of armed periaguas.

The resolution and sagacity with which the Indian behaved upon this occasion, had they been exerted on a more extensive object than retrieving the freedom of a single family, might perhaps have immortalised the exploit, and have given him a rank amongst the illustrious names of antiquity. Indeed, his late masters did so much justice to his merit as to own that it was a most gallant enterprise, and that they were grieved they had ever been necessitated, by their attention to their own safety, to abridge the liberty of a person of whose prudence and courage they had now such a distinguished proof. And as it was supposed by some of them that he still continued in the woods in the neighbourhood of the port, where it was feared he might suffer for want of provisions, they easily prevailed upon the master to leave a quantity of such food as they thought would be most agreeable to him, in a particular part where they imagined he would be likely to find it; and there was reason to conjecture that this piece of humanity was not altogether useless to him, for on visiting the place some time after, it was found that the provision was gone, and in a manner that made them conclude it had fallen into his hands.

But, however, though many of them were satisfied that this Indian still continued near them, yet others would needs conclude that he was gone to the Island of Chiloe, where they feared he would alarm the Spaniards, and would soon return with a force sufficient to surprise the pink. And on this occasion the master of the pink was prevailed on to omit firing the evening gun; for it must be remembered (and there is a particular reason hereafter for attending to this circumstance) that the master, from an ostentatious imitation of the practice of men-of-war, had hitherto fired a gun every evening at the setting of the watch. This, he pretended, was to awe the enemy, if there was any within hearing, and to convince them that the pink was always on her guard; but it being now represented to him that his great security was his concealment, and that the evening-gun might possibly discover him and serve to guide the enemy to him, he was prevailed on, as has been mentioned, to omit it for the future. And his crew being now well refreshed, and their wood and water sufficiently replenished, he, in a few days after the escape of the Indian, put to sea, and had a fortunate passage to the rendezvous at the Island of Juan Fernandez, where he arrived on the 16th of August, as has been already mentioned in the preceding Chapter.

This vessel, the Anna pink, was, as I have observed, the last that joined the Commodore at Juan Fernandez. The remaining ships of the squadron were the Severn, the Pearl, and the Wager store-ship. The Severn and Pearl parted company with the squadron off Cape Noir, and, as we afterwards learned, put back to the Brazils; so that of all the ships which came into the South Seas the Wager, Captain Cheap, was the only one that was missing. This ship had on board some field-pieces mounted for land service, together with some cohorn mortars, and several kinds of artillery, stores, and tools, intended for the operations on shore. And, therefore, as the enterprise on Baldivia had been resolved on for the first undertaking of the squadron, Captain Cheap was extremely solicitous that these materials, which were in his custody, might be ready before Baldivia; that if the squadron should possibly rendezvous there (as he knew not the condition they were then reduced to) no delay nor disappointment might be imputed to him.

But whilst the Wager, with these views, was making the best of her way to her first rendezvous off the Island of Socoro, whence (as there was little probability of meeting any of the squadron there) she proposed to steer directly for Baldivia, she made the land on the 14th of May, about the Latitude of 47° S.; and the captain exerting himself on this occasion, in order to get clear of it, he had the misfortune to fall down the after-ladder, and thereby dislocated his shoulder, which rendered him incapable of acting. This accident, together with the crazy condition of the ship, which was little better than a wreck, prevented her from getting off to sea, and entangled her more and more with the land; so that the next morning, at daybreak, she struck on a sunken rock, and soon after bilged, and grounded between two small islands, at about a musket-shot from the shore. In this situation the ship continued entire a long time, so that all the crew had it in their power to get safe on shore; but a general confusion taking place, numbers of them, instead of consulting their safety, or reflecting on their calamitous condition, fell to pillaging the ship, arming themselves with the first weapons that came to hand, and threatening to murder all who should oppose them. This frenzy was greatly heightened by the liquors they found on board, with which they got so extremely drunk, that some of them tumbling down between decks, were drowned as the water flowed in, being incapable of getting up and retreating to other places where the water had not yet entered. And the captain, having done his utmost to get the whole crew on shore, was at last obliged to leave these mutineers behind him, and to follow his officers and such as he had been able to prevail on; but he did not fail to send back the boats to persuade those who remained to have some regard to their preservation, though all his efforts were for some time without success. However, the weather next day proving stormy, and there being great danger of the ship's parting, they[1] began to be alarmed with the fears of perishing, and were desirous of getting to land; but it seems their madness had not yet left them, for the boat not appearing to fetch them off so soon as they expected, they at last pointed a four-pounder, which was on the quarter-deck, against the hut where they knew the captain resided on shore, and fired two shots, which passed but just over it.

From this specimen of the behaviour of part of the crew, it will not be difficult to frame some conjecture of the disorder and anarchy which took place when they at last got all on shore. For the men conceived that by the loss of the ship the authority of the officers was at an end; and, they being now on a desolate coast, where scarcely any other provisions could be got except what should be saved out of the wreck, this was another insurmountable source of discord. For as the working upon the wreck, and the securing the provisions, so that they might be preserved for future exigencies as much as possible, and the taking care that what was necessary for immediate subsistence might be sparingly and equally distributed, were matters not to be brought about but by discipline and subordination; the mutinous disposition of the people, stimulated by the impulses of immediate hunger, rendered every regulation made for this purpose ineffectual. So that there were continual concealments, frauds, and thefts, which animated each man against his fellow, and produced infinite feuds and contests. And hence there was constantly kept on foot a perverse and malevolent turn of tem-

---

[1] Ed. 1776: "The refractory part of the crew."

per, which rendered them utterly ungovernable.[1]

But besides these heart-burnings, occasioned by petulance and hunger, there was another important point, which set the greatest part of the people at variance with the captain. This was their differing with him in opinion on the measures to be pursued in the present exigency; for the captain was determined, if possible, to fit up the boats in the best manner he could, and to proceed with them to the northward. For having with him above 100[2] men in health, and having got some fire-arms and ammunition from the wreck, he did not doubt but they could master any Spanish vessel they should meet with in those seas; and he thought he could not fail of meeting with one in the neighbourhood of Chiloe or Baldivia, in which, when he had taken her, he intended to proceed to the rendezvous at Juan Fernandez; and he further insisted, that should they meet with no prize by the way, yet the boats alone would easily carry them there. But this was a scheme that, however prudent, was no ways relished by the generality of his people; for, being quite jaded with the [fatigues,] distresses, and dangers they had already run through, they could not think of prosecuting an enterprise further which had hitherto proved so disastrous; and therefore the common resolution was to lengthen the long-boat, and with that and the rest of the boats, to steer to the southward, to pass through the Straits of Magellan, and to range along the east side of South America, till they should arrive at Brazil, where they doubted not to be well received, and to procure a passage to Great Britain. This project was at first sight infinitely more hazardous and tedious than what was proposed by the captain; but as it had the air of returning home, and flattered them with the hopes of bringing them once more to their native country, this circumstance alone rendered them inattentive to all its inconveniences, and made them adhere to it with insurmountable obstinacy; so that the captain himself, though he never changed his opinion, was yet obliged to give way to the torrent, and in appearance to acquiesce in this resolution, whilst he endeavoured underhand to give it all the obstruction he could, particularly in the lengthening of the long-boat, which he contrived should be of such a size, that though it might serve to carry them to Juan Fernandez, would yet, he hoped, appear incapable of so long a navigation as that to the coast of Brazil.

But the captain, by his steady opposition at first to this favourite project, had much embittered the people against him; to which, likewise, the following unhappy accident greatly contributed. There was a midshipman, whose name was Cozens, who had appeared the foremost in all the refractory proceedings of the crew. He had involved himself in brawls with most of the officers who had adhered to the captain's authority, and had even treated the captain himself with great abuse and insolence. As his turbulence and brutality grew every day more and more intolerable, it was not in the least doubted but there were some violent measures in agitation in which Cozens was engaged as the ringleader; for which reason the captain, and those about him, constantly kept themselves on their guard. But at last the purser, having, by the captain's order, stopped the allowance of a fellow who would not work, Cozens, though the man did not com-

[1] Sir John Barrow, in his "Life of Anson," states that "it was in consequence of the mutinous and bad conduct of the shipwrecked seamen of the Wager, that Anson, in 1748, when he had the management of the Admiralty, in the absence of the Duke of Bedford and Lord Sandwich, got an Act passed (21 George II.) for extending the discipline of the navy to the crews of His Majesty's ships, wrecked, lost, or taken, and continuing to them their wages upon certain conditions."

[2] Ed. 1776: "Above 200."

plain to him, intermeddled in the affair with great eagerness; and grossly insulting the purser, who was then delivering our provisions just by the captain's tent, and was himself sufficiently violent, the purser, enraged by his scurrility, and perhaps piqued by former quarrels, cried out—"A mutiny!" adding "that the dog had pistols," and then himself fired a shot at Cozens, which, however, missed him. But the captain, on this outcry and the report of the pistol, rushed out of his tent; and, not doubting but it had been fired by Cozens as the commencement of a mutiny, he immediately shot him in the head without further deliberation, and though he did not kill him on the spot, yet the wound proved mortal, and he died about fourteen days after.

This incident, however displeasing to the people, did yet for a considerable time awe them to their duty, and rendered them more submissive to the captain's authority. But at last, when towards the middle of October the long-boat was nearly completed, and they were preparing to put to sea, the additional provocation he gave them by covertly traversing their project of proceeding through the Straits of Magellan, and their fears that he might at length engage a party sufficient to overturn this favourite measure, made them resolve to make use of the death of Cozens as a reason for depriving him of his command, under pretence of carrying him a prisoner to England to be tried for murder; and he was accordingly confined under a guard. But they never intended to carry him with them, as they too well knew what they had to apprehend on their return to England if their commander should be present to confront them; and therefore, when they were just ready to put to sea, they set him at liberty, leaving him, and the few who chose to take their fortunes with him, no other embarkation but the yawl; to which the barge was afterwards added by the people on board her being prevailed on to return back.

When the ship was wrecked, there remained alive on board the Wager near 130 persons; of these, above thirty died during their stay upon the place, and near eighty went off in the long-boat and the cutter to the southward: so that there remained with the captain, after their departure, no more than nineteen persons, which, however, was as many as the barge and the yawl—the only embarkations left them—could well carry off. It was on the 13th of October, five months after the shipwreck, that the long-boat, converted into a schooner, weighed and stood to the southward, giving the captain, who, with Lieutenant Hamilton of the land forces, and the surgeon, was then on the beach, three cheers at their departure. It was the 29th of January following before they arrived at Rio Grande on the coast of Brazil; and having, by various accidents, left about twenty of their people on shore at the different places they touched at; and a greater number having perished by hunger during the course of their navigation, there were no more than thirty of them left when they arrived in that port. Indeed the undertaking of itself was a most extraordinary one, for, not to mention the length of the run, the vessel was scarcely able to contain the number that first put to sea in her; and their stock of provisions (being only what they had saved out of the ship) was extremely slender; and the cutter, the only boat they had with them, soon broke away from the stern and was staved to pieces; so that when their provision and their water failed them, they had frequently no means of getting on shore to search for a fresh supply.

When the long-boat and cutter were gone, the captain and those who were left with him proposed to pass to the northward in the barge and yawl; but the weather was so bad, and the difficulty of subsisting so great, that it was two months after the departure of the long-boat before he was able to put to sea. It seems the place where the Wager was cast away was not a part of the continent, as was first imagined, but an island at some distance from the main, which afforded no other sorts of provision but shell-fish and a few herbs; and as the greatest part of what they had got from the ship was carried off in the long-boat, the captain and his people were often in great necessity, especially as they chose to preserve what little sea-provisions remained for their store when they should go to the northward. During their residence at this island, which was by the seamen denominated Wager Island,[1] they had now and then a straggling canoe or two of the Indians, which came and bartered their fish and other provisions with our people. This was indeed some little succour, and at another season might perhaps have been greater; for as there were several Indian huts on the shore, it was supposed that in some years, during the height of summer, many of these savages might resort thither to fish. And, from what has been related in the account of the Anna pink, it should seem to be the general practice of those Indians to frequent this coast in the summer time for the benefit of fishing, and to retire in the winter into a better climate more to the northward. And on this mention of the Anna pink, I cannot but observe how much it is to be lamented that the Wager's people had no knowledge of her being so near them on the coast; for as she was not above thirty leagues distant from them, and came into their neighbourhood about the same time the Wager was lost, and was a fine roomy ship, she could easily have taken them all on board and have carried them to Juan Fernandez. Indeed, I suspect she was still nearer to them than what is here estimated, for several of the Wager's people, at different times, heard the report of a cannon, which I conceive could be no other than the evening gun fired from the Anna pink, especially as what was heard at Wager Island was about the same time of the day. But to return to Captain Cheap.

Upon the 14th of December the captain and his people embarked in the barge and the yawl in order to proceed to the northward, taking on board with them all the provisions they could amass from the wreck of the ship; but they had scarcely been an hour at sea when the wind began to blow hard, and the sea ran so high that they were obliged to throw the greatest part of their provisions overboard to avoid immediate destruction. This was a terrible misfortune in a part of the world where food is so difficult to be got; however, they still persisted in their design, putting on shore as often as they could to seek subsistence. But, about a fortnight after, another dreadful accident befell them, for the yawl sank at an anchor, and one of the men in her was drowned; and as the barge was incapable of carrying the whole company, they were now reduced to the hard necessity of leaving four marines behind them on that desolate shore. But they still kept on their course to the northward, struggling with their disasters, and greatly delayed by the perverseness of the winds and the frequent interruptions which their search after food occasioned; till at last, about the end of January, having made three unsuccessful attempts to double a headland which they supposed to be what the Spaniards called Cape Tres Montes, it was unanimously resolved to give over this expedition, the difficulties of which appeared insuperable, and to return again to Wager Island, where they got back about the middle of February, quite disheartened and dejected with their reiterated disappointments, and almost perishing with hunger and fatigue.

However, on their return they had the good luck to meet with several pieces of beef which had been washed out of the ship, and were swimming in the sea. This was a most seasonable relief to them after the hardships they had endured; and, to complete

[1] A small island just north of Wellington Island on the western coast of Patagonia, and divided from the peninsula of Tres Montes, lying to the northward still, by the Gulf of Penas.

their good fortune, there came in a short time two canoes of Indians, amongst whom was a native of Chiloe who spoke a little Spanish; and the surgeon who was with Captain Cheap understanding that language, he made a bargain with the Indian, that if he would carry the captain and his people to Chiloe in the barge he should have her, and all that belonged to her, for his pains. Accordingly, on the 6th of March, the eleven persons, to which the company was now reduced, embarked in the barge on this new expedition; but after having proceeded for a few days, the captain and four of his principal officers being on shore, the six, who together with an Indian remained in the barge, put off with her to sea and did not return.

By this means there were left on shore Captain Cheap, Mr Hamilton, lieutenant of marines; the Honourable Mr Byron[1] and Mr Campbell, midshipmen; and Mr Elliot, the surgeon. One would have thought their distresses had long before this time been incapable of augmentation, but they found, on reflection, that their present situation was much more dismaying than anything they had yet gone through, being left on a desolate coast without any provision, or the means of procuring any, for their arms, ammunition, and every conveniency they were masters of, except the tattered habits they had on, were all carried away in the barge. But when they had sufficiently revolved in their own minds the various circumstances of this unexpected calamity, and were persuaded that they had no relief to hope for, they perceived a canoe at a distance, which proved to be that of the Indian who had undertaken to carry them to Chiloe, he and his family being then on board it. He made no difficulty of coming to them, for it seems he had left Captain Cheap and his people a little before to go a-fishing, and had in the meantime committed them to the care of the other Indian, whom the sailors had carried to sea in the barge. But when he came on shore and found the barge gone and his companion missing, he was extremely concerned, and could with difficulty be persuaded that the other Indian was not murdered; but being at last satisfied with the account that was given him, he still undertook to carry them to the Spanish settlements, and (as the Indians are well skilled in fishing and fowling) to procure them provisions by the way.

About the middle of March, Captain Cheap and the four who were left with him set out for Chiloe, the Indian having procured a number of canoes, and got many of his neighbours together for that purpose. Soon after they embarked, Mr Elliot the surgeon died, so that there now remained only four of the whole company. At last, after a very complicated passage by land and water, Captain Cheap, Mr Byron, and Mr Campbell, arrived in the beginning of June at the Island of Chiloe, where they were received by the Spaniards with great humanity; but, on account of some quarrel among the Indians, Mr Hamilton did not get thither till two months after. Thus, above a twelvemonth after the loss of the Wager, ended this fatiguing peregrination, which by a variety of misfortunes had diminished the company from twenty to no more than four, and those too, brought so low, that had their distresses continued but a few days longer, in all probability none of them would have survived. For the captain himself was with difficulty recovered; and the rest were so reduced by the severity of the weather, their labour, and their want of all kinds of necessaries, that it was wonderful how they supported themselves so long. After some stay at

[1] The Honourable John Byron, who left a well-written narrative of his sufferings and adventures; as Commodore, he commanded an expedition of discovery to the southern parts of South America in 1764-1766, and circumnavigated the globe partly in the same track as his former commander. He afterwards rose to the rank of Admiral, and survived till 1798. Lord Byron, the poet, was his grandson.

Chiloe, the captain and the three who were with him were sent to Valparaiso, and thence to Santiago, the capital of Chili, where they continued above a year: but on the advice of a cartel being settled betwixt Great Britain and Spain, Captain Cheap, Mr Byron, and Mr Hamilton were permitted to return to Europe on board a French ship. The other midshipman, Mr Campbell, having changed his religion whilst at Santiago, chose to go back [overland] to Buenos Ayres with Pizarro and his officers, with whom he went afterwards to Spain on board the Asia; and there having failed in his endeavours to procure a commission from the Court of Spain, he returned to England, and attempted to get reinstated in the British navy; and has since published a narration of his adventures, in which he complains of the injustice that had been done him, and strongly disavows his ever being in the Spanish service. But as the change of his religion, and his offering himself to the Court of Spain (though not accepted), are matters which, he is conscious, are capable of being incontestably proved; on these two heads he has been entirely silent. And now, after this account of the accidents which befell the Anna pink, and the catastrophe of the Wager, I shall again resume the thread of our own story.

CHAPTER IV.

About a week after the arrival of our victualler, the Trial sloop, that had been sent to the Island of Mas-a-fuera, returned to an anchor at Juan Fernandez, after having been round that island without meeting any part of our squadron. . . .

The latter part of the month of August was spent in unloading the provisions from the Anna pink; and here we had the mortification to find that great quantities of our provisions, as bread, rice, groats, &c., were decayed and unfit for use. This was owing to the water the pink had made by her working and straining in bad weather; for thereby several of her casks had rotted, and her bags were soaked through.[1] . . . . The thorough refitting of the Anna pink, proposed by the carpenters, was, in our present situation, impossible to be complied with, as all the plank and iron in the squadron was insufficient for that purpose. And now the master, finding his own sentiments confirmed by the opinion of all the carpenters, offered a petition to the Commodore in behalf of his owners, desiring that, since it appeared he was incapable of leaving the island, Mr Anson would please to purchase the hull and furniture of the pink for the use of the squadron. Hereupon the Commodore ordered an inventory to be taken of every particular belonging to the pink, with its just value; and as by this inventory it appeared that there were many stores which would be useful in refitting the other ships, and which were at present very scarce in the squadron by reason of the great quantities that had been already expended, he agreed with Mr Gerard to purchase the whole together for £300. The pink being thus broken up, Mr Gerard, with the hands belonging to the pink, were sent on board the Gloucester; as that ship had buried the greatest number of men in proportion to her complement. But afterwards one or two of them were received on board the Centurion on their own petition, they being extremely averse to sailing in the same ship with their old master, on account of some particular ill-usage they conceived they had suffered from him.

This transaction brought us down to the beginning of September, and our people by this time were so far recovered of the scurvy that there was little danger of burying any more at present; and therefore I shall now sum up the total of our loss since our departure from England, the better to convey some idea of our past suffer-

[1] The Anna pink was here discharged from the service of the squadron, and on examination was found to be un-seaworthy.

ings and of our present strength. We had buried on board the Centurion since our leaving St Helens 292, and had now remaining on board 214. This will doubtless appear a most extraordinary mortality; but yet on board the Gloucester it had been much greater, for out of a much smaller crew than ours they had buried the same number, and had only eighty-two remaining alive. It might be expected that on board the Trial the slaughter would have been the most terrible, as her decks were almost constantly knee-deep in water; but it happened otherwise, for she escaped more favourably than the rest, since she only buried forty-two, and had now thirty-nine remaining alive. The havoc of this disease had fallen still severer on the invalids and marines than on the sailors; for on board the Centurion, out of fifty invalids and seventy-nine marines there remained only four invalids, including officers, and eleven marines; and on board the Gloucester every invalid perished, and out of forty-eight marines only two escaped. From this account it appears that the three ships together departed from England with 961 men on board, of whom 626 were dead before this time; so that the whole of our remaining crews, which were now to be distributed amongst three ships, amounted to no more than 335 men and boys, a number greatly insufficient for manning the Centurion alone, and barely capable of navigating all the three with the utmost exertion of their strength and vigour. This prodigious reduction of our men was still the more terrifying as we were hitherto uncertain of the fate of Pizarro's squadron, and had reason to suppose that some part of it at least had got round into these seas. Indeed we were satisfied from our own experience that they must have suffered greatly in their passage; but then every port in the South Seas was open to them, and the whole power of Chili and Peru would doubtless be united in refreshing and refitting them, and recruiting the numbers they had lost. Besides, we had some obscure knowledge of a force to be fitted out from Callao; and, however contemptible the ships and sailors of this part of the world may have been generally esteemed, it was scarcely possible for anything bearing the name of a ship of force to be feebler or less considerable than ourselves. And had there been nothing to be apprehended from the naval power of the Spaniards in this part of the world, yet our enfeebled condition would nevertheless give us the greatest uneasiness, as we were incapable of attempting any of their considerable places; for the risking of twenty men, weak as we then were, was risking the safety of the whole. So that we conceived we should be necessitated to content ourselves with what few prizes we could pick up at sea before we were discovered, after which we should in all probability be obliged to depart with precipitation, and esteem ourselves fortunate to regain our native country, leaving our enemies to triumph on the inconsiderable mischief they had received from a squadron whose equipment had filled them with such dreadful apprehensions. This was a subject on which we had reason to imagine the Spanish ostentation would remarkably exert itself; though the causes of our disappointment and their security were neither to be sought for in their valour nor our misconduct. Such were the desponding reflections which at that time arose on the review and comparison of our remaining strength with our original numbers. Indeed our fears were far from being groundless or disproportioned to our feeble and almost desperate situation. It is true the final event proved more honourable than we had foreboded; but the intermediate calamities did likewise greatly surpass our most gloomy apprehensions, and could they have been predicted to us at this Island of Juan Fernandez, they would doubtless have appeared insurmountable.

In the beginning of September, as has been already mentioned, our men were tolerably well recovered; and now the time of navigation in this climate drawing near, we exerted ourselves in getting our ships in readiness for the sea. We converted the foremast of the victualler into a mainmast for the Trial sloop; and, still flattering ourselves with the possibility of the arrival of some other ships of our squadron, we intended to leave the mainmast of the victualler to make a mizzenmast for the Wager. Thus all hands being employed in forwarding our departure, we on the 8th, about eleven in the morning, espied a sail to the NE., which continued to approach us till her courses appeared even with the horizon. In this interval we all had hopes she might prove one of our own squadron; but at length, finding she steered away to the eastward, without hauling in for the island, we concluded she must be a Spaniard. And now great disputes were set on foot about the possibility of her having discovered our tents on shore, some of us strongly insisting that she had doubtless been near enough to have perceived something that had given her a jealousy of an enemy, which had occasioned her standing to the eastward without hauling in; but, leaving these contests to be settled afterwards, it was resolved to pursue her; and the Centurion being in the greatest forwardness, we immediately got all our hands on board, set up our rigging, bent our sails, and by five in the afternoon got under sail. We had at this time very little wind, so that all the boats were employed to tow us out of the bay; and even what wind there was lasted only long enough to give us an offing of two or three leagues, when it flattened to a calm. The night coming on, we lost sight of the chase, and were extremely impatient for the return of daylight, in hopes to find that she had been becalmed as well as we; though I must confess that her greater distance from the land was a reasonable ground for suspecting the contrary, as we indeed found in the morning, to our great mortification; for though the weather continued perfectly clear, we had no sight of the ship from the mast-head. But as we were now satisfied that it was an enemy, and the first we had seen in these seas, we resolved not to give over the search lightly; and a small breeze springing up from the WNW., we got up our top-gallant masts and yards, set all the sails, and steered to the SE., in hopes of retrieving our chase, which we imagined to be bound to Valparaiso. We continued on this course all that day and the next; and then, not getting sight of our chase, we gave over the pursuit, conceiving that by that time she must in all probability have reached her port.

And now we prepared to return to Juan Fernandez, and hauled up to the SW. with that view, having but very little wind till the 12th, when, at three in the morning, there sprang up a fresh gale from the WSW., and we tacked and stood to the NW.; and at daybreak we were agreeably surprised with the sight of a sail on our weather-bow, between four and five leagues distant. On this we crowded all the sail we could, and stood after her, and soon perceived it not to be the same ship we originally gave chase to. She at first bore down upon us, showing Spanish colours, and making a signal as to her consort; but observing that we did not answer her signal, she instantly luffed close to the wind and stood to the southward. Our people were now all in spirits, and put the ship about with great alacrity; and as the chase appeared to be a large ship, and had mistaken us for her consort, we conceived that she was a man-of-war, and probably one of Pizarro's squadron. This induced the Commodore to order all the officers' cabins to be knocked down and thrown overboard, with several casks of water and provisions which stood between the guns; so that we had soon a clear ship, ready for an engagement. About 9 o'clock we had thick, hazy weather, and a shower of rain, during which we lost sight of the chase; and we were apprehensive, if the weather should continue, that by going upon the other tack, or by some other artifice, she might escape us; but it clearing up in less than an hour, we found that we had both weathered and fore-reached upon her considerably, and now we were near enough to discover that she was only a merchantman, without so much as a single tier of guns. About half-an-hour after twelve, being then within a reasonable distance of her, we fired four shot amongst her rigging; on which they lowered their topsails and bore down to us, but in very great confusion, their top-gallant-sails and stay-sails all fluttering in the wind. This was owing to their having let run their sheets and halyards just as we fired at them; after which not a man amongst them had courage enough to venture aloft (for there the shot had passed but just before) to take them in.

As soon as the vessel came within hail of us, the Commodore ordered them to bring to under his lee-quarter, and then hoisted out the boat and sent Mr Saumarez, his first lieutenant, to take possession of the prize, with directions to send all the prisoners on board the Centurion, but first the officers and passengers. When Mr Saumarez came on board them, they received him at the side with the strongest tokens of the most abject submission; for they were all of them (especially the passengers, who were twenty-five in number) extremely terrified, and under the greatest apprehensions of meeting with very severe and cruel usage. But the lieutenant endeavoured, with great courtesy, to dissipate their fright, assuring them that their fears were altogether groundless, and that they would find a generous enemy in the Commodore, who was not less remarkable for his lenity and humanity than for his resolution and courage. The prisoners who were first sent on board the Centurion informed us that our prize was called Nuestra Señora del Monte Carmelo, and was commanded by Don Manuel Zamorra. Her cargo consisted chiefly of sugar, and great quantities of blue cloth made in the province of Quito, somewhat resembling our English coarse broad-cloths, but inferior to them. They had, besides, several bales of a coarser sort of cloth, of different colours, somewhat like Colchester baize, called by them Pannia da Tierra, with a few bales of cotton, and tobacco, which though strong was not ill flavoured. These were the principal goods on board her; but we found, besides, what was to us much more valuable than the rest of the cargo. This was some trunks of wrought plate, and twenty-three serons[1] of dollars, each weighing upwards of 200 lbs. avoirdupois. The ship's burthen was about 450 tons; she had fifty-three sailors on board, both whites and blacks. She came from Callao, and had been twenty-seven days at sea before she fell into our hands. She was bound to the port of Valparaiso, in the kingdom of Chili,[2] and proposed to have returned thence loaded with corn and Chili wine, some gold, dried beef, and small cordage, which at Callao they convert into larger rope. Our prize had been built upwards of thirty years; yet as they lie in harbour all the winter months, and the climate is favourable, they esteemed it no very great age. Her rigging was very indifferent, as were likewise her sails, which were made of cotton. She had only three 4-pounders, which were altogether unserviceable, their carriages being scarcely able to support them; and

[1] A seron or seroon is a species of packet made and used in Spanish America, consisting of a piece of raw bullock's hide, with the hair on, formed while wet into the shape of a small trunk, and sewed together. In Kerr's Collection of Voyages, the quantity of dollars taken on this occasion is estimated at between £70,000 and £80,000.

[2] Thomas says that those ships annually trade to Valparaiso, exchanging silver in return for gold and coin, the latter being very scarce in Peru. Some of the prisoners said that if the ship had been taken on the return from Chili to Peru, the captors would have found in her as much gold as they had now found silver.

there were no small arms on board, except a few pistols belonging to the passengers. The prisoners informed us that they left Callao in company with two other ships, which they had parted with some days before, and that at first they conceived us to be one of their company; and by the description we gave them of the ship we had chased from Juan Fernandez, they assured us she was of their number, but that the coming in sight of that island was directly repugnant to the merchants' instructions, who had expressly forbid it, as knowing that if any English squadron was in those seas, the Island of Fernandez was most probably the place of their rendezvous.

And now, after this short account of the ship and her cargo, it is necessary that I should relate the important intelligence which we met with on board her, partly from the information of the prisoners, and partly from the letters and papers which fell into our hands. We here first learned with certainty the force and destination of that squadron which cruised off Madeira at our arrival there, and afterwards chased the Pearl in our passage to Port St Julian. This we now knew was a squadron composed of five large Spanish ships, commanded by Admiral Pizarro, and purposely fitted out to traverse our designs, as has been already more amply related in the Third Chapter of the First Book. And we had, at the same time, the satisfaction to find that Pizarro, after his utmost endeavours to gain his passage into these seas, had been forced back again into the River of Plate, with the loss of two of his largest ships; and besides this disappointment of Pizarro, which, considering our great debility, was no unacceptable intelligence, we further learned that an embargo had been laid upon all shipping in these seas by the Viceroy of Peru, in the month of May preceding, on a supposition that about that time we might arrive upon the coast. But on the account sent overland by Pizarro of his own distresses, part of which they knew we must have encountered, as we were at sea during the same time, and on their having no news of us in eight months after we were known to set sail from St Catherine's, they were fully persuaded that we were either shipwrecked, or had perished at sea, or at least had been obliged to put back again; for it was conceived impossible for any ships to continue at sea during so long an interval: and therefore, on the application of the merchants, and the firm persuasion of our having miscarried, the embargo had been lately taken off.[1]

This last article made us flatter ourselves that, as the enemy was still a stranger to our having got round Cape Horn, and the navigation of

[1] Thomas makes a curiously different report of Pizarro's despatch, and one much more flattering to English pride. Pizarro, he says, told the Viceroy of Peru "that, though he himself had been forced back in such a miserable condition, not having above 80 or 100 of his men living, and his ships in so ill a state, that, till sufficient reinforcements could come to him from Old Spain, he could not possibly come into those seas, yet as the English were a stubborn and resolute people, and daring enough to persist obstinately in the most desperate undertakings, he did believe some of us might possibly get round; but as he experimentally knew what of necessity we must have suffered in that dreadful passage, he made no doubt but we should be in a very weak and defenceless condition; he therefore advised the Viceroy to fit out all the strength of shipping he could, and send them to cruise at the Island of Juan Fernandez, where we must of necessity touch to refresh our people, and to repair our ships; and further advised, that, in case of meeting us, they should not stand to fight or cannonade at a distance, in which possibly we might have the advantage, or make our escape, but should board us at once sword in hand; which must, if well executed, in our weak condition, infallibly prove the means of taking us."

these seas was restored, we might meet with some considerable captures, and might thereby indemnify ourselves for the incapacity we were now under of attempting any of their considerable settlements on shore. And thus much we were certain of, from the information of our prisoners, that whatever our success might be as to the prizes we might light on, we had nothing to fear, weak as we were, from the Spanish force in this part of the world; though we discovered that we had been in most imminent peril from the enemy when we least apprehended it, and when our other distresses were at the greatest height. For we learned from the letters on board, that Pizarro, in the express he despatched to the Viceroy of Peru after his return to the River of Plate, had intimated to him that it was possible some part at least of the English squadron might get round; but that, as he was certain from his own experience that if they did arrive in those seas it must be in a very weak and defenceless condition, he advised the Viceroy, in order to be secure at all events, to fit out what ships of force he had, and send them to the southward, where in all probability they would intercept us singly and before we had an opportunity of touching anywhere for refreshment; in which case he doubted not but we should prove an easy conquest. The Viceroy of Peru approved of this advice, and immediately fitted out four ships of force from Callao; one of 50 guns, two of 40 guns, and one of 24 guns. Three of them were stationed off the Port of Conception, and one of them at the Island of Juan Fernandez; and in these stations they continued cruising for us till the 6th of June, when, not seeing anything of us, and conceiving it to be impossible that we could have kept the seas so long, they quitted their cruise and returned to Callao, fully satisfied that we had either perished or at least had been driven back. As the time of their quitting their station was but a few days before our arrival at the Island of Fernandez, it is evident that had we made that island on our first search for it, without hauling in for the main to secure our easting (a circumstance which at that time we considered as very unfortunate to us, on account of the numbers which we lost by our longer continuance at sea), had we, I say, made the island on the 28th of May, when we first expected to see it, and were in reality very near it, we had doubtless fallen in with some part of the Spanish squadron; and in the distressed condition we were then in, the meeting with a healthy, well-provided enemy was an incident that could not but have been perplexing, and might perhaps have proved fatal, not only to us [in the Centurion], but to the Trial, the Gloucester, and the Anna pink, which separately joined us, and which were each of them less capable than we were of making any considerable resistance. I shall only add, that these Spanish ships sent out to intercept us had been greatly shattered by a storm during their cruise; and that, after their arrival at Callao, they had been laid up. And our prisoners assured us, that whenever intelligence was received at Lima of our being in these seas, it would be at least two months before this armament could be again fitted out.

The whole of this intelligence was as favourable as we in our reduced circumstances could wish for. And now we were fully satisfied as to the broken jars, ashes, and fish-bones, which we had observed at our first landing at Juan Fernandez; these things being doubtless the relics of the cruisers stationed off that port. Having thus satisfied ourselves in the material articles, and having got on board the Centurion most of the prisoners and all the silver, we at eight in the same evening made sail to the northward, in company with our prize, and at six the next morning discovered the Island of Juan Fernandez, where the next day both we and our prize came to an anchor. And here I cannot omit one remarkable incident which occurred when the prize and her crew came into the bay, where the rest of the squadron lay. The Spaniards in the Carmelo had been sufficiently informed of the distresses we had gone through, and were greatly surprised that we had ever surmounted them: but when they saw the Trial sloop at anchor, they were still more astonished that, after all our fatigues, we had the industry (besides refitting our other ships) to complete such a vessel in so short a time, they taking it for granted that she had been built upon the spot. And it was with great difficulty they were prevailed on to believe that she came from England with the rest of the squadron; they at first insisting that it was impossible such a bauble as that could pass round Cape Horn, when the best ships of Spain were obliged to put back.

By the time we arrived at Juan Fernandez, the letters found on board our prize were more minutely examined; and it appearing from them and from the accounts of our prisoners that several other merchantmen were bound from Callao to Valparaiso, Mr Anson despatched the Trial sloop the very next morning to cruise off the last-mentioned port, reinforcing her with ten hands from on board his own ship. Mr Anson likewise resolved, on the intelligence recited above, to separate the ships under his command, and employ them in distinct cruises, as he thought that by this means we should not only increase our chance for prizes, but that we should likewise run less risk of alarming the coast and of being discovered. And now, the spirits of our people being greatly raised, and their despondency dissipated by this earnest of success, they forgot all their past distresses, and resumed their wonted alacrity, and laboured indefatigably in completing our water, receiving our lumber, and preparing to take our farewell of the island. But as these occupations took us up four or five days, with all our industry, the Commodore in that interval directed that the guns belonging to the Anna pink, being four 6-pounders, four 4-pounders, and two swivels, should be mounted on board the Carmelo, our prize. And having sent on board the Gloucester six passengers and twenty-three seamen[1] to assist in navigating the ship, he directed Captain Mitchel to leave the island as soon as possible, the service requiring the utmost despatch, ordering him to proceed to the Latitude of 5° S., and there to cruise off the high land of Paita,[2] at such a distance from shore as should prevent his being discovered. On this station he was to continue till he should be joined by the Commodore, which would be whenever it should be known that the Viceroy had fitted out the ships at Callao, or on Mr Anson's receiving any other intelligence that should make it necessary to unite our strength. These orders being delivered to the captain of the Gloucester, and all our business completed, we on the Saturday following, being the 19th of September, weighed our anchor, in company with our prize, and got out of the bay, taking our last leave of the Island of Juan Fernandez, and steering to the eastward, with an intention of joining the Trial sloop in her station off Valparaiso.

CHAPTER V.

ALTHOUGH the Centurion, with her prize the Carmelo, weighed from the Bay of Juan Fernandez on the 19th of September, leaving the Gloucester at anchor behind her, yet, by the irregularity and fluctuation of the winds in the offing, it was the 22d of the same month, in the evening, before we lost sight of the island; after which we continued our course to the eastward, in order to reach our station and to join the Trial off Valparaiso. The next night the weather proved squally, and we split our main-topsail, which we handed for the pre-

[1] Selected from among the prisoners for their strength or their knowledge of seamanship.

[2] Where the vessels trading between Lima and Panama generally touched to deliver part of their cargoes for dispersion through the inland parts of Peru.

sent, but got it repaired, and set it again the next morning. And now, on the 24th, a little before sunset, we saw two sail to the eastward, on which our prize stood directly from us, to avoid giving any suspicion of our being cruisers; whilst we in the meantime made ourselves ready for an engagement, and steered towards the two ships we had discovered, with all our canvas. We soon perceived that one of these, which had the appearance of being a very stout ship, made directly for us, whilst the other kept at a very great distance. By 7 o'clock we were within pistol-shot of the nearest, and had a broadside ready to pour into her, the gunners having their matches in their hands, and only waiting for orders to fire; but as we knew it was now impossible for her to escape us, Mr Anson, before he permitted them to fire, ordered the master to hail the ship in Spanish; on which the commanding officer on board her, who proved to be Mr Hughes, lieutenant of the Trial, answered us in English, and informed us that she was a prize taken by the Trial a few days before, and that the other sail at a distance was the Trial herself, disabled in her masts. We were soon after joined by the Trial, and Captain Saunders, her commander, came on board the Centurion. He informed the Commodore that he had taken this ship the 18th instant; that she was a prime sailer, and had cost him thirty-six hours' chase before he could come up with her; that for some time he gained so little upon her that he began to despair of taking her; and the Spaniards, though alarmed at first with seeing nothing but a cloud of sail in pursuit of them, the Trial's hull being so low in the water that no part of it appeared, yet knowing the goodness of their ship, and finding how little the Trial neared them, they at length laid aside their fears, and recommending themselves to the blessed Virgin for protection, began to think themselves secure. And indeed their success was very near doing honour to their Ave Marias; for altering their course in the night, and shutting up their windows to prevent any of their lights from being seen, they had some chance of escaping. But a small crevice in one of the shutters rendered all their invocations ineffectual; for through this crevice the people on board the Trial perceived a light, which they chased till they arrived within gunshot; and then Captain Saunders alarmed them unexpectedly with a broadside, when they flattered themselves they were got out of his reach. However, for some time after they still kept the same sail abroad, and it was not observed that this first salute had made any impression on them; but just as the Trial was preparing to repeat her broadside, the Spaniards crept from their holes, lowered their sails, and submitted without any opposition. She was one of the largest merchantmen employed in those seas, being about 600 tons burthen, and was called the [Nuestra Señora de] Arranzazu. She was bound from Callao to Valparaiso, and had much the same cargo with the Carmelo we had taken before, except that her silver amounted only to about £5000 sterling.

But to balance this success, we had the misfortune to find that the Trial had sprung her mainmast, and that her main-topmast had come by the board; and as we were all of us standing to the eastward the next morning, with a fresh gale at S., she had the additional ill-luck to spring her foremast; so that now she had not a mast left on which she could carry sail. These unhappy incidents were still [further] aggravated by the impossibility we were just then under of assisting her; for the wind blew so hard, and raised such a hollow sea that we could not venture to hoist out our boat, and consequently could have no communication with her; so that we were obliged to lie to for the greatest part of forty-eight hours to attend her, as we could have no thought of leaving her to herself in her present unhappy situation. And as an accumulation to our misfortunes, we were all the while driving to the leeward of our station, at the very time when, by our intelligence, we had reason to expect several of the enemy's ships would appear upon the coast, who would now gain the port of Valparaiso without obstruction. And I am verily persuaded that the embarrassment we received from the dismasting of the Trial, and our absence from our intended station occasioned thereby, deprived us of some very considerable captures.[1]

The weather proving somewhat more moderate on the 27th, we sent our boat for the captain of the Trial, who, when he came on board us, produced an instrument, signed by himself and all his officers, representing that the sloop, besides being dismasted, was so very leaky in her hull, that even in moderate weather it was necessary to keep the pumps constantly at work, and that they were then scarcely sufficient to keep her free; so that in the late gale, though they had all been engaged at the pumps by turns, yet the water had increased upon them; and, upon the whole, they apprehended her to be at present so very defective, that if they met with much bad weather they must all inevitably perish, and therefore they petitioned the Commodore to take some measures for their future safety. But the refitting of the Trial, and the repairing of her defects, was an undertaking that in the present conjuncture greatly exceeded his power; for we had no masts to spare her, we had no stores to complete her rigging, nor had we any port where she might be hove down and her bottom examined. Besides, had a port and proper requisites for this purpose been in our possession, yet it would have been extreme imprudence, in so critical a conjuncture, to have loitered away so much time as would have been necessary for these operations. The Commodore therefore had no choice left him but that of taking out her people and destroying her; but at the same time, as he conceived it necessary for his Majesty's service to keep up the appearance of our force, he appointed the Trial's prize (which had been often employed by the Viceroy of Peru as a man-of-war) to be a frigate in his Majesty's service, manning her with the Trial's crew, and giving new commissions to the captain and all the inferior officers accordingly. This new frigate, when in the Spanish service, had mounted 32 guns; but she was now to have only 20, which were the 12 that were on board the Trial, and 8 that had belonged to the Anna pink. When this affair was thus far regulated, Mr Anson gave orders to Captain Saunders to put it in execution, directing him to take out of the sloop the arms, stores, ammunition, and everything that could be of any use to the other ships, and then to scuttle her and sink her. And after Captain Saunders had seen her destroyed, he was to proceed with his new frigate (to be called the Trial's prize) and to cruise off the high land of Valparaiso, keeping it from him NNW., at the distance of twelve or fourteen leagues. For as all ships bound from Valparaiso to the northward steer that course, Mr Anson proposed by this means to stop any intelligence that might be despatched to Callao of two of their ships being missing, which might give them apprehensions of the English squadron being in their neighbourhood. The Trial's prize was to continue on this station twenty-four days, and if not joined by the Commodore at the expiration of that term, she was then to proceed down the coast to Pisco, or Nasca,[1] where she would be certain

[1] Thomas, with regard to the disabled condition of the Trial, says: "This was a great destruction, for now we had intelligence by the Trial's prize that there were many ships at sea richly laden, and that they had no apprehensions of being attacked by us, having received intelligence that our squadron was either put back or destroyed. In the course, therefore, of the forty-eight hours we were detained in waiting upon the Trial, I am persuaded we missed the taking many valuable prizes."

[1] Pisco town and bay are about 120

to meet with Mr Anson. The Commodore likewise ordered Lieutenant Saumarez, who commanded the Centurion's prize, to keep company with Captain Saunders, both to assist him in unloading the sloop, and also that, by spreading in their cruise, there might be less danger of any of the enemy's ships slipping by unobserved. These orders being despatched, the Centurion parted from them at eleven in the evening on the 27th of September, directing her course to the southward, with a view of cruising for some days to the windward of Valparaiso.

And now by this disposition of our ships we flattered ourselves that we had taken all the advantages of the enemy that we possibly could with our small force, since our disposition was doubtless the most prudent that could be projected. For as we might suppose the Gloucester by this time to be drawing near her station off the high land of Paita, we were enabled by our separate stations, to intercept all vessels employed either betwixt Peru and Chili to the southward, or betwixt Panama and Peru to the northward. Since the principal trade from Peru to Chili being carried on to the port of Valparaiso, the Centurion cruising to the windward of Valparaiso would in all probability meet with them, as it is the constant practice of those ships to fall in with the coast to the windward of that port. And the Gloucester would, in like manner, be in the way of the trade bound from Panama or the northward to any part of Peru; since the high land off which she was stationed is constantly made by all ships in that voyage. And whilst the Centurion and Gloucester were thus situated for interrupting the enemy's trade, the Trial's prize and Centurion's prize were as conveniently stationed for preventing all intelligence, by intercepting all ships bound from Valparaiso to the northward; for it was on board these vessels that it was to be feared some account of us might possibly be sent to Peru.

But the most prudent dispositions carry with them only a probability of success, and can never ensure its certainty; since those chances, which it was reasonable to overlook in deliberations, are sometimes of most powerful influence in execution. Thus, in the present case, the distress of the Trial, and the quitting our station to assist her (events which no degree of prudence could either foresee or obviate) gave an opportunity to all the ships bound to Valparaiso to reach that port without molestation during this unlucky interval; so that though, after leaving Captain Saunders, we were very expeditious in regaining our station, where we got the 29th at noon,[1] yet in plying on and off till the 6th of October we had not the good fortune to discover a sail of any sort.[2] And then, having lost all hopes

or 130 miles south-east from Lima; Nasca Point is about 100 miles in the same direction from Pisco.

[1] Thomas, who frequently differs in date from Mr Walter, says that "on the 30th we saw the main land of Chili. This day we began to exercise our people with small arms, which was the first time we had done it since we came into those seas, and which we continued at all proper opportunities during the voyage."

[2] Thomas here notices a dissension among the ships' companies, of which Mr Walter, with an obvious official bias, says not a word: "On the 5th, the Commodore being informed that there were murmurings amongst the people, because the prize-money was not immediately divided, ordered the articles of war to be read; and after that remonstrated to them on the danger of mutiny, and said he had heard the reason of their discontent, but assured them their properties were secured by act of parliament as firmly as any one's own inheritance, and that the money, plate, &c., were weighed and marked in public; so that any capable person, if he pleased, might take an inventory of the whole. He then read an account of the particulars, and told them they might (if they pleased) make choice of any per-

of making any advantage by a longer stay, we made sail to the leeward of the port, in order to join our prizes; but when we arrived on the station appointed for them, we did not meet with them, though we continued there four or five days. We supposed that some chase had occasioned their leaving their station, and therefore we proceeded down the coast to the high land of Nasca, where Captain Saunders was directed to join us. Here we arrived on the 21st, and were in great expectation of meeting with some of the enemy's ships on the coast, as both the accounts of former voyages and the information of our prisoners assured us that all ships bound to Callao constantly make this land, to prevent the danger of running to the leeward of the port. But notwithstanding the advantages of this station we saw no sail till the 2d of November, when two ships appeared in sight together; we immediately gave them chase, but soon perceived that they were the Trial's and Centurion's prizes. As they had the wind of us, we brought to and waited their coming up, when Captain Saunders came on board us, and acquainted the Commodore that he had cleared the Trial pursuant to his orders, and having scuttled her he remained by her till she sank, but that it was the 4th of October before this was effected; for there ran so large and hollow a sea, that the sloop, having neither masts nor sails to steady her, rolled and pitched so violently that it was impossible for a boat to lie alongside of her for the greatest part of the time. And during this attendance on the sloop they were all driven so far to the northwest, that they were afterwards obliged to stretch a long way to the westward to regain the ground they had lost, which was the reason that we had not met with them on their station as we expected. We found they had not been more fortunate in their cruise than we were, for they had seen no vessel since they separated from us.

The little success we all had, and our certainty that had any ships been stirring in these seas for some time past we must have met with them, made us believe that the enemy at Valparaiso, on missing the two ships we had taken, had suspected us to be in the neighbourhood, and had consequently laid an embargo on all the trade in the southern ports. We likewise apprehended that they might by this time be fitting out the men-of-war at Callao, for we knew that it was no uncommon thing for an express from Valparaiso to reach Lima in twenty-nine or thirty days, and it was now more than fifty since we had taken our first prize. These apprehensions of an embargo along the coast, and of the equipment of the Spanish squadron at Callao, determined the Commodore to hasten down to the leeward of Callao, and to join Captain Mitchel (who was stationed off Paita) as soon as possible, that, our strength being united, we might be prepared to give the ships from Callao a warm reception if they dared to put to sea. With this view we bore away the same afternoon, taking particular care to keep at such a distance from the shore that there might be no danger of our being discovered from thence; for we knew that all the country ships were commanded, under the severest penalty, not to sail by the port of Callao without stopping; and as this order was constantly complied with, we should undoubtedly be known for enemies if we were seen to act contrary to it. In this new navigation, not being certain whether we might not meet the Spanish squadron in our route, the Commodore took on board the Centurion part of his crew with which he had formerly manned the Carmelo. And now, standing to the northward, we, before night came on, had a view of the small island called St Gallan,[1] which bore from us NNE. half E.,

son to take an inventory for them, or buy their parts. This spread a visible joy, and gave content to every one."

[1] Just to the southward of the well-known Chincha Islands, in the opening of Pisco Bay.

about seven leagues distant. This island lies in the Latitude of about 14° S., and about five miles to the northward of a high land called Morro Viejo, or the Old Man's Head. I mention this island and the high land near it more particularly, because between them is the most eligible station on that coast for cruising upon the enemy, as all ships bound to Callao, whether from the northward or the southward, run well in with the land in this part. By the 5th of November, at three in the afternoon, we were advanced within view of the high land of Barranca, lying in the Latitude of 10° 36′ S., bearing from us NE. by E., distant eight or nine leagues; and an hour and a half afterwards we had the satisfaction we had so long wished for, of seeing a sail. She first appeared to leeward, and we all immediately gave her chase; but the Centurion so much outsailed the two prizes, that we soon ran them out of sight, and gained considerably on the chase. However, night coming on before we came up with her, we about 7 o'clock lost sight of her, and were in some perplexity what course to steer; but at last Mr Anson resolved, as we were then before the wind, to keep all his sails set, and not to change his course. For though we had no doubt but the chase would alter her course in the night; yet, as it was uncertain what tack she would go upon, it was thought more prudent to keep on our course, as we must by this means unavoidably near her, than to change it on conjecture, when if we should mistake we must infallibly lose her. Thus, then, we continued the chase about an hour and half in the dark, some one or other on board us constantly imagining they discerned her sails right ahead of us; but at last Mr Brett, then our second lieutenant, did really discover her about four points on the larboard-bow, steering off to the seaward. We immediately clapped the helm a-weather, and stood for her, and in less than an hour came up with her; and having fired fourteen shots at her, she struck. Our third lieutenant, Mr Dennis, was sent in the boat with sixteen men to take possession of the prize, and to return the prisoners to our ship. This ship was named the Santa Teresa de Jesus, built at Guayaquil, of about 300 tons burthen, and was commanded by Bartolome Urrunaga, a Biscayer. She was bound from Guayaquil to Callao; her loading consisted of timber, cacao, cocoa-nuts, tobacco, hides, Pito thread (which is very strong, and is made of a species of grass), Quito cloth, wax, &c. The specie on board her was inconsiderable, being principally small silver money, and not amounting to more than £170 sterling. It is true, her cargo was of great value, could we have disposed of it; but the Spaniards having strict orders never to ransom their ships, all the goods that we took in these seas, except what little we had occasion for ourselves, were of no advantage to us. Indeed, though we could make no profit thereby ourselves, it was some satisfaction to us to consider that it was so much really lost to the enemy, and that the despoiling them was no contemptible branch of that service in which we were now employed by our country.[1]

---

[1] Thomas gives a quaint and amusing account of the use the squadron made of those stores so much despised by the Chaplain: "The 7th, we were employed in getting aboard several necessary stores, as planks, cordage, and the like, for the use of our squadron. The 9th, we brought from on board the Teresa ten serons of cocoa, one of wax, and 180 fathom of three and a half rope. The 10th, we brought from on board our first prize, the Carmelo, the following goods, viz., cloth, two bales; baize, five ditto; sugar, 182 loaves; straw mats, two; tar, one skin; raisins, three bales; indigo, four serons; cotton cloth, one bale; hats, two cases and twenty-five loose ones: skins, one parcel; chocolate, one bag; camlet, one bale and two parcels; silks, one box; lead, four pigs; and combs, one small parcel." After such an enumeration, one is better able to

Besides our prize's crew, which amounted to forty-five hands, there were on board her ten passengers, consisting of four men and three women, who were natives of the country, born of Spanish parents, and three black female slaves that attended them. The women were a mother and her two daughters, the eldest about twenty-one, and the youngest about fourteen. It is not to be wondered at that women of these years should be excessively alarmed at falling into the hands of an enemy whom, from the former outrages of the Buccaneers, and by the artful insinuations of their priests, they had been taught to consider as the most terrible and brutal of all mankind. These apprehensions, too, were in the present instance exaggerated by the singular beauty of the youngest of the women, and the riotous disposition which they might well expect to find in a set of sailors that had not seen a woman for near a twelvemonth. Full of these terrors, the women all hid themselves when our officer went on board; and, when they were found out, it was with great difficulty that he could persuade them to approach the light. However, he soon satisfied them, by the humanity of his conduct and his assurances of their future security and honourable treatment, that they had nothing to fear; and the Commodore being informed of the matter, sent directions that they should be continued on board their own ship, with the use of the same apartments, and with all the other conveniences they had enjoyed before, giving strict orders that they should receive no kind of inquietude or molestation whatever. And that they might be the more certain of having these orders complied with, or of complaining if they were not, the Commodore permitted the pilot, who in Spanish ships is generally the second person on board, to stay with them as their guardian and protector. He was particularly chosen for this purpose by Mr Anson, as he seemed to be extremely interested in all that concerned the women, and had at first declared that he was married to the youngest of them; though it afterwards appeared, both from the information of the rest of the prisoners and other circumstances, that he had asserted this with a view the better to secure them from the insults they expected on their first falling into our hands. By this compassionate and indulgent behaviour of the Commodore, the consternation of our female prisoners entirely subsided, and they continued easy and cheerful during the whole time they were with us, as I shall have occasion to mention more particularly hereafter.

I have before observed, that at the beginning of this chase the Centurion ran her two consorts out of sight, for which reason we lay by all the night, after we had taken the prize, for Captain Saunders and Lieutenant Saumarez to join us, firing guns and making false fires every half hour, to prevent their passing us unobserved; but they were so far astern that they neither heard nor saw any of our signals, and were not able to come up with us till broad daylight. When they had joined us, we proceeded together to the northward, being now four sail in company. We here found the sea, for many miles round us, of a beautiful red colour: this, upon examination, we imputed to an immense quantity of spawn spread upon its surface; and taking up some of the water in a wine-glass, it soon changed from a dirty aspect to a clear crystal, with only some red globules of a slimy nature floating on the top. And now, having a supply of timber on board our new prize, the Commodore ordered our boats to be repaired, and a swivel gun-stock to be fixed in the bow, both of the barge and pinnace, in order to increase their force, in case we should be obliged to have recourse to them for boarding ships or for any attempts on shore. As we stood from hence to the northward nothing remarkable occurred for two or three days, though

understand how ships in those days could keep at sea for years without their crews losing the habits and semblance of civilisation.

we spread our ships in such a manner that it was not probable any vessel of the enemy could escape us. In our run along this coast we generally observed that there was a current which set us to the northward at the rate of ten or twelve miles each day. And now, being in about 8° of S. Latitude, we began to be attended with vast numbers of flying fish and bonitos, which were the first we saw after our departure from the coast of Brazil. But it is remarkable, that on the east side of South America they extended to a much higher latitude than they do on the west side; for we did not lose them on the coast of Brazil till we approached the southern tropic. The reason for this diversity is doubtless the different degrees of heat obtaining in the same latitude on different sides of that continent.[1]

On the 10th of November we were three leagues south of the southernmost Island of Lobos, lying in the Lat. of 60° 27′ S. There are two islands of this name: this, called Lobos de la Mar; and another, which lies to the northward of it, very much resembling it in shape and appearance, and often mistaken for it, called Lobos de Tierra. We were now drawing near to the station appointed to the Gloucester; for which reason, fearing to miss her, we made an easy sail all night. The next morning, at daybreak, we saw a ship in shore, and to windward, plying up to the coast. She had passed by us with the favour of the night, and we, soon perceiving her not to be the Gloucester, got our tacks on board and gave her chase; but it proving very little wind, so that neither of us could make much way, the Commodore ordered the barge, his pinnace, and the Trial's pinnace, to be manned and armed, and to pursue the chase and board her. Lieutenant Brett, who commanded the barge, came up with her first, about 9 o'clock, and running alongside of her, he fired a volley of small shot between the masts, just over the heads of the people on board, and then instantly entered with the greatest part of his men; but the enemy made no resistance, being sufficiently frightened by the dazzling of the cutlasses, and the volley they had just received. Lieutenant Brett ordered the sails to be trimmed, and bore down to the Commodore, taking up in his way the two pinnaces. When he was arrived within about four miles of us, he put off in the barge, bringing with him a number of the prisoners, who had given him some material intelligence which he was desirous the Commodore should be acquainted with as soon as possible. On his arrival we learned that the prize was called Nuestra Señora del Carmen, of about 270 tons burthen; she was commanded by Marcos Morena,[2] a native of Venice, and had on board forty-three mariners. She was deep laden with steel, iron, wax, pepper, cedar, plank, snuff, rosaries, European bale goods, powder-blue, cinnamon, Romish indulgences, and other species of merchandise. And though this cargo, in our present circumstances, was but of little value to us, yet with respect to the Spaniards it was the most considerable capture that fell into our hands in this part of the world; for it amounted to upwards of 400,000 dollars prime cost at Panama. This ship was bound to Callao, and had stopped at Paita in her passage to take in a recruit of water and provisions, and had not left that place above twenty-four hours before she fell into our hands.

I have mentioned that Mr Brett had received some important intelligence from the prisoners, which he endeavoured to acquaint the Commodore with immediately. The first person he received it from (though upon further examination it was con-

[1] Here we omit a long digression "on the heat and cold of different climates, and on the varieties which occur in the same place in different parts of the year, and in different places lying in the same degrees of latitude."

[2] Or Marco Marina.

firmed by the other prisoners) was one John Williams, an Irishman, whom he found on board the Spanish vessel. Williams was a Papist, who worked his passage from Cadiz, and had travelled over all the kingdom of Mexico as a pedlar. He pretended that by this business he had got 4000 or 5000 dollars; but that he was embarrassed by the priests, who knew he had money, and was at last stripped of all he had. He was, indeed, at present all in rags, being but just got out of Paita gaol, where he had been confined for some misdemeanour; he expressed great joy upon seeing his countrymen, and immediately informed them, that a few days before a vessel came into Paita, where the master of her informed the Governor that he had been chased in the offing by a very large ship, which, from her size, and the colour of her sails, he was persuaded must be one of the English squadron. This we then conjectured to have been the Gloucester, as we afterwards found it was. The Governor, upon examining the master, was fully satisfied of his relation, and immediately sent away an express to Lima to acquaint the Viceroy therewith; and the royal officer residing at Paita, being apprehensive of a visit from the English, was busily employed in removing the King's treasure, and his own, to Piura, a town within land about fourteen leagues distant. We further learned from our prisoners, that there was a very considerable sum of money,[1] belonging to some merchants at Lima, that was now lodged at the customhouse at Paita; and that this was intended to be shipped on board a vessel which was then in the port of Paita, and was preparing to sail with the utmost expedition, being bound for the Bay of Sonsonnate, on the coast of Mexico, in order to purchase a part of the cargo of the Manilla ship. This vessel at Paita was esteemed a prime sailer, and had just received a new coat of tallow on her bottom; and, in the opinion of the prisoners, she might be able to sail the succeeding morning.

[1] According to Thomas's account, 400,000 dollars.

The character they gave us of this vessel, on which the money was to be shipped, left us little reason to believe that our ship, which had been in the water near two years, could have any chance of coming up with her, if we once suffered her to escape out of the port. And therefore, as we were now discovered, and the coast would be soon alarmed, and as our cruising in these parts any longer would answer no purpose, the Commodore resolved to surprise the place, having first minutely informed himself[2] of its strength and condition, and being fully satisfied that there was little danger of losing many of our men in the attempt. This surprise of Paita, besides the treasure it promised us, and its being the only enterprise it was in our power to undertake, had these other advantages attending it, that we should in all probability supply ourselves with great quantities of live provision, of which we were at this time in want. And we should likewise have an opportunity of setting our prisoners on shore, who were now very numerous, and made a greater consumption of our food than our stock that remained was capable of furnishing long. In all these lights, the attempt was a most eligible one, and what our necessities, our situation, and every prudential consideration prompted us to. How it succeeded, and how far it answered our expectations, shall be the subject of the following Chapter.

CHAPTER VI.

THE town of Paita is situated in the Latitude of 50° 12′ S., in a most barren soil, composed only of sand and slate; the extent of it is but small, containing in all less than 200 families. The houses are only groundfloors, the walls built of split cane and

[2] By examining the prisoners.

mud, and the roofs thatched with leaves. These edifices, though extremely slight, are abundantly sufficient for a climate where rain is considered as a prodigy, and is not seen in many years; so that it is said that a small quantity of rain falling in this country in the year 1728, it ruined a great number of buildings, which mouldered away, and, as it were, melted before it. The inhabitants of Paita are principally Indians and black slaves, or at least a mixed breed, the whites being very few. The port of Paita, though in reality little more than a bay, is esteemed the best on that part of the coast, and is indeed a very secure and commodious anchorage. It is greatly frequented by all vessels coming from the north, since it is here only that the ships from Acapulco, Sonsonnate, Realejo, and Panama can touch and refresh in their passage to Callao; and the length of these voyages (the wind for the greatest part of the year being full against them) renders it impossible to perform them without calling upon the coast for a recruit of fresh water. It is true, Paita is situated on so parched a spot that it does not itself furnish a drop of fresh water, or any kind of greens or provisions, except fish and a few goats; but there is an Indian town called Colan, about two or three leagues distant to the northward, whence water, maize, greens, fowls, &c., are brought to Paita on *balsas*, or floats, for the convenience of the ships that touch here; and cattle are sometimes brought from Piura, a town which lies about fourteen leagues up in the country. The water brought from Colan is whitish, and of a disagreeable appearance, but is said to be very wholesome; for it is pretended by the inhabitants that it runs through large woods of sarsaparilla, and that it is sensibly impregnated therewith. This port of Paita, besides furnishing the northern trade bound to Callao with water and necessaries, is the usual place where passengers from Acapulco or Panama, bound to Lima, disembark; for as it is 200 leagues from hence to Callao, the port of Lima, and as the wind is generally contrary, the passage by sea is very tedious and fatiguing; but by land there is a tolerably good road parallel to the coast, with many stations and villages for the accommodation of travellers. The town of Paita is itself an open place; its sole protection and defence is [a small fort or redoubt near the shore of the bay]. It was of consequence to us to be well informed of the fabric and strength of this fort; and by the examination of our prisoners we found that there were eight pieces of cannon mounted in it, but that it had neither ditch nor outwork, being only surrounded by a plain brick wall; and that the garrison consisted of only one weak company, but the town itself might possibly arm 300 men more.

Mr Anson, having informed himself of the strength of the place, resolved to attempt it that very night.[1] We were then about twelve leagues distant from the shore, far enough to prevent our being discovered, yet not so far but that, by making all the sail we could, we might arrive in the bay with our ships in the night. However, the Commodore prudently considered that this would be an improper method of proceeding, as our ships, being such large bodies, might be easily discovered at a distance even in the night, and might thereby alarm the inhabitants and give them an opportunity of removing their valuable effects. He therefore, as the strength of the place did not require our whole force, resolved to attempt it with our boats only, ordering the eighteen-oared barge, and our own and the Trial's pinnaces, on that service; and having picked out fifty-eight men to man them, well provided with arms and ammunition, he gave the command of the expedition to Lieutenant Brett, and gave him his necessary orders. And the better to prevent the disappointment and confusion which might arise from the darkness of the night, and the ignorance of the streets and passages of the place; two of the Spanish pilots

[1] The 12th of November 1741.

were ordered to attend the lieutenant, and to conduct him to the most convenient landing-place, and were afterwards to be his guides on shore. And that we might have the greater security for their faithful behaviour on this occasion, the Commodore took care to assure all our prisoners, that if the pilots acted properly they should all of them be released and set on shore at this place; but in case of any misconduct or treachery, he threatened them that the pilots should be instantly shot, and that he would carry all the rest of the Spaniards who were on board him prisoners to England. So that the prisoners themselves were interested in our success; and therefore we had no reason to suspect our conductors either of negligence or perfidy. And on this occasion I cannot but remark a singular circumstance of one of the pilots employed by us in this business. It seems (as we afterwards learned) he had been taken by Captain Clipperton above twenty years before, and had been forced to lead Clipperton and his people to the surprise of Truxillo, a town within land to the southward of Paita, where, however, he contrived to alarm his countrymen, and to save them, though the place was taken. Now that the only two attempts on shore, which were made at so long an interval from each other, should be guided by the same person, and he, too, a prisoner both times, and forced upon the employ contrary to his inclination, is an incident so very extraordinary that I could not help taking notice of it.

During our preparations, the ships themselves stood towards the port with all the sail they could make, being secure that we were yet at too great a distance to be seen. But about 10 o'clock at night, the ships being then within five leagues of the place, Lieutenant Brett, with the boats under his command, put off, and arrived at the mouth of the bay without being discovered; but no sooner had he entered it than some of the people on board a vessel riding at anchor there perceived him, who instantly put off in their boat, rowing towards the fort, shouting and crying, "The English! The English dogs!" by which the whole town was suddenly alarmed; and our people soon observed several lights hurrying backwards and forwards in the fort, and other marks of the inhabitants being in great motion. Lieutenant Brett on this encouraged his men to pull briskly up to the shore that they might give the enemy as little time as possible to prepare for their defence. However, before our boats could reach the shore, the people in the fort had got ready some of their cannon, and pointed them towards the landing-place; and though in the darkness of the night it might be well supposed that chance had a greater share than skill in their direction, yet the first shot passed extremely near one of the boats, whistling just over the heads of the crew. This made our people redouble their efforts, so that they had reached the shore, and were in part disembarked, by the time the second gun fired. As soon as our men landed, they were conducted by one of the Spanish pilots to the entrance of a narrow street, not above fifty yards distant from the beach, where they were covered from the fire of the fort; and being formed in the best manner the shortness of the time would allow, they immediately marched for the parade, which was a large square at the end of this street, the fort being one side of the square and the Governor's house another. In this march (though performed with tolerable regularity) the shouts and clamours of threescore sailors who had been confined so long on shipboard, and were now for the first time on shore in an enemy's country—joyous as they always are when they land, and animated besides in the present case with the hopes of an immense pillage—the huzzas, I say, of this spirited detachment, joined with the noise of their drums, and favoured by the night, had augmented their numbers, in the opinion of the enemy, to at least 300: by which persuasion the inhabitants were so greatly intimidated that they were much more solicitous about the means of their flight than of their resistance. So that though upon entering the parade our people received a volley from the merchants who owned the treasure then in the town, and who, with a few others, had ranged themselves in a gallery that ran round the Governor's house, yet that post was immediately abandoned upon the first fire made by our people, who were thereby left in quiet possession of the parade.

On this success Lieutenant Brett divided his men into two parties, ordering one of them to surround the Governor's house, and if possible to secure the Governor, whilst he himself with the other marched to the fort with an intent to force it. But, contrary to his expectation, he entered it without opposition;[1] for the enemy, on his approach, abandoned it, and made their escape over the walls. By this means the whole place was mastered in less than a quarter of an hour's time from the first landing, with no other loss than that of one man killed on the spot, and two wounded;[2] one of whom was the Spanish pilot of the Teresa, who received a slight bruise by a ball which grazed on his wrist. Indeed, another of the company, the Honourable Mr Keppel, son to the Earl of Albemarle, had a very narrow escape; for having on a jockey cap, one side of the peak was shaved off close to his temple by a ball, which, however, did him no other injury. And now Lieutenant Brett, after this success, placed a guard at the fort, and another at the Governor's house, and appointed sentinels at all the avenues of the town, both to prevent any surprise from the enemy, and to secure the effects in the place from being embezzled. And this being done, his next care was to seize on the custom-house where the treasure lay, and to examine if any of the inhabitants remained in the town, that he might know what further precautions it was necessary to take. But he soon found that the numbers left behind were no ways formidable: for the greatest part of them (being in bed when the place was surprised) had run away with so much precipitation, that they had not given themselves time to put on their clothes.[3] And in this precipitate rout the Governor was not the last to secure himself, for he fled betimes, half-naked, leaving his wife, a young lady of about seventeen years of age to whom he had been married but three or four days, behind him; though she too was afterwards carried off in her shift by a couple of sentinels, just as the detachment ordered to invest the house arrived before it. This escape of the Governor was an unpleasing circumstance, as Mr Anson had particularly recommended it to Lieutenant Brett to secure his person if possible, in hopes that by that means

[1] "On our getting possession of the castle," says Thomas, "our commanding officer very inconsiderately ordered the guns to be thrown over the walls, which accordingly was executed; but some time after, reflecting on the ill-consequence which might attend that proceeding, he ordered two of them to be got up and remounted."

[2] In Thomas's narrative we are told more particularly: "We lost one man, Peter Obrian the Commodore's steward, who was shot through the breast by a musket ball; and had two wounded, to wit, Arthur Lusk, a quarter-master, and the Spanish pilot of the Teresa, whom we had made use of as a guide; and I have had it reported from several officers then on shore, that our men ran to the attack, and fired in so irregular a manner, that it was, and still remains a doubt, whether those were not shot by our people rather than by the enemy."

[3] "These people," says Thomas contemptuously enough, "having enjoyed a long peace, and being enervated by the luxury so customary in those parts, their arms in a bad condition, and no person of experience or courage to head them, it is no wonder that they made so small a resistance, and were all driven out of the town in less than half-an-hour by only forty-nine men."

we might be able to treat for the ransom of the place; but it seems his alertness rendered it impossible to seize him. The few inhabitants who remained were confined in one of the churches under a guard, except some stout Negroes who were found in the place; these, instead of being shut up, were employed the remaining part of the night to assist in carrying the treasure from the custom-house and other places to the fort: however, there was care taken that they should be always attended by a file of musketeers.

The transporting the treasure from the custom-house to the fort was the principal occupation of Mr Brett's people after he had got possession of the place. But the sailors, while they were thus employed, could not be prevented from entering the houses which lay near them, in search of private pillage. And, the first things which occurred to them being the clothes which the Spaniards in their flight had left behind them, and which, according to the custom of the country, were most of them either embroidered or laced, our people eagerly seized these glittering habits, and put them on over their own dirty trousers and jackets; not forgetting, at the same time, the tie or bag-wig, and laced hat, which were generally found with the clothes. When this practice was once begun, there was no preventing the whole detachment from imitating it; and those who came latest into the fashion, not finding men's clothes sufficient to equip themselves, were obliged to take up with women's gowns and petticoats, which (provided there was finery enough) they made no scruple of putting on and blending with their own greasy dress. So that, when a party of them thus ridiculously metamorphosed first appeared before Mr Brett, he was extremely surprised at their appearance, and could not immediately be satisfied they were his own people.

These were the transactions of our detachment on shore at Paita the first night: and now to return to what was done on board the Centurion in that interval. I must observe, that after the boats were gone off we lay by till 1 o'clock in the morning, and then, supposing our detachment to be near landing, we made an easy sail for the bay. About seven in the morning we began to open the bay, and soon after we had a view of the town; and though we had no reason to doubt of the success of the enterprise, yet it was with great joy that we first discovered an infallible signal of the certainty of our hopes: this was by means of our perspectives, for through them we saw an English flag hoisted on the flagstaff of the fort, which to us was an incontestible proof that our people had got possession of the town. We plied into the bay with as much expedition as the wind, which then blew off shore, would permit us, and at eleven the Trial's boat came on board us, laden with dollars and church-plate; and the officer who commanded her informed us of the preceding night's transactions, such as we have already related them. About two in the afternoon we came to an anchor in ten fathoms and a half, at a mile and a half distance from the town, and were consequently near enough to have a more immediate intercourse with those on shore. And now we found that Mr Brett had hitherto gone on in collecting and removing the treasure without interruption; but that the enemy had rendezvoused from all parts of the country on a hill at the back of the town, where they made no inconsiderable appearance: for, amongst the rest of their force, there were 200 horse seemingly very well armed and mounted, and, as we conceived, properly trained and regimented, being furnished with trumpets, drums, and standards. These troops paraded about the hill with great ostentation, sounding their military music, and practising every art to intimidate us (as our numbers on shore were by this time not unknown to them), in hopes that we might be induced by our fears to abandon the place before the pillage was completed. But we were not so ignorant as to believe that this body of horse, which seemed to be what the enemy principally depended on, would dare to venture in streets and among houses, even had their numbers been three times as great; and therefore, notwithstanding their menaces, we went on, as long as the daylight lasted, calmly, in sending off the treasure, and in employing the boats to carry on board the refreshments, such as hogs, fowls, &c., which we found here in great abundance. But at night, to prevent any surprise, the Commodore sent on shore a reinforcement, who posted themselves in all the streets leading to the parade; and for their greater security they traversed the streets with barricades six feet high: and the enemy continuing quiet all night, we at daybreak returned again to our labour of loading the boats and sending them off.

By this time we were convinced of what consequence it would have been to us had fortune seconded the prudent views of the Commodore, by permitting us to have secured the Governor. For we found in the place many storehouses full of valuable effects, which were useless to us at present, and such as we could not find room for on board. But had the Governor been in our power, he would in all probability have treated for a ransom, which would have been extremely advantageous both to him and us; Whereas he being now at liberty, and having collected all the force of the country for many leagues round, and having even got a body of militia from Piura, he was so elated with his numbers, and so fond of his new military command, that he seemed not to trouble himself about the fate of his government. So that though Mr Anson sent several messages to him by the inhabitants who were in our power, desiring him to enter into a treaty for the ransom of the town and goods, giving him at the same time an intimation that he should be far from insisting on a rigorous equivalent, but perhaps might be satisfied with some live cattle and a few necessaries for the use of the squadron, and assuring him too, that if he would not condescend at least to treat, he would set fire to the town and all the warehouses: yet the governor was so imprudent and arrogant, that he despised all these reiterated applications, and did not deign even to return the least answer to them.

On the second day of our being in possession of the place, several Negro slaves deserted from the enemy on the hill, and, coming into the town, voluntarily entered into our service. One of these was well known to a gentleman on board, who remembered him formerly at Panama. And the Spaniards without the town being in extreme want of water, many of their slaves crept into the place by stealth, and carried away several jars of water to their masters on the hill; and though some of them were seized by our men in the attempt, yet the thirst amongst the enemy was so pressing,[1] that they continued this practice till we left the place. And now, on this second day, we were assured both by the deserters and by these prisoners we took, that the Spaniards on the hill, who were by this time increased to a formidable number, had resolved to storm the town and fort the succeeding night; and that one Gordon, a Scotch Papist, and captain of a ship in those seas, was to have the command of this enterprise. But we, notwithstanding, continued sending off our boats, and prosecuted our work without the least hurry or precipitation till the evening; and then a reinforcement was again sent on shore by the Commodore, and Lieutenant Brett doubled his guards at each of the barricades; and our posts being

[1] Thomas says: "The country thereabouts being for many miles round quite barren and sandy, without either water or any other thing necessary for life, and the nearest town to them, named as I think Santa Cruz, whence relief might be got, being a day and a half or two days' journey off, the people who had left the town were in a starving condition, and we had melancholy accounts of several dying among them for want chiefly of water during our small stay."

connected by means of sentinels placed within call of each other, and the whole being visited by frequent rounds, attended with a drum, these marks of our vigilance, which the enemy could not be ignorant of, as they could doubtless hear the drum, if not the calls of the sentinels; these marks, I say, of our vigilance and of our readiness to receive them, cooled their resolution, and made them forget the vaunts of the preceding day; so that we passed the second night with as little molestation as we had done the first.

We had finished sending the treasure on board the Centurion the evening before; so that the third morning, being the 15th of November, the boats were employed in carrying off the most valuable part of the effects that remained in the town.[1] And the Commodore intending to sail this day, he about 10 o'clock, pursuant to his promise, sent all his prisoners, amounting to eighty-eight, on shore, giving orders to Lieutenant Brett to secure them in one of the churches under a strict guard till he was ready to embark his men. Mr Brett was at the same time ordered to set the whole town on fire,[2] except the two churches (which by good fortune stood at some distance from the other houses), and then he was to abandon the place and to come on board. These orders were punctually complied with; for Mr Brett immediately set his men to work to distribute pitch, tar, and other combustibles (of which great quantities were found here) into houses situated in different streets of the town; so that, the place being fired in many quarters at the same time, the destruction might be more violent and sudden, and the enemy, after our departure, might not be able to extinguish it. These preparations being made, he in the next place ordered the cannon which he found in the fort, to be nailed up;[3] and then, setting fire to those houses which were most windward, he collected his men, and marched towards the beach, where the boats waited to carry them off. And the part of the beach where he intended to embark being an open place without the town, the Spaniards on the hill, perceiving he was retreating, resolved to try if they could not precipitate his departure, and thereby lay some foundation for their future boasting. And for this purpose a small squadron of their horse, consisting of about sixty, picked out as I suppose for this service, marched down the hill with much seeming resolution; so that, had we not been prepossessed with a juster opinion of their prowess, we might have suspected that, now we were on the open beach with no advantage of situation, they would certainly have charged us. But we presumed (and we were not mistaken) that this was mere ostentation; for, notwithstanding the pomp and parade they advanced with, Mr Brett had no sooner ordered his men to halt and face about, but the enemy stopped their career, and never dared to advance a step farther.

When our people were arrived at their boats, and were ready to go on board, they were for some time delayed by missing one of their number; but being unable, by their mutual

[1] "Which," by Thomas's account, "chiefly consisted of rich brocades, laced cloths, bales of fine linens and woollens, britannias, stays, and the like; together with a great number of hogs, some sheep and fowls, cases of Spanish brandies and wines, a great quantity of onions, olives, sweet-meats, and many other things too tedious to name, all which the sailors hoped would have been equally divided among the ships' companions, but they found themselves disappointed."

[2] The burning of Paita, inflicting cruel injury not on the Spanish Government but on an unoffending and industrious community, has been generally censured as a violation of the laws of civilised warfare. Earl Stanhope, usually slow to blame, says the act "has imprinted a deep blot on the glory of Lord Anson's expedition."

[3] Spiked.

inquiries amongst each other, to inform themselves where he was left, or by what accident he was detained, they, after considerable delay, resolved to get into their boats and to put off without him. And the last man was actually embarked, and the boats just putting off, when they heard him calling to them to take him in. The town was by this time so thoroughly on fire, and the smoke covered the beach so effectually, that they could scarcely see him, though they heard his voice. The lieutenant instantly ordered one of the boats to his relief, which found him up to the chin in water, for he had waded as far as he durst, being extremely frightened with the apprehensions of falling into the hands of an enemy, enraged, as they doubtless were, with the pillage and destruction of their town. On inquiring into the cause of his staying behind, it was found that he had taken that morning too large a dose of brandy, which had thrown him into so sound a sleep, that he did not awake till the fire came near enough to scorch him. He was strangely amazed on first opening his eyes, to see the place all on a blaze on one side, and several Spaniards and Indians not far from him on the other. The greatness and suddenness of his fright instantly reduced him to a state of sobriety, and gave him sufficient presence of mind to push through the thickest of the smoke, as the likeliest means to escape the enemy; and making the best of his way to the beach, he ran as far into the water as he durst (for he could not swim) before he ventured to look back. . . .

By the time our people had taken their comrade out of the water, and were making the best of their way for the squadron, the flames had taken possession of every part of the town, and had got such hold, both by means of combustibles that had been distributed for that purpose, and by the slightness of the materials of which the houses were composed and their aptitude to take fire, that it was sufficiently apparent no efforts of the enemy (though they flocked down in great numbers) could possibly put a stop to it, or prevent the entire destruction of the place, and all the merchandise contained therein.

Our detachment under Lieutenant Brett having safely joined the squadron, the Commodore prepared to leave the place the same evening. He found, when he first came into the bay, six vessels of the enemy at anchor; one of which was the ship which, according to our intelligence, was to have sailed with the treasure to the coast of Mexico, and which, as we were persuaded she was a good sailer, we resolved to take with us. The others were two snows, a bark, and two row-galleys of thirty-six oars a-piece; these last, as we were afterwards informed, with many others of the same kind built at different ports, were intended to prevent our landing in the neighbourhood of Callao; for the Spaniards, on the first intelligence of our squadron and its force, expected that we would attempt the city of Lima. The Commodore, having no occasion for these other vessels, had ordered the masts of all five of them to be cut away on his first arrival; and now, at his leaving the place, they were towed out of the harbour, and scuttled and sunk; and the command of the remaining ship, called the Solidad, being given to Mr Hughes, the lieutenant of the Trial, who had with him a crew of ten men to navigate her, the squadron towards midnight weighed anchor and sailed out of the bay, being now augmented to six sail, that is, the Centurion, and the Trial prize, together with the Carmelo, the Teresa, the Carmen, and our last acquired vessel, the Solidad.

And now, before I entirely quit the account of our transactions at this place, it may not, perhaps, be improper to give a succinct relation of the booty we made here, and of the loss the Spaniards sustained. I have before observed that there were great quantities of valuable effects in the town; but, as the greatest part of them were what we could neither dispose of nor carry away, the total amount of this merchandise can only be rudely guessed at. But the Spaniards, in the representations they made to the Court of Madrid (as we were afterwards assured), estimated their whole loss at a million and a half of dollars; and when it is considered that no small part of the goods we burned there were of the richest and most expensive species, as broadcloths, silks, cambrics, velvets, &c., I cannot but think their valuation sufficiently moderate. As to our part, our acquisition, though inconsiderable in comparison of what we destroyed, was yet in itself far from despicable; for the wrought plate, dollars, and other coin which fell into our hands, amounted to upwards of £30,000 sterling, besides several rings, bracelets, and jewels, whose intrinsic value we could not then determine; and over and above all this, the plunder which became the property of the immediate captors was very great; so that upon the whole it was by much the most important booty we made upon that coast.

There remains, before I take leave of this place, another particularity to be mentioned, which, on account of the great honour which our national character in those parts has thence received, and the reputation which our Commodore in particular has thereby acquired, merits a distinct and circumstantial discussion. It has been already related that all the prisoners taken by us in our preceding prizes were put on shore and discharged at this place; amongst which there were some persons of considerable distinction, particularly a youth of about seventeen years of age, son of the Vice-President of the Council of Chili. As the barbarity of the Buccaneers, and the artful use the [Spanish] ecclesiastics had made of it, had filled the natives of those countries with the most terrible ideas of the English cruelty, we always found our prisoners, at their first coming on board us, to be extremely dejected and under great horror and anxiety. In particular, this youth, whom I last mentioned, having never been from home before, lamented his captivity in the most moving manner, regretting in very plaintive terms his parents, his brothers, his sisters, and his native country, of all which he was fully persuaded he had taken his last farewell, believing that he was now devoted for the remaining part of his life to an abject and cruel servitude; nor was he singular in his fears, for his companions on board, and indeed all the Spaniards that came into our power, had the same desponding opinion of their situation. Mr Anson constantly exerted his utmost endeavours to efface these inhuman impressions they had received of us; always taking care that as many of the principal people among them as there was room for should dine at his table by turns; and giving the strictest orders, too, that they should at all times, and in every circumstance, be treated with the utmost decency and humanity. But, notwithstanding this precaution, it was generally observed that for the first day or two they did not quit their fears, but suspected the gentleness of their usage to be only preparatory to some unthought-of calamity. However, being confirmed by time, they grew perfectly easy in their situation, and remarkably cheerful, so that it was often disputable whether or no they considered their being detained by us as a misfortune. For the youth I have above mentioned, who was near two months on board us, had at last so far conquered his melancholy surmises, and had taken such an affection to Mr Anson, and seemed so much pleased with the manner of life, totally different from all he had ever seen before, that it is doubtful to me whether, if his own opinion had been taken, he would not have preferred a voyage to England in the Centurion to the being set on shore at Paita, where he was at liberty to return to his country and his friends.

This conduct of the Commodore to his prisoners, which was continued without interruption or deviation, gave them all the highest idea of his humanity and benevolence, and induced them likewise (as mankind are fond of forming general opinions) to entertain very favourable thoughts of the whole English nation. But whatever they might be disposed to think of Mr Anson before the taking of the Teresa, their veneration for him was prodigiously increased by his conduct towards those women whom (as I have already mentioned) he took in that vessel. For the leaving them in the possession of their apartments, the strict orders given to prevent all his people on board from approaching them, and the permitting the pilot to stay with them as their guardian, were measures that seemed so different from what might be expected from an enemy and an heretic, that the Spaniards on board, though they had themselves experienced his beneficence, were surprised at this new instance of it; and the more so, as all this was done without his ever having seen the women, though the two daughters were both esteemed handsome, and the youngest was celebrated for her uncommon beauty. The women themselves, too, were so sensible of the obligations they owed him for the care and attention with which he had protected them, that they absolutely refused to go on shore at Paita till they had been permitted to wait on him on board the Centurion, to return him thanks in person. Indeed, all the prisoners left us with the strongest assurances of their grateful remembrance of his uncommon treatment. A Jesuit, in particular, whom the Commodore had taken, and who was an ecclesiastic of some distinction, could not help expressing himself with great thankfulness for the civilities he and his countrymen had found on board, declaring that he should consider it as his duty to do Mr Anson justice at all times; adding, that his usage of the men prisoners was such as could never be forgotten, and such as he could never fail to acknowledge and recite upon all occasions; but that his behaviour to the women was so extraordinary, and so extremely honourable, that he doubted all the regard due to his own ecclesiastical character would be scarcely sufficient to render it credible. And, indeed, we were afterwards informed that both he and the rest of our prisoners had not been silent on this head, but had, both at Lima and other places, given the greatest encomiums to our Commodore; the Jesuit in particular, as we were told, having on his account interpreted in a lax and hypothetical sense that article of his Church which asserts the impossibility of heretics being saved.

And let it not be imagined that the impression which the Spaniards hence received to our advantage is a matter of small import; for, not to mention several of our countrymen who have already felt the good effects of these prepossessions, the Spaniards are a nation whose good opinion of us is doubtless of more consequence than that of all the world besides. Not only as the commerce we have formerly carried on with them, and perhaps may again hereafter, is so extremely valuable, but also as the transacting it does so immediately depend on the honour and good faith of those who are entrusted with its management. But, however, [even] had no national conveniencies attended it, the Commodore's equity and good temper would not less have deterred him from all tyranny and cruelty to those whom the fortune of war had put into his hands. I shall only add, that by his constant attachment to these humane and prudent maxims he has acquired a distinguished reputation amongst the Creole Spaniards which is not confined merely to the coast of the South Seas, but is extended through all the Spanish settlements in America: so that his name is frequently to be met with in the mouths of[1] most of the Spanish inhabitants of that prodigious empire.[2]

[1] Ed. 1776: "Was universally mentioned with honour and applause by."

[2] Byron, who met, during his residence as a prisoner on parole in Chili, some of the released captives, says: "They all spoke in the highest terms of the kind treatment they had re-

CHAPTER VII.

When we got under sail from the road of Paita (which, as I have already observed, was about midnight on the 16th of November) we stood to the westward; and in the morning the Commodore gave orders that the whole squadron should spread themselves, in order to look out for the Gloucester; for we now drew near to the station where Captain Mitchel had been directed to cruise, and hourly expected to get sight of him, but the whole day passed without seeing him.

And now a jealousy which had taken its rise at Paita, between those who had been ordered on shore for the attack and those who had continued on board, grew to such a height, that the Commodore, being made acquainted with it, thought it necessary to interpose his authority to appease it.[1]

The ground of this animosity was the plunder gotten at Paita, which those who had acted on shore had appropriated to themselves, and considered it as a reward for the risks they had run and the resolution they had shown in that service. But those who had remained on board considered this as a very partial and unjust procedure, urging that, had it been left to their choice, they should have preferred the acting on shore to the continuing on board; that their duty, while their comrades were on shore, was extremely fatiguing, for besides the labour of the day they were constantly under arms all night to secure the prisoners, whose numbers exceeded their own, and of whom it was then necessary to be extremely watchful, to prevent any attempts they might have formed in that critical conjuncture; that upon the whole it could not be denied but that the presence of a sufficient force on board was as necessary to the success of the enterprise, as the action of the others on shore; and therefore those who had continued on board insisted that they could not be deprived of their share of the plunder, without manifest injustice. These were the contests amongst our men, which were carried on with great heat on both sides; and though the plunder in question was a very trifle in comparison of the treasure taken in the place (in which there was no doubt but those on board had an equal right), yet as the obstinacy of sailors is not always regulated by the importance of the matter in dispute, the Commodore thought it necessary to put a stop to this ferment betimes. And accordingly, the morning after our leaving Paita, he ordered all hands upon the quarter-deck, where, addressing himself to those who had been detached on shore, he commended their behaviour, and thanked them for their services on that occasion; but then, representing to them the reasons urged by those who had continued on board for an equal distribution of the plunder, he told them that he thought these reasons very conclusive, and that the expectations of their comrades were

---

ceived; and some of them told us they were so happy on board the Centurion, that they would not have been sorry if the Commodore had taken them with him to England." Still more remarkable, however, is Captain Basil Hall's testimony, in his "South America:" "Lord Anson's proceedings are still traditionally known at Paita; and it is curious to observe that the kindness with which that sagacious officer invariably treated his Spanish prisoners is, at the distance of eighty years, better known and more dwelt upon by the inhabitants of Paita than the capture and wanton destruction of the town."

[1] Thomas tells a very different story about this division of the spoil: "The 22d, a division was made of the plunder of Paita, and the Commodore not appearing in that affair, it was done at the pleasure, and to the entire satisfaction of five or six (no doubt) very disinterested officers; and, indeed, most things of this nature, during the course of the voyage being managed with the same discretion and honour, no room was left for complaining of particular partialities."

justly founded; and therefore he ordered, that not only the men, but all the officers likewise, who had been employed in taking the place, should produce the whole of their plunder immediately upon the quarter-deck; and that it should be impartially divided amongst the whole crew, in proportion to each man's rank and commission. And to prevent those who had been in possession of the plunder from murmuring at this diminution of their share, the Commodore added, that as an encouragement to others who might be hereafter employed on like services, he would give his entire share to be distributed amongst those who had been detached for the attack of the place. Thus this troublesome affair, which, if permitted to have gone on, might perhaps have been attended with mischievous consequences, was by the Commodore's prudence soon appeased, to the general satisfaction of the ship's company; not but there were some few whose selfish dispositions were uninfluenced by the justice of this procedure, and who were incapable of discerning the force of equity, however glaring, when it tended to deprive them of any part of what they had once got into their hands.

This important business employed the best part of the day after we came from Paita. And now at night, having no sight of the Gloucester, the Commodore ordered the squadron to bring to, that we might not pass her in the dark. The next morning we again looked out for her, and at ten we saw a sail, to which we gave chase; and at two in the afternoon we came near enough to her to discover her to be the Gloucester, with a small vessel in tow. About an hour after we were joined by them, and then we learned that Captain Mitchel, in the whole time of his cruise, had only taken two prizes, one of them being a small snow,[1] whose cargo consisted chiefly of wine, brandy, and olives in jars, with about £7000 in specie;[2] and the other a large boat or launch which the Gloucester's barge came up with near the shore. The prisoners on board this vessel alleged that they were very poor, and that their loading consisted only of cotton, though the circumstances in which the barge surprised them seemed to insinuate that they were more opulent than they pretended to be, for the Gloucester's people found them at dinner upon pigeon-pie served up in silver dishes. However, the officer who commanded the barge having opened several of the jars on board to satisfy his curiosity, and finding nothing in them but cotton, he was inclined to believe the account the prisoners gave him; but the cargo being taken into the Gloucester, and there examined more strictly, they were agreeably surprised to find that the whole was a very extraordinary piece of false package, and that there was concealed amongst the cotton, in every jar, a considerable quantity of double doubloons and dollars to the amount, in the whole, of near £12,000. This treasure was going to Paita, and belonged to the same merchants who were the proprietors of the greatest part of the money we had taken there; so that, had this boat escaped the Gloucester, it is probable her cargo would have fallen into our hands. Besides these two prizes which we have mentioned, the Gloucester's people told us that they had been in sight of two or three other ships of the enemy, which had escaped them; and one of them we had reason to believe, from some of our intelligence, was of an immense value.

Being now joined by the Gloucester and her prize, it was resolved that we

---

[1] Called the Del Oro.

[2] Thomas says: "On board this prize of the Gloucester were two horses, which being, I suppose, fat, and probably better food than their salt beef or pork, they killed and eat them; and this, I imagine, gave ground to that fiction which one of the spurious accounts of our voyage has given, of our eagerly hunting and eating wild horses, whereas in reality we never saw nor heard of a wild horse during our voyage."

should stand to the northwards, and make the best of our way either to Cape St Lucas, in California, or to Cape Corrientes on the coast of Mexico. Indeed the Commodore, when at Juan Fernandez, had determined with himself to touch in the neighbourhood of Panama, and to endeavour to get some correspondence overland with the fleet under the command of Admiral Vernon. For, when we departed from England, we left a large force at Portsmouth, which was intended to be sent to the West Indies, there to be employed in an expedition against some of the Spanish settlements.[1] And Mr Anson taking it for granted that this enterprise had succeeded, and that Porto Bello perhaps might be then garrisoned by British troops, he hoped that on his arrival at the Isthmus he should easily procure an intercourse with our countrymen on the other side, either by the Indians, who were greatly disposed in our favour, or even by the Spaniards themselves, some of whom, for proper rewards, might be induced to carry on this intelligence, which, after it was once begun, might be continued with very little difficulty. So that Mr Anson flattered himself that he might by this means have received a reinforcement of men from the other side, and that, by settling a prudent plan of operations with our commanders in the West Indies, he might have taken even Panama itself, which would have given to the British nation the possession of that Isthmus, whereby we should have been in effect masters of all the treasures of Peru, and should have had in our hands an equivalent for any demands, however extraordinary, which we might have been induced to have made on either of the branches of the House of Bourbon. Such were the projects which the Commodore resolved in his thoughts at the Island of Juan Fernandez, notwithstanding the feeble condition to which he was then reduced. And indeed, had the success of our force in the West Indies been answerable to the general expectation, it cannot be denied but these views would have been the most prudent that could have been thought of. But in examining the papers which were found on board the Carmelo, the first prize we took, we learned (though I then omitted to mention it) that our attempt against Carthagena had failed, and that there was no probability that our fleet in that part of the world would engage in any new enterprise that would at all facilitate this plan. And therefore Mr Anson gave over all hopes of being reinforced across the Isthmus, and consequently had no inducement at present to proceed to Panama, as he was incapable of attacking the place; and there was great reason to believe that by this time there was a general embargo on all the coast.

The only feasible measure, then, which was left us, was to get as soon as possible to the southern parts of California, or to the adjacent coast of Mexico, there to cruise for the Manilla galleon, which we knew was now at sea, bound to the port of Acapulco. And we doubted not to get on that station time enough to intercept her, for this ship does not [usually] arrive at Acapulco till towards the middle of January, and we were now but in the middle of November, and did not conceive that our passage thither would cost us above a month or five weeks; so that we imagined we had near twice as much time as was necessary for our purpose. . . .

Having determined to go to Quibo, we directed our course to the northward, being eight sail in company, and consequently having the appearance of a very formidable fleet; and on the 19th, at daybreak, we discovered Cape Blanco, bearing SSE. half E., seven miles distant. This cape lies in the Latitude of 4° 15′ S., and is always made by ships bound either to windward or to leeward, so that off this cape is a most excellent station to cruise upon the enemy. By this time we found that our last prize, the Solidad, was far from answering the character given her of a good sailer; and she and the Santa Teresa delay-

[1] See Note 3, page 15.

ing us considerably, the Commodore ordered them both to be cleared of everything that might prove useful to the rest of the ships, and then to be burned. Having given proper instructions, and a rendezvous to the Gloucester and the other prizes, we proceeded in our course for Quibo; and on the 22d, in the morning, saw the Island of Plata,[1] bearing E., distant four leagues. Here one of our prizes was ordered to stand close in with it, both to discover if there were any ships between that island and the continent, and likewise to look out for a stream of fresh water which was reported to be there, and which would have saved us the trouble of going to Quibo; but she returned without having seen any ship or finding any water. At three in the afternoon, Point Manta bore SE. by E., seven miles distant; and there being a town of the same name in the neighbourhood, Captain Mitchel took this opportunity of sending away several of his prisoners from the Gloucester in the Spanish launch. The boats were now daily employed in distributing provisions on board the Trial and other prizes to complete their stock for six months; and that the Centurion might be the better prepared to give the Manilla ship (one of which we were told was of an immense size) a warm reception, the carpenters were ordered to fix eight stocks in the main and fore tops, which were properly fitted for the mounting of swivel guns.

On the 25th we had a sight of the Island of Gallo, bearing ESE. half E., four leagues distant; and hence we crossed the Bay of Panama with a NW. course, hoping that this would have carried us in a direct line to the Island of Quibo. But we afterwards found that we ought to have stood more to the westward; for the winds in a short time began to incline to that quarter, and made it difficult for us to gain the island.

On the 27th, Captain Mitchel having finished the clearing of his largest prize, she was scuttled and set on fire; but we still consisted of five ships and were fortunate enough to find them all good sailers, so that we never occasioned any delay to each other. Being now in a rainy climate, which we had been long disused to, we found it necessary to calk the sides of the Centurion, to prevent the rain-water from running into her. On the 3d of December we had a view of the Island of Quibo; the east end of which then bore from us NNW., four leagues distant, and the Island of Quicara WNW., at about the same distance. Here we struck ground with sixty-five fathoms of line, and found the bottom to consist of grey sand with black specks. When we had thus got sight of the land, we found the wind to hang westerly; and therefore, night coming on, we thought it advisable to stand off till morning, as there are said to be some shoals in the entrance of the channel. At six the next morning, Point Mariato bore NE. half N., three or four leagues distant. In weathering this point all the squadron, except the Centurion, were very near it; and the Gloucester, being the leewardmost ship, was forced to tack and stand to the southward, so that we lost sight of her. At nine, the Island Sebaco bore NW. by N., four leagues distant; but the wind still proving unfavourable, we were obliged to ply on and off for the succeeding twenty-four hours, and were frequently taken aback. However, at eleven the next morning the wind happily settled in the SSW., and we bore away for the SSE. end of the island, and about three in the afternoon entered Canal Bueno, passing round a shoal which stretches off about two miles from the south point of the island. This Canal Bueno, or Good Channel, is at least six miles in breadth; and as we had the wind large, we kept in a good depth of water, generally from twenty-eight to thirty-three fathoms, and came not within a mile and a half distance of the breakers; though in all proba-

[1] So called, it is said, because here Sir Francis Drake divided the treasure he had captured in the South Seas.

bility, if it had been necessary, we might have ventured much nearer without incurring the least danger. At seven in the evening we came to an anchor in thirty-three fathoms muddy ground; the south point of the island bearing SE. by S., a remarkable high part of the Island W. by N., and the Island Sebaco E. by N.

CHAPTER VIII.

THE next morning, after our coming to an anchor, an officer was despatched on shore to discover the watering-place, who having found it, returned before noon; and then we sent the long-boat for a load of water, and at the same time we weighed and stood farther in with our ships. At two we came again to an anchor in twenty-two fathoms, with a bottom of rough gravel intermixed with broken shells, the watering-place now bearing from us NW. half N., only three quarters of a mile distant. This Island of Quibo is extremely convenient for wooding and watering; for the trees grow close to the high-water mark, and a large rapid stream of fresh water runs over the sandy beach into the sea: so that we were little more than two days in laying in all the wood and water we wanted. . . .

Whilst the ship continued here at anchor, the Commodore, attended by some of his officers, went in a boat to examine a bay which lay to the northward; and they afterwards ranged all along the eastern side of the island. And in the places where they put on shore in the course of his expedition, they generally found the soil to be extremely rich, and met with great plenty of excellent water. In particular, near the NE. point of the island they discovered a natural cascade which surpassed, as they conceived, everything of this kind which human art or industry has hitherto produced. It was a river of transparent water, about forty yards wide, which ran down a declivity of near 150 yards in length. The channel it ran in was very irregular; for it was entirely formed of rock, both its sides and bottom being made up of large detached blocks; and by these the course of the water was frequently interrupted: for in some places it ran sloping with a rapid but uniform motion, while in other parts it tumbled over the ledges of rocks with a perpendicular descent. All the neighbourhood of this stream was a fine wood; and even the huge masses of rock which overhung the water, and which, by their various projections, formed the inequalities of the channel, were covered with lofty forest trees. Whilst the Commodore, and those who were with him, were attentively viewing this place, and remarking the different blendings of the water, the rocks, and the wood, there came in sight (as it were with an intent still to heighten and animate the prospect) a prodigious flight of macaws, which, hovering over this spot, and often wheeling and playing on the wing about it, afforded a most brilliant appearance by the glittering of the sun on their variegated plumage; so that some of the spectators cannot refrain from a kind of transport when they recount the complicated beauties which occurred in this extraordinary water-fall.

In this expedition, which the boat made along the eastern side of the island, though they met with no inhabitants, yet they saw many huts upon the shore, and great heaps of shells of fine mother-of-pearl scattered up and down in different places. These were the remains left by the pearl fishers from Panama, who often frequent this place in the summer season; for the pearl oysters, which are to be met with everywhere in the Bay of Panama, are so plenty at Quibo, that by advancing a very little way into the sea, you might stoop down and reach them from the bottom. They are usually very large, and out of curiosity we opened some of them with a view of tasting them, but we found them extremely tough and unpalatable. . . .

Though the pearl oyster was incapable of being eaten, yet the sea at this place furnished us with another dainty in the greatest plenty and perfection. This was the turtle, of which we took here what quantity we pleased. There are generally reckoned four species of turtle, that is, the trunk turtle, the loggerhead, the hawksbill, and the green turtle. The two first are rank and unwholesome; the hawksbill (which furnishes the tortoise-shell) is but indifferent food, though better than the other two; but the green turtle is generally esteemed, by the greatest part of those who are acquainted with its taste, to be the most delicious of all eatables; and that it is a most wholesome food we are amply convinced by our own experience. For we fed on this last species, or the green turtle, for near four months, and consequently, had it been in any degree noxious, its ill effects could not possibly have escaped us. . . .

In three days' time we had completed our business at this place, and were extremely impatient to put to sea, that we might arrive in time enough on the coast of Mexico to intercept the Manilla galleon. But the wind being contrary detained us a night, and the next day, when we got into the offing (which we did through the same channel by which we entered) we were obliged to keep hovering about the island in hopes of getting sight of the Gloucester, which, as I have in the last Chapter mentioned, was separated from us on our first arrival. It was the 9th of December, in the morning, when we put to sea; and continuing to the southward of the island, looking out for the Gloucester, we, on the 10th, at five in the afternoon, discerned a small sail to the northward of us, to which we gave chase, and coming up with her took her. She proved to be a bark from Panama, bound to Cheripe, an inconsiderable village on the continent, and was called the Jesu Nazereno. She had nothing on board but some oakum, about a ton of rock salt, and between £30 and £40 in specie, most of it consisting of small silver money intended for purchasing a cargo of provisions at Cheripe.

On the 12th of December we were at last relieved from the perplexity we had suffered by the separation of the Gloucester; for on that day she joined us, and informed us that in tacking to the southward, on our first arrival, she had sprung her fore-top-mast, which had disabled her from working to windward, and prevented her from joining us sooner. And now we scuttled and sunk the Jesu Nazareno, the prize we took last; and having the greatest impatience to get into a proper station for the galleon, we stood all together to the westward, leaving the Island of Quibo (notwithstanding all the impediments we met with) in about nine days after our first coming in sight of it.

CHAPTER IX.

On the 12th of December we stood from Quibo to the westward; and the same day the Commodore delivered fresh instructions to the captains of the men-of-war, and the commanders of our prizes, appointing them the rendezvous they were to make, and the courses they were to steer in case of a separation. And first they were directed to use all possible despatch in getting to the northward of the harbour of Acapulco, where they were to endeavour to fall in with the land between the Latitudes of 18° and 19°; thence they were to beat up the coast, at eight or ten leagues' distance from the shore, till they came abreast of Cape Corrientes, in the Latitude of 20° 20′. When they arrived there, they were to continue cruising on that station till the 14th of February; and then they were to proceed to the middle island of the Tres Marias, in the Latitude of 21° 25′, bearing from Cape Corrientes NW. by N., twenty-five leagues distant. And if at this island they did not meet the Commodore, they were there to recruit their wood and water, and then to make the best of their way to the

Island of Macao, on the coast of China. These orders being distributed to all the ships, we had little doubt of arriving soon upon our intended station, as we expected, upon increasing our offing from Quibo, to fall in with the regular trade-wind. But, to our extreme vexation, we were baffled for near a month, either with tempestuous weather from the western quarter, or with dead calms and heavy rains, attended with a sultry air; so that it was the 25th of December before we got a sight of the Island of Cocos, which, by our reckoning was only 100 leagues from the continent; and we had the mortification to make so little way that we did not lose sight of it again in five days. This island we found to be in the Latitude of 5° 20′ N. It has a high hummock towards the western part, which descends gradually, and at last terminates in a low point to the eastward. From the Island of Cocos we stood W. by N., and were till the 9th of January in running 100 leagues more. We had at first flattered ourselves that the uncertain weather and western gales we met with were owing to the neighbourhood of the continent, from which, as we got more distant, we expected every day to be relieved by falling in with the eastern trade-wind. But as our hopes were so long baffled, and our patience quite exhausted, we began at length to despair of succeeding in the great purpose we had in view, that of intercepting the Manilla galleon; and this produced a general dejection amongst us, as we had at first considered this project as almost infallible, and had indulged ourselves in the most boundless hopes of the advantages we should thence receive. However, our despondency was at last somewhat alleviated by a favourable change of the wind; for on the 9th of January a gale for the first time sprung up from the NE., and on this we took the Carmelo in tow, as the Gloucester did the Carmen, making all the sail we could to improve the advantage, for we still suspected that it was only a temporary gale, which would not last long; but the next day we had the satisfaction to find that the wind did not only continue in the same quarter, but blew with so much briskness and steadiness, that we now no longer doubted of its being the true trade-wind. And as we advanced apace towards our station, our hopes began to revive, and our former despair by degrees gave place to more sanguine prejudices; for though the customary season of the arrival of the galleon at Acapulco was already elapsed, yet we were by this time unreasonable enough to flatter ourselves that some accidental delay might, for our advantage, lengthen out her passage beyond its usual limits.

When we got into the trade-wind, we found no alteration in it till the 17th of January, when we were advanced to the Latitude of 12° 50′; but on that day it shifted to the westward of N. This change we imputed to our having hauled up too soon, though we then esteemed ourselves full seventy leagues from the coast, which plainly shows that the trade-wind does not take place but at a considerable distance from the continent. After this the wind was not so favourable to us as it had been; however, we still continued to advance, and on the 26th of January, being then to the northward of Acapulco, we tacked and stood to the eastward, with a view of making the land. In the preceding fortnight we caught some turtle on the surface of the water, and several dolphins, bonitos, and albicores. One day, as one of the sail-makers' mates was fishing from the end of the jib-boom, he lost his hold, and dropped into the sea; and the ship, which was then going at the rate of six or seven knots, went directly over him. But, as we had the Carmelo in tow, we instantly called out to the people on board her, who threw him over several ends of ropes, one of which he fortunately caught hold of, and twisting it round his arm, they hauled him into the ship without his having received any other injury than a wrench in his arm, of which he soon recovered.

When, on the 26th of January, we stood to the eastward, we expected by our reckonings to have fallen in with the land on the 28th; but though the weather was perfectly clear, we had no sight of it at sunset, and therefore we continued on our course, not doubting but we should see it by the next morning. About ten at night we discovered a light on the larboard-bow, bearing from us NNE. The Trial's prize, too, which was about a mile ahead of us, made a signal at the same time for seeing a sail; and as we had none of us any doubt but what we saw was a ship's light, we were all extremely animated with a firm persuasion that it was the Manilla galleon, which had been so long the object of our wishes. And what added to our alacrity was our expectation of meeting with two of them instead of one, for we took it for granted that the light in view was carried in the top of one ship for a direction to her consort. We immediately cast off the Carmelo, and pressed forward with all our canvas, making a signal for the Gloucester to do the same. Thus we chased the light, keeping all our hands at their respective quarters, under an expectation of engaging in the next half hour, as we sometimes conceived the chase to be about a mile distant, and at other times to be within reach of our guns; and some on board us positively averred that besides the light they could plainly discern her sails. The Commodore himself was so fully persuaded that we should be soon alongside of her, that he sent for his first lieutenant, who commanded between decks, and directed him to see all the great guns loaded with two round-shot for the first broadside, and after that with one round-shot and one grape; strictly charging him, at the same time, not to suffer a gun to be fired till he, the Commodore, should give orders, which he informed the lieutenant would not be till we arrived within pistol-shot of the enemy. In this constant and eager attention we continued all night, always presuming that another quarter of an hour would bring us up with this Manilla ship, whose wealth, with that of her supposed consort, we now estimated by round millions. But when the morning broke, and daylight came on, we were most strangely and vexatiously disappointed by finding that the light which had occasioned all this bustle and expectancy was only a fire on the shore. Indeed, the circumstances of this deception are so extraordinary as to be scarcely credible; for by our run during the night, and the distance of the land in the morning, there was no doubt to be made but this fire, when we first discovered it, was about twenty-five leagues from us: and yet I believe there was no person on board who doubted of its being a ship's light, or of its being near at hand. It was, indeed, upon a very high mountain, and continued burning for several days afterwards; it was not a volcano, but rather, as I suppose, stubble or heath set on fire for some purpose of agriculture.

At sun-rising, after this mortifying delusion, we found ourselves about nine leagues off the land, which extended from the NW. to E. half N. On this land we observed two remarkable hummocks, such as are usually called paps, which bore N. from us; these a Spanish pilot and two Indians, who were the only persons amongst us that pretended to have traded in this part of the world, affirmed to be over the harbour of Acapulco. Indeed, we very much doubted their knowledge of the coast; for we found these paps to be in the Latitude of 17° 56′, whereas those over Acapulco are said to be in 17° only; and we afterwards found our suspicions of their skill to be well-grounded.[1] However, they were very confident, and assured us that the height of the mountains was itself an infallible mark of the harbour; the coast, as they pretended (though falsely) being generally low to the eastward and westward of it.

And now, being in the track of the Manilla galleon, it was a great doubt with us (as it was near the end of January) whether she was or was not

[1] See Dampier's description of the place, Chapter IX.

arrived. But, examining our prisoners about it, they assured us that she was sometimes known to come in after the middle of February; and they endeavoured to persuade us that the fire we had seen on shore was a proof that she was as yet at sea, it being customary, as they said, to make use of these fires as signals for her direction when she continued longer out than ordinary. On this information, strengthened by our propensity to believe them in a matter which so pleasingly flattered our wishes, we resolved to cruise for her for some days; and we accordingly spread our ships at the distance of twelve leagues from the coast, in such a manner that it was impossible she should pass us unobserved. However, not seeing her soon, we were at intervals inclined to suspect that she had gained her port already; and as we now began to want a harbour to refresh our people, the uncertainty of our present situation gave us great uneasiness, and we were very solicitous to get some positive intelligence, which might either set us at liberty to consult our necessities, if the galleon was arrived, or might animate us to continue on our present cruise with cheerfulness, if she was not. With this view the Commodore, after examining our prisoners very particularly, resolved to send a boat, under colour of the night, into the harbour of Acapulco, to see if the Manilla ship was there or not, one of the Indians being very positive that this might be done without the boat itself being discovered. To execute this project, the barge was despatched the 6th of February, with a sufficient crew and two officers, who took with them a Spanish pilot, and the Indian who had insisted on the practicability of this measure, and had undertaken to conduct it. Our barge did not return to us again till the 11th, when the officers acquainted Mr Anson that, agreeable to our suspicion, there was nothing like a harbour in the place where the Spanish pilots had at first asserted Acapulco to lie; that, when they had satisfied themselves in this particular, they steered to the eastward in hopes of discovering it, and had coasted along shore thirty-two leagues; that in this whole range they met chiefly with sandy beaches of a great length, over which the sea broke with so much violence that it was impossible for a boat to land; that at the end of their run they could just discover two paps at a very great distance to the eastward, which from their appearance and their latitude they concluded to be those in the neighbourhood of Acapulco; but that, not having a sufficient quantity of fresh water and provision for their passage thither and back again, they were obliged to return to the Commodore, to acquaint him with their disappointment. On this intelligence we all made sail to the eastward, in order to get into the neighbourhood of that port; the Commodore resolving to send the barge a second time upon the same enterprise when we were arrived within a moderate distance. And the next day, which was the 12th of February, we being by that time considerably advanced, the barge was again despatched, and particular instructions given to the officers to preserve themselves from being seen from the shore. On the 13th, we espied a high land to the eastward, which we first imagined to be that over the harbour of Acapulco; but we afterwards found that it was the high land of Seguateneo,[1] where there is a small harbour of which we shall have occasion to make more ample mention hereafter. And now, having waited six days without any news of our barge, we began to be uneasy for her safety; but on the seventh day, that is, on the 19th of February, she returned. The officers informed the Commodore that they had discovered the harbour of Acapulco, which they esteemed to bear from us ESE. at least fifty leagues distant; that on the 17th, about two in the morning, they were got within the island that lies at the mouth of the harbour, and yet neither the Spanish pilot nor the Indian who were with them could give them any information where they

[1] Chequetan; see Chapter XII.

then were; but that, while they were lying upon their oars in suspense what to do, being ignorant that they were then at the very place they sought for, they discerned a small light upon the surface of the water, on which they instantly plied their paddles, and moving as silently as possible towards it, they found it to be in a fishing canoe, which they surprised, with three Negroes that belonged to it. It seems the Negroes at first attempted to jump overboard, and being so near the land, they would easily have swam on shore; but they were prevented by presenting a piece at them, on which they readily submitted, and were taken into the barge. The officers further added, that they had immediately turned the canoe adrift against the face of a rock, where it would inevitably be dashed to pieces by the fury of the sea; this they did to deceive those who perhaps might be sent from the town to search after the canoe; for, upon seeing several pieces of a wreck, they would immediately conclude that the people on board her had been drowned, and would have no suspicion of their having fallen into our hands. When the crew of the barge had taken this precaution, they exerted their utmost strength in pulling out to sea, and by dawn of day had gained such an offing as rendered it impossible for them to be seen from the coast.

And now having got the three Negroes in our possession, who were not ignorant of the transactions at Acapulco, we were soon satisfied about the most material points which had long kept us in suspense. And on examination we found that we were indeed disappointed in our expectation of intercepting the galleon before her arrival at Acapulco; but we learned other circumstances which still revived our hopes, and which, we then conceived, would more than balance the opportunity we had already lost. For though our Negro prisoners informed us that the galleon arrived at Acapulco on our 9th of January, which was about twenty days before we fell in with this coast, yet they at the same time told us that the galleon had delivered her cargo and was taking in water and provisions for her return, and that the Viceroy of Mexico had by proclamation fixed her departure from Acapulco to the 14th of March, N.S. This last news was most joyfully received by us, as we had no doubt but she must certainly fall into our hands, and as it was much more eligible to seize her on her return than it would have been to have taken her before her arrival, as the specie for which she had sold her cargo, and which she would now have on board, would be prodigiously more to be esteemed by us than the cargo itself, great part of which would have perished on our hands; and no part of it could have been disposed of by us at so advantageous a mart as Acapulco.

Thus we were a second time engaged in an eager expectation of meeting with this Manilla ship, which, by the fame of its wealth, we had been taught to consider as the most desirable prize that was to be met with in any part of the globe. As all our future projects will be in some sort regulated with a view to the possession of this celebrated galleon, and as the commerce which is carried on by means of these vessels between the city of Manilla and the port of Acapulco is perhaps the most valuable, in proportion to its quantity, of any in the known world, I shall endeavour in the ensuing Chapter to give as distinct an account as I can of all the particulars relating thereto; both as it is a matter in which I conceive the public to be in some degree interested, and as I flatter myself that, from the materials which have fallen into my hands, I am enabled to describe it with more distinctness than has hitherto been done, at least in our language.

CHAPTER X.[1]

THOUGH Spain did not [by the voyage of Magellan] acquire the property of

[1] The historical portion of this

any of the Spice Islands, yet the discovery made, in his expedition, of the Philippine Islands, was thought too considerable to be neglected, for these were not far distant from those places which produced spices, and were very well situated for the Chinese trade, and for the commerce of other parts of India; and therefore a communication was soon established and carefully supported between these islands and the Spanish colonies on the coast of Peru. So that the city of Manilla (which was built on the Island of Luconia, the chief of the Philippines) soon became the mart for all Indian commodities, which were brought up by the inhabitants, and were annually sent to the South Seas to be there vended on their account; and the returns of this commerce to Manilla being principally made in silver, the place by degrees grew extremely opulent and considerable, and its trade so far increased as to engage the attention of the Court of Spain, and to be frequently controlled and regulated by royal edicts.

In the infancy of this trade, it was carried on from the port of Callao to the city of Manilla, in which voyage the trade-wind continually favoured them; so that, notwithstanding these places were distant between three and four thousand leagues, yet the voyage was often made in little more than two months. But then the return from Manilla was extremely troublesome and tedious, and is said to have sometimes taken them up above a twelvemonth, which, if they pretended to ply up within the limits of the trade-wind, is not at all to be wondered at; and it is asserted that in their first voyages they were so imprudent and unskilful as to attempt this course. However, that route was soon laid aside by the advice, as it is said, of a Jesuit, who persuaded them to steer to the northward till they got clear of the trade-winds, and then, by the favour of the westerly winds, which generally prevail in high latitudes, to stretch away for the coast of California. This has been the practice for at least 160 years past; for Sir Thomas Cavendish, in the year 1586, engaged off the south end of California a vessel bound from Manilla to the American coast.[1] And it was in compliance with this new plan of navigation, and to shorten the run both backwards and forwards, that the staple[2] of this commerce to and from Manilla was removed from Callao, on the coast of Peru, to the port of Acapulco, on the coast of Mexico, where it continues fixed at this time.

The trade carried on from Manilla to China, and different parts of India, is principally for such commodities as are intended to supply the kingdoms of Mexico and Peru. These are, spices; all sorts of Chinese silks and manufactures, particularly silk stockings, of which I have heard that no less than 50,000 pairs were the usual number shipped on board the annual ship; vast quantities of Indian stuffs—as calicoes and chintzes, which are much worn in America; together with other minuter articles—as goldsmiths' work, &c., which is principally done at the city of Manilla itself by the Chinese; for it is said there are at least 20,000 Chinese who constantly reside there, either as servants, manufacturers, or brokers. All these different commodities are collected at Manilla, thence to be transported annually in one or more ships to the port of Acapulco. But this trade to Acapulco is not laid open to all the inhabitants of Manilla, but is confined by very particular regulations, somewhat analogous to those by which the

Chapter, relating to the old feud between the Spanish and Portuguese on the score of their discoveries, and to the origin of the trade, has been left out.

[1] The Santa Anna, of 700 tons, the Admiral of the South Seas, bearing a cargo valued at 122,000 pesos. From Cape St Lucas, after capturing the Manilla ship, Cavendish sailed to the Ladrones in forty-five days.

[2] The place of trade established by decree or ordinance.

trade of the register ships from Cadiz to the West Indies is restrained. The ships employed herein are found by the King of Spain, who pays the officers and crew; and the tonnage is divided into a certain number of bales, all of the same size. These are distributed amongst the convents at Manilla, but principally to the Jesuits, as a donation for the support of their missions for the propagation of the Catholic faith; and these convents have hereby a right to embark such a quantity of goods on board the Manilla ship as the tonnage of their bales amounts to; or, if they choose not to be concerned in trade themselves, they have the power of selling this privilege to others. And as the merchants to whom they grant their shares are often unprovided of a stock, it is usual for the convents to lend them considerable sums of money on bottomry. The trade is by the royal edicts limited to a certain value, which the annual cargo ought not to exceed. Some Spanish manuscripts I have seen mention this limitation to be 600,000 dollars; but the annual cargo does certainly surpass this sum; and though it may be difficult to fix its exact value, yet from many comparisons I conclude that the return cannot be greatly short of 3,000,000 dollars.

This trade from Manilla to Acapulco and back again is usually carried on in one or at most two annual ships, which set sail from Manilla about July, arrive at Acapulco in the December, January, or February following, and, having there disposed of their effects, return for Manilla some time in March, where they generally arrive in June; so that the whole voyage takes up very near an entire year. For this reason, though there is often no more than one ship employed at a time, yet there is always one ready for the sea when the other arrives; and therefore the commerce at Manilla are provided with three or four stout ships, that, in case of any accident, the trade may not be suspended. The largest of these ships, whose name I have not learned, is described as little less than one of our first-rate men-of-war, and indeed she must be of an enormous size, for it is known, that when she was employed with other ships from the same port to cruise for our China trade, she had no less than 1200 men on board. Their other ships, though far inferior in bulk to this, are yet stout, large vessels, of the burthen of 1200 tons and upwards, and usually carry from 350 to 600 hands, passengers included, with fifty odd guns. As these are all King's ships, commissioned and paid by him, there is usually one of the captains who is styled the General, and who carries the royal standard of Spain at the main-topgallant masthead.

The ship having received her cargo on board, and being fitted for the sea, generally weighs from the mole of Cabite[1] about the middle of July, taking the advantage of the westerly monsoon, which then sets in, to carry them to sea. The getting through the Boccadero[2] to the eastward must be a troublesome navigation; and, in fact, it is sometimes the end of August before they get clear of the land. When they have got through this passage, and are clear of the islands, they stand to the northward of the east, in order to get into the Latitude of thirty odd degrees, when they expect to meet with westerly winds, before which they run away for the coast of California.[3] It is most remarkable, that by the concurrent testimony of all the Spanish navigators, there is not one port, nor even a tolerable road, as yet found out betwixt the Philippine Islands and the coast of California and Mexico; so that from the time the Manilla ship first loses

---

[1] The port of Manilla, about two leagues to the southward of the city.

[2] Luzon, or Luconia, is separated from Mindoro by the strait of that name, about five miles broad; and from Samar by the "Embocadero de San Bernardino," the common passage for vessels navigating the Pacific on their way to China.

[3] Compare Dampier's account of the navigation in Chapter IX.

sight of land, she never lets go her anchor till she arrives on the coast of California, and very often not till she gets to its southernmost extremity. And therefore, as this voyage is rarely of less than six months' continuance, and the ship is deep laden with merchandise and crowded with people, it may appear wonderful how they can be supplied with a stock of fresh water for so long a time; and indeed their method of procuring it is extremely singular.[1] . . .

The Manilla ship, having stood so far to the northward as to meet with a westerly wind, stretches away nearly in the same latitude for the coast of California; and when she has run into the Longitude of 96° from Cape Espiritu Santo, she generally meets with a plant floating on the sea, which, being called *porra*,[2] by the Spaniards, is, I presume, a species of sea-leek. On the sight of this plant they esteem themselves sufficiently near the Californian shore, and immediately stand to the southward; and they rely so much on this circumstance, that on the first discovery of the plant the whole ship's company chant a solemn Te Deum, esteeming the difficulties and hazards of their passage to be now at an end; and they constantly correct their longitude thereby, without ever coming within sight of land. After falling in with these signs, as they denominate them, they steer to the southward, without endeavouring to fall in with the coast till they have run into lower latitude; for as there are many islands and some shoals adjacent to California, the extreme caution of the Spanish navigators makes them very apprehensive of being engaged with the land. However, when they draw near its southern extremity, they venture to haul in, both for the sake of making Cape St Lucas to ascertain their reckoning, and also to receive intelligence from the Indian inhabitants whether or no there are any enemies on the coast; and this last circumstance, which is a particular article in the captain's instructions, makes it necessary to mention the late proceedings of the Jesuits amongst the Californian Indians.

Since the first discovery of California there have been various wandering missionaries who have visited it at different times, though to little purpose; but of late years the Jesuits, encouraged and supported by a large donation from the Marquis da Valero, a most munificent bigot, have fixed themselves upon the place and have established a very considerable mission. Their principal settlement lies just within Cape St Lucas, where they have collected a great number of savages, and have endeavoured to inure them to agriculture and other mechanic arts. And their efforts have not been altogether ineffectual; for they have planted vines at their settlements with very good success, so that they already make a considerable quantity of wine, resembling in flavour the inferior sorts of Madeira, which begins to be esteemed in the neighbouring kingdom of Mexico. The Jesuits, then, being thus firmly rooted on California, they have already extended their jurisdiction quite across the country from sea to sea, and are endeavouring to spread their influence farther to the northward, with which view they have made several expeditions up the gulf between California and Mexico, in order to discover the nature of the adjacent countries, all

---

[1] In allusion to the custom of the Spaniards in the South Seas carrying a great quantity of water jars hung on the shrouds and stays of the vessel, and in this way conserving the water during the voyage. They depended for a fresh supply on the rains which fell, and which they caught in mats hung all over the deck, from which it was led into the jars by means of split bamboos.

[2] "Puerro" is the Spanish for leek; but "porra" is a word, though generally used in a tropical sense, sufficiently near the other to have been quite honestly used in Anson's time to serve the same meaning; and "porreta" signifies the green leaf of onions or garlick.

which they hope hereafter to bring under their power. And being thus occupied in advancing the interests of their Society, it is no wonder if some share of attention is engaged about the security of the Manilla ship, in which their convents at Manilla are so deeply concerned. For this purpose there are refreshments, as fruits, wine, water, &c., constantly kept in readiness for her; and there is besides care taken at Cape St Lucas to look out for any ship of the enemy which might be cruising there to intercept her; this being a station where she is constantly expected, and where she has been often waited for and fought with, though generally with little success. In consequence, then, of the measures mutually settled between the Jesuits of Manilla and their brethren at California, the captain of the galleon is ordered to fall in with the land to the northward of Cape St Lucas, where the inhabitants are directed, on sight of the vessel, to make the proper signals with fires; and on discovering these fires the captain is to send his launch on shore with twenty men, well-armed, who are to carry with them the letters from the convents at Manilla to the Californian missionaries, and are to bring back the refreshments which will be prepared for them, and likewise intelligence whether or no there are any enemies on the coast. And if the captain finds, from the account which is sent him, that he has nothing to fear, he is directed to proceed for Cape St Lucas, and thence to Cape Corrientes; after which he is to coast it along for the port of Acapulco.

The most usual time of the arrival of the galleon at Acapulco is towards the middle of January; but this navigation is so uncertain that she sometimes gets in a month sooner, and at other times has been detained at sea above a month longer. The port of Acapulco is by much the securest and finest in all the northern parts of the Pacific Ocean; being, as it were, a basin surrounded with very high mountains: but the town is a most wretched place, and extremely unhealthy, for the air about it is so pent up by the hills, that it has scarcely any circulation. The place is, besides, destitute of fresh water, except what is brought from a considerable distance; and is in all respects so inconvenient, that except at the time of the mart, whilst the Manilla galleon is in the port, it is almost deserted. When the galleon arrives in this port, she is generally moored on its western side, and her cargo is delivered with all possible expedition. And now the town of Acapulco, from almost a solitude, is immediately thronged with merchants from all parts of the kingdom of Mexico. The cargo being landed and disposed of, the silver and the goods intended for Manilla are taken on board, together with provisions and water, and the ship prepares to put to sea with the utmost expedition. There is indeed no time to be lost; for it is an express order to the captain to be out of the port of Acapulco, on his return, before the first day of April, N.S.

And having mentioned the goods intended for Manilla, I must observe that the principal return is always made in silver, and consequently the rest of the cargo is but of little account; the other articles, besides the silver, being some cochineal, and a few sweetmeats, the produce of the American settlements, together with European millinery ware for the women at Manilla, and some Spanish wines, such as tent and sherry, which are intended for the use of their priests in the administration of the sacrament. And this difference in the cargo of the ship to and from Manilla occasions a very remarkable variety in the manner of equipping the ship for these two different voyages. For the galleon, when she sets sail from Manilla, being deep laden with a variety of bulky goods, has not the conveniency of mounting her lower tier of guns, but carries them in her hold till she draws near Cape St Lucas, and is apprehensive of an enemy. Her hands, too, are as few as is consistent with the safety of the ship, that she may be less pestered with the stowage of provisions. But, on her return from Acapulco, as her cargo lies in less room, her lower tier is (or ought to be) always mounted before she leaves the port; and her crew is augmented with a supply of sailors, and with one or two companies of foot, which are intended to reinforce the garrison at Manilla. And there being, besides, many merchants who take their passage to Manilla on board the galleon, her whole number of hands on her return is usually little short of six hundred, all which are easily provided for by reason of the small stowage necessary for the silver.

The galleon being thus fitted for her return, the captain, on leaving the port of Acapulco, steers for the Latitude of 13° or 14°, and runs on that parallel till he gets sight of the Island of Guam, one of the Ladrones. In this run the captain is particularly directed to be careful of the shoals of St Bartholomew, and of the Island of Gasparico.[1] He is also told in his instructions that, to prevent his passing the Ladrones in the dark, there are orders given that through all the month of June fires shall be lighted every night on the highest part of Guam and Rota, and kept in till the morning. At Guam there is a small Spanish garrison, (as will be more particularly mentioned hereafter) purposely intended to secure that place for the refreshment of the galleon, and to yield her all the assistance in their power. However, the danger of the road at Guam is so great, that though the galleon is ordered to call there, yet she rarely stays above a day or two; but getting her water and refreshments on board as soon as possible, she steers away directly for Cape Espiritu Santo, on the Island of Samal.[2] Here the captain is again ordered to look out for signals; and he is told that sentinels will be posted not only on that cape, but likewise in Catanduanas, Butusan, Birriborongo, and on the Island of Batan. These sentinels are instructed to make a fire when they discover the ship, which the captain is carefully to observe. For if, after this first fire is extinguished, he perceives that four or more are lighted up again, he is then to conclude that there are enemies on the coast; and on this he is immediately to endeavour to speak with the sentinel on shore, and to procure from him more particular intelligence of their force, and of the station they cruise in; pursuant to which he is to regulate his conduct, and to endeavour to gain some secure port amongst those islands, without coming in sight of the enemy; and in case he should be discovered when in port, and should be apprehensive of an attack, he is then to land his treasure, and to take some of his artillery on shore for its defence, not neglecting to send frequent and particular accounts to the city of Manilla of all that passes. But if, after the first fire on shore, the captain observes that two others only are made by the sentinels, he is then to conclude that there is nothing to fear; and he is to pursue his course without interruption, and to make the best of his way to the port of Cabite, which is the port to the city of Manilla, and the constant station for all the ships employed in this commerce to Acapulco.

---

[1] In Anson's Chart San Bartolomeo is laid down as a considerable island, in about Latitude 13° N., Longitude 159° E. The position nearly corresponds with that of some of the smaller islands, north of Torres, belonging to the Caroline group. Gaspar Rico, not shown in Anson's Chart, is in about Latitude 12° 30′ N., Longitude 171° 30′ E. But the two islands specially signalled out for caution are no more than a speck among the hundreds of isles which for fully thirty degrees of longitude the Centurion passed to the northward in her voyage to the Ladrones.

[2] Or Samar; an island of considerable size, lying to the north of Mindanao, about the centre of the Archipelago, with its point farthest advanced towards the east.

CHAPTER XI.

I HAVE already mentioned in the ninth Chapter, that the return of our barge from the port of Acapulco, where she had surprised three Negro fishermen, gave us inexpressible satisfaction; as we learned from our prisoners that the galleon was then preparing to put to sea, and that her departure was fixed, by an edict of the Viceroy of Mexico, to the 14th of March N.S., that is, to the 3d of March according to our reckoning. What related to this Manilla ship being the matter to which we were most attentive, it was necessarily the first article of our examination; but having satisfied ourselves upon this head, we then indulged our curiosity in inquiring after other news; when the prisoners informed us, that they had received intelligence at Acapulco of our having plundered and burned the town of Paita; and that on this occasion the Governor of Acapulco had augmented the fortifications of the place, and had taken several precautions to prevent us from forcing our way into the harbour; that in particular he had placed a guard on the island which lies at the harbour's mouth, and that this guard had been withdrawn but two nights before the arrival of our barge: so that had the barge succeeded in her first attempt, or had she arrived at the port the second time two days sooner, she could scarcely have avoided being seized on, or if she had escaped it must have been with the loss of the greatest part of her crew, as she would have been under the fire of the guard before she had known her danger.

And now, on the 1st of March, we made the high lands usually called the paps, over Acapulco, and got with all possible expedition into the situation prescribed by the Commodore's orders. The distribution of our squadron on this occasion, both for the intercepting the galleon and for the avoiding a discovery from the shore, was so very judicious that it well merits to be distinctly described. The order of it was thus: The Centurion brought the paps over the harbour to bear NNE., at fifteen leagues' distance, which was a sufficient offing to prevent our being seen by the enemy. To the westward of the Centurion there was stationed the Carmelo, and to the eastward were the Trial prize, the Gloucester, and the Carmen. These were all ranged in a circular line, and each ship was three leagues distant from the next; so that the Carmelo and the Carmen, which were the two extremes, were twelve leagues distant from each other. And as the galleon could without doubt be discerned at six leagues' distance from either extremity, the whole sweep of our squadron, within which nothing could pass undiscovered, was at least twenty-four leagues in extent; and yet we were so connected by our signals as to be easily and speedily informed of what was seen in any part of the line. And to render this disposition still more complete, and to prevent even the possibility of the galleon's escaping us in the night, the two cutters belonging to the Centurion and the Gloucester were both manned and sent in-shore, and were ordered to lie all day at the distance of four or five leagues from the entrance of the port, where, by reason of their smallness, they could not possibly be discovered; but in the night they were directed to stand nearer to the harbour's mouth, and, as the light of the morning came on, they were to return back again to their day posts. When the cutters should first discover the Manilla ship one of them was ordered to return to the squadron, and to make a signal whether the galleon stood to the eastward or to the westward; whilst the other was to follow the galleon at a distance, and, if it grew dark, was to direct the squadron in their chase by showing false fires.

Besides the care we had taken to prevent the galleon from passing by us unobserved, we had not been inattentive to the means of engaging her to advantage when we came up with her; for, considering the thinness of our hands, and the vaunting accounts given by the Spaniards of her size, her guns, and her strength, this was a consideration not to be neglected. As we supposed that none of our ships but the Centurion and the Gloucester were capable of lying alongside of her, we took on board the Centurion all the hands belonging to the Carmelo and the Carmen, except what were just sufficient to navigate those ships; and Captain Saunders was ordered to send from the Trial prize ten Englishmen and as many Negroes to reinforce the crew of the Gloucester. And for the encouragement of our Negroes, of whom we had a considerable number on board, we promised them that on their good behaviour they should all have their freedom; and as they had been almost every day trained to the management of the great guns for the two preceding months, they were very well qualified to be of service to us; and from their hopes of liberty, and in return for the usage they had met with amongst us, they seemed disposed to exert themselves to the utmost of their power.

And now, being thus prepared for the reception of the galleon, we expected with the utmost impatience the so often-mentioned 3d of March, the day fixed for her departure. And on that day we were all of us most eagerly engaged in looking out towards Acapulco: and we were so strangely prepossessed with the certainty of our intelligence, and with an assurance of her coming out of port, that some or other on board us were constantly imagining that they discovered one of our cutters returning with a signal. But to our extreme vexation both this day and the succeeding night passed over without any news of the galleon. However, we did not yet despair, but were all heartily disposed to flatter ourselves that some unforeseen accident had intervened which might have put off her departure for a few days; and suggestions of this kind occurred in plenty, as we knew that the time fixed by the Viceroy for her sailing was often prolonged on the petition of the merchants of Mexico. Thus we kept up our hopes, and did not abate of our vigilance; and as the 7th of March was Sunday, the beginning of Passion Week, which is observed by the Papists with great strictness and a total cessation from all kinds of labour, so that no ship is permitted to stir out of port during the whole week, this quieted our apprehensions for some days, and disposed us not to expect the galleon till the week following. On the Friday in this week our cutters returned to us, and the officers on board them were very confident that the galleon was still in port, for that she could not possibly have come out but they must have seen her. On the Monday morning succeeding Passion Week—that is, on the 15th of March—the cutters were again despatched to their old station, and our hopes were once more indulged in as sanguine prepossessions as before; but in a week's time our eagerness was greatly abated, and a general dejection and despondency took place in its room. It is true there were some few amongst us who still kept up their spirits, and were very ingenious in finding out reasons to satisfy themselves that the disappointment we had hitherto met with had only been occasioned by a casual delay of the galleon, which a few days would remove, and not by a total suspension of her departure for the whole season. But these speculations were not relished by the generality of our people; for they were persuaded that the enemy had by some accident discovered our being upon the coast, and had therefore laid an embargo on the galleon till the next year. And indeed this persuasion was but too well founded; for we afterwards learned that our barge, when sent on the discovery of the port of Acapulco, had been seen from the shore, and that this circumstance (no embarkations but canoes ever frequenting that coast) was to them a sufficient proof of the neighbourhood of our squadron; on which they stopped the galleon till the succeeding year. . . .

When we had taken up the cutters,

all the ships being joined, the Commodore made a signal to speak with their commanders; and upon inquiry into the stock of fresh water remaining on board the squadron, it was found to be so very slender, that we were under a necessity of quitting our station to procure a fresh supply. And consulting what place was the properest for this purpose, it was agreed that the harbour of Seguataneo or Chequetan, being the nearest to us, was on that account the most eligible; and it was therefore immediately resolved to make the best of our way thither. And that, even while we were recruiting our water, we might not totally abandon our views upon the galleon, which perhaps, upon certain intelligence of our being employed at Chequetan, might venture to slip out to sea, our cutter, under the command of Mr Hughes, the lieutenant of the Trial prize, was ordered to cruise off the port of Acapulco for twenty-four days; that, if the galleon should set sail in that interval, we might be speedily informed of it. In pursuance of these resolutions, we endeavoured to ply to the westward, to gain our intended port, but were often interrupted in our progress by calms and adverse currents. In these intervals we employed ourselves in taking out the most valuable part of the cargoes of the Carmelo and Carmen prizes, which two ships we intended to destroy as soon as we had tolerably cleared them. By the 1st of April we were so far advanced towards Seguataneo, that we thought it expedient to send out two boats, that they might range along the coast, and discover the watering-place. They were gone some days, and, our water being now very short, it was a particular felicity to us that we met with daily supplies of turtle; for had we been entirely confined to salt provisions we must have suffered extremely in so warm a climate. Indeed, our present circumstances were sufficiently alarming, and gave the most considerate amongst us as much concern as any of the numerous perils we had hitherto encountered; for our boats, as we conceived by their not returning, had not as yet discovered a place proper to water at, and by the leakage of our casks and other accidents we had not ten days' water on board the whole squadron; so that, from the known difficulty of procuring water on this coast, and the little reliance we had on the Buccaneer writers (the only guides we had to trust to), we were apprehensive of being soon exposed to a calamity, the most terrible of any in the long, disheartening catalogue of the distresses of a sea-faring life.

But these gloomy suggestions were soon happily ended; for our boats returned on the 5th of April, having discovered a place proper for our purpose about seven miles to the westward of the rocks of Seguataneo, which, by the description they gave of it, appeared to be the port called by Dampier the harbour of Chequetan. The success of our boats was highly agreeable to us; and they were ordered out again the next day to sound the harbour and its entrance, which they had represented as very narrow. At their return they reported the place to be free from any danger; so that on the 7th we stood in, and that evening came to an anchor in eleven fathoms. The Gloucester came to an anchor at the same time with us; but the Carmelo and the Carmen having fallen to leeward, the Trial prize was ordered to join them, and to bring them in, which in two or three days she effected. Thus, after a four months' continuance at sea from the leaving of Quibo, and having but six days' water on board, we arrived in the harbour of Chequetan.

CHAPTER XII.

The harbour of Chequetan lies in the Latitude of 17° 36′ N., and is about thirty leagues to the westward of Acapulco. It is easy to be discovered by any ship that will keep well in with the land, especially by such as range down coast from Acapulco, and will attend to the following particulars.

There is a beach of sand, which extends eighteen leagues from the harbour of Acapulco to the westward, against which the sea breaks with such violence that it is impossible to land in any part of it; but yet the ground is so clean, that ships, in the fair season, may anchor in great safety at the distance of a mile or two from the shore. The land adjacent to this beach is generally low, full of villages, and planted with a great number of trees; and on the tops of some small eminences there are several look-out towers, so that the face of the country affords a very agreeable prospect. . . .

And on this occasion I cannot help mentioning another adventure which happened to some of our people in the Bay of Petaplan, as it may help to give the reader a just idea of the temper of the inhabitants of this part of the world. Some time after our arrival at Chequetan, Lieutenant Brett was sent by the Commodore, with two of our boats under his command, to examine the coast to the eastward, particularly to make observations on the bay and watering-place of Petaplan. As Mr Brett, with one of the boats, was preparing to go on shore towards the hill of Petaplan, he, accidentally looking across the bay, perceived on the opposite strand three small squadrons of horse parading upon the beach, and seeming to advance towards the place where he proposed to land. On sight of this, he immediately put off the boat, though he had but sixteen men with him, and stood over the bay towards them; and he soon came near enough to perceive that they were mounted on very sightly horses, and were armed with carbines and lances. On seeing him make towards them, they formed upon the beach, and seemed resolved to dispute his landing, firing several distant shots at him as he drew near; till at last, the boat being arrived within a reasonable distance of the most advanced squadron, Mr Brett ordered his people to fire, upon which this resolute cavalry instantly ran in great confusion into the wood through a small opening. In this precipitate flight one of their horses fell down and threw his rider; but whether he was wounded or not we could not learn, for both man and horse soon got up again and followed the rest into the wood. In the meantime the other two squadrons, who were drawn up at a great distance behind, out of the reach of our shot, were calm spectators of the rout of their comrades, for they had halted on our first approach, and never advanced afterwards. It was doubtless fortunate for our people that the enemy acted with so little prudence, and exerted so little spirit; for had they concealed themselves till our men had landed, it is scarcely possible but the whole boat's crew must have fallen into their hands, since the Spaniards were not much short of 200, and the whole number with Mr Brett only amounted to sixteen. However, the discovery of so considerable a force collected in this Bay of Petaplan obliged us constantly to keep a boat or two before it; for we were apprehensive that the cutter, which we had left to cruise off Acapulco, might on her return be surprised by the enemy, if she did not receive timely information of her danger. . . .

CHAPTER XIII.

THE next morning after our coming to an anchor in the harbour of Chequetan, we sent about ninety of our men well armed on shore, forty of whom were ordered to march into the country, as has been mentioned, and the remaining fifty were employed to cover the watering-place and to prevent any interruption from the natives. Here we completed the unloading of the Carmelo and Carmen, which we had begun at sea—at least we took out of them the indigo, cacao, and cochineal, with some iron for ballast, which were all the goods we intended to preserve, though they did not amount to a tenth of their cargoes. Here, too, it was agreed after a mature consultation to destroy the Trial's prize, as well as the Carmelo and Carmen, whose fate had been before resolved on. Indeed the ship was in good repair and fit for the sea; but as the whole numbers on board our squadron did not amount to the complement of a fourth-rate man-of-war, we found it was impossible to divide them into three ships without rendering them incapable of navigating in safety in the tempestuous weather we had reason to expect on the coast of China, where we supposed we should arrive about the time of the change of the monsoons. These considerations determined the Commodore to destroy the Trial prize, and to reinforce the Gloucester with the greatest part of her crew. And in consequence of this resolve, all the stores on board the Trial prize were removed into the other ships; and the prize herself, with the Carmelo and Carmen, were prepared for scuttling with all the expedition we were masters of. But the great difficulties we were under in laying in a store of water (which have been already touched on), together with the necessary repairs of our rigging and other unavoidable occupations, took us up so much time, and found us such unexpected employment, that it was near the end of April before we were in a condition to leave the place.

During our stay here there happened an incident which, as it proved the means of convincing our friends in England of our safety, which for some time they had despaired of and were then in doubt about, I shall beg leave particularly to recite. I have observed in the preceding Chapter that from this harbour of Chequetan there was but one pathway, which led through the woods into the country. This we found much beaten, and were thence convinced that it was well known to the inhabitants. As it passed by the spring-head, and was the only avenue by which the Spaniards could approach us, we, at some distance beyond the spring-head, felled several large trees, and laid them one upon the other across the path; and at this barricade we constantly kept a guard; and we, besides, ordered our men employed in watering to have their arms ready, and, in case of any alarm, to march instantly to this spot. And though our principal intention was to prevent our being disturbed by any sudden attack of the enemy's horse, yet it answered another purpose which was not in itself less important—this was to hinder our own people from straggling singly into the country, where we had reason to believe they would be surprised by the Spaniards, who would doubtless be extremely solicitous to pick up some of them in hopes of getting intelligence of our future designs. To avoid this inconvenience, the strictest orders were given to the sentinels to let no person whatever pass beyond their post. But, notwithstanding this precaution, we missed one Lewis Leger, who was the Commodore's cook; and as he was a Frenchman, and suspected to be a Papist, it was by some imagined that he had deserted with a view of betraying all that he knew to the enemy; but this appeared by the event to be an ill-grounded surmise, for it was afterwards known that he had been taken by some Indians, who carried him prisoner to Acapulco, whence he was transferred to Mexico, and then to Vera Cruz, where he was shipped on board a vessel bound to Old Spain. And the vessel being obliged by some accident to put into Lisbon, Leger escaped on shore, and was by the British Consul sent thence to England, where he brought the first authentic account of the safety of the Commodore, and of what he had done in the South Seas. The relation he gave of his own seizure was, that he had rambled into the woods at some distance from the barricade where he had first attempted to pass, but had been stopped and threatened to be punished—that his principal view was to get a quantity of limes for his master's store; and that in this occupation he was surprised unawares by four Indians, who stripped him naked, and carried him in that condition to Acapulco, exposed to the scorching heat of the sun, which at that time of the year shone with its greatest violence. And afterwards at Mexico, his treatment in prison was sufficiently severe, and the whole course of his captivity was a continued instance of the hatred which the Spaniards bear to all those who endeavour to disturb them in the peaceable possession of the coasts of the South Seas. Indeed, Leger's fortune was, upon the whole, extremely singular; for after the hazards he had run in the Commodore's squadron, and the severities he had suffered in his long confinement amongst the enemy, a more fatal disaster attended him on his return to England. For though, when he arrived in London, some of Mr Anson's friends interested themselves in relieving him from the poverty to which his captivity had reduced him, yet he did not long enjoy the benefit of their humanity, for he was killed in an insignificant night brawl, the cause of which could scarcely be discovered.

And here I must observe that though the enemy never appeared in sight during our stay in this harbour, yet we perceived that there were large parties of them encamped in the woods about us; for we could see their smokes, and could thence determine that they were posted in a circular line surrounding us at a distance; and just before our coming away they seemed, by the increase of their fires, to have received a considerable reinforcement.

Towards the latter end of April, the unloading of our three prizes, our wooding and watering, and, in short, all our proposed employments at the harbour of Chequetan were completed; so that on the 27th the Trial's prize, the Carmelo and the Carmen—all which we here intended to destroy—were towed on shore and scuttled, and a quantity of combustible materials were distributed in their upper works; and next morning the Centurion and the Gloucester weighed anchor, but as there was but little wind, and that not in their favour, they were obliged to warp out of the harbour. When they had reached the offing, one of the boats was despatched back again to set fire to our prize, which was accordingly executed. And a canoe was left fixed to a grapnel in the middle of the harbour, with a bottle in it well corked, enclosing a letter to Mr Hughes, who commanded the cutter which was ordered to cruise before the port of Acapulco when we came off that station. And on this occasion I must mention more particularly than I have yet done the views of the Commodore in leaving the cutter before that port.

When we were necessitated to make for Chequetan to take in our water, Mr Anson considered that our being in that harbour would soon be known at Acapulco; and therefore he hoped, that on the intelligence of our being employed in port, the galleon might put to sea, especially as Chequetan is so very remote from the course generally steered by the galleon. He therefore ordered the cutter to cruise twenty-four days off the port of Acapulco; and her commander was directed, on perceiving the galleon under sail, to make the best of his way to the Commodore at Chequetan. As the Centurion was doubtless a much better sailer than the galleon, Mr Anson, in this case, resolved to have got to sea as soon as possible, and to have pursued the galleon across the Pacific Ocean; and supposing he should not have met with her in his passage (which, considering that he would have kept nearly the same parallel, was not very improbable) yet he was certain of arriving off Cape Espiritu Santo, on the Island of Samal, before her; and that being the first land she makes on her return to the Philippines, we could not have failed to have fallen in with her by cruising a few days in that station. But the Viceroy of Mexico ruined this project by keeping the galleon in the port of Acapulco all that year.

The letter left in the canoe for Mr Hughes, the commander of the cutter (the time of whose return was now considerably elapsed), directed him to go back immediately to his former station before Acapulco, where he would find Mr Anson, who resolved to cruise for him there for a certain number of days; after which it was added, that the Commodore would return to the southward to join the rest of the squadron. This last article was inserted to deceive the Spaniards, if they got possession of the canoe (as we afterwards learned they did), but could not impose on Mr Hughes, who well knew that the Commodore had no squadron to join, nor any intention of steering back to Peru.

Being now in the offing of Chequetan, bound across the vast Pacific Ocean in our way to China, we were impatient to run off the coast as soon as possible; for as the stormy season was approaching apace, and as we had no further views in the American seas, we had hoped that nothing would have prevented us from standing to the westward the moment we got out of the harbour of Chequetan. And it was no small mortification to us that our necessary employment there had detained us so much longer than we expected; and now we were further detained by the absence of the cutter, and the standing towards Acapulco in search of her. Indeed, as the time of her cruise had been expired for near a fortnight, we suspected that she had been discovered from the shore, and that the Governor of Acapulco had thereupon sent out a force to seize her, which, as she carried but six hands, was no very difficult enterprise. However, this being only conjecture, the Commodore as soon as he was got clear of the harbour of Chequetan, stood along the coast to the eastward in search of her. And to prevent her from passing by us in the dark, we brought to every night, and the Gloucester, whose station was a league within us towards the shore, carried a light, which the cutter could not but perceive if she kept along shore, as we supposed she would do; and as a further security, the Centurion and the Gloucester alternately showed two false fires every half-hour.

By Sunday, the 2d of May, we were advanced within three leagues of Acapulco; and having seen nothing of our boat, we gave her over for lost, which, besides the compassionate concern for our shipmates, and for what it was apprehended they might have suffered, was in itself a misfortune which in our present scarcity of hands we were all greatly interested in. For the crew of the cutter, consisting of six men and the lieutenant, were the very flower of our people, purposely picked out for this service, and known to be, every one of them, of tried and approved resolution, and as skilful seamen as ever trod a deck. However, as it was the general belief among us that they were taken and carried into Acapulco, the Commodore's prudence suggested a project which we hoped would recover them. This was founded on our having many Spanish and Indian prisoners in our possession, and a number of sick Negroes, who could be of no service to us in the navigating of the ship. The Commodore therefore wrote a letter the same day to the Governor of Acapulco, telling him that he would release them all, provided the Governor returned the cutter's crew; and the letter was despatched the same afternoon by a Spanish officer, of whose honour we had a good opinion, and who was furnished with a launch belonging to one of our prizes, and a crew of six other prisoners, who all gave their parole for their return. The officer, too, besides the Commodore's letter, carried with him a joint petition signed by all the rest of the prisoners, beseeching his Excellency to acquiesce in the terms proposed for their liberty. From a consideration of the number of our prisoners, and the quality of some of them, we did not doubt but the Governor would readily comply with Mr Anson's proposal; and therefore we kept plying on and off the whole night, intending to keep well in with the land, that we might receive an answer at the limited time, which was the next day, being Monday. But both on the Monday and Tuesday we were driven so far off shore that we could not hope to receive any answer; and on the Wednesday morning we found ourselves fourteen leagues from the harbour of Acapulco; but, as the wind was now favourable, we pressed forwards with all our sail, and did not doubt of getting in with the land in a few hours.

Whilst we were thus standing in, the man at the mast-head called out that he saw a boat under sail at a considerable distance to the south-eastward. This we took for granted was the answer of the Governor to the Commodore's message, and we instantly edged towards it; but when we drew nearer we found to our unspeakable joy that it was our own cutter. While she was still at a distance, we imagined that she had been discharged out of the port of Acapulco by the Governor; but when she drew nearer, the wan and meagre countenances of the crew, the length of their beards, and the feeble and hollow tone of their voices, convinced us that they had suffered much greater hardships than could be expected from even the severities of a Spanish prison. They were obliged to be helped into the ship, and were immediately put to bed; and with rest, and nourishing diet, which they were plentifully supplied with from the Commodore's table, they recovered their health and vigour apace. And now we learned that they had kept the sea the whole time of their absence, which was above six weeks; that when they finished their cruise before Acapulco, and had just begun to ply to the westward in order to join the squadron, a strong adverse current had forced them down the coast to the eastward in spite of all their efforts; that at length, their water being all expended, they were obliged to search the coast farther on to the eastward, in quest of some convenient landing-place, where they might get a fresh supply; that in this distress they ran upwards of eighty leagues to leeward, and found everywhere so large a surf, that there was not the least possibility of their landing; that they passed some days in this dreadful situation, without water, and having no other means left them to allay their thirst than sucking the blood of the turtle which they caught; and at last, giving up all hopes of relief, the heat of the climate, too, augmenting their necessities, and rendering their sufferings insupportable, they abandoned themselves to despair, fully persuaded that they should perish by the most terrible of all deaths. But they were soon after happily relieved by a most unexpected incident, for there fell so heavy a rain, that by spreading their sails horizontally, and by putting bullets in the centre of them to draw them to a point, they caught as much water as filled all their casks; immediately upon this fortunate supply, they stood to the westward in quest of the Commodore; and, being now luckily favoured by a strong current, they joined us in less than fifty hours from the time they stood to the westward, after having been absent from us full forty-three days. Those who have an idea of the inconsiderable size of a cutter belonging to a sixty-gun ship (being only an open boat about twenty-two feet in length), and who will attend to the various accidents to which she was exposed during a six weeks' continuance alone in the open ocean, on so impracticable and dangerous a coast, will readily own that her return to us at last, after all the difficulties which she actually experienced, and the hazards to which she was each hour exposed, may be considered as little short of miraculous. I cannot finish the article of this cutter without remarking how little reliance navigators ought to have on the accounts of the Buccaneer writers. For though, in this run of hers eighty leagues to the eastward of Acapulco, she found no place where it was possible for a boat to land, yet those writers have not been ashamed to feign harbours and convenient watering-places within these limits, thereby exposing such as should confide in their relations to the risk of being destroyed by thirst.

And now, having received our cutter, the sole object of our coming a second time before Acapulco, the Commodore resolved not to lose a moment's time longer, but to run off the coast with the utmost expedition; both as the stormy season on the coast of Mexico was now approaching apace, and as we were apprehensive of having the westerly monsoon to struggle with when we came upon the coast of China. And therefore he no longer stood towards Acapulco, as he now wanted no answer from the Governor; but yet he resolved not to deprive his prisoners of the liberty which he had promised them, so that they were all immediately embarked in two launches which belonged to our prizes, those from the Centurion in one launch, and those from the Gloucester in the other. The launches were well equipped with masts, sails, and oars; and, lest the wind might prove unfavourable, they had a stock of water and provisions put on board them sufficient for fourteen days. There were discharged thirty-nine persons from on board the Centurion, and eighteen from the Gloucester, the greatest part of them Spaniards, the rest Indians and sick Negroes; but as our crews were very weak, we kept the Mulattoes and some of the stoutest of the Negroes, with a few Indians, to assist us, but we dismissed every Spanish prisoner whatever.[1] We have since learned that these two launches arrived safe at Acapulco, where the prisoners could not enough extol the humanity with which they had been treated; and that the Governor, before their arrival, had returned a very obliging answer to the Commodore's letter, and had attended it with a present of two boats laden with the choicest refreshments and provisions which were to be got at Acapulco; but that these boats, not having found our ships, were at length obliged to put back again after having thrown all their provisions overboard in a storm which threatened their destruction.

The sending away our prisoners was our last transaction on the American coast, for no sooner had we parted with them than we and the Gloucester made sail to the SW., proposing to get a good offing from the land, where we hoped in a few days to meet with the regular trade-wind, which the accounts of former navigators had represented as much brisker and steadier in this ocean than in any other part of the globe. For it has been esteemed no uncommon passage to run from hence to the easternmost parts of Asia in two months, and we flattered ourselves that we were as capable of making an expeditious passage as any ships that had ever run this course before us; so that we hoped soon to gain the coast of China, for which we were now bound. And conformable to the general idea of this navigation given by former voyagers, we considered it as free from all kinds of embarrassment of bad weather, fatigue, or sickness: and consequently we undertook it with alacrity, especially as it was no contemptible step towards our arrival at our native country, for which many of us by this time began to have great longings. Thus, on the 6th of May, we for the last time lost sight of the mountains of Mexico, persuaded that in a few weeks we should arrive at the River of Canton in China, where we expected to meet with many English ships and numbers of our countrymen, and hoped to enjoy the advantages of an amicable, well-frequented port, inhabited by a polished people, and abounding with the conveniences and indulgences of a civilised life, blessings which now for near twenty months had never been once in our power.[2]

[1] "About four in the evening they left us," says Thomas, "having first, though enemies, observed the custom of seafaring people at parting, and wished us a prosperous voyage."

[2] In the original, a Fourteenth Chapter of Book II. is devoted to a disquisition, entitled "A brief Account of what might have been expected from our squadron had it arrived in the South Seas in good time;" but apart from the unprofitablene[illegible]most our

BOOK III.

CHAPTER I.

WHEN, on the 6th of May 1742, we left the coast of America, we stood to the SW. with a view of meeting with the NE. trade-wind, which the accounts of former writers made us expect at seventy or eighty leagues' distance from the land. We had, besides, another reason for standing to the southward, which was the getting into the Latitude of 13° or 14° N., that being the parallel where the Pacific Ocean is most usually crossed, and consequently where the navigation is esteemed the safest. This last purpose we had soon answered, being in a day or two sufficiently advanced to the south. At the same time, we were also farther from the shore than we had presumed was necessary for falling in with the trade-wind; but in this particular we were most grievously disappointed, for the wind still continued to the westward, or at best variable. As the getting into the NE. trade was to us a matter of the last consequence, we stood more to the southward, and made many experiments to meet with it; but all our efforts were for a long time unsuccessful, so that it was seven weeks from our leaving the coast before we got into the true trade-wind. This was an interval in which we believed we should well nigh have reached the easternmost parts of Asia; but we were so baffled with the contrary and variable winds which for all that time perplexed us, that we were not as yet advanced above a fourth part of the way. The delay alone would have been a sufficient mortification, but there were other circumstances attending it which rendered this situation not less terrible, and our apprehensions perhaps still greater, than in any of our past distresses. For our two ships were by this time extremely crazy; and many days had not passed before we discovered a spring in the foremast of the Centurion, which rounded about twenty-six inches of its circumference, and which was judged to be at least four inches deep. And no sooner had our carpenters secured this with fishing it, but the Gloucester made a signal of distress; and we learned that she had a dangerous spring in her mainmast twelve feet below the trussel-trees,[1] so that she could not carry any sail upon it. Our carpenters, on a strict examination of this mast, found it so very rotten and decayed that they judged it necessary to cut it down as low as it appeared to have been injured, and by this it was reduced to nothing but a stump, which served only as a step to the topmast. These accidents augmented our delay and occasioned us great anxiety about our future security, for on our leaving the coast of Mexico the scurvy had begun to make its appearance again amongst our people, though from our departure from Juan Fernandez we had till then enjoyed a most uninterrupted state of health. We too well knew the effects of this disease from our former fatal experience, to suppose that anything but a speedy passage could secure the greater part of our crew from perishing by it; and as, after being seven weeks at sea, there did not appear any reasons that could persuade us we were nearer the

cussing a probability subject to so many conditions that the Narrator could not contemplate, it will appear to the reader that quite enough has been said, both in the opening Chapter and throughout the whole narrative, to show wherein and how the squadron came short of its intents.

[1] Or trestle-trees; "two strong pieces of timber placed horizontally and fore-and-aft on opposite sides of a mast-head, to support the cross-trees and top, and also for the fid of the mast above to rest on."

trade-wind than when we first set out, there was no ground for us to suppose but our passage would prove at least three times as long as we at first expected; and consequently we had the melancholy prospect either of dying by the scurvy or perishing with the ship for want of hands to navigate her. Indeed some amongst us were at first willing to believe that in this warm climate, so different from what we felt in passing round Cape Horn, the violence of this disease and its fatality might be in some degree mitigated, as it had not been unusual to suppose that its particular virulence in that passage was in a great measure owing to the severity of the weather. But the havoc of the distemper in our present circumstances soon convinced us of the falsity of this speculation, as it likewise exploded some other opinions which usually pass current about the cause and nature of this disease.[1] Our surgeon[2] (who, during our passage round Cape Horn, had ascribed the mortality we suffered to the severity of the climate) exerted himself in the present run to the utmost, and at last declared that all his measures were totally ineffectual, and did not in the least avail his patients.[3]

When we reached the trade-wind, and it settled between the north and the east, yet it seldom blew with so much strength but the Centurion might have carried all her small sails abroad with the greatest safety; so that now, had we been a single ship, we might have run down our longitude apace, and have reached the Ladrones soon enough to have recovered great numbers of our men who afterwards perished. But the Gloucester, by the loss of her mainmast, sailed so very heavily that we had seldom any more than our topsails set, and yet were frequently obliged to lie to for her; and, I conceive, that in the whole we lost little less than a month by our attendance upon her, in consequence of the various mischances she encountered. In all this run, it was remarkable that we were rarely many days together without seeing great numbers of birds, which is a proof that there are many islands, or at least rocks, scattered all along at no very considerable distance from our track.[4] Some indeed there are

---

[1] Some observations on the general medical treatment of the disease, and on the effect of certain specifics tried on some of the crew, have been here omitted. Speaking of the scurvy, Thomas strives to remove the prevalent notion that it attacks none but the lazy; whereas experience in the voyage proved the direct contrary, the most laborious, active, stirring persons being oftenest seized with the disease, and the continuance of their labour, instead of curing, only helped to kill them the sooner. Nor, he adds, does the scurvy generally incline people to indolence till it has come to such a height that at the least motion the sufferer is ready to faint.

[2] Mr Henry Ettrick, originally of the Wager; he succeeded Mr Thomas Walter, the first surgeon of the Centurion, who died off the coast of Brazil.

[3] About the middle of June, Thomas remarks that abundance of scorbutic symptoms, such as blackness in the skin, hard nodes in the flesh, shortness of breath, and a general lassitude and weakness of all the parts, began to prevail almost universally among the people. Towards the end of July he writes: "About this time our people began to die very fast, and I believe above five parts out of six of the ship's company were ill and expected to follow in a short time. Those whose breath was anyways affected, dropped off immediately; but those who were attacked first in the more remote parts of the body, languished generally a month or six weeks, the distemper advancing in the meantime towards the lungs by a very regular and sensible approach."

[4] More recent discoveries have fully borne out this sagacious conjecture. Thomas records, early in July: "We had, not only now, but for almost our

marked in Spanish charts; but the frequency of the birds seems to evince that there are many more than have been hitherto discovered; for the greatest part of the birds, we observed, were such as are known to roost on shore; and the manner of their appearance sufficiently made out that they came from some distant haunt every morning, and returned thither again in the evening; for we never saw them early or late, and the hour of their arrival and departure gradually varied, which we supposed was occasioned by our running nearer their haunts or getting farther from them.

The trade-wind continued to favour us without any fluctuation from the end of June till towards the end of July. But on the 26th of July, being then, as we esteemed, about 300 leagues distant from the Ladrones, we met with a westerly wind, which did not come about again to the eastward in four days' time. This was a most dispiriting incident, as it at once damped all our hopes of speedy relief, especially, too, as it was attended with a vexatious accident to the Gloucester; for in one part of these four days the wind flattened to a calm, and the ships rolled very deep, by which means the Gloucester's fore-cap split, and her topmast came by the board and broke her foreyard directly in the slings.[1] As she was hereby rendered incapable of making any sail for some time, we were obliged, as soon as a gale sprung up, to take her in tow; and near twenty of the healthiest and ablest of our seamen were taken from the business of our own ship, and were employed for eight or ten days together on board the Gloucester in repairing her damages. But these things, mortifying as we thought them, were but the beginning of our disasters; for scarce had our people finished their business in the Gloucester, before we met with a most violent storm in the western board, which obliged us to lie to. In the beginning of this storm our ship sprung a leak, and let in so much water, that all our people, officers included, were employed continually in working the pumps. And the next day we had the vexation to see the Gloucester

whole passage, abundance of birds of prey, also flying fish, which are their proper food, and vast quantities of skip-jacks, albicores, &c., whereof we took a great number, which contributed much to our refreshment after the loss of the tortoises, that generally leave all ships about twenty or thirty leagues off the land. I think this the more worthy of notice, because Dampier, Rogers, Cook, Cowley, and most other voyagers, some of whom have been not only once, but several times on this voyage, have reported that they never saw a fish or fowl in this whole run. For my part, I readily believe and conclude, that this difference in our observations and accounts is really occasioned by the different seasons of the year in which we happened to perform this passage; it being a known truth, and confirmed by the experience of thousands in all ages, that most fish have their different seasons for their different rendezvouses. The 10th, we saw three gannets, or, as they call them in Scotland, solan geese, being, by what I can learn from the most intelligent of that nation whom I have conversed with, and who often have opportunity to observe them in several different parts, of one and the same species."

[1] Anson records in his Official Report: "On the 15th of June the Gloucester found her mainmast sprung at the head, which, upon examination, was discovered to be entirely rotten. On the 29th of July the Gloucester carried away her foretopmast and foreyard. My ship's company are now miserably afflicted with the scurvy, the ship very leaky, the men and officers that were well being only able to make one spell at the pump." "This is all," observes Sir John Barrow, "that Anson says of the second attack of this afflicting malady; but, coming from the Commodore, it speaks volumes."

with her topmast once more by the board; and whilst we were viewing her with great concern for this new distress, we saw her maintop-mast, which had hitherto served as a jury mainmast, share the same fate. This completed our misfortunes, and rendered them without resource; for we knew the Gloucester's crew were so few and feeble that without our assistance they could not be relieved; and our sick were now so far increased, and those that remained in health so continually fatigued with the additional duty of our pumps, that it was impossible for us to lend them any aid. Indeed we were not as yet fully apprised of the deplorable situation of the Gloucester's crew; for when the storm abated (which during its continuance prevented all communication with them) the Gloucester bore up under our stern, and Captain Mitchel informed the Commodore that besides the loss of his masts, which was all that had appeared to us, the ship had then no less than seven feet of water in her hold, although his officers and men had been kept constantly at the pump for the last twenty-four hours.

This last circumstance was indeed a most terrible accumulation to the other extraordinary distresses of the Gloucester, and required, if possible, the most speedy and vigorous assistance, which Captain Mitchel begged the Commodore to send him. But the debility of our people, and our own immediate preservation, rendered it impossible for the Commodore to comply with his request. All that could be done was to send our boat on board for a more particular condition of the ship; and it was soon suspected that the taking her people on board us, and then destroying her, was the only measure that could be prosecuted in the present emergency, both for the security of their lives and of our own. Our boat soon returned with a representation of the state of the Gloucester, and of her several defects, signed by Captain Mitchel and all his officers; by which it appeared that she had sprung a leak by the sternpost being loose and working with every roll of the ship, and by two beams amidships being broken in the orlop, no part of which the carpenters reported was possible to be repaired at sea; that both officers and men had worked twenty-four hours at the pump without intermission, and were at length so fatigued that they could continue their labour no longer, but had been forced to desist, with seven feet of water in the hold, which covered their casks, so that they could neither come at fresh water nor provision; that they had no mast standing except the foremast, the mizzenmast, and the mizzentop-mast, nor had they any spare masts to get up in the room of those they had lost; that the ship was besides extremely decayed in every part, for her knees and clamps were all worked quite loose, and her upper works in general were so loose that the quarter-deck was ready to drop down; and that her crew was greatly reduced, for there remained alive on board her no more than seventy-seven men, eighteen boys, and two prisoners, officers included; and that of this whole number only sixteen men and eleven boys were capable of keeping the deck, and several of these very infirm.

The Commodore, on the perusal of this melancholy representation, presently ordered them a supply of water and provisions, of which they seemed to be in immediate want, and at the same time sent his own carpenter on board them to examine into the truth of every particular; and it being found, on the strictest inquiry, that the preceding account was in no instance exaggerated, it plainly appeared that there was no possibility of preserving the Gloucester any longer, as her leaks were irreparable, and the united hands on board both ships, capable of working, would not be able to free her, even if our own ship should not employ any part of them. What then could be resolved on, when it was the utmost we ourselves could do to manage our own pumps? Indeed, there was no room for deliberation;

the only step to be taken was the saving the lives of the few that remained on board the Gloucester, and getting out of her as much as was possible before she was destroyed. And therefore the Commodore immediately sent an order to Captain Mitchel, as the weather was now calm and favourable, to send his people on board the Centurion as expeditiously as he could, and to take out such stores as he could get at whilst the ship could be kept above water. And as our leak required less attention whilst the present easy weather continued, we sent our boats, with as many men as we could spare, to Captain Mitchel's assistance.

The removing the Gloucester's people on board us, and the getting out such stores as could most easily be come at, gave us full employment for two days. Mr Anson was extremely desirous to have got two of her cables and an anchor, but the ship rolled so much, and the men were so excessively fatigued, that they were incapable of effecting it; nay, it was even with the greatest difficulty that the prize-money which the Gloucester had taken in the South Seas was secured and sent on board the Centurion. However, the prize-goods on board her, which amounted to several thousand pounds in value, and were principally the Centurion's property, were entirely lost; nor could any more provision be got out than five casks of flour, three of which were spoiled by the salt water. Their sick men, amounting to near seventy, were removed into boats with as much care as the circumstances of that time would permit; but three or four of them expired as they were hoisting them into the Centurion.

It was the 15th of August, in the evening, before the Gloucester was cleared of everything that was proposed to be removed; and though the hold was now almost full of water, yet as the carpenters were of opinion that she might still swim for some time if the calm should continue and the water become smooth, she was set on fire; for we knew not how near we might now be to the Island of Guam, which was in the possession of our enemies, and the wreck of such a ship would have been to them no contemptible acquisition. When she was set on fire, Captain Mitchel and his officers left her, and came on board the Centurion; and we immediately stood from the wreck, not without some apprehensions (as we had now only a light breeze) that, if she blew up soon, the concussion of the air might damage our rigging; but she fortunately burned, though very fiercely, the whole night, her guns firing successively as the flames reached them. And it was six in the morning, when we were about four leagues distant, before she blew up; the report she made upon this occasion was but a small one, but there was an exceeding black pillar of smoke, which shot up into the air to a very considerable height. Thus perished his Majesty's ship the Gloucester. . . .

The 23d, at daybreak, we were cheered with the discovery of two islands in the western board. This gave us all great joy, and raised our drooping spirits; for before this a universal dejection had seized us, and we almost despaired of ever seeing land again. The nearest of these islands we afterwards found to be Anatacan. We judged it to be full fifteen leagues from us, and it seemed to be high land, though of an indifferent length. The other was the Island of Serigan, and had rather the appearance of a high rock than a place we could hope to anchor at. We were extremely impatient to get in with the nearest island, where we expected to meet with anchoring ground, and an opportunity of refreshing our sick; but the wind proved so variable all day, and there was so little of it, that we advanced towards it but slowly. However, by the next morning we were got so far to the westward that we were in view of a third island, which was that of Paxaros, though marked in the chart only as a rock. This was small and very low land, and we had passed within less than a mile of it in the night without seeing it. And now at noon, being within four miles of the Island of Anatacan, the boat was sent away to examine the anchoring ground and the produce of the place; and we were not a little solicitous for her return, as we then conceived our fate to depend upon the report we should receive; for the other two islands were obviously enough incapable of furnishing us with any assistance, and we knew not then that there were any others which we could reach. In the evening the boat came back, and the crew informed us that there was no place for a ship to anchor, the bottom being everywhere foul ground, and all, except one small spot, not less than fifty fathoms in depth; that on that spot there was thirty fathoms, though not above half-a-mile from the shore; and that the bank was steep and could not be depended on. They further told us that they had landed on the island, but with some difficulty, on account of the greatness of the swell; that they found the ground was everywhere covered with a kind of cane or rush; but that they met with no water, and did not believe the place to be inhabited, though the soil was good, and abounded with groves of cocoa-nut trees.

This account of the impossibility of anchoring at this island occasioned a general melancholy on board, for we considered it as little less than the prelude to our destruction; and our despondency was increased by a disappointment we met with the succeeding night; for, as we were plying under topsails, with an intention of getting nearer to the island and of sending our boat on shore to load with cocoa-nuts for the refreshment of our sick, the wind proved squally, and blew so strong off shore that we were driven so far to the southward that we dared not to send off our boat. And now the only possible circumstance that could secure the few that remained alive from perishing was the accidental falling in with some other of the Ladrone Islands better prepared for our accommodation; and as our knowledge of these islands was extremely imperfect, we were to trust entirely to chance for our guidance; only, as they are all of them usually laid down near the same meridian, and we had conceived those we had already seen to be part of them, we concluded to stand to the southward as the most probable means of falling in with the next. Thus, with the most gloomy persuasion of our approaching destruction, we stood from the Island of Anatacan, having all of us the strongest apprehensions (and those not ill founded) either of dying of the scurvy or of perishing with the ship, which, for want of hands to work her pumps, might in a short time be expected to founder.

CHAPTER II.

It was the 26th of August 1742, in the morning, when we lost sight of Anatacan. The next morning we discovered three other islands to the eastward, which were from ten to fourteen leagues from us. These were, as we afterwards learned, the islands of Saypan, Tinian, and Aguigan. We immediately steered towards Tinian, which was the middlemost of the three; but had so much of calms and light airs, that though we were helped forwards by the currents, yet next day at daybreak we were at least five leagues distant from it. However, we kept on our course, and about ten in the morning we perceived a proa under sail to the southward, between Tinian and Aguigan. As we imagined from hence that these islands were inhabited, and knew that the Spaniards had always a force at Guam, we took the necessary precautions for our own security and for preventing the enemy from taking advantage of our present wretched circumstances, of which they would be sufficiently informed by the manner of our working the ship. We therefore mustered all our hands who were capable of standing to their arms, and loaded our upper and quarter deck guns with grape shot; and that we might the more readily procure some intelligence of the state of these islands, we showed Spanish colours and hoisted a red flag at the foretop-mast head, to give our ship the appearance of the Manilla galleon, hoping thereby to decoy some of the inhabitants on board us. Thus preparing ourselves, and standing towards the land, we were near enough at three in the afternoon to send the cutter in-shore to find out a proper berth for the ship; and we soon perceived that a proa came off the shore to meet the cutter, fully persuaded, as we afterwards found, that we were the Manilla ship. As we saw the cutter returning back with the proa in tow, we immediately sent the pinnace to receive the proa and the prisoners, and to bring them on board, that the cutter might proceed on her errand. The pinnace came back with a Spaniard and four Indians,[1] who were the people taken in the proa. The Spaniard was immediately examined as to the produce and circumstances of this Island of Tinian, and his account of it surpassed even our most sanguine hopes; for he informed us that it was uninhabited, which, in our present defenceless condition, was an advantage not to be despised, especially as it wanted but few of the conveniences that could be expected in the most cultivated country; for he assured us that there was great plenty of very good water, and that there were an incredible number of cattle, hogs, and poultry running wild on the island, all of them excellent in their kind; that the woods produced sweet and sour oranges, limes, lemons, and cocoa-nuts in great plenty, besides a fruit peculiar to these islands (called by Dampier bread-fruit[2]); that, from the quantity and goodness of the provisions produced here, the Spaniards at Guam made use of it as a store for supplying the garrison; that he himself was a sergeant of that garrison, and was sent here with twenty-two Indians to jerk beef, which he was to load for Guam on board a small bark of about fifteen tons, which lay at anchor near the shore.

This account was received by us with inexpressible joy. Part of it we were ourselves able to verify on the spot, as we were by this time near enough to discover several numerous herds of cattle feeding in different places of the island; and we did not anyways doubt the rest of his relation, as the appearance of the shore prejudiced us greatly in its favour, and made us hope that not only our necessities might be there fully relieved and our diseased recovered, but that, amidst those pleasing scenes which were then in view, we might procure ourselves some amusement and relaxation after the numerous fatigues we had undergone. For the prospect of the country did by no means resemble that of an uninhabited and uncultivated place, but had much more the air of a magnificent plantation, where large lawns and stately woods had been laid out together with great skill, and where the whole had been so artfully combined, and so judiciously adapted to the slopes of the hills and the inequalities of the ground, as to produce a most striking effect, and to do honour to the invention of the contriver. Thus (an event not unlike what we had already seen) we were forced upon the most desirable and salutary measures by accidents which at first sight we considered as the greatest of misfortunes; for had we not been driven by the contrary winds and currents to the northward of our course (a circumstance which at that time gave us the most terrible apprehensions), we should in all probability never have arrived at this delightful island, and consequently we should have missed that place where alone all our wants could be most amply relieved, our sick recovered, and our enfeebled crew once

---

[1] Thomas says: "One of those Indians was a carpenter by trade, and his father was one of the principal builders at Manilla. This young man having been ill used by the Governor at Guam, voluntarily entered with us, and became one of our carpenter's crew, and proved a very useful handy fellow."

[2] In Chapter X.

more refreshed and enabled to put again to sea.

The Spanish sergeant, from whom we received the account of the island, having informed us that there were some Indians on shore under his command employed in jerking beef, and that there was a bark at anchor to take it on board, we were desirous if possible to prevent the Indians from escaping, who doubtless would have given the Governor of Guam intelligence of our arrival; and we therefore immediately despatched the pinnace to secure the bark, which the sergeant told us was the only embarkation on the place. And then, about eight in the evening, we let go our anchor, in twenty-two fathoms, and though it was almost calm, and whatever vigour and spirit was to be found on board was doubtless exerted to the utmost on this pleasing occasion, when, after having kept the sea for some months, we were going to take possession of this little paradise, yet we were full five hours in furling our sails. It is true, we were somewhat weakened by the crews of the cutter and pinnace which were sent on shore; but it is not less true that, including those absent with the boats and some Negro and Indian prisoners, all the hands we could muster capable of standing at a gun amounted to no more than seventy-one, most of which number too were incapable of duty; but on the greatest emergencies this was all the force we could collect, in our present enfeebled condition, from the united crews of the Centurion, the Gloucester, and the Trial, which, when we departed from England, consisted altogether of near 1000 hands.

When we had furled our sails, the remaining part of the night was allowed to our people for their repose, to recover them from the fatigue they had undergone; and in the morning a party was sent on shore well-armed, of which I myself was one, to make ourselves masters of the landing-place, as we were not certain what opposition might be made by the Indians on the island. We landed without difficulty, for the Indians having perceived, by our seizure of the bark the night before, that we were enemies, they immediately fled into the woody parts of the island. We found on shore many huts which they had inhabited, and which saved us both the time and trouble of erecting tents. One of these huts, which the Indians made use of for a store-house, was very large, being twenty yards long and fifteen broad; this we immediately cleared of some bales of jerked beef which we found in it, and converted it into an hospital for our sick, who, as soon as the place was ready to receive them, were brought on shore, being in all 128. Numbers of these were so very helpless, that we were obliged to carry them from the boats to the hospital upon our shoulders, in which humane employment (as before at Juan Fernandez) the Commodore himself and every one of his officers were engaged without distinction;[1] and, notwithstanding the great debility and other dying aspects of the greatest part of our sick, it is almost incredible how soon they began to feel the salutary influence of the land. For though we buried twenty-one men on this and the preceding day, yet we did not lose above ten men more during our whole two months' stay here; and in general our diseased received so much benefit from the fruits of the island, particularly the fruits of the acid kind, that in a week's time there were but few who were not so far recovered as to be able to move about without help.

This island [of Tinian] lies in Latitude 50° 8′ N., and Longitude from Acapulco 114° 50′ W. Its length is about twelve miles, and its breadth about half as much; it extending from the SSW. to NNE. The soil is everywhere dry and healthy, and somewhat sandy, which, being less disposed than other soils to a rank

---

[1] "And indeed," says Thomas, "they were almost the only persons on board capable of performing this service; the healthiest seamen being so much enfeebled, that they had but just strength enough left to help themselves."

and over-luxuriant vegetation, occasions the meadows and the bottoms of the woods to be much neater and smoother than is customary in hot climates. The land rises by easy slopes, from the very beach where we watered, to the middle of the island; though the general course of its ascent is often interrupted and traversed by gentle descents and valleys; and the inequalities that are formed by the different combinations of these gradual swellings of the ground are most beautifully diversified with large lawns, which are covered with a very fine trefoil, intermixed with a variety of flowers, and are skirted by woods of tall and well-spread trees, most of them celebrated either for their aspect or their fruit. The turf of the lawns is quite clean and even, and the bottoms of the woods in many places clear of all bushes and underwoods; and the woods themselves usually terminate on the lawns with a regular outline, not broken nor confused with straggling trees, but appearing as uniform as if laid out by art. . . .

I must now observe that all these advantages were greatly enhanced by the healthiness of its climate, by the almost constant breezes which prevail there, and by the frequent showers which fall, and which, though of a very short and almost momentary duration, are extremely grateful and refreshing, and are perhaps one cause of the salubrity of the air and of the extraordinary influence it was observed to have upon us in increasing and invigorating our appetites and digestion. This was so remarkable, that those amongst our officers who were at all other times spare and temperate eaters, who besides a slight breakfast made but one moderate repast a-day, were here in appearance transformed into gluttons; for instead of one reasonable flesh-meal, they were now scarcely satisfied with three, and each of them so prodigious in quantity as would at another time have produced a fever or a surfeit. And yet our digestion so well corresponded with the keenness of our appetites, that we were neither disordered nor even loaded by this repletion; for after having, according to the custom of the island, made a large beef breakfast, it was not long before we began to consider the approach of dinner as a very desirable though somewhat tardy incident. . . .

Our first undertaking after our arrival was the removal of our sick on shore, as has been mentioned. Whilst we were thus employed, four of the Indians on shore, being part of the Spanish sergeant's detachment, came and surrendered themselves to us; so that with those we took in the proa, we had now eight of them in our custody. One of the four who submitted undertook to show us the most convenient place for killing cattle, and two of our men were ordered to attend him on that service; but one of them unwarily trusting the Indian with his firelock and pistol, the Indian escaped with them into the woods. His countrymen, who remained behind, were apprehensive of suffering for this perfidy of their comrade, and therefore begged leave to send one of their own party into the country, who they engaged should both bring back the arms and persuade the whole detachment from Guam to submit to us. The Commodore granted their request, and one of them was despatched on this errand, who returned next day and brought back the firelock and pistol, but assured us he had met with them in a pathway in the wood, and protested that he had not been able to meet with any one of his countrymen. This report had so little the air of truth, that we suspected there was some treachery carrying on; and therefore, to prevent any future communication amongst them, we immediately ordered all the Indians who were in our power on board the ship, and did not permit them to return any more on shore.

Towards the middle of September several of our sick were tolerably recovered by their residence on shore; and, on the 12th of September all those who were so far relieved since their arrival as to be capable of doing duty were sent on board the ship. And then the Commodore, who was himself ill of the scurvy, had a tent erected for him on shore, where he went with the view of staying a few days for the recovery of his health; being convinced, by the general experience of his people, that no other method but living on the land was to be trusted to for the removal of this dreadful malady. The place where his tent was pitched on this occasion, was near the well whence we got all our water, and was indeed a most elegant spot. As the crew on board were now reinforced by the recovered hands returned from the island, we began to send our casks on shore to be fitted up, which till now could not be done, for the coopers were not well enough to work. We likewise weighed our anchors, that we might examine our cables, which we suspected had by this time received considerable damage. And as the new moon was now approaching, when we apprehended violent gales, the Commodore, for our greater security, ordered that part of the cables next to the anchors to be armed with the chains of the fire-grapnels; and they were besides cackled twenty fathoms from the anchors, and seven fathoms from the service, with a good rounding of a 4½-inch hawser; and to all these precautions we added that of lowering the main and fore yards close down, that in case of blowing weather the wind might have less power upon the ship to make her ride a-strain.

Thus effectually prepared, as we conceived, we expected the new moon, which was the 18th of September; and riding safe that and the three succeeding days (though the weather proved very squally and uncertain), we flattered ourselves (for I was then on board) that the prudence of our measures had secured us from all accidents. But on the 22d the wind blew from the eastward with such fury that we soon despaired of riding out the storm; and therefore we should have been extremely glad that the Commodore and the rest of our people on shore, which were the greatest part of our hands, had been on board with us, since our only hopes of safety seemed to depend on our putting immediately to sea. But all communication with the shore was now effectually cut off, for there was no possibility that a boat could live, so that we were necessitated to ride it out till our cables parted. Indeed it was not long before this happened, for the small bower parted at five in the afternoon, and the ship swung off to the best bower; and as the night came on, the violence of the wind still increased. But, notwithstanding its inexpressible fury, the tide ran with so much rapidity as to prevail over it; for the tide, having set to the northward in the beginning of the storm, turned suddenly to the southward about six in the evening, and forced the ship before it in despite of the storm, which blew upon the beam. And now the sea broke most surprisingly all around us, and a large tumbling swell threatened to poop us; the long-boat, which was at this time moored astern, was on a sudden canted so high that it broke the transom of the Commodore's gallery, whose cabin was on the quarter-deck, and would doubtless have risen as high as the taffrail had it not been for this stroke which stove the boat all to pieces; but the poor boat-keeper, though extremely bruised, was saved almost by miracle. About eight the tide slackened, but the wind did not abate; so that at eleven the best bower cable, by which alone we rode, parted. Our sheet anchor, which was the only one we had left, was instantly cut from the bow; but before it could reach the bottom we were driven from twenty-two into thirty-five fathoms; and after we had veered away one whole cable, and two-thirds of another, we could not find ground with sixty fathoms of line. This was a plain indication that the anchor lay near the edge of the bank, and could not hold us [long].

In this pressing danger, Mr Saumarez, our first lieutenant, who now commanded on board, ordered several guns to be fired and lights to be shown, as a signal to the Commodore of our distress; and in a short time after, it being then about 1 o'clock, and the night excessively dark, a strong gust, attended with rain and lightning, drove us off the bank and forced us out to sea, leaving behind us on the island Mr Anson, with many more of our officers, and great part of our crew, amounting in the whole to 113 persons. Thus were we all, both at sea and on shore, reduced to the utmost despair by this catastrophe; those on shore conceiving they had no means left them ever to leave the island, and we on board utterly unprepared to struggle with the fury of the seas and winds we were now exposed to, and expecting each moment to be our last.

CHAPTER III.

The storm which drove the Centurion to sea blew with too much turbulence to permit either the Commodore or any of the people on shore from hearing the guns which she fired as signals of distress, and the frequent glare of the lightning had prevented the explosions from being observed: so that when at daybreak it was perceived from the shore that the ship was missing, there was the utmost consternation amongst them. For much the greatest part of them immediately concluded that she was lost, and entreated the Commodore that the boat might be sent round the island to look for the wreck; and those who believed her safe had scarcely any expectation that she would ever be able to make the island again; for the wind continued to blow strong at east, and they knew how poorly she was manned and provided for struggling with so tempestuous a gale. And if the Centurion was lost, or should be incapable of returning, there appeared in either case no possibility of their ever getting off the island; for they were at least 600 leagues from Macao, which was their nearest port; and they were masters of no other vessel than the small Spanish bark, of about fifteen tons, which they seized at their first arrival, and which would not even hold a fourth part of their number. And the chance of their being taken off the island by the casual arrival of any other ship was altogether desperate, as perhaps no European ship had ever anchored here before, and it were madness to expect that like incidents should send another here in 100 ages to come; so that their desponding thoughts could only suggest to them the melancholy prospect of spending the remainder of their days on this island, and bidding adieu for ever to their country, their friends, their families, and all their domestic endearments. Nor was this the worst they had to fear: for they had reason to expect that the Governor of Guam, when he should be informed of their situation, might send a force sufficient to overpower them and to remove them to that island; and then the most favourable treatment they could hope for would be to be detained prisoners for life; since, from the known policy and cruelty of the Spaniards in their distant settlements, it was rather to be expected that the Governor, if he once had them in his power, would make their want of commissions (all of them being on board the Centurion) a pretext for treating them as pirates, and for depriving them of their lives with infamy.[1]

In the midst of these gloomy reflections Mr Anson had doubtless his share of disquietude, but he always kept up his usual composure and steadiness; and having soon projected

---

[1] As in 1575, John Oxenham, or Oxnam, who had accompanied Drake in his expedition to the West Indies, was put to death by the Governor of Panama, with all his companions, because he had undertaken a daring but ultimately calamitous privateering expedition without any commission from his sovereign. Oxenham was the first Englishman who ever navigated the Pacific Ocean.

a scheme for extricating himself and his men from their present anxious situation, he first communicated it to some of the most intelligent persons about him; and having satisfied himself that it was practicable, he then endeavoured to animate his people to a speedy and vigorous prosecution of it. With this view he represented to them how little foundation there was for their apprehensions of the Centurion's being lost; that he should have hoped they had been all of them better acquainted with sea affairs than to give way to the impression of so chimerical a fright, and that he doubted not but, if they would seriously consider what such a ship was capable of enduring, they would confess that there was not the least probability of her having perished; that he was not without hopes that she might return in a few days, but if she did not, the worst that could be supposed was that she was driven so far to the leeward of the island that she could not regain it, and that she would consequently be obliged to bear away for Macao on the coast of China; that, as it was necessary to be prepared against all events, he had, in this case, considered of a method of carrying them off the island, and joining their old ship the Centurion again at Macao; that this method was to haul the Spanish bark on shore, to saw her asunder, and to lengthen her twelve feet, which would enlarge her to near forty tons burthen, and would enable her to carry them all to China; that he had consulted the carpenters, and they had agreed that this proposal was very feasible, and that nothing was wanting to execute it but the united resolution and industry of the whole body. He added that for his own part he would share the fatigue and labour with them, and would expect no more from any man than what he, the Commodore himself, was ready to submit to; and concluded with representing to them the importance of saving time, and that, in order to be the better prepared for all events, it was necessary to set to work immediately and to take it for granted that the Centurion would not be able to put back (which was indeed the Commodore's secret opinion); since, if she did return, they should only throw away a few days' application; but, if she did not, their situation, and the season of the year, required their utmost despatch.

These remonstrances, though not without effect, did not immediately operate so powerfully as Mr Anson could have wished. He indeed raised their spirits by showing them the possibility of their getting away, of which they had before despaired; but then, from their confidence of this resource; they grew less apprehensive of their situation, gave a greater scope to their hopes, and flattered themselves that the Centurion would return and prevent the execution of the Commodore's scheme, which they could easily foresee would be a work of considerable labour. By this means it was some days before they were all of them heartily engaged in the project; but at last, being in general convinced of the impossibility of the ship's return, they set themselves zealously to the different tasks allotted them, and were as industrious and as eager as their commander could desire,[1] punctually assembling at daybreak at the rendezvous, whence they were distributed to their different employments, which they followed with unusual vigour till night came on.

And here I must interrupt the course of this transaction for a moment to relate an incident which for some time gave Mr Anson more concern than all the preceding disasters. A few days after the ship was driven off, some of the people on shore cried out, "A sail!" This spread a general joy, every one supposing that it was the ship returning; but presently a second sail was descried, which quite de-

[1] And the Commodore, Thomas says, encouraged their diligence by his example; for being always at work by daybreak himself, it was thought a disgrace to be idle when their chief was employed.

stroyed their conjecture, and made it difficult to guess what they were. The Commodore eagerly turned his glass towards them, and saw they were two boats; on which it immediately occurred to him that the Centurion was gone to the bottom, and that these were her two boats coming back with the remains of her people; and this sudden and unexpected suggestion wrought on him so powerfully that, to conceal his emotion, he was obliged (without speaking to any one) instantly to retire to his tent, where he passed some bitter moments in the firm belief that the ship was lost, and that now all his views of further distressing the enemy, and of still signalising his expedition by some important exploit, were at an end. But he was soon relieved from these disturbing thoughts by discovering that the two boats in the offing were Indian proas; and perceiving that they stood towards the shore, he directed every appearance that could give them any suspicion to be removed, and concealed his people in the adjacent thickets, prepared to secure the Indians when they should land. But after the proas had stood in within a quarter of a mile of the land, they suddenly stopped short, and, remaining there motionless for near two hours, they then made sail again and stood to the southward.

But to return to the projected enlargement of the bark. If we examine how they were prepared for going through with this undertaking, on which their safety depended, we shall find that, independent of other matters which were of as much importance, the lengthening of the bark alone was attended with great difficulty. Indeed, in a proper place, where all the necessary materials and tools were to be had, the embarrassment would have been much less; but some of these tools were to be made, and many of the materials were wanting; and it required no small degree of invention to supply all these deficiencies. And when the hull of the bark should be completed, this was but one article; and there were many others of equal weight, which were to be well considered. These were the rigging it, the victualling it, and lastly, the navigating it for the space of six or seven hundred leagues, through unknown seas, where no one of the company had ever passed before. In some of these particulars such obstacles occurred, that without the intervention of very extraordinary and unexpected accidents the possibility of the whole enterprise would have fallen to the ground, and their utmost industry and efforts must have been fruitless.[1] . . .

And now, all these obstacles being in some degree removed (which were always as much as possible concealed from the vulgar,[2] that they might not grow remiss with the apprehension of labouring to no purpose), the work proceeded very successfully and vigorously. The necessary ironwork was in great forwardness, and the timbers and planks (which, though not the most exquisite performances of the sawyer's art, were yet sufficient for the purpose) were all prepared; so that on the 6th of October, being the fourteenth day from the departure of the ship, they hauled the bark on shore, and, on the two succeeding days she was sawn asunder (though with great care not to cut her planks), and her two parts were separated the proper distance from each other; and, the ma-

---

[1] Both carpenters and smiths were here hard at work in the enlargement of the bark, Anson himself lending a hand in the sawing of trees into plank. When the equipment of the vessel was being proceeded with, they made the disheartening discovery that they were without a compass by which to steer, but in about eight days from the departure of the Centurion, they were relieved from their perplexity, by the discovery of a small one in a chest belonging to the Spanish bark.

[2] From the general knowledge of the company on shore. Thomas says, "The alacrity with which the business was carried on left no room for reflection among the common sailors, though their superiors were not without their fears."

terials being all ready beforehand, they the next day, being the 9th of October, went on with great despatch in their proposed enlargement of her. And by this time they had all their future operations so fairly in view, and were so much masters of them, that they were able to determine when the whole would be finished, and had accordingly fixed the 5th of November for the day of their putting to sea. But their projects and labours were now drawing to a speedier and happier conclusion; for on the 11th of October, in the afternoon, one of the Gloucester's men, being upon a hill in the middle of the island, perceived the Centurion at a distance, and running down with his utmost speed towards the landing-place, he in the way saw some of his comrades, to whom he hallooed out with great ecstasy, "The ship! The ship!" This being heard by Mr Gordon, a lieutenant of marines, who was convinced by the fellow's transport that his report was true, Mr Gordon ran towards the place where the Commodore and his people were at work, and being fresh and in breath easily outstripped the Gloucester's man, and got before him to the Commodore; who, on hearing this happy and unexpected news, threw down his axe with which he was then at work, and by his joy broke through for the first time the equable and unvaried character which he had hitherto preserved. The others who were with him instantly ran down to the seaside in a kind of frenzy, eager to feast themselves with a sight they had so ardently wished for, and of which they had now for a considerable time despaired. By five in the evening the Centurion was visible in the offing to them all; and, a boat being sent off with eighteen men to reinforce her, and with fresh meat and fruits for the refreshment of her crew, she the next afternoon happily came to an anchor in the road, where the Commodore immediately came on board her, and was received by us with the sincerest and heartiest acclamations. For from the following short recital of the fears, the dangers, and fatigues we in the ship underwent during our nineteen days' absence from Tinian, it may be easily conceived that a harbour, refreshments, repose, and the joining of our Commander and shipmates, were not less pleasing to us than our return was to them.

CHAPTER IV.

The Centurion being now once more safely arrived at Tinian, to the mutual respite of the labours of our divided crew, it is high time that the reader, after the relation already given of the projects and employment of those left on shore, should be apprised of the fatigues and distresses to which we, who were driven off to sea, were exposed during the long interval of nineteen days that we were absent from the island.

It has been already mentioned, that it was the 22d of September, about 1 o'clock in an extreme dark night, when by the united violence of a prodigious storm, and an exceeding rapid tide, we were driven from our anchors and forced to sea. Our condition then was truly deplorable; we were in a leaky ship, with three cables in our hawses, to one of which hung our only remaining anchor; we had not a gun on board lashed, nor a port barred in; our shrouds were loose, and our topmasts unrigged, and we had struck our fore and main yards close down before the storm came on, so that there were no sails we could set except our mizzen. In this dreadful extremity we could muster no more strength on board to navigate the ship than 108 hands, several Negroes and Indians included. This was scarcely the fourth part of our complement; and of these the greater number were either boys, or such as, being lately recovered from the scurvy, had not yet arrived at half their former vigour. No sooner were we at sea, but by the violence of the storm, and the working of the ship, we made a great quantity of water through our hawse-holes, ports, and scuppers, which, added to the constant effect of our leak, rendered our pumps alone a sufficient employment for us all. But though this leakage, by being a short time neglected, would inevitably end in our destruction, yet we had other dangers then impending, which occasioned this to be regarded as a secondary consideration only. For we all imagined that we were driving directly on the neighbouring island of Aguigan, which was about two leagues distant; and as we had lowered our main and fore yards close down, we had no sails we could set but the mizzen, which was altogether insufficient to carry us clear of this instant peril. We therefore immediately applied ourselves to work, endeavouring by the utmost of our efforts to heave up the main and fore yards, in hopes that, if we could but be enabled to make use of our lower canvas, we might possibly weather the island, and thereby save ourselves from this impending shipwreck. But after full three hours' ineffectual labour the jeers broke, and the men being quite jaded, we were obliged by mere debility to desist, and quietly to expect our fate, which we then conceived to be unavoidable. For we imagined ourselves by this time to be driven just upon the shore, and the night was so extremely dark, that we expected to discover the island no otherwise than by striking upon it; so that the belief of our destruction, and the uncertainty of the point of time when it would take place, occasioned us to pass several hours under the most serious apprehensions, that each succeeding moment would send us to the bottom. Nor did these continued terrors of instantly striking and sinking end but with the daybreak; when we with great transport perceived that the island we had thus dreaded was at a considerable distance, and that a strong northern current had been the cause of our preservation.

The turbulent weather which forced us from Tinian did not begin to abate till three days after; and then we swayed up the foreyard, and began to heave up the mainyard, but the jeers broke and killed one of our men, and prevented us at that time from proceeding. The next day, being the 26th of September, was a day of most severe fatigue to us all; for it must be remembered that in these exigencies no rank or office exempted any person from the manual application and bodily labour of a common sailor. The business of this day was no less than an attempt to heave up the sheet-anchor, which we had hitherto dragged at our bows with two cables an end. This was a work of great importance to our future preservation; for, not to mention the impediment to our navigation, and the hazard it would be to our ship if we attempted to make sail with the anchor in its present situation, we had this most interesting consideration to animate us, that it was the only anchor we had left, and, without securing it, we should be under the utmost difficulties and hazards whenever we made the land again; and therefore, being all of us fully apprised of the consequence of this enterprise, we laboured at it with the severest application for full twelve hours, when we had indeed made a considerable progress, having brought the anchor in sight. But it then growing dark, and we being excessively fatigued, we were obliged to desist, and to leave the work unfinished till the next morning, when, by the benefit of a night's rest, we completed it, and hung the anchor at our bow.

It was the 27th of September, in the morning, that is, five days after our departure, when we thus secured our anchor; and the same day we got up our mainyard. And having now conquered in some degree the distress and disorder which we were necessarily involved in at our first driving out to sea, and being enabled to make use of our canvas, we set our courses, and for the first time stood to the eastward, in hopes of regaining the Island of Tinian, and joining our Commodore in a few days; for we were then, by our accounts, only forty-seven leagues to the south-west of Tinian, so that on the 1st day of October, having then run the distance necessary for making the island according to our reckoning, we were in full expectation of seeing it; but we were unhappily disappointed, and were thereby convinced that a current had driven us to the westward. And as we could not judge how much we might hereby have deviated, and consequently how long we might still expect to be at sea, we had great apprehensions that our stock of water might prove deficient; for we were doubtful about the quantity we had on board, and found many of our casks so decayed as to be half leaked out. However, we were delivered from our uncertainty the next day, by having a sight of the Island of Guam, by which we discovered that the currents had driven us forty-four leagues to the westward of our accounts. This sight of land having satisfied us of our situation, we kept plying to the eastward, though with excessive labour; for the wind continuing fixed in the eastern board, we were obliged to tack often, and our crew were so weak, that without the assistance of every man on board, it was not in our power to put the ship about. This severe employment lasted till the 11th of October, being the nineteenth day from our departure, when, arriving in the offing of Tinian, we were reinforced from the shore, as has been already mentioned; and on the evening of the same day we, to our inexpressible joy, came to an anchor in the road, thereby procuring to our shipmates on shore, as well as to ourselves, a cessation from the fatigues and apprehensions which this disastrous incident had given rise to.

CHAPTER V.

When the Commodore came on board the Centurion on her return to Tinian, as already mentioned, he resolved to stay no longer at the island than was absolutely necessary to complete our stock of water, a work which we immediately set ourselves about. But the loss of our long-boat, which was staved against our poop when we were driven out to sea, put us to great inconveniences in getting our water on board, for we were obliged to raft off all our casks, and the tide ran so strong that, besides the frequent delays and difficulties it occasioned, we more than once lost the whole raft. Nor was this our only misfortune, for on the 14th of October, being but the third day after our arrival, a sudden gust of wind brought home our anchor, forced us off the bank, and drove the ship out to sea a second time. The Commodore, it is true, and the principal officers, were now on board; but we had near seventy men on shore who had been employed in filling our water and procuring provisions. These had with them our two cutters, but as they were too many for the cutters to bring off at once, we sent the eighteen-oared barge to assist them, and at the same time made a signal for all that could to embark. The two cutters soon came off to us full of men, but forty of the company who were employed in killing cattle in the wood and in bringing them down to the landing-place were left behind; and though the eighteen-oared barge was left for their conveyance, yet, as the ship soon drove to a considerable distance, it was not in their power to join us. However, as the weather was favourable, and our crew was now stronger than when we were first driven out, we in about five days' time returned again to an anchor at Tinian, and relieved those we had left behind us from their second fears of being deserted by their ship.

On our arrival we found that the Spanish bark, the old object of their hopes, had undergone a new metamorphosis. For those we had left on shore began to despair of our return, and conceiving that the lengthening the bark as formerly proposed was both a toilsome and unnecessary measure, considering the small number they consisted of, they had resolved to join her again and to restore her to her first state; and in this scheme they had made some progress, for they had brought the two parts together, and would have soon completed her had not our coming back put a period to their labours and disquietudes. These people we had left behind informed us that, just before we were seen in the offing, two proas had stood in very near the shore, and had continued there for some time; but on the appearance of our ship they crowded away, and were presently out of sight. And on this occasion I must mention an incident which, though it happened during the first absence of the ship, was then omitted to avoid interrupting the course of the narration.

It has been already observed that a part of the detachment sent to this island under the command of the Spanish sergeant lay concealed in the woods, and we were the less solicitous to find them out as our prisoners all assured us that it was impossible for them to get off, and consequently that it was impossible for them to send any intelligence about us to Guam. But when the Centurion drove out to sea and left the Commodore on shore, he one day, attended by some of his officers, endeavoured to make the tour of the island. In this expedition, being on a rising ground, they perceived in the valley beneath them the appearance of a small thicket which, by observing more nicely, they found had a progressive motion; this at first surprised them, but they soon discovered that it was no more than several large cocoa bushes which were dragged along the ground by persons concealed beneath them. They immediately concluded that these were some of the sergeant's party (which was indeed true), and therefore the Commodore and his people made after them in hopes of finding out their retreat. The Indians soon perceived they were discovered, and hurried away with precipitation; but Mr Anson was so near them that he did not lose sight of them till they arrived at their cell, which he and his officers entering found to be abandoned, there being a passage from it down a precipice contrived for the convenience of flight. They found here an old firelock or two, but no other arms. However, there was a great quantity of provisions, particularly salted spareribs of pork, which were excellent; and from what our people saw here, they concluded that the extraordinary appetite which they had found at this island was not confined to themselves alone; for it being about noon, the Indians had laid out a very plentiful repast, considering their numbers, and had their bread-fruit and cocoa-nuts prepared ready for eating, and in a manner which plainly evinced that with them too a good meal was neither an uncommon nor an unheeded article. The Commodore having in vain endeavoured to discover the path by which the Indians had escaped, he and his officers contented themselves with sitting down to the dinner which was thus luckily fitted to their present appetites; after which they returned back to their old habitation, displeased at missing the Indians, as they hoped to have engaged them in our service if they could have had any conference with them. But, notwithstanding what our prisoners had asserted, we were afterwards assured that these Indians were carried off to Guam long before we left the place.

On our coming to an anchor again, after our second driving off to sea, we laboured indefatigably in getting in our water;[1] and having by the 20th of October completed it to fifty tuns, which we supposed would be sufficient for our passage to Macao, we on the next day sent one of each mess on shore to gather as large a quantity of oranges, lemons, cocoa-nuts, and other

[1] "In which service," says the useful Thomas, "two of our men employed in the well unfortunately perished; for the sides of the well being loose earth, by the carelessness of those above in not properly attending the filling, the bank gave way by the weight of a heavy cask, and both that and the bank fell in upon them together."

fruits of the island, as they possibly could for the use of themselves and messmates when at sea. And these purveyors returning on board us on the evening of the same day, we then set fire to the bark and proa, hoisted in our boats, and got under sail, steering away for the south end of the Island of Formosa, and taking our leave for the third and last time of the Island of Tinian, an island which, whether we consider the excellence of its productions, the beauty of its appearance, the elegance of its woods and lawns, the healthiness of its air, or the adventures it gave rise to, may in all these views be truly styled romantic.[1]

CHAPTER VI.

I HAVE already mentioned that on the 21st of October, in the evening, we took our leave of the Island of Tinian, steering the proper course for Macao in China. The eastern monsoon was now, we reckoned, fairly settled; and we had a constant gale blowing right upon our stern; so that we generally ran from forty to fifty leagues a day. But we had a large hollow sea pursuing us, which occasioned the ship to labour much; whence we received great damage in our rigging, which was grown very rotten, and our leak was augmented; but happily for us our people were now in full health, so that there were no complaints of fatigue, but all went through their attendance on the pumps, and every other duty of the ship, with ease and cheerfulness.

Having now no other but our sheet-anchor left, except our prize anchors, which were stowed in the hold, and were too light to be depended on, we were under great concern how we should manage on the coast of China, where we were all entire strangers, and where we should doubtless be frequently under the necessity of coming to an anchor. Our sheet-anchor being obviously much too heavy for a coasting anchor, it was at length resolved to fix two of our largest prize anchors into one stock, and to place between their shanks two guns, four-pounders, which was accordingly executed, and it was to serve as a best bower. And a third prize-anchor being in like manner joined with our stream-anchor, with guns between them, we thereby made a small bower, so that, besides our sheet-anchor, we had again two others at our bows, one of which weighed 3900 and the other 2900 pounds.

The 3d of November, about three in the afternoon, we saw an island, which at first we imagined to be the Island of Botel Tobago Xima, but on our nearer approach we found it to be much smaller than that is usually represented; and about an hour after we saw another island five or six miles farther to the westward. As no chart, nor any journal we had seen, took notice of any other island to the eastward of Formosa than Botel Tobago Xima; and as we had no observation of our latitude at noon, we were in some perplexity, being apprehensive that an extraordinary current had driven us into the neighbourhood of the Bashee Islands; and therefore when night came on we brought to, and continued in this posture till the next morning, which, proving dark and cloudy, for some time prolonged our uncertainty; but it cleared up about nine o'clock, when we again discerned the two islands above mentioned; we then pressed forwards to the westward, and by eleven got a sight of the southern part of the Island of Formosa. This satisfied us that the second island we saw was Botel Tobago Ximà, and the first a small island or rock lying five or six miles due east from it, which not being mentioned by any of our books or charts, was the occasion of our

---

[1] A description of the Ladrones, and of the wonderfully fast-sailing proas of the inhabitants, is omitted. Dampier gives an interesting account of Guam and of the proas in his Tenth Chapter.

fears. While we were passing by these rocks of Vele Rete, there was an outcry of fire on the forecastle; this occasioned a general alarm, and the whole crew instantly flocked together in the utmost confusion, so that the officers found it difficult for some time to appease the uproar. But having at last reduced the people to order, it was perceived that the fire proceeded from the furnace; and pulling down the brick-work, it was extinguished with great facility, for it had taken its rise from the bricks, which, being overheated, had begun to communicate the fire to the adjacent wood-work. In the evening we were surprised with a view of what we at first sight conceived to have been breakers, but on a stricter examination we found them to be only a great number of fires on the Island of Formosa. These, we imagined, were intended by the inhabitants of that island as signals for us to touch there; but that suited not our views, we being impatient to reach the port of Macao as soon as possible. From Formosa we steered WNW., and sometimes still more northerly, proposing to fall in with the coast of China to the eastward of Pedro Blanco; for the rock so called is usually esteemed an excellent direction for ships bound to Macao. We continued this course till the following night, and then frequently brought to, to try if we were in soundings; but it was the 5th of November, at nine in the morning, before we struck ground, and then we had forty-two fathoms, and a bottom of grey sand mixed with shells. When we had got about twenty miles farther WNW., we had thirty-five fathoms, and the same bottom, from whence our soundings gradually decreased from thirty-five to twenty-five fathoms; but soon after, to our great surprise, they jumped back again to thirty fathoms. This was an alteration we could not very well account for,[1] since all the charts laid down regular soundings everywhere to the northward of Pedro Blanco; and for this reason we kept a very careful look-out, and altered our course to NNW., and having run thirty-five miles in this direction our soundings again gradually diminished to twenty-two fathoms, and we at last, about midnight, got sight of the mainland of China, bearing N. by W., four leagues distant.

We then brought the ship to, with her head to the sea, proposing to wait for the morning; and before sunrise we were surprised to find ourselves in the midst of an incredible number of fishing-boats, which seemed to cover the surface of the sea as far as the eye could reach. I may well style their number incredible, since I cannot believe, upon the lowest estimate, that there were so few as 6000; most of them manned with five hands, and none with less than three. Nor was this swarm of fishing vessels peculiar to this spot; for, as we ran on to the westward, we found them as abundant on every part of the coast. We at first doubted not but we should procure a pilot from them to carry us to Macao; but though many of them came close to the ship, and we endeavoured to tempt them by showing them a number of dollars, a most alluring bait for Chinese of all ranks and professions, yet we could not entice them on board us; though I presume the only difficulty was their not comprehending what we wanted them to do, for we could have no communication with them but by signs. Indeed we often pronounced the word Macao; but this we had reason to suppose they understood in a different sense; for in return they sometimes held up fish to us, and we afterwards learned that the Chinese name for fish is of a somewhat similar sound. But what surprised us most was the inattention and want of curiosity which we observed in this herd of fishermen. A ship like ours had doubtless never been in those seas before; perhaps there might not be one amongst all the Chinese employed in this fishery who had ever seen any European vessel; so that we might reasonably

[1] In recent maps a sandbank is laid down at this part of the Centurion's course.

have expected to have been considered by them as a very uncommon and extraordinary object. But though many of their vessels came close to the ship, yet they did not appear to be at all interested about us, nor did they deviate in the least from their course to regard us; which insensibility, especially in maritime persons, about a matter in their own profession, is scarcely to be credited, did not the general behaviour of the Chinese in other instances furnish us with continual proofs of a similar turn of mind. It may perhaps be doubted whether this cast of temper be the effect of nature or education; but in either case it is an incontestible symptom of a mean and contemptible disposition, and is alone a sufficient confutation of the extravagant panegyrics which many hypothetical writers have bestowed on the ingenuity and capacity of this nation.[1] Not being able to procure any information from the Chinese fishermen about our proper course to Macao, it was necessary for us to rely entirely on our own judgment; and concluding from our latitude, which was 22° 42′ N., and from our soundings, which were only seventeen or eighteen fathoms, that we were yet to the eastward of Pedro Blanco, we stood to the westward.

It was on the 5th of November at midnight when we first made the coast of China; and the next day about 2 o'clock, as we were standing to the westward within two leagues of the coast, and still surrounded by fishing-vessels in as great numbers as at first, we perceived that a boat ahead of us waved a red flag, and blew a horn. This we considered as a signal made to us either to warn us of some shoal or to inform us that they would supply us with a pilot, and in this belief we immediately sent our cutter to the boat to know their intentions; but we were soon made sensible of our mistake, and found that this boat was the commodore of the whole fishery, and that the signal she had made was to order them all to leave off fishing and to return in shore, which we saw them instantly obey. On this disappointment we kept on our course, and soon after passed by two very small rocks which lay four or five miles distant from the shore; but night came on before we got sight of Pedro Blanco, and we therefore brought to till the morning, when we had the satisfaction to discover it. It is a rock of a small circumference, but of a moderate height, and both in shape and colour resembles a sugar-loaf, and is about seven or eight miles from the shore. We passed within a mile and a half of it, and left it between us and the land, still keeping on to the westward; and the next day, being the 7th, we were abreast of a chain of islands which stretched from east to west. These, as we afterwards found, were called the Islands of Lema; they are rocky and barren, and are in all, small and great, fifteen or sixteen; and there are besides a great number of other islands between them and the mainland of China. These islands we left on the starboard side, passing within four miles of them, where we had twenty-four fathoms water. We were still surrounded by fishing-boats; and we once more sent the cutter on board one of them to endeavour to procure a pilot, but could not prevail; however, one of the Chinese directed us by signs to sail round the westernmost of the islands or rocks of Lema, and then to haul up. We followed this direction, and in the evening came to an anchor in eighteen fathoms.

After having continued at anchor all night, we on the 9th, at four in the morning, sent our cutter to sound the channel where we proposed to pass; but before the return of the cutter a Chinese pilot put on board

[1] Mr Walter evidently was strongly prejudiced against the Chinese; but he knew too little about them to trace the conduct of the fishermen to what was probably its true source—the contempt of the people for everything foreign, and the exclusive policy of the authorities, under whose vigilant control the fishermen obviously plied their trade.

us, and told us in broken Portuguese he would carry us to Macao for thirty dollars. These were immediately paid him, and we then weighed and made sail; and soon after several other pilots came on board us, who, to recommend themselves, produced certificates from the captains of several ships they had piloted in; but we continued the ship under the management of the Chinese who came first on board. By this time we learned that we were not far distant from Macao, and that there were in the River of Canton, at the mouth of which Macao lies, eleven European ships, of which four were English. Our pilot carried us between the Islands of Bamboo and Cabouce; but the winds hanging in the northern board, and the tides often setting strongly against us, we were obliged to come frequently to an anchor, so that we did not get through between the two islands till the 12th of November at two in the morning. In passing through, our depth of water was from twelve to fourteen fathoms, and as we still steered on NW. half W., between a number of other islands, our soundings underwent little or no variation till towards the evening, when they increased to seventeen fathoms, in which depth (the wind dying away) we anchored not far from the Island of Lantoon, which is the largest of all this range of islands. At seven in the morning we weighed again, and steering WSW. and SW. by W., we at 10 o'clock happily anchored in Macao road, in five fathoms water, the city of Macao bearing W. by N., three leagues distant; the peak of Lantoon E. by N., and the Grand Ladrone S. by E., each of them about five leagues distant. Thus, after a fatiguing cruise of above two years' continuance, we once more arrived in an amicable port in a civilised country, where the conveniences of life were in great plenty; where the naval stores, which we now extremely wanted, could be in some degree procured; where we expected the inexpressible satisfaction of receiving letters from our relations and friends; and where our countrymen who were lately arrived from England would be capable of answering the numerous inquiries we were prepared to make both about public and private occurrences, and to relate to us many particulars which, whether of importance or not, would be listened to by us with the utmost attention, after the long suspension of our correspondence with our country to which the nature of our undertaking had hitherto subjected us.

CHAPTER VII.

THE city of Macao, in the road of which we came to an anchor on the 12th of November, is a Portuguese settlement situated in an island at the mouth of the River of Canton. It was formerly a very rich and populous city, and capable of defending itself against the power of the adjacent Chinese Governors, but at present it is much fallen from its ancient splendour; for though it is inhabited by Portuguese, and has a Governor nominated by the King of Portugal, yet it subsists merely by the sufferance of the Chinese, who can starve the place and dispossess the Portuguese whenever they please. This obliges the Governor of Macao to behave with great circumspection, and carefully to avoid every circumstance that may give offence to the Chinese. The River of Canton, at the mouth of which this city lies, is the only Chinese port frequented by European ships; and this river is indeed a more commodious harbour on many accounts than Macao. But the peculiar customs of the Chinese, only adapted to the entertainment of trading ships, and the apprehensions of the Commodore lest he should embroil the East India Company with the Regency of Canton if he should insist on being treated upon a different footing than the merchantmen, made him resolve to go first to Macao before he ventured into the port of Canton. Indeed, had not this reason prevailed with him, he himself had nothing to fear; for it is certain that he might have entered the port of Canton, and might have continued there as long as he pleased, and afterwards have left it again, although the whole power of the Chinese empire had been brought together to oppose him.

The Commodore, not to depart from his usual prudence, no sooner came to an anchor in Macao road than he despatched an officer with his compliments to the Portuguese Governor of Macao, requesting his Excellency by the same officer to advise him in what manner it would be proper to act to avoid offending the Chinese, which, as there were then four of our ships in their power at Canton, was a matter worthy of attention. The difficulty which the Commodore principally apprehended related to the duty usually paid by all ships in the River of Canton, according to their tonnage. For as men-of-war are exempted in every foreign harbour from all manner of port charges, the Commodore thought it would be derogatory to the honour of his country to submit to this duty in China; and therefore he desired the advice of the Governor of Macao, who, being a European, could not be ignorant of the privileges claimed by a British man-of-war, and consequently might be expected to give us the best lights for avoiding this perplexity. Our boat returned in the evening with two officers sent by the Governor, who informed the Commodore that it was the Governor's opinion that if the Centurion ventured into the River of Canton, the duty would certainly be demanded; and therefore, if the Commodore approved of it, he would send him a pilot who should conduct us into another safe harbour, called the Typa,[1] which was every way commodious for careening the ship (an operation we were resolved to begin upon as soon as possible), and where the above-mentioned duty would in all probability be never asked for.

This proposal the Commodore agreed to, and in the morning we weighed anchor, and, under the direction of the Portuguese pilot, steered for the intended harbour. As we entered two islands, which form the eastern passage to it, we found our soundings decreased to three fathoms and a half. But the pilot assuring us that this was the least depth we should meet with, we continued our course, till at length the ship stuck fast in the mud, with only eighteen feet water abaft; and, the tide of ebb making, the water sewed[2] to sixteen feet, but the ship remained perfectly upright. We then sounded all round us, and finding the water deepened to the northward, we carried out our small bower with two hawsers an-end, and at the return of the tide of flood hove the ship afloat; and a small breeze springing up at the same instant, we set the fore-topsail, and slipping the hawser ran into the harbour, where we moored in about five fathoms water. This harbour of the Typa is formed by a number of islands, and is about six miles distant from Macao. Here we saluted the Castle of Macao with eleven guns, which were returned by an equal number.

The next day the Commodore paid a visit in person to the Governor, and was saluted at his landing by eleven guns, which were returned by the Centurion. Mr Anson's business in this visit was to solicit the Governor to grant us a supply of provisions, and to furnish us with such stores as were necessary to refit the ship. The Governor seemed really inclined to do us all the service he could, and assured the Commodore, in a friendly manner, that he would privately give us all the assistance in his power; but he, at the same time, frankly owned that he dared not openly furnish us with anything we demanded, unless we first procured an order for it from the Viceroy of Canton; for that he neither received provisions for his garrison, nor any other necessaries,

[1] The Island of Typa, directly to the south of Macao, at the mouth of the river.

[2] Sank away from the ship with the out-going tide.

but by permission from the Chinese Government; and as they took care only to furnish him from day to day, he was indeed no other than their vassal, whom they could at all times compel to submit to their own terms, only by laying an embargo on his provisions.

On this declaration of the Governor, Mr Anson resolved himself to go to Canton to procure a license from the Viceroy, and he accordingly hired a Chinese boat for himself and his attendants; but just as he was ready to embark, the "hoppo," or Chinese custom-house officer at Macao, refused to grant a permit to the boat, and ordered the watermen not to proceed at their peril. The Commodore at first endeavoured to prevail with the "hoppo" to withdraw his injunction and to grant a permit; and the Governor of Macao employed his interest with the "hoppo" to the same purpose. Mr Anson, seeing the officer inflexible, told him the next day, that if he longer refused to grant the permit, he would man and arm his own boats to carry him thither; asking the "hoppo," at the same time, who he imagined would dare to oppose him. This threat immediately brought about what his entreaties had laboured for in vain; the permit was granted, and Mr Anson went to Canton. On his arrival there, he consulted with the supercargoes and officers of the English ships how to procure an order from the Viceroy for the necessaries he wanted; but in this he had reason to suppose that the advice they gave him, though doubtless well intended, was yet not the most prudent; for as it is the custom with these gentlemen never to apply to the supreme magistrate himself, whatever difficulties they labour under, but to transact all matters relating to the Government by the mediation of the principal Chinese merchants, Mr Anson was advised to follow the same method upon this occasion; the English promising (in which they were doubtless sincere) to exert all their interest to engage the merchants in his favour. And when the Chinese merchants were applied to, they readily undertook the management of it, and promised to answer for its success; but after near a month's delay, and reiterated excuses, during which interval they pretended to be often upon the point of completing the business, they at last (being pressed, and measures being taken for delivering a letter to the Viceroy) threw off the mask, and declared they neither had applied to the Viceroy, nor could they, for he was too great a man, they said, for them to approach on any occasion. And not contented with having themselves thus grossly deceived the Commodore, they now used all their persuasion with the English at Canton to prevent them from intermeddling with anything that regarded him, representing to them that it would in all probability embroil them with the Government, and occasion them a great deal of unnecessary trouble; which groundless insinuations had, indeed, but too much weight with those they were applied to.

It may be difficult to assign a reason for this perfidious conduct of the Chinese merchants. Interest, indeed, is known to exert a boundless influence over the inhabitants of that empire; but how their interest could be affected in the present case is not easy to discover, unless they apprehended that the presence of a ship of force might damp their Manilla trade, and therefore acted in this manner with a view of forcing the Commodore to Batavia; but it might be as natural in this light to suppose that they would have been eager to have got him despatched. I therefore rather impute their behaviour to the unparalleled pusillanimity of the nation, and to the awe they are under of the Government; for as such a ship as the Centurion, fitted for war only, had never been seen in those parts before, she was the horror of these dastards, and the merchants were in some degree terrified even with the idea of her, and could not think of applying to the Viceroy (who is doubtless fond of all opportunities of fleecing them) without representing to themselves the pretences which a hungry and tyrannical magistrate might possibly find for censuring their intermeddling in so unusual a transaction, in which he might pretend the interest of the State was immediately concerned. However, be this as it may, the Commodore was satisfied that nothing was to be done by the interposition of the merchants, as it was on his pressing them to deliver a letter to the Viceroy that they had declared they durst not intermeddle, and had confessed that notwithstanding all their pretences of serving him, they had not yet taken one step towards it. Mr Anson therefore told them that he would proceed to Batavia and refit his ship there; but informed them, at the same time, that this was impossible to be done unless he was supplied with a stock of provisions sufficient for his passage. The merchants on this undertook to procure him provisions, but assured him that it was what they durst not engage in openly, but proposed to manage it in a clandestine manner, by putting a quantity of bread, flour, and other provision on board the English ships, which were now ready to sail; and these were to stop at the mouth of the Typa, where the Centurion's boats were to receive it. This article, which the merchants represented as a matter of great favour, being settled, the Commodore, on the 16th of December, returned from Canton to the ship, seemingly resolved to proceed to Batavia to refit as soon as he should get his supplies of provision on board.

But Mr Anson (who never intended going to Batavia) found, on his return to the Centurion, that her mainmast was sprung in two places, and that the leak was considerably increased; so that, upon the whole, he was fully satisfied that though he should lay in a sufficient stock of provisions, yet it would be impossible for him to put to sea without refitting. For, if he left the port with his ship in her present condition, she would be in the utmost danger of foundering; and therefore, notwithstanding the difficulties he had met with, he resolved at all events to have her hove down before he left Macao. He was fully convinced, by what he had observed at Canton, that his great caution not to injure the East India Company's affairs, and the regard he had shown to the advice of their officers, had occasioned all his embarrassments. For he now saw clearly, that if he had at first carried his ship into the River of Canton, and had immediately applied himself to the mandarins, who are the chief officers of State, instead of employing the merchants to apply for him, he would in all probability have had all his requests granted, and would have been soon despatched. He had already lost a month by the wrong measures he had been put upon, but he resolved to lose as little more time as possible; and therefore, the 17th of December, being the next day after his return from Canton, he wrote a letter to the Viceroy of that place, acquainting him that he was commander-in-chief of a squadron of his Britannic Majesty's ships of war, which had been cruising for two years past in the South Seas against the Spaniards, who were at war with the King his master; that, in his way back to England, he had put into the port of Macao, having a considerable leak in his ship, and being in great want of provisions, so that it was impossible for him to proceed on his voyage till his ship was repaired, and he was supplied with the necessaries he wanted; that he had been at Canton in hopes of being admitted to a personal audience of his Excellency, but, being a stranger to the customs of the country, he had not been able to inform himself what steps were necessary to be taken to procure such an audience; and therefore was obliged to apply to him in this manner, to desire his Excellency to give orders for his being permitted to employ carpenters and proper workmen to refit his ship, and to furnish himself with provisions and stores, thereby to enable him to pursue his voyage to

Great Britain with this monsoon; hoping at the same time that these orders would be issued with as little delay as possible, lest it might occasion his loss of the season, and he might be prevented from departing till the next winter.

This letter was translated into the Chinese language, and the Commodore delivered it himself to the "hoppo" or chief officer of the Emperor's customs at Macao, desiring him to forward it to the Viceroy of Canton with as much expedition as he could. The officer at first seemed unwilling to take charge of it, and raised many difficulties about it, so that Mr Anson suspected him of being in league with the merchants of Canton, who had always shown a great apprehension of the Commodore's having any immediate intercourse with the Viceroy or mandarins; and therefore the Commodore, with some resentment, took back his letter from the "hoppo," and told him he would immediately send an officer with it to Canton in his own boat, and would give him positive orders not to return without an answer from the Viceroy. The "hoppo" perceiving the Commodore to be in earnest, and fearing to be called to an account for his refusal, begged to be entrusted with the letter, and promised to deliver it and to procure an answer as soon as possible. And now it was soon seen how justly Mr Anson had at last judged of the proper manner of dealing with the Chinese; for this letter was written but the 17th of December, as has been already observed, and on the 19th in the morning a mandarin of the first rank, who was Governor of the city of Janson, together with two mandarins of an inferior class, and a great retinue of officers and servants, having with them eighteen half-galleys decorated with a great number of streamers, and furnished with music, and full of men, came to grapnel ahead of the Centurion; whence the mandarin sent a message to the Commodore, telling him that he (the mandarin) was ordered by the Viceroy of Canton, to examine the condition of the ship, and desiring the ship's boat might be sent to fetch him on board. The Centurion's boat was immediately despatched, and preparations were made for receiving him; for a hundred of the most sightly of the crew were uniformly dressed in the regimentals of the marines, and were drawn up under arms on the main-deck, against his arrival. When he entered the ship he was saluted by the drums, and what other military music there was on board; and passing by the new-formed guard, he was met by the Commodore on the quarter-deck, who conducted him to the great cabin. Here the mandarin explained his commission, declaring that his business was to examine all the particulars mentioned in the Commodore's letter to the Viceroy, and to confront them with the representation that had been given of them; that he was particularly instructed to inspect the leak, and had for that purpose brought with him two Chinese carpenters; and that, for the greater regularity and despatch of his business, he had every head of inquiry separately written down on a sheet of paper, with a void space opposite to it where he was to insert such information and remarks thereon as he could procure by his own observation.

This mandarin appeared to be a person of very considerable parts, and endowed with more frankness and honesty than is to be found in the generality of the Chinese. After the proper inquiries had been made, particularly about the leak, which the Chinese carpenters reported to be as dangerous as it had been represented, and consequently that it was impossible for the Centurion to proceed to sea without being refitted, the mandarin expressed himself satisfied with the account given in the Commodore's letter. And this magistrate, as he was more intelligent than any other person of his nation that came to our knowledge, so likewise was he more curious and inquisitive, viewing each part of the ship with particular attention, and appearing greatly surprised at the largeness of the lower-deck guns, and at the weight and size of the shot. The Commodore, observing his astonishment, thought this a proper opportunity to convince the Chinese of the prudence of granting him a speedy and ample supply of all he wanted. With this view he told the mandarin, and those who were with him, that besides the demands he made for a general supply, he had a particular complaint against the proceedings of the custom-house of Macao; that at his first arrival the Chinese boats had brought on board plenty of greens, and variety of fresh provisions for daily use, for which they had always been paid to their full satisfaction, but that the custom-house officers at Macao had soon forbid them, by which means he was deprived of those refreshments which were of the utmost consequence to the health of his men after their long and sickly voyage; that as they, the mandarins, had informed themselves of his wants, and were eyewitnesses of the force and strength of his ship, they might be satisfied it was not for want of power to supply himself that he desired the permission of the Government to purchase what provisions he stood in need of; that they must be convinced that the Centurion alone was capable of destroying the whole navigation of the port of Canton, or of any other port in China, without running the least risk from all the force the Chinese could collect; that it was true this was not the manner of proceeding between nations in friendship with each other, but it was likewise true that it was not customary for any nation to permit the ships of their friends to starve and sink in their ports, when those friends had money to supply their wants, and only desired liberty to lay it out; that they must confess he and his people had hitherto behaved with great modesty and reserve, but that, as his wants were each day increasing, hunger would at last prove too strong for any restraint, and necessity was acknowledged in all countries to be superior to every other law, and therefore it could not be expected that his crew would long continue to starve in the midst of that plenty to which their eyes were every day witnesses. To this the Commodore added (though perhaps with a less serious air) that if by the delay of supplying him with fresh provisions his men should be reduced to the necessity of turning cannibals, and preying upon their own species, it was easy to be foreseen that, independent of their friendship to their comrades, they would in point of luxury prefer the plump, well-fed Chinese to their own emaciated shipmates. The first mandarin acquiesced in the justness of this reasoning, and told the Commodore that he should that night proceed for Canton; that on his arrival a council of mandarins would be summoned, of which he himself was a member, and that by being employed in the present commission he was of course the Commodore's advocate; that, as he was fully convinced of the urgency of Mr Anson's necessity, he did not doubt but on his representation the council would be of the same opinion, and that all that was demanded would be amply and speedily granted. And with regard to the Commodore's complaint of the custom-house of Macao, he undertook to rectify that immediately by his own authority; for, desiring a list to be given him of the quantity of provision necessary for the expense of the ship for a day, he wrote a permit under it, and delivered it to one of his attendants, directing him to see that quantity sent on board early every morning; and this order from that time forward was punctually complied with.

When this weighty affair was thus in some degree regulated, the Commodore invited him and his two attendant mandarins to dinner, telling them at the same time that if his provision, either in kind or quantity, was not what they might expect, they must thank themselves for having confined him to so hard an allowance. One of his dishes was beef, which the Chinese all dislike, though Mr Anson was not apprised of it; this seems to be derived from the Indian superstition, which for some ages past has made a great progress in China.

However, his guests did not entirely fast; for the three mandarins completely finished the white part of four large fowls. But they were extremely embarrassed with their knives and forks, and were quite incapable of making use of them; so that, after some fruitless attempts to help themselves, which were sufficiently awkward, one of the attendants was obliged to cut their meat in small pieces for them. But whatever difficulty they might have in complying with the European manner of eating, they seemed not to be novices in drinking. The Commodore excused himself in this part of the entertainment, under the pretence of illness; but there being another gentleman present, of a florid and jovial complexion, the chief mandarin clapped him on the shoulder, and told him by the interpreter that certainly he could not plead sickness, and therefore insisted on his bearing him company; and that gentleman perceiving that after they had despatched four or five bottles of Frontiniac, the mandarin still continued unruffled, he ordered a bottle of citron-water to be brought up, which the Chinese seemed much to relish; and this being near finished they arose from table, in appearance cool and uninfluenced by what they had drunk. And the Commodore having, according to custom, made the mandarin a present, they all departed in the same vessels that brought them.

After their departure the Commodore with great impatience expected the resolution of the council, and the necessary licenses for his refitment. For it must be observed, as has already appeared from the preceding narration, that he could neither purchase stores nor necessaries with his money, nor did any kind of workmen dare to engage themselves to work for him, without the permission of the Government first obtained. And in the execution of these particular injunctions the magistrates never fail of exercising great severity, they, notwithstanding the fustian eulogiums bestowed upon them by the Catholic missionaries and their European copiers, being composed of the same fragile materials with the rest of mankind, and often making use of the authority of the law not to suppress crimes, but to enrich themselves by the pillage of those who commit them. For capital punishments are rare in China, the effeminate genius of the nation, and their strong attachment to lucre, disposing them rather to make use of fines; and hence arises no inconsiderable profit to those who compose their tribunals. Consequently prohibitions of all kinds, particularly such as the alluring prospect of great profit may often tempt the subject to infringe, cannot but be favourite institutions in such a government.

Some time before this, Captain Saunders took his passage to England on board a Swedish ship, and was charged with despatches from the Commodore; and soon after, in the month of December, Captain Mitchel, Colonel Cracherode, and Mr Tassel, one of the agent-victuallers, with his nephew, Mr Charles Harriot, embarked on board some of our Company's ships; and I, having obtained the Commodore's leave to return home, embarked with them. I must observe, too (having omitted it before), that whilst we lay here at Macao we were informed by some of the officers of our Indiamen that the Severn and Pearl, the two ships of our squadron which had separated from us off Cape Noir, were safely arrived at Rio Janeiro, on the coast of Brazil; and it was with great joy we received the news, after the strong persuasion which had so long prevailed amongst us, of their having both perished.

Notwithstanding the favourable disposition of the mandarin Governor of Janson at his leaving Mr Anson, several days were elapsed before he had any advice from him, and Mr Anson was privately informed there were great debates in council upon his affair; partly, perhaps, owing to its being so unusual a case, and in part to the influence, as I suppose, of the intrigues of the French at Canton. For they had a countryman and fast friend residing on the spot, who spoke the language very well, and was not unacquainted with the venality of the Government, nor with the persons of several of the magistrates, and consequently could not be at a loss for means of traversing the assistance desired by Mr Anson. And this opposition by the French was not merely the effect of national prejudice or contrariety of political interests, but was in good measure owing to their vanity, a motive of much more weight with the generality of mankind than any attachment to the public service of their community. For the French pretending their Indiamen to be men-of-war, their officers were apprehensive that any distinction granted to Mr Anson, on account of his bearing the King's commission, would render them less considerable in the eyes of the Chinese, and would establish a prepossession at Canton in favour of ships of war, by which they, as trading vessels, would suffer in their importance; and I wish the affectation of endeavouring to pass for men-of-war, and the fear of sinking in the estimation of the Chinese if the Centurion was treated in a different manner from themselves, had been confined to the officers of the French ships only.[1] However, notwithstanding all these obstacles, it should seem that the representation of the Commodore to the mandarins of the facility with which he could right himself, if justice were denied him, had at last its effect; for on the 6th of January, in the morning, the Governor of Janson, the Commodore's advocate, sent down the Viceroy of Canton's warrant for the refitment of the Centurion, and for supplying her people with all they wanted; and next day a number of Chinese smiths and carpenters went on board to agree for all the work by the great.[2] They demanded at first to the amount of £1000 sterling for the necessary repairs of the ship, the boats, and the masts. This the Commodore seemed to think an unreasonable sum, and endeavoured to persuade them to work by the day; but that proposal they would not hearken to, so it was at last agreed that the carpenters should have to the amount of about £600 for their work, and that the smiths should be paid for their iron work by weight, allowing them at the rate of £3 a hundred nearly for the small work, and 46s. for the large. . . .

It was the beginning of April before they had new-rigged the ship, stowed their provisions and water on board, and fitted her for the sea; and before this time the Chinese grew very uneasy, and extremely desirous that she should be gone; either not knowing, or pretending not to believe, that this was a point the Commodore was as eagerly set on as they could be. On the 3d of April two mandarin boats came on board from Macao to urge his departure; and this having been often done before, though there had been no pretence to suspect Mr Anson of any affected delays, he at this last message answered them in a determined tone, desiring them to give him no further trouble, for he would go when he thought proper and not before. On this rebuke the Chinese (though it was not in their power to compel him to be gone) immediately prohibited all provisions from being carried on board him, and took such care that their injunctions should be complied with, that from that time forwards nothing could be purchased at any rate whatever.

On the 6th of April the Centurion weighed from the Typa, and warped to the southward, and by the 15th she was got into Macao road, completing her water as she passed along, so that there remained now very few articles more to attend to; and her whole business being finished by the 19th, she at three in the afternoon of that day weighed and made sail, and stood to sea.

CHAPTER VIII.

The Commodore was now got to sea, with his ship very well refitted, his

[1] Glancing, apparently, at the jealousies of the English merchants.

[2] In the lump, or for the whole job.

stores replenished, and an additional stock of provisions on board. His crew, too, was somewhat reinforced; for he had entered twenty-three men during his stay at Macao, the greatest part of which were Lascars or Indian sailors, and some few Dutch.[1] He gave out at Macao that he was bound to Batavia, and thence to England; and though the western monsoon was now set in, when that passage is considered as impracticable, yet by the confidence he had expressed in the strength of his ship and the dexterity of his people he had persuaded not only his own crew, but the people at Macao likewise, that he proposed to try this unusual experiment; so that there were many letters put on board him by the inhabitants of Canton and Macao for their friends at Batavia.

But his real design was of a very different nature; for he knew that instead of one annual ship from Acapulco to Manilla there would be this year, in all probability, two, since by being before Acapulco, he had prevented one of them from putting to sea the preceding season. He therefore resolved to cruise for these returning vessels off Cape Espiritu Santo, on the Island of Samal, which is the first land they always make in the Philippine Islands. And as June is generally the month in which they arrive there, he doubted not but he should get to his intended station in time enough to intercept them. It is true, they were said to be stout vessels, mounting 44 guns a-piece, and carrying above 500 hands, and might be expected to return in company; and he himself had but 227 hands on board, of which near thirty were boys. But this disproportion of strength did not deter him, as he knew his ship to be much better fitted for a sea engagement than theirs, and as he had reason to expect that his men would exert themselves in the most extraordinary manner when they had in view the immense wealth of these Manilla galleons.

This project the Commodore had resolved on in his own thoughts ever since his leaving the coast of Mexico; and the greatest mortification which he received from the various delays he had met with in China was his apprehension lest he might be thereby so long retarded as to let the galleons escape him. Indeed at Macao it was incumbent on him to keep these views extremely secret, for there being a great intercourse and a mutual connection of interests between that port and Manilla, he had reason to fear that if his designs were discovered, intelligence would be immediately sent to Manilla, and measures would be taken to prevent the galleons from falling into his hands. But being now at sea, and entirely clear of the coast, he summoned all his people on the quarter-deck, and informed them of his resolution to cruise for the two Manilla ships, of whose wealth they were not ignorant. He told them he should choose a station where he could not fail of meeting with them; and though they were stout ships, and full-manned, yet, if his own people behaved with their accustomed spirit, he was certain he should prove too hard for them both, and that one of them at least could not fail of becoming his prize. He further added, that many ridiculous tales had been propagated about the strength of the sides of these ships, and their being impenetrable to cannon-shot; that these fictions had been principally invented to palliate the cowardice of those who had formerly engaged them; but he hoped there were none of those present weak enough to give credit to so absurd a story. For his own part, he did assure them upon his word that, whenever he met with them, he would fight them so near that they should find his bullets, instead of being stop-

[1] Yet the ship's company was deplorably far short of her requirements; for Anson says in his Official Report: "The number of men I have now borne is 201, amongst which are included all the officers and boys which I had out of the Gloucester, Trial prize, and Anna pink, so that I have not before the mast more than forty-five able seamen."

ped by one of their sides, should go through them both.

This speech of the Commodore's was received by his people with great joy, for no sooner had he ended than they expressed their approbation, according to naval custom, by three strenuous cheers, and all declared their determination to succeed or perish whenever the opportunity presented itself. And now their hopes, which since their departure from the coast of Mexico had entirely subsided, were again revived; and they all persuaded themselves that, notwithstanding the various casualties and disappointments they had hitherto met with, they should yet be repaid the price of their fatigues, and should at last return home enriched with the spoils of the enemy. For, firmly relying on the assurances of the Commodore that they should certainly meet with the vessels, they were all of them too sanguine to doubt a moment of mastering them; so that they considered themselves as having them already in their possession. And this confidence was so universally spread through the whole ship's company that, the Commodore having taken some Chinese sheep to sea with him for his own provision, and one day inquiring of his butcher why for some time past he had seen no mutton at his table, asking him if all the sheep were killed, the butcher very seriously replied that there were indeed two sheep left; but that if his honour would give him leave, he proposed to keep those for the entertainment of the General of the galleons.

When the Centurion left the port of Macao she stood for some days to the westward; and on the 1st of May they saw part of the Island of Formosa, and, standing thence to the southward, they on the 4th of May were in the latitude of the Bashee Islands, as laid down by Dampier; but they suspected his account of inaccuracy, as they found that he had been considerably mistaken in the latitude of the south end of Formosa. For this reason they kept a good look-out, and about seven in the evening discovered from the masthead five small islands, which were judged to be the Bashees, and they had afterwards a sight of Botel Tobago Xima. By this means they had an opportunity of correcting the position of the Bashee Islands, which had been hitherto laid down twenty-five leagues too far to the westward; for by their observations they esteemed the middle of these islands to be in 21° 4′ N., and to bear from Botel Tobago Xima SSE., twenty leagues distant, that island itself being in 21° 57′ N. After getting a sight of the Bashee Islands they stood between the S. and SW. for Cape Espiritu Santo, and the 20th of May at noon they first discovered that cape, which, about 4 o'clock, they brought to bear SSW., about eleven leagues distant. It appeared to be of a moderate height, with several round hummocks on it. As it was known that there were sentinels placed upon this cape to make signals to the Acapulco ship when she first falls in with the land, the Commodore immediately tacked, and ordered the top-gallant sails to be taken in to prevent being discovered; and this being the station in which it was resolved to cruise for the galleons, they kept the cape between the S. and W., and endeavoured to confine themselves between the Latitude of 12° 50′ and 13° 5′, the cape itself lying, by their observations, in 12° 40′ N. and 4° of E. Longitude from Botel Tobago Xima.

It was the last of May, by the foreign style,[1] when they arrived off this Cape; and the month of June, by the same style, being that in which the Manilla ships are usually expected, the Centurion's people were now waiting each hour with the utmost impatience for the happy crisis which was to balance the account of all their past calamities. As from this time there was but small employment for the crew, the Commodore ordered them almost every day to be exercised in the management of the great guns, and in the use of their small arms. This had been his practice, more or less, at all

[1] New Style.

convenient seasons during the whole course of his voyage; and the advantages which he received from it in his engagement with the galleon were an ample recompense for all his care and attention. [The men] were taught the shortest method of loading with cartridges, and were constantly trained to fire at a mark, which was usually hung at the yard-arm, and some little reward was given to the most expert. The whole crew, by this management, were rendered extremely skilful, quick in loading, all of them good marksmen, and some of them most extraordinary ones; so that I doubt not but, in the use of small arms, they were more than a match for double their number who had not been habituated to the same kind of exercise.

It was the last of May, N.S., as has been already said, when the Centurion arrived off Cape Espiritu Santo, and consequently the next day began the month in which the galleons were to be expected. The Commodore therefore made all necessary preparations for receiving them, having hoisted out his long-boat, and lashed her alongside, that the ship might be ready for engaging if they fell in with the galleons in the night. All this time, too, he was very solicitous to keep at such a distance from the cape as not to be discovered; but it has been since learned that, notwithstanding his care, he was seen from the land; and advice of him was sent to Manilla, where it was at first disbelieved; but on reiterated intelligence (for it seems he was seen more than once) the merchants were alarmed, and the Governor was applied to, who undertook (the commerce[1] supplying the necessary sums) to fit out a force consisting of two ships of 32 guns, one of 20 guns, and two sloops of 10 guns each, to attack the Centurion on her station. And some of these vessels did actually weigh with this view; but the principal ship not being ready, and the monsoon being against them, the commerce and the Government disagreed, and the enterprise was laid aside. This frequent discovery of the Centurion from the shore was somewhat extraordinary, for the pitch of the cape is not high, and she usually kept from ten to fifteen leagues distant; though once, indeed, by an indraught of the tide as was supposed, they found themselves in the morning within seven leagues of the land.

As the month of June advanced, the expectancy and impatience of the Commodore's people each day increased. And I think no better idea can be given of their great eagerness on this occasion than by copying a few paragraphs from the journal of an officer who was then on board, as it will, I presume, be a more natural picture of the full attachment of their thoughts to the business of their cruise than can be given by any other means. The paragraphs I have selected, as they occur in order of time, are as follow:

"May 31.—Exercising our men at their quarters, in great expectation of meeting with the galleons very soon; this being the 11th of June their style.

"June 3.—Keeping in our stations and looking out for the galleons.

"June 5.—Begin now to be in great expectations, this being the middle of June their style.

"June 11.—Begin to grow impatient at not seeing the galleons.

"June 13.—The wind having blown fresh easterly for the forty-eight hours past gives us great expectations of seeing the galleons soon.

"June 15.—Cruising on and off, and looking out strictly.

"June 19.—This being the last day of June, N.S., the galleons, if they arrive at all, must appear soon."

From these samples it is sufficiently evident, how completely the treasure of the galleons had engrossed their imagination, and how anxiously they passed the latter part of their cruise, when the certainty of the arrival of

[1] That is, the commercial community collectively; as "trade" is used, early in the narrative, to signify the collection of merchant ships sailing under convoy.

these vessels was dwindled down to probability only, and that probability became each hour more and more doubtful. However, on the 20th of June, O.S., being just a month from their arrival on their station, they were relieved from this state of uncertainty when, at sunrise, they discovered a sail from the masthead in the SE. quarter.[1] On this a general joy spread through the whole ship; for they had no doubt but this was one of the galleons, and they expected soon to see the other. The Commodore instantly stood towards her, and at half-an-hour after seven they were near enough to see her from the Centurion's deck; at which time the galleon fired a gun, and took in her topgallant sails, which was supposed to be a signal to her consort to hasten her up; and therefore the Centurion fired a gun to leeward, to amuse her. The Commodore was surprised to find that in all this time the galleon did not change her course, but continued to bear down upon him; for he hardly believed, what afterwards appeared to be the case, that she knew his ship to be the Centurion and resolved to fight him.

About noon the Commodore was little more than a league distant from the galleon, and could fetch her wake, so that she could not now escape; and, no second ship appearing, it was concluded that she had been separated from her consort. Soon after, the galleon hauled up her foresail, and brought to under topsails, with her head to the northward, hoisting Spanish colours, and having the standard of Spain flying at the topgallant-masthead. Mr Anson in the meantime had prepared all things for an engagement on board the Centurion, and had taken all possible care both for the most effectual exertion of his small strength, and for avoiding the confusion and tumult too frequent in actions of this kind. He picked out about thirty of his choicest hands and best marksmen, whom he distributed into his tops, and who fully answered his expectation by the signal services they performed. As he had not hands enough remaining to quarter a sufficient number to each great gun in the customary manner, he therefore, on his lower tier, fixed only two men to each gun, who were to be solely employed in loading it, whilst the rest of his people were divided into different gangs of ten or twelve men each, who were constantly moving about the decks, to run out and fire such guns as were loaded. By this management he was enabled to make use of all his guns; and, instead of firing broadsides with intervals between them, he kept up a constant fire without intermission, whence he doubted not to procure very signal advantages. For it is common with the Spaniards to fall down upon the decks when they see a broadside preparing, and to continue in that posture till it is given; after which they rise again, and, presuming the danger to be for some time over, work their guns, and fire with great briskness, till another broadside is ready: but the firing gun by gun, in the manner directed by the Commodore, rendered this practice of theirs impossible.

The Centurion being thus prepared, and nearing the galleon apace, there happened, a little after noon, several squalls of wind and rain, which often obscured the galleon from their sight; but whenever it cleared up they observed her resolutely lying to; and, towards 1 o'clock, the Centurion hoisted her broad pendant and colours, she being then within gun-shot of the enemy. And the Commodore observing the Spaniards to have neglected clearing their ship till that time, as he then saw them throwing overboard cattle and lumber, he gave orders to fire upon them with the chase guns, to embarrass them in their work, and prevent them from completing it, though his general directions had been not to engage till they were

[1] Thomas commemorates the name of Mr Charles Proby, midshipman, as having been the first on board to discover the long-looked-for treasure-ship.

within pistol-shot. The galleon returned the fire with two of her stern-chasers;[1] and the Centurion getting her spritsail-yard fore and aft, that if necessary she might be ready for boarding, the Spaniards in a bravado rigged their spritsail-yard fore and aft likewise. Soon after, the Centurion came abreast of the enemy within pistol-shot, keeping to the leeward, with a view of preventing them from putting before the wind and gaining the port of Jalapay, from which they were about seven leagues distant. And now the engagement began in earnest, and for the first half-hour Mr Anson overreached the galleon, and lay on her bow; where by the great wideness of his ports he could traverse almost all his guns upon the enemy, whilst the galleon could only bring a part of hers to bear. Immediately on the commencement of the action, the mats with which the galleon had stuffed her netting took fire, and burned violently, blazing up half as high as the mizzentop. This accident (supposed to be caused by the Centurion's wads) threw the enemy into great confusion, and at the same time alarmed the Commodore, for he feared lest the galleon should be burned, and lest he himself too might suffer by her driving on board him. But the Spaniards at last freed themselves from the fire, by cutting away the netting, and tumbling the whole mass which was in flames into the sea. But still the Centurion kept her first advantageous position, firing her cannon with great regularity and briskness, whilst at the same time the galleon's decks lay open to her topmen, who having at their first volley driven the Spaniards from their tops, made prodigious havoc with their small arms, killing or wounding every officer but one that ever appeared on the quarter-deck, and wounding in particular the General of the galleon himself. And though the Centurion, after the first half-hour, lost her original situation, and was close alongside the galleon, and the enemy continued to fire briskly for near an hour longer, yet at last the Commodore's grape-shot swept their decks so effectually, and the number of their slain and wounded was so considerable, that they began to fall into great disorder; especially as the General, who was the life of the action, was no longer capable of exerting himself. Their embarrassment was visible from on board the Commodore. For the ships were so near, that some of the Spanish officers were seen running about with great assiduity to prevent the desertion of their men from their quarters. But all their endeavours were in vain, for after having, as a last effort, fired five or six guns with more judgment than usual, they gave up the contest; and, the galleon's colours being singed off the ensign staff in the beginning of the engagement, she struck the standard at her maintop-gallant masthead, the person who was employed to do it having been in imminent peril of being killed, had not the Commodore, who perceived what he was about, given express orders to his people to desist from firing.

Thus was the Centurion possessed of this rich prize, amounting in value to near a million and a half of dollars. She was called the Nuestra Señora de Cabadonga, and was commanded by the General Don Jeronimo de Montero, a Portuguese by birth, and the most approved officer for skill and courage of any employed in that service. The galleon was much larger than the Centurion, had 550 men and 36 guns mounted for action, besides twenty-eight pidreroes in her gunwale, quarters, and tops, each of which carried a four-pound ball. She was very well furnished with small arms, and was particularly provided against boarding, both by her close quarters, and by a strong net-work of two inch rope which was laced over her waist and was defended by half pikes.[3] She

[1] "One of which," Thomas records, "carried away one of our fore shrouds and our forestay tackle."

[3] Placed in the manner of *chevaux de frise.*

had sixty-seven killed in the action and eighty-four wounded; whilst the Centurion had only two killed, and a lieutenant and sixteen wounded, all of whom but one recovered: of so little consequence are the most destructive arms in untutored and unpractised hands.[1]

The treasure thus taken by the Centurion having been for at least eighteen months the great object of their hopes, it is impossible to describe the transport on board when, after all their reiterated disappointments, they at last saw their wishes accomplished. But their joy was near being suddenly damped by a most tremendous incident: for no sooner had the galleon struck, than one of the lieutenants, coming to Mr Anson to congratulate him on his prize, whispered him at the same time that the Centurion was dangerously on fire near the powder-room. The Commodore received this dreadful news without any apparent emotion, and, taking care not to alarm his people, gave the necessary orders for extinguishing it, which was happily done in a short time, though its appearance at first was extremely terrible. It seems some cartridges had been blown up by accident between decks, whereby a quantity of oakum in the after hatchway, near the after powder-room, was set on fire; and the great smother and smoke of the oakum occasioned the apprehension of a more extended and mischievous fire. At the same instant, too, the galleon fell on board the Centurion on the starboard quarter, but she was cleared without doing or receiving any considerable damage.

[1] Thomas, who was one of the party sent on board as prize crew, says: "I had heard we had killed them sixty men, and wounded as many more, and expected to have seen the horrid spectacle of mangled limbs, dead carcasses, and decks covered with blood; but no such spectacle appeared; a party having been properly stationed, during the time of action, to wash away the blood, and to throw the dead overboard."

The Commodore made his first lieutenant, Mr Saumarez, captain of this prize, appointing her a post-ship in his Majesty's service. Captain Saumarez, before night, sent on board the Centurion all the Spanish prisoners but such as were thought the most proper to be retained to assist in navigating the galleon. And now the Commodore learned from some of these prisoners that the other ship, which he had kept in the port of Acapulco the preceding year, instead of returning in company with the present prize, as was expected, had set sail from Acapulco alone much sooner than usual, and had in all probability got into the port of Manilla long before the Centurion arrived off Cape Espiritu Santo; so that Mr Anson, notwithstanding his present success, had great reason to regret his loss of time at Macao, which prevented him from taking two rich prizes instead of one.[1]

The Commodore, when the action was ended, resolved to make the best of his way with his prize for the River of Canton, being in the meantime fully employed in securing his prisoners, and in removing the treasure from on board the galleon into the Centurion. The last of these operations was too important to be postponed; for as the navigation to Canton was through seas but little known, and where, from the season of the year, much bad weather might be expected, it was of great conse-

[1] Among the prisoners, we are told by Thomas, was "an old gentleman, Governor of Guam, who was going to Manilla to renew his commission, and who had scarce mounted the Centurion's side before he was received with open arms by Mr Crooden, captain of marines, who thirty-six years before, at the battle of Almanza, had been his prisoner, and honourably used by him. These two renewed their old acquaintance, and Captain Crooden had a long-wished-for opportunity of returning the favours he had formerly received, and which he gratefully remembered."

quence that the treasure should be sent on board the Centurion; which ship, by the presence of the Commander-in-chief, the greater number of her hands, and her other advantages, was doubtless much safer against all the casualties of winds and seas than the galleon. And the securing the prisoners was a matter of still more consequence, as not only the possession of the treasure, but the lives of the captors depended thereon. This was indeed an article which gave the Commodore much trouble and disquietude, for they were above double the number of his own people; and some of them, when they were brought on board the Centurion, and had observed how slenderly she was manned, and the large proportion which the striplings bore to the rest, could not help expressing themselves with great indignation to be thus beaten by a handful of boys. The method which was taken to hinder them from rising was by placing all but the officers and the wounded in the hold, where, to give them as much air as possible, two hatchways were left open; but then (to avoid all danger whilst the Centurion's people should be employed upon the deck) there was a square partition of thick planks, made in the shape of a funnel, which inclosed each hatchway on the lower deck, and reached to that directly over it on the upper deck. These funnels served to communicate the air to the hold better than could have been done without them, and, at the same time, added greatly to the security of the ship; for they being seven or eight feet high, it would have been extremely difficult for the Spaniards to have clambered up; and, still to augment that difficulty, four swivel-guns loaded with musket-bullets were planted at the mouth of each funnel, and a sentinel with lighted match constantly attended, prepared to fire into the hold amongst them in case of any disturbance. Their officers, who amounted to seventeen or eighteen, were all lodged in the first lieutenant's cabin, under a constant guard of six men; and the General, as he was wounded, lay in the Commodore's cabin with a sentinel always with him; and they were all informed that any violence or disturbance would be punished with instant death. And that the Centurion's people might be at all times prepared, if notwithstanding these regulations any tumult should arise, the small arms were constantly kept loaded in a proper place, whilst all the men went armed with cutlasses and pistols; and no officer ever pulled off his clothes, and when he slept had always his arms lying ready by him.

These measures were obviously necessary, considering the hazards to which the Commodore and his people would have been exposed had they been less careful. Indeed the sufferings of the poor prisoners, though impossible to be alleviated, were much to be commiserated; for the weather was extremely hot, the stench of the hold loathsome beyond all conception, and their allowance of water but just sufficient to keep them alive, it not being practicable to spare them more than at the rate of a pint a-day for each, the crew themselves having only an allowance of a pint and a half. All this considered, it was wonderful that not a man of them died during their long confinement, except three of the wounded, who died the same night they were taken; though it must be confessed that the greatest part of them were strangely metamorphosed by the heat of the hold. For when they were first taken they were sightly, robust fellows; but when, after above a month's imprisonment, they were discharged in the River of Canton, they were reduced to mere skeletons, and their air and looks corresponded much more to the conception formed of ghosts and spectres than to the figure and appearance of real men.

Thus employed in securing the treasure and the prisoners, the Commodore, as has been said, stood for the River of Canton, and on the 30th of June, at six in the evening, got sight of Cape Delangano,[1] which then bore W. ten leagues distant; and the next day he made the Bashee Islands, and the wind being so far to the northward that it was difficult to weather them, it was resolved to stand through between Grafton and Monmouth Islands, where the passage seemed to be clear; but in getting through, the sea had a very dangerous aspect, for it rippled and foamed as if it had been full of breakers, which was still more terrible, as it was then night. But the ships got through very safe (the prize always keeping a-head), and it was found that the appearance which had alarmed them had been occasioned only by a strong tide. I must here observe, that though the Bashee Islands are usually reckoned to be no more than five, yet there are many more lying about them to the westward, which, as the channels amongst them are not at all known, makes it advisable for ships rather to pass to the northward or southward than through them; and indeed the Commodore proposed to have gone to the northward, between them and Formosa, had it been possible for him to have weathered them. From hence the Centurion steering the proper course for the River of Canton, she, on the 8th of July, discovered the Island of Supata, the westernmost of the Lema Islands. This island they made to be 139 leagues distant from Grafton Island, and to bear from it N. 82°, 37° W.; and on the 11th, having taken on board two Chinese pilots, one for the Centurion and the other for the prize, they came to an anchor off the city of Macao.

By this time the particulars of the cargo of the galleon were well ascertained, and it was found that she had on board 1,313,843 pieces of eight, and 35,682 oz. of virgin silver, besides some cochineal and a few other commodities, which however were but of small account in comparison of the specie. And this being the Commodore's last prize, it hence appears that all the treasure taken by the Centurion was not much short of £400,000, independent of the ships and merchandise which she either burnt or destroyed, and which by the most reasonable estimation could not amount to so little as £600,000 more; so that the whole loss of the enemy by our squadron did doubtless exceed a million sterling. To which if there be added the great expense of the Court of Spain in fitting out Pizarro, and in paying the additional charges in America incurred on our account, together with the loss of their men-of-war, the total of all these articles will be a most exorbitant sum, and is the strongest conviction of the utility of this expedition, which, with all its numerous disadvantages, did yet prove so extremely prejudicial to the enemy. . . .

[1] Cape Engano, near the north-western extremity of the Island of Luconia or Luzon.

CHAPTER IX.

THE Commodore, having taken pilots on board, proceeded with his prize for the River of Canton, and on the 14th of July came to an anchor short of the Bocca Tigris, which is a narrow passage forming the mouth of that river. This entrance he proposed to stand through the next day, and to run up as far as Tiger Island, which is a very safe road, secured from all winds. But whilst the Centurion and her prize were thus at anchor, a boat with an officer came off from the mandarin commanding the forts at Bocca Tigris, to examine what the ships were and whence they came. Mr Anson informed the officer that his ship was a ship of war belonging to the King of Great Britain, and that the other in company with him was a prize he had taken; that he was going into Canton River to shelter himself against the hurricanes which were then coming on; and that as soon as the monsoon shifted he should proceed for England. The officer then desired an account of what men, guns, and ammunition were on board, a list of all which, he said, was to be sent to the Government of Canton. But when these articles were repeated to him, particularly when he was told that there were in the Centurion 400 firelocks and between 300 and 400 barrels of powder, he shrugged up his shoulders and seemed to be terrified with the bare recital, saying that no ships ever came into Canton River armed in that manner; adding, that he durst not set down the whole of this force, lest it should too much alarm the Regency. After he had finished his inquiries, and was preparing to depart, he desired to leave the two custom-house officers behind him; on which the Commodore told him, that though as a man-of-war he was prohibited from trading, and had nothing to do with customs or duties of any kind, yet for the satisfaction of the Chinese he would permit two of their people to be left on board, who might themselves be witnesses how punctually he should comply with his instructions. The officer seemed amazed when Mr Anson mentioned being exempted from all duties, and told him that the Emperor's duty must be paid by all ships that came into his ports. And it is supposed that on this occasion private directions were given by him to the Chinese pilot not to carry the Commodore through the Bocca Tigris, which makes it necessary more particularly to describe that entrance. . . .

On the 16th of July the Commodore sent his second lieutenant to Canton with a letter to the Viceroy, informing him of the reason of the Centurion's putting into that port; and that the Commodore himself soon proposed to repair to Canton to pay a visit to the Viceroy. The lieutenant was very civilly received, and was promised that an answer should be sent to the Commodore the next day. In the meantime Mr Anson gave leave to several of the officers of the galleon to go to Canton, they engaging their parole to return in two days. When these prisoners got to Canton, the Regency sent for them and examined them, inquiring particularly by what means they had fallen into Mr Anson's power. And on this occasion the prisoners were honest enough to declare that, as the Kings of Great Britain and Spain were at war, they had proposed to themselves the taking of the Centurion, and had bore down upon her with that view, but that the event had been contrary to their hopes; however, they acknowledged that they had been treated by the Commodore much better than they believed they should have treated him had he fallen into their hands. This confession from an enemy had great weight with the Chinese, who till then, though they had revered the Commodore's power, had yet suspected his morals, and had considered him rather as a lawless freebooter than as one commissioned by the State for the revenge of public injuries. But they now changed their opinion, and regarded him as a more important person, to which perhaps the vast treasure of his prize might not a little contribute, the acquisition of wealth being a matter greatly adapted to the estimation and reverence of the Chinese nation. ·

In this examination of the Spanish prisoners, though the Chinese had no reason in the main to doubt the account which was given them, yet there were two circumstances which appeared to them so singular as to deserve a more ample explanation. One of them was, the great disproportion of men between the Centurion and the galleon; the other was the humanity with which the people of the galleon were treated after they were taken. The mandarins therefore asked the Spaniards how they came to be overpowered by so inferior a force, and how it happened, since the two nations were at war, that they were not put to death when they came into the hands of the English. To the first of these inquiries the Spaniards replied, that though they had more hands than the Centurion, yet she, being intended solely for war, had a great superiority in the size of her guns, and in many other articles, over the galleon, which was a vessel fitted out principally for traffic. And as to the second question, they told the Chinese that amongst the nations of Europe it was not customary to put to death those who submitted, though they readily owned that the Commodore, from the natural bias of his temper, had treated both them, and their countrymen who had formerly been in his power, with very unusual courtesy, much beyond what they could have expected, or than was required by the customs established between nations at war with each other. These replies fully satisfied the Chinese, and at the same time wrought very powerfully in the Commodore's favour.

On the 20th of July, in the morning, three mandarins, with a great number of boats and a vast retinue, came on board the Centurion, and delivered to the Commodore the Viceroy of Canton's order for a daily supply of provisions, and for pilots to carry the ships up the river as far as the second bar; and at the same time they delivered him a message from the Viceroy in answer to the letter sent to Canton. The substance of the message was, that the Viceroy desired to be excused from receiving the Commodore's visit during the then excessive hot weather, because the assembling the mandarins and soldiers necessary to that ceremony would prove extremely inconvenient and fatiguing; but that in September, when the weather would be more temperate, he should be glad to see both the Commodore himself and the English captain of the other ship that was with him. As Mr Anson knew that an express had been despatched to the Court at Pekin with an account of the Centurion and her prize being arrived in the River of Canton, he had no doubt but the principal motive for putting off this visit was, that the Regency at Canton might gain time to receive the Emperor's instructions about their behaviour on this unusual affair.

When the mandarins had delivered their message, they began to talk to the Commodore about the duties to be paid by his ships; but he immediately told them that he would never submit to any demand of that kind; that as he neither brought any merchandise thither, nor intended to carry any away, he could not be reasonably deemed to be within the meaning of the Emperor's orders, which were doubtless calculated for trading vessels only; adding that no duties were ever demanded of men-of-war by nations accustomed to their reception, and that his master's orders expressly forbade him from paying any acknowledgment for his ships anchoring in any port whatever. The mandarins being thus cut short on the subject of the duty, they said they had another matter to mention, which was the only remaining one they had in charge; this was a request to the Commodore, that he would release the prisoners he had taken on board the galleon; for that the Viceroy of Canton apprehended the Emperor, his master, might be displeased if he should be informed that persons who were his allies, and carried on a great commerce with his subjects, were under confinement in his dominions. Mr Anson was himself extremely desirous to get rid of the Spaniards, having on his first arrival sent about 100 of them to Macao, and those who remained, which were near 400 more, were on many accounts a great encumbrance to him. However, to enhance the favour, he at first raised some difficulties; but, permitting himself to be prevailed on, he at last told the mandarins, that to show his readiness to oblige the Viceroy, he would release the prisoners, whenever they, the Chinese, would send boats to fetch them off. This matter being adjusted, the mandarins departed; and on the 28th of July, two Chinese junks were sent from Canton to take on board the prisoners, and to carry them to Macao. And the Commodore, agreeable to his promise, dismissed them all, and ordered his purser to send with them eight days' provision for their subsistence during their sailing down the river.[1] This being

[1] Thomas, who was one of the prize

despatched; the Centurion and her prize came to her moorings above the second bar, where they proposed to continue till the monsoon shifted.

Though the ships, in consequence of the Viceroy's permit, found no difficulty in purchasing provisions for their daily consumption, yet it was impossible for the Commodore to proceed to England without laying in a large quantity both of provisions and stores for his use during the voyage. The procuring this supply was attended with much embarrassment; for there were people at Canton who had undertaken to furnish him with biscuit and whatever else he wanted; and his linguist,[1] towards the middle of September, had assured him from day to day that all was ready, and would be sent on board him immediately. But a fortnight being elapsed, and nothing being brought, the Commodore sent to Canton to inquire more particularly into the reasons of this disappointment, and he had soon the vexation to be informed that the whole was an illusion; that no order had been procured from the Viceroy to furnish him with his sea stores, as had been pretended; that there was no biscuit baked, nor any one of the articles in readiness which had been promised him; nor did it appear that the contractors had taken the least step to comply with their agreement. This was most disagreeable news, and made it suspected that the furnishing the Centurion for her return to Great Britain might prove a more troublesome matter than had been hitherto imagined; especially, too, as the month of September was nearly elapsed without Mr Anson's having received any message from the Viceroy of Canton.

And here, perhaps, it might be expected that some satisfactory account should be given of the motives of the Chinese for this faithless procedure. But as I have already in a former Chapter[2] made some kind of conjectures about a similar event, I shall not repeat them again in this place, but shall observe that, after all, it may, perhaps, be impossible for a European, ignorant of the customs and manners of that nation, to be fully apprised of the real incitements to this behaviour.[3] Indeed, thus much may undoubtedly be asserted, that in ar-

---

crew, tells a somewhat ugly story of the Spanish "General" now set at liberty. He was not only allowed the use of his own cabin till he should be recovered of his wound, but obtained the services of an English surgeon; Anson, at the same time, sending an officer to demand his commission. The General made the officer look in a box in the locker of his private cabin, where he said the commission would be found, along with a sword-belt studded with diamonds of great value; and when the box was found empty, the Spaniard averred that some of the English, rummaging in his cabin, must have stolen and secreted the contents. Despite the non-production of his commission, the General received the most humane and liberal treatment, being allowed at his departure to carry off several chests and trunks unsearched which he claimed as his private property, though he had many valuable ventures concealed which should have been given up as fair and lawful prize. Persisting to the last in the theft of his commission and sword-belt, he brought down on the prize crew a heavy and undeserved punishment; for Anson, on their arrival in the Canton River, absolutely prohibited their intercourse with the natives, that the thief might have no chance of parting with his booty undiscovered. Thomas, however, was afterwards told at Macao by an Irish priest, that the General had both his commission and his sword-belt; that he had made no secret of his fraud; and that he had offered the diamonds—which were made up in the belt only by way of a blind—among the merchants at Macao for sale.

[1] Interpreter.

[2] Chapter VII. of this Book, page 136.

[3] Thomas says: "We could no

tifice, falsehood, and an attachment to all kinds of lucre, many of the Chinese are difficult to be paralleled by any other people; but then the combination of these talents, and the manner in which they are applied in particular emergencies, are often beyond the reach of a foreigner's penetration; so that though it may be falsely concluded that the Chinese had some interest in thus amusing the Commodore, yet it may not be easy to assign the individual views by which they were influenced. . . .

It were endless to recount all the artifices, extortions, and frauds which were practised on the Commodore and his people by this interested race. The method of buying all things in China being by weight, the tricks made use of by the Chinese to increase the weight of the provision they sold to the Centurion were almost incredible. One time, a large quantity of fowls and ducks being bought for the ships' use, the greatest part of them presently died. This alarmed the people on board with the apprehension that they had been killed by poison; but on examination it appeared that it was only owing to their being crammed with stones and gravel to increase their weight, the quantity thus forced into most of the ducks being found to amount to ten ounces in each. The hogs, too, which were bought ready killed of the Chinese butchers, had water injected into them for the same purpose, so that a carcase hung up all night for the water to drain from it, has lost above a stone of its weight; and when, to avoid this cheat, the hogs were bought alive, it was found that the Chinese gave them salt to increase their thirst, and having by this means excited them to drink great quantities of water, they then took measures to prevent them from discharging it again by urine, and sold the tortured animal in this inflated state. When the Commodore first put to sea from Macao, they practised an artifice of another kind; for as the Chinese never object to the eating of any food that dies of itself, they took care, by some secret practices, that great part of his live sea-store should die in a short time after it was put on board, hoping to make a second profit of the dead carcases, which they expected would be thrown overboard; and two-thirds of the hogs dying before the Centurion was out of sight of land, many of the Chinese boats followed her only to pick up the carrion. These instances may serve as a specimen of the manners of this celebrated nation, which is often recommended to the rest of the world as a pattern of all kinds of laudable qualities.

The Commodore, towards the end of September, having found out (as has been said) that those who had contracted to supply him with sea-provisions and stores had deceived him, and that the Viceroy had not sent to him according to his promise, he saw it would be impossible for him to surmount the embarrassment he was under without going himself to Canton and visiting the Viceroy. And therefore, on the 27th of September, he sent a message to the mandarin who attended the Centurion, to inform him that he, the Commodore, intended on the 1st of October to proceed in his boat to Canton; adding, that the day after he got there he should notify his arrival to the Viceroy, and should desire him to fix a time for his audience; to which the mandarin returned no other answer than that he would acquaint the Viceroy with the Commodore's intentions. In the meantime all things were prepared for this expedition, and the boat's crew in particular, which Mr Anson proposed to take with him, were clothed in a uniform dress, resembling that of the watermen on the Thames. They were in number eighteen and a coxswain; they had scarlet jackets and blue silk waist-

otherwise account for this faithless procedure of the Chinese, than by supposing they meant to starve us into a compliance with their accustomed demands for port charges, with which the Commodore was determined never to acquiesce."

coats, the whole trimmed with silver buttons, and with silver badges on their jackets and caps. As it was apprehended, and even asserted, that the payment of the customary duties for the Centurion and her prize would be demanded by the Regency of Canton, and would be insisted on previous to the granting a permission for victualling the ship for her future voyage, the Commodore, who was resolved never to establish so dishonourable a precedent, took all possible precautions to prevent the Chinese from facilitating the success of their unreasonable pretensions by having him in their power at Canton. And therefore, for the security of his ship and the great treasure on board her, he appointed his first lieutenant, Mr Brett,[1] to be captain of the Centurion under him, giving him proper instructions for his conduct; directing him particularly, if he, the Commodore, should be detained at Canton on account of the duties in dispute, to take out the men from the Centurion's prize and to destroy her; and then to proceed down the river through the Bocca Tigris, with the Centurion alone, and to remain without that entrance till he received further orders from Mr Anson.

These necessary steps being taken, which were not unknown to the Chinese, it should seem as if their deliberations were in some sort embarrassed thereby. It is reasonable to imagine that they were in general very desirous of getting the duties to be paid them; not perhaps solely in consideration of the amount of those dues, but to keep up their reputation for address and subtlety, and to avoid the imputation of receding from claims on which they had already so frequently insisted. However, as they now foresaw that they had no other method of succeeding than by violence, and that even against this the Commodore was prepared, they were at last disposed, I conceive, to let the affair drop, rather than entangle themselves in a hostile measure which they found would only expose them to the risk of having the whole navigation of their port destroyed, without any certain prospect of gaining their favourite point thereby. However, though there is reason to imagine that these were their thoughts at that time, yet they could not depart at once from the evasive conduct to which they had hitherto adhered. For when the Commodore, on the morning of the 1st of October, was preparing to set out for Canton, his linguist came to him from the mandarin who attended his ship, to tell him that a letter had been received from the Viceroy of Canton, desiring the Commodore to put off his going thither for two or three days. But, in the afternoon of the same day, another linguist came on board, who with much seeming fright, told Mr Anson that the Viceroy had expected him up that day, that the council was assembled, and the troops had been under arms to receive him; and that the Viceroy was highly offended at the disappointment, and had sent the Commodore's linguist to prison chained, supposing that the whole had been owing to the linguist's negligence. This plausible tale gave the Commodore great concern, and made him apprehend that there was some treachery designed him, which he could not yet fathom; and though it afterwards appeared that the whole was a fiction, not one article of it having the least foundation, yet (for reasons best known to themselves) this falsehood was so well supported by the artifices of the Chinese merchants at Canton, that three days afterwards the Commodore received a letter signed by all the supercargoes of the English ships then at that place, expressing their great uneasiness at what had happened, and intimating their fears that some insult would be offered to his boat if he came thither before the Viceroy was fully satisfied about the mistake. To this letter Mr Anson replied that he did not believe there had been any mistake, but was persuaded it was a forgery of the Chinese to prevent his visiting the Viceroy; that therefore

[1] Afterwards Sir Percy Brett.

he would certainly come up to Canton on the 13th of October, confident that the Chinese would not dare to offer him an insult, as well knowing it would be properly returned.

On the 13th of October, the Commodore continuing firm to his resolution, all the supercargoes of the English, Danish, and Swedish ships came on board the Centurion, to accompany him to Canton, for which place he set out in his barge the same day, attended by his own boats and by those of the trading ships, which on this occasion came to form his retinue; and as he passed by Wampo,[1] where the European vessels lay, he was saluted by all of them but the French; and in the evening he arrived safely at Canton.

CHAPTER X.

When the Commodore arrived at Canton he was visited by the principal Chinese merchants, who affected to appear very much pleased that he had met with no obstruction in getting thither, and who thence pretended to conclude that the Viceroy was satisfied about the former mistake, the reality of which they still insisted on; they added that as soon as the Viceroy should be informed that Mr Anson was at Canton (which they promised should be done the next morning), they were persuaded a day would be immediately appointed for the visit which was the principal business that had brought the Commodore thither.

The next day the merchants returned to Mr Anson, and told him that the Viceroy was then so fully employed in preparing his despatches for Pekin, that there was no getting admittance to him for some days; but that they had engaged one of the officers of his court to give them information as soon as he should be at leisure, when they proposed to notify Mr Anson's arrival, and to endeavour to fix the day of audience. The Commodore was by this time too well acquainted with their artifices not to perceive that this was a falsehood; and had he consulted only his own judgment he would have applied directly to the Viceroy by other hands. But the Chinese merchants had so far prepossessed the supercargoes of our ships with chimerical fears, that they were extremely apprehensive of being embroiled with the Government, and of suffering in their interest, if those measures were taken which appeared to Mr Anson at that time to be the most prudential; and therefore, lest the malice and double-dealing of the Chinese might have given rise to some sinister incident which would be afterwards laid at his door, he resolved to continue passive as long as it should appear that he lost no time by thus suspending his own opinion. With this view he promised not to take any immediate step himself for getting admittance to the Viceroy, provided the Chinese with whom he contracted for provisions would let him see that his bread was baked, his meat salted, and his stores prepared with the utmost despatch. But if, by the time when all was in readiness to be shipped off (which it was supposed would be in about forty days), the merchants should not have procured the Viceroy's permission, then the Commodore proposed to apply for it himself. These were the terms Mr Anson thought proper to offer to quiet the uneasiness of the supercargoes; and notwithstanding the apparent equity of the conditions, many difficulties and objections were urged, nor would the Chinese agree to them till the Commodore had consented to pay for every article he bespoke before it was put in hand. However, at last the contract being passed, it was some satisfaction to the Commodore to be certain that his preparations were now going on; and, being himself on the spot, he took care to hasten them as much as possible.

During this interval, in which the stores and provisions were getting ready, the merchants continually entertained Mr Anson with accounts of

[1] Whampoa.

their various endeavours to get a license from the Viceroy, and their frequent disappointments, which to him was now a matter of amusement, as he was fully satisfied there was not one word of truth in anything they said. But when all was completed, and wanted only to be shipped, which was about the 24th of November, at which time, too, the NE. monsoon was set in, he then resolved to apply himself to the Viceroy to demand an audience, as he was persuaded that without this ceremony the procuring a permission to send his stores on board would meet with great difficulty. On the 24th of November, therefore, Mr Anson sent one of his officers to the mandarin who commanded the guard of the principal gate of the city of Canton with a letter directed to the Viceroy. When this letter was delivered to the mandarin, he received the officer who brought it very civilly, and took down the contents of it in Chinese, and promised that the Viceroy should be immediately acquainted with it; but told the officer it was not necessary for him to wait for an answer, because a message would be sent to the Commodore himself. On this occasion Mr Anson had been under great difficulties about a proper interpreter to send with his officer, as he was well aware that none of the Chinese usually employed as linguists could be relied on; but he at last prevailed with Mr Flint, an English gentleman belonging to the factory, who spoke Chinese perfectly well, to accompany his officer. This person, who upon this occasion and many others was of singular service to the Commodore, had been left at Canton, when a youth, by the late Captain Rigby. The leaving him there to learn the Chinese language was a step taken by that captain merely from his own persuasion of the great advantages which the East India Company might one day receive from an English interpreter; and though the utility of this measure has greatly exceeded all that was expected from it, yet I have not heard that it has been to this day imitated, but we imprudently choose (except in this single instance) to carry on the vast transactions of the port of Canton either by the ridiculous jargon of broken English which some few of the Chinese have learned, or by the suspected interpretation of the linguists of other nations.

Two days after the sending the above-mentioned letter, a fire broke out in the suburbs of Canton. On the first alarm, Mr Anson went thither with his officers and his boats' crew to assist the Chinese. When he came there, he found that it had begun in a sailor's shed, and that by the slightness of the buildings and the awkwardness of the Chinese it was getting head apace. But he perceived that by pulling down some of the adjacent sheds it might easily be extinguished; and particularly observing that it was running along a wooden cornice which would soon communicate it to a great distance, he ordered his people to begin with tearing away that cornice. This was presently attempted, and would have been soon executed, but in the meantime he was told, that as there was no mandarin there to direct what was to be done, the Chinese would make him (the Commodore) answerable for whatever should be pulled down by his orders. On this his people desisted, and he sent them to the English factory to assist in securing the Company's treasure and effects, as it was easy to foresee that no distance was a protection against the rage of such a fire, where so little was done to put a stop to it; for all this time the Chinese contented themselves with viewing it, and now and then holding one of their idols near it, which they seemed to expect should check its progress. However, at last a mandarin came out of the city, attended by four or five hundred firemen; these made some feeble efforts to pull down the neighbouring houses, but by this time the fire had greatly extended itself, and was got amongst the merchants' warehouses; and the Chinese firemen, wanting both skill and spirit, were incapable of checking its violence, so that its fury increased upon them, and it was feared the whole city would be destroyed. In this general confusion the Viceroy himself came thither, and the Commodore was sent to and was entreated to afford his assistance, being told that he might take any measures he should think most prudent in the present emergency. And now he went thither a second time, carrying with him about forty of his people, who upon this occasion exerted themselves in such a manner as in that country was altogether without example.[1] For they were rather animated than deterred by the flames and falling buildings amongst which they wrought; so that it was not uncommon to see the most forward of them tumble to the ground on the roofs and amidst the ruins of houses which their own efforts brought down with them. By their boldness and activity the fire was soon extinguished, to the amazement of the Chinese; and the building being all on one floor, and the materials slight, the seamen, notwithstanding their daring behaviour, happily escaped with no other injuries than some considerable bruises. The fire, though at last thus luckily extinguished, did great mischief during the time it continued; for it consumed an hundred shops and eleven streets full of warehouses, so that the damage amounted to an immense sum; and one of the Chinese merchants, well known to the English, whose name was Succoy, was supposed for his own share to have lost near £200,000 sterling. It raged, indeed, with unusual violence, for in many of the warehouses there were large quantities of camphor, which greatly added to its fury, and produced a column of exceeding white flame, which shot up into the air to such a prodigious height that the flame itself was plainly seen on board the Centurion, though she was thirty miles distant.

Whilst the Commodore and his people were labouring at the fire, and the terror of its becoming general still possessed the whole city, several of the most considerable Chinese merchants came to Mr Anson to desire that he would let each of them have one of his soldiers (for such they styled his boat's crew from the uniformity of their dress) to guard their warehouses and dwelling-houses, which, from the known dishonesty of the populace, they feared would be pillaged in the tumult. Mr Anson granted them this request; and all the men that he thus furnished to the Chinese behaved greatly to the satisfaction of their employers, who afterwards highly applauded their great diligence and fidelity. By this means the resolution of the English at the fire, and their trustiness and punctuality elsewhere, was the subject of general conversation amongst the Chinese; and the next morning, many of the principal inhabitants waited on the Commodore to thank him for his assistance, frankly owning to him that they could never have extinguished the fire of themselves, and that he had saved their city from being totally consumed. And soon after a message came to the Commodore from the Viceroy, appointing the 30th of November for his audience, which sudden resolution of the Viceroy, in a matter that had been so long agitated in vain, was also owing to the signal services performed by Mr Anson and his people at the fire, of which the Viceroy himself had been in some measure an eye-witness. The fixing this business of the audience was, on all accounts, a circumstance which Mr Anson was much pleased with, as he was satisfied that the Chinese Government would not have determined this point without having agreed among themselves to give up their pretensions to the duties they claimed, and to grant him all he could reasonably ask; for, as they well knew the Commodore's sentiments, it would have been a piece of imprudence not consistent with the refined cunning of

[1] Thomas enthusiastically says, that "they in sight of the whole city performed such daring, and, to the people who beheld them, such astonishing feats, that they looked upon them as salamanders, and cried out that they could live in fire."

the Chinese to have admitted him to an audience only to have contested with him. And, therefore, being himself perfectly easy about the result of his visit, he made all necessary preparations against the day, and engaged Mr Flint, whom I have mentioned before, to act as interpreter in the conference, who in this affair, as in all others, acquitted himself much to the Commodore's satisfaction, repeating with great boldness, and, doubtless, with exactness, all that was given in charge, a part which no Chinese linguist would ever have performed with any tolerable fidelity.

At 10 o'clock in the morning, on the day appointed, a mandarin came to the Commodore to let him know that the Viceroy was ready to receive him, on which the Commodore and his retinue immediately set out. And as soon as he entered the outer gate of the city, he found a guard of 200 soldiers drawn up ready to attend him; these conducted him to the great parade before the Emperor's palace, where the Viceroy then resided. In this parade a body of troops, to the number of 10,000, were drawn up under arms, and made a very fine appearance, being all of them new clothed for this ceremony; and Mr Anson and his retinue having passed through the middle of them, he was then conducted to the great hall of audience, where he found the Viceroy seated under a rich canopy in the Emperor's chair of state, with all his Council of Mandarins attending. Here there was a vacant seat prepared for the Commodore, in which he was placed on his arrival. He was ranked the third in order from the Viceroy, there being above him only the head of the law and of the treasury, who in the Chinese Government take place of all military officers. When the Commodore was seated, he addressed himself to the Viceroy by his interpreter, and began with reciting the various methods he had formerly taken to get an audience; adding, that he imputed the delays he had met with to the insincerity of those he had employed, and that he had therefore no other means left than to send, as he had done, his own officer with a letter to the gate. On the mention of this, the Viceroy stopped the interpreter, and bid him assure Mr Anson that the first knowledge they had of his being at Canton was from that letter. Mr Anson then proceeded, and told him that the subjects of the King of Great Britain trading to China, had complained to him (the Commodore) of the vexatious impositions both of the merchants and inferior custom-house officers, to which they were frequently necessitated to submit, by reason of the difficulty of getting access to the mandarins, who alone could grant them redress; that it was his (Mr Anson's) duty, as an officer of the King of Great Britain, to lay before the Viceroy these grievances of the British subjects, which he hoped the Viceroy would take into consideration, and would give orders that for the future there should be no just reason for complaint. Here Mr Anson paused, and waited some time in expectation of an answer, but nothing being said, he asked his interpreter if he was certain the Viceroy understood what he had urged; the interpreter told him he was certain it was understood, but he believed no reply would be made to it. Mr Anson then represented to the Viceroy the case of the ship Haslingfield, which, having been dismasted on the coast of China, had arrived in the River of Canton but a few days before. The people on board this vessel had been great sufferers by the fire; the captain in particular had all his goods burned, and had lost besides, in the confusion, a chest of treasure of 4500 taels, which was supposed to be stolen by the Chinese boatmen. Mr Anson therefore desired that the captain might have the assistance of the Government, as it was apprehended the money could never be recovered without the interposition of the mandarins; and to this request the Viceroy made answer, that in settling the Emperor's customs for that ship, some abatement should be made in consideration of her losses.

And now, the Commodore having despatched the business with which the officers of the East India Company had entrusted him, he entered on his own affairs; acquainting the Viceroy that the proper season was now set in for returning to Europe, and that he waited only for a license to ship off his provisions and stores, which were all ready; and that, as soon as this should be granted to him, and he should have got his necessaries on board, he intended to leave the River of Canton, and to make the best of his way for England. The Viceroy replied to this, that the license should be immediately issued, and that everything should be ordered on board the following day. And finding that Mr Anson had nothing further to insist on, the Viceroy continued the conversation for some time, acknowledging in very civil terms how much the Chinese were obliged to him for his signal services at the fire, and owning that he had saved the city from being destroyed; and then, observing that the Centurion had been a good while on their coast, he closed his discourse by wishing the Commodore a good voyage to Europe. After which, the Commodore, thanking him for his civility and assistance, took his leave.

As soon as the Commodore was out of the hall of audience, he was much pressed to go into a neighbouring apartment, where there was an entertainment provided; but finding on inquiry that the Viceroy himself was not to be present, he declined the invitation, and departed, attended in the same manner as at his arrival; only at his leaving the city he was saluted by three guns, which are as many as in that country are ever fired on any ceremony. Thus the Commodore, to his great joy, at last finished this troublesome affair, which for the preceding four months had given him great disquietude. Indeed, he was highly pleased with procuring a license for the shipping of his stores and provisions; for thereby he was enabled to return to Great Britain with the first of the monsoon, and to prevent all intelligence of his being expected. But this, though a very important point, was not the circumstance which gave him the greatest satisfaction; for he was more particularly attentive to the authentic precedent established on this occasion, by which his Majesty's ships of war are for the future exempted from all demands of duty in any of the ports of China.

In pursuance of the promises of the Viceroy, the provisions were begun to be sent on board the day after the audience, and four days after, the Commodore embarked at Canton for the Centurion; and on the 7th of December, the Centurion and her prize unmoored, and stood down the river, passing through the Bocca Tigris on the 10th. And on this occasion I must observe, that the Chinese had taken care to man the two forts on each side of that passage with as many men as they could well contain, the greatest part of them armed with pikes and matchlock muskets. These garrisons affected to show themselves as much as possible to the ships, and were doubtless intended to induce Mr Anson to think more reverently than he had hitherto done of the Chinese military power. For this purpose they were equipped with much parade, having a great number of colours exposed to view; and on the castle in particular there were laid considerable heaps of large stones, and a soldier of unusual size, dressed in very sightly armour, stalked about on the parapet with a battle-axe in his hand endeavouring to put on as important and martial an air as possible, though some of the observers on board the Centurion shrewdly suspected, from the appearance of his armour, that instead of steel, it was composed only of a particular kind of glittering paper.[1]

[1] We omit Mr Walter's strictures on the merely imitative genius, the bad government, and the pusillanimity and military weakness of the Chinese; strictures founded admittedly on very partial information, and

The Commodore, on the 12th of December, anchored before the town of Macao. Whilst the ships lay here, the merchants of Macao finished their agreement for the galleon, for which they had offered 6000 dollars; this was much short of her value, but the impatience of the Commodore to get to sea, to which the merchants were no strangers, prompted them to insist on so unequal a bargain. Mr Anson had learned enough from the English at Canton to conjecture, that the war betwixt Great Britain and Spain was still continued, and that probably the French might engage in the assistance of Spain before he could arrive in Great Britain; and therefore, knowing that no intelligence could get to Europe of the prize he had taken, and the treasure he had on board, till the return of the merchantmen from Canton, he was resolved to make all possible expedition in getting back, that he might be himself the first messenger of his own good fortune, and might thereby prevent the enemy from forming any projects to intercept him. For these reasons, he to avoid all delay accepted of the sum offered for the galleon; and she being delivered to the merchants the 15th of December 1743, the Centurion the same day got under sail on her return to England. And on the 3d of January she came to an anchor at Prince's Island in the Straits of Sunda, and continued there wooding and watering till the 8th; when she weighed and stood for the Cape of Good Hope, where on the 11th of March she anchored in Table Bay.

Here the Commodore continued till the beginning of April, highly delighted with the place, which by its extraordinary accommodations, the healthiness of its air, and the picturesque appearance of the country, all enlivened by the addition of a civilised colony, was not disgraced in an imaginary comparison with the valleys of Juan Fernandez and the lawns of Tinian. During his stay he entered about forty new men; and having, by the 3d of April 1744, completed his water and provison, he on that day weighed and put to sea. The 19th of the same month they saw the Island of St Helena, which, however, they did not touch at, but stood on their way; and on the 10th of June, being then in soundings, they spoke with an English ship from Amsterdam bound for Philadelphia, whence they received the first intelligence of a French war. The 12th they got sight of the Lizard; and the 15th in the evening, to their infinite joy, they came safe to an anchor at Spithead. But that the signal perils which had so often threatened them in the preceding part of the enterprise might pursue them to the very last, Mr Anson learned on his arrival that there was a French fleet of considerable force cruising in the Chops of the Channel; which, by the account of their position, he found the Centurion had run through, and had been all the time concealed by a fog. Thus was this expedition finished, when it had lasted three years and nine months; after having, by its event, strongly evinced this important truth: That though prudence, intrepidity, and perseverance united are not exempted from the blows of adverse fortune, yet in a long series of transactions they usually rise superior to its power, and in the end rarely fail of proving successful.

---

stamped with an almost venomous spirit of prejudice.

END OF ANSON'S VOYAGE.

www.ingramcontent.com/pod-product-compliance
Lightning Source LLC
Chambersburg PA
CBHW031053110726
47900CB00003B/908
9783744722285